Some modern methods of organic synthesis

SOME MODERN METHODS OF ORGANIC SYNTHESIS

W. CARRUTHERS

Chemistry Department
University of Exeter

THIRD EDITION

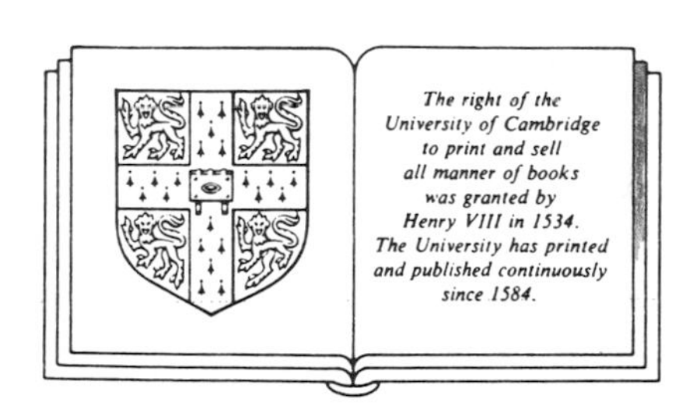
The right of the
University of Cambridge
to print and sell
all manner of books
was granted by
Henry VIII in 1534.
The University has printed
and published continuously
since 1584.

CAMBRIDGE UNIVERSITY PRESS
Cambridge
New York Port Chester
Melbourne Sydney

Published by the Press Syndicate of the University of Cambridge
The Pitt Building, Trumpington Street, Cambridge CB2 1RP
40 West 20th Street, New York, NY 10011, USA
10 Stamford Road, Oakleigh, Melbourne 3166, Australia

First published 1971
Second edition 1978
Reprinted 1985
Third edition 1986
Reprinted 1987 (with corrections), 1988, 1989

Reprinted in Great Britain by Billings & Sons Ltd, Worcester

British Library cataloguing in publication data

Carruthers, W.
Some modern methods of organic synthesis.—
3rd ed.—(Cambridge texts in chemistry and
biochemistry)
1. Chemistry, Organic—Synthesis
I. Title
547'.2 QD262

Library of Congress cataloging in publication data

Carruthers, W.
Some modern methods of organic synthesis.
(Cambridge texts in chemistry and biochemistry)
Includes index.
1. Chemistry, Organic—Synthesis. I. Title.
II. Series.
QD262.C33 1986 547'.2 85-21270

ISBN 0 521 32234 0 hard covers
ISBN 0 521 31117 9 paperback
(second edition ISBN 0 521 21715 6 hard covers
ISBN 0 521 29241 7 paperback)

J.W.A.

To D.H.R.B.

Contents

Preface to the first edition

This book is addressed principally to advanced undergraduates and to graduates at the beginning of their research careers, and aims to bring to their notice some of the reactions used in modern organic syntheses. Clearly, the whole field of synthesis could not be covered in a book of this size, even in a cursory manner, and a selection has had to be made. This has been governed largely by consideration of the usefulness of the reactions, their versatility and, in some cases, their selectivity.

A large part of the book is concerned with reactions which lead to the formation of carbon–carbon single and double bonds. Some of the reactions discussed, such as the alkylation of ketones and the Diels–Alder reaction, are well established reactions whose scope and usefulness has increased with advancing knowledge. Others, such as those involving phosphorus ylids, organoboranes and new organometallic reagents derived from copper, nickel, and aluminium, have only recently been introduced and add powerfully to the resources available to the synthetic chemist. Other reactions discussed provide methods for the functionalisation of unactivated methyl and methylene groups through intramolecular attack by free radicals at unactivated carbon–hydrogen bonds. The final chapters of the book are concerned with the modification of functional groups by oxidation and reduction, and emphasise the scope and limitations of modern methods, particularly with regard to their selectivity.

Discussion of the various topics is not exhaustive. My object has been to bring out the salient features of each reaction rather than to provide a comprehensive account. In general, reaction mechanisms are not discussed except in so far as is necessary for an understanding of the course or stereochemistry of a reaction. In line with the general policy in the series references have been kept to a minimum. Relevant reviews are noted but, for the most part, references to the original literature are given only for points of outstanding interest and for very recent work. Particular reference is made here to the excellent book by H. O. House, *Modern Synthetic Reactions* which has been my guide at several points and on which I have tried to build, I feel all too inadequately.

I am indebted to my friend and colleague, Dr K. Schofield, for much helpful comment and careful advice which has greatly assisted me in writing the book.

W. CARRUTHERS
26 October 1970

Preface to the third edition

The general plan of this third edition follows that of the earlier editions, but the opportunity has been taken to bring the book up to date as far as possible and to take account of advances in knowledge and of new synthetic methods which have come into use since publication of the second edition. Perhaps the most striking trend in synthesis since then has been the development of highly stereoselective reactions and their application in complex syntheses. These reactions include the stereoselective alkylation of carbonyl compounds, stereoselective aldol condensations and stereoselective oxidations, epoxidations and reductions, and these are among the topics discussed in this edition. New methods for the stereoselective formation of carbon–carbon double bonds, and modern applications of the Diels–Alder reaction, particularly its use in the control of stereochemistry in the synthesis of natural products, and the related class of 1,3-dipolar cyclo-addition reactions are also considered. Other sections of the book are concerned with the increasingly important application in synthesis of organo-metallic reagents, including organoboranes and organosilanes and reagents derived from copper, nickel and palladium, and with the continuing interest in selective reactions at unactivated carbon–hydrogen bonds.

The book is addressed principally to advanced undergraduates and to graduates at the beginning of their research careers, and my aim has been to bring out the salient features of the reactions and reagents rather than to provide a comprehensive account. Reaction mechanisms are not discussed except in so far as is necessary for an understanding of the course or stereochemistry of a reaction. To prevent the book from becoming too big some material of less immediate interest which appeared in earlier editions has been excised from this one. Discussion of new reactions is supported by references.

W. CARRUTHERS

May 1985

1 Formation of carbon–carbon single bonds

In spite of the fundamental importance in organic synthesis of the formation of carbon–carbon single bonds there are comparatively few general methods available for effecting this process, and fewer still which proceed in good yield under mild conditions. Many of the most useful procedures involve carbanions, themselves derived from organometallic compounds, or from compounds containing 'activated' methyl or methylene groups. They include reactions which proceed by attack of the carbanion on a carbonyl or conjugated carbonyl group, as in the Grignard reaction, the aldol and Claisen ester condensations and the Michael reaction, and reactions which involve nucleophilic displacement at a saturated carbon atom, as in the alkylation of ketones and the coupling reactions of some organometallic compounds. Other reactions employed in the formation of carbon–carbon bonds involve carbonium ions and pericyclic processes and recently free-radial reactions have been finding useful application. Examples of all of these procedures will be discussed in this chapter.

1.1. Alkylation: importance of enolate anions

It is well known that certain unsaturated groups attached to a saturated carbon atom render hydrogen atoms attached to that carbon relatively acidic, so that the compound can be converted into an anion on treatment with an appropriate base. Table 1.1, taken from House (1965), shows the pK_a values for some compounds of this type and for some common solvents and reagents.

The acidity of the C—H bonds in these compounds is due to a combinaion of the inductive electron-withdrawing effect of the unsaturated groups and resonance stabilisation of the anion formed by removal of a proton (1.1). Not all groups are equally effective in 'activating' a neighbouring CH_2 or CH_3; nitro is the most powerful of the common groups and thereafter the series follows the approximate order $NO_2 >$ $COR > SO_2R > CO_2R > CN > C_6H_5$. Two activating groups reinforce

Table 1.1. *Approximate acidities of active methylene compounds and other common reagents*

Compound	pK_a	Compound	pK_a
$CH_3CO_2\underline{H}$	5	$C_6H_5COC\underline{H}_3$	19
$C\underline{H}_2(CN)CO_2C_2H_5$	9	$CH_3COC\underline{H}_3$	20
$C\underline{H}_2(CO.CH_3)_2$	9	$C\underline{H}_3SO_2C\underline{H}_3$	~23
$C\underline{H}_3NO_2$	10	$C\underline{H}_3CO_2C_2H_5$	~24
$CH_3COC\underline{H}_2CO_2C_2H_5$	11	$C\underline{H}_3CO_2H$	~24
$C\underline{H}_2(CO_2C_2H_5)_2$	13	$C\underline{H}_3CN$	~25
$CH_3O\underline{H}$	16	$C_6H_5N\underline{H}_2$	~30
$C_2H_5O\underline{H}$	18	$(C_6H_5)_3C\underline{H}$	~40
$(CH_3)_3CO\underline{H}$	19	$C\underline{H}_3SOC\underline{H}_3$	~40

(Acidic hydrogen atoms are underlined)

H. O. House, *Modern synthetic reactions*, copright 1972, W. A. Benjamin, Inc. Menlo Park, California.

each other, as can be seen by comparing diethyl malonate (p$K_a \approx 13$) with ethyl acetate (p$K_a \approx 24$). Acidity is also increased slightly by electron-withdrawing substituents (e.g. sulphide), and decreased by alkyl groups, so that diethyl methylmalonate, for example, has a slightly less acidic C—H group than diethyl malonate itself.

$$CH_3-\overset{+}{N}(=O)O^- \xrightarrow{\text{base}} \left[{}^-CH_2-\overset{+}{N}(=O)O^- \leftrightarrow CH_2=\overset{+}{N}(O^-)O^- \right]$$

$$H_2C(COOC_2H_5)_2 \longrightarrow \left[{}^-CH(COOC_2H_5)_2 \leftrightarrow CH(=C(O^-)OC_2H_5)(COOC_2H_5) \right] \qquad (1.1)$$

By far the most important activating groups in synthesis are the carbonyl and carboxylic ester groups. Removal of a proton from the α-carbon atom of a carbonyl compound with base gives the corresponding enolate anion, and it is these anions which are involved in base-catalysed condensation reactions of carbonyl compounds, such as the aldol condensation, and in bimolecular nucleophilic displacements (alkylations) (1.2).

The enolate anions should be distinguished from the enols themselves, which are always present in equilibrium with the carbonyl compound. Most monoketones and esters contain only small amounts of enol (<1

$$\mathrm{R{-}CH_2{-}CO{-}R'} \xrightleftharpoons{\text{base (slow)}} \mathrm{R{-}\bar{C}H{-}CO{-}R'} \leftrightarrow \mathrm{R{-}CH{=}\overset{\overset{\displaystyle O^-}{|}}{C}{-}R'}$$

$$\mathrm{-\overset{\overset{\displaystyle O^-}{|}}{C}{=}\underset{|}{C}H \quad \gt C{=}O \rightleftharpoons -\overset{\overset{\displaystyle O}{\|}}{C}{-}\overset{|}{C}H{-}\overset{|}{\underset{|}{C}}{-}O^-} \qquad (1.2)$$

$$\mathrm{-\overset{\overset{\displaystyle O^-}{|}}{C}{=}\underset{|}{C}H \quad -\overset{|}{\underset{|}{C}}{-}X \rightarrow -\overset{\overset{\displaystyle O}{\|}}{C}{-}\overset{|}{C}H{-}\overset{|}{\underset{|}{C}}{-} + X^-}$$

per cent) at equilibrium, but with 1,2- and 1,3-dicarbonyl compounds much higher amounts of enol (>50 per cent) may be present. In the presence of acid catalysts monoketones may be converted largely into the enol form, and enols are concerned in many acid-catalysed condensations of carbonyl compounds (1.3).

$$\mathrm{R{-}CH_2{-}CO{-}R'} \rightleftharpoons \mathrm{R{-}CH{=}\overset{\overset{\displaystyle OH}{|}}{C}{-}R'} \qquad (1.3)$$

The formation of the enolate anion results from an equilibrium reaction between the carbonyl compound and the base. A competing equilibrium involves the enolate anion and the solvent. Thus, with diethyl malonate in solvent SolH in presence of base B^-, we have

$$\begin{aligned} \mathrm{CH_2(CO_2C_2H_5)_2 + B^-} &\rightleftharpoons \mathrm{{}^-CH(CO_2C_2H_5)_2 + BH} \\ \mathrm{{}^-CH(CO_2C_2H_5)_2 + SolH} &\rightleftharpoons \mathrm{CH_2(CO_2C_2H_5)_2 + Sol^-,} \end{aligned} \qquad (1.4)$$

and to ensure an adequate concentration of the enolate anion at equilibrium clearly both the solvent and the conjugate acid of the base must be much weaker acids than the active methylene compound. The correct choice of base and solvent is thus of great importance if the subsequent alkylation, or other, reaction is to be successful. Reactions must normally be effected under anhydrous conditions since water is a much stronger acid than the usual activated methylene compounds and, if present, would instantly protonate any carbanion produced. Another point of importance is that the solvent must not be a much stronger acid than the conjugate acid of the base, otherwise the equilibrium

$$\mathrm{B^- + SolH \rightleftharpoons BH + Sol^-} \qquad (1.5)$$

will lie too far to the right and lower the concentration of B^-. For example, sodamide can be used as base in liquid ammonia or in benzene, but, obviously, not in ethanol. Base–solvent combinations commonly used to convert active methylene compounds into the corresponding anions include sodium methoxide, sodium ethoxide and sodium or potassium t-butoxide in solution in the corresponding alcohol, or as suspensions in ether, benzene or dimethoxyethane. Potassium t-butoxide is a particularly useful reagent, since it is a poor nucleophile and its solutions in different solvents have widely different basic strengths; it is most active in solution in dry dimethyl sulphoxide (Pearson and Buehler, 1974). Metallic sodium or potassium, or sodium hydride, in suspension in benzene, ether or dimethoxyethane, sodamide in suspension in an inert solvent or in solution in liquid ammonia, and solutions of sodium or potassium triphenylmethyl in ether or benzene have also been used with the less 'active' compounds.

For many purposes, however, these traditional bases have now been superseded by the lithium salts of certain sterically hindered secondary amines, particularly lithium diisopropylamide and lithium 2,2,6,6-tetramethylpiperidide (Olofson and Dougherty, 1973) or the alkali metal salts of bis(trimethylsilyl)amine, $HN(SiMe_3)_2$ (Colvin, 1978; Smith and Richmond, 1983). These strong amide bases are only weakly nucleophilic, so that they do not themselves attack susceptible functional groups, and they have the added advantage that they are soluble in non-polar, even hydrocarbon, solvents. The insolubility of the traditional bases in most common organic solvents seriously limits their usefulness.

1.2. Alkylation of relatively acidic methylene groups

In order to effect a reasonably rapid reaction it is, of course, necessary to have a high concentration of the appropriate carbanion. Because of their relatively high acidity (see Table 1.1) compounds in which a C—H bond is activated by a nitro group or by two or more carbonyl, ester or cyano groups can be converted largely into their anions with a comparatively weak base such as a solution of sodium ethoxide in ethanol. An alternative procedure is to prepare the enolate in benzene or ether, using finely divided sodium or potassium metal or sodium hydride, which react irreversibly with compounds containing active methylene groups with formation of the metal salt and evolution of hydrogen. β-Diketones can often be converted into their enolates with alkali metal hydroxides or carbonates in aqueous alcohol or acetone.

Much faster alkylation of enolate anions can often be achieved in dimethylformamide, dimethyl sulphoxide, 1,2-dimethoxyethane or hexamethylphosphoramide than in the usual protic solvents. This appears to be due to the fact that the former solvents do not solvate the enolate anion and thus do not diminish its reactivity as a nucleophile. At the same time they are able to solvate the cation, separating it from the cation–enolate ion pair and leaving a relatively free enolate ion which would be expected to be a more reactive nucleophile than the ion pair (Parker, 1962). Reactions effected with aqueous alkali as base are often improved in the presence of a phase-transfer catalyst such as a tetra-alkylammonium salt (cf. Makosza and Jończyk, 1976).

Alkylation of enolate anions is readily effected with alkyl halides or other alkylating agents. Both primary and secondary alkyl, allyl or benzyl halides may be used successfully, but with tertiary halides poor yields of alkylated product often result because of competing dehydrohalogenation of the halide. It is often advantageous to proceed by way of the toluene-*p*-sulphonate or methanesulphonate rather than a halide. The sulphonates are excellent alkylating agents, and can usually be obtained from the alcohol in a pure condition more readily than the corresponding halides. Epoxides have also been used as alkylating agents, generally reacting at the less substituted carbon atom. Attack of the enolate anion on the alkylating agent takes place by an S_N2 pathway and thus results in inversion of configuration at the carbon atom of the alkylating agent.

$$p\text{-}CH_3C_6H_4SO_2O\text{-(steroid, H}_3C\text{)} \xrightarrow[C_2H_5OH]{CH_2(CO_2C_2H_5)_2,\ C_2H_5ONa} \text{(steroid, H}_3C\text{, H)}\text{-}CH(CO_2C_2H_5)_2 \quad (1.6)$$

With secondary and tertiary allylic halides or sulphonates, reaction of an enolate anion may give mixtures of products formed by competing attack at the α- and γ-positions (1.7).

$$C_2H_5\text{—}\underset{}{\overset{Cl}{\overset{|}{C}}}H\text{—}CH{=}CH_2 \xrightarrow[C_2H_5ONa,\ C_2H_5OH]{CH_2(CO_2C_2H_5)_2}$$

$$C_2H_5\text{—}\overset{CH(CO_2C_2H_5)_2}{\overset{|}{C}}H\text{—}CH{=}CH_2 + \underset{(10\%\ \text{of product})}{C_2H_5CH{=}CHCH_2CH(CO_2C_2H_5)_2} \quad (1.7)$$

A difficulty sometimes encountered in the alkylation of active methylene compounds is the formation of unwanted dialkylated products. During the alkylation of diethyl sodiomalonate, the monoalkyl derivative formed initially is in equilibrium with its anion as indicated in the first equation of (1.8). In ethanol solution, dialkylation does not take place to any appreciable extent because ethanol is sufficiently acidic to reduce the concentration of the anion of the alkyl derivative, but not that of the more acidic diethyl malonate itself, to a very low value.

$$\begin{aligned} RCH(CO_2C_2H_5)_2+\bar{C}H(CO_2C_2H_5)_2 &\rightleftharpoons R\bar{C}(CO_2C_2H_5)_2+CH_2(CO_2C_2H_5)_2 \\ R\bar{C}(CO_2C_2H_5)_2+C_2H_5OH &\rightleftharpoons RCH(CO_2C_2H_5)_2+C_2H_5O^- \end{aligned} \quad (1.8)$$

However, replacement of ethanol by an inert solvent favours dialkylation, and dialkylation also becomes a more serious problem with the more acidic alkylcyanoacetic esters, and in alkylations with very reactive compounds such as allyl or benzyl halides or sulphonates.

Dialkylation may, of course, be effected deliberately if required by carrying out two successive operations, using either the same or a different alkylating agent in the two steps. Thus, alkylation with $\alpha\omega$-polymethylene dihalides, and intramolecular alkylation of ω-haloalkylmalonic esters provides a useful route to three- to seven-membered ring compounds. Non-cyclic products are frequently formed at the same time by competing intermolecular reactions and conditions have to be carefully chosen to suppress their formation (1.9).

$$Br(CH_2)_3Br + CH_2(CO_2C_2H_5)_2 \xrightarrow[C_2H_5OH]{NaOC_2H_5} \text{(cyclobutane-1,1-diyl)}(CO_2C_2H_5)_2 + CH(CO_2C_2H_5)_2\text{–(CH}_2)_3\text{–}CH(CO_2C_2H_5)_2 \quad (1.9)$$

Under ordinary conditions aryl or vinyl halides do not react with enolate anions, although aryl halides with strongly electronegative substituents in the *ortho* and *para* positions do. 2,4-Dinitrochlorobenzene, for example, with ethyl cyanoacetate gives ethyl (2,4-dinitrophenyl)cyanoacetate in 90 per cent yield by an addition–elimination, not an S_N2, pathway. Unactivated aryl halides may also react with enolates under more vigorous conditions. Reaction of bromobenzene with diethyl malonate, for example, takes place readily in presence of an excess of sodium amide in liquid ammonia, to give diethyl phenylmalonate in 50 per cent yield. The reaction is not a direct nucleophilic displacement, however, but takes place by an elimination–addition sequence in which benzyne is an intermediate (1.10). Similar reactions can be effected intramolecularly and provide a good route to some cyclic systems (1.11). Vinyl derivatives of active methylene compounds can be obtained indirectly from alkylidene derivatives by moving the double bond out of conjugation as illustrated in (1.12). Kinetically controlled alkylation of the delocalised anion takes place at the α-carbon atom to give the

$$C_6H_5Br \xrightarrow[\text{liq. } NH_3]{NaNH_2} \text{(1)} \xrightarrow{\bar{C}H(CO_2C_2H_5)_2} \text{C}_6\text{H}_4^-\text{–}CH(CO_2C_2H_5)_2 \rightarrow C_6H_5\bar{C}(CO_2C_2H_5)_2 \xrightarrow{H_3O^+} C_6H_5CH(CO_2C_2H_5)_2 \quad (1.10)$$

$\beta\gamma$-unsaturated compound directly. A similar course is followed in the kinetically controlled protonation of such anions.

$$\text{Br-indoline (}CH_3O_2C\text{ side chain, N–}COCH_3\text{)} \xrightarrow{KNH_2,\ NH_3} \text{aryne (}CH_3O_2\bar{C}\text{ side chain, N–}COCH_3\text{)} \rightarrow \text{tricyclic product (}CH_3O_2C\text{, N–}COCH_3\text{)} \quad (1.11)$$

$$(C_2H_5)_2C{=}C(CN)(CO_2C_2H_5) \xrightarrow[C_2H_5OH]{C_2H_5ONa} \left[CH_3\bar{C}H{-}C(C_2H_5){=}C(CN)(CO_2C_2H_5) \leftrightarrow CH_3CH{=}C(C_2H_5){-}\bar{C}(CN)(CO_2C_2H_5)\right] \quad (1.12)$$

$$\xrightarrow{H_3O^+} CH_3CH{=}C(C_2H_5){-}CH(CN)(CO_2C_2H_5)$$

$$\xrightarrow{CH_3I} CH_3CH{=}C(C_2H_5){-}C(CH_3)(CN)(CO_2C_2H_5)$$

A wasteful side reaction which frequently occurs in the alkylation of 1,3-dicarbonyl compounds is the formation of the *O*-alkylated product. Thus, reaction of the sodium salt of cyclohexan-1,3-dione with butyl bromide gives 37 per cent of 1-butoxycyclohexen-3-one and only 15 per cent of 2-butylcyclohexan-1,3-dione. In general, however, *O*-alkylation competes significantly with *C*-alkylation only with reactive methylene compounds in which the equilibrium concentration of enol is relatively high, as in 1,3-dicarbonyl compounds and phenols. Phenols, of course, generally undergo predominant *O*-alkylation.

Alkylation of malonic ester, cyanacetic ester and β-keto esters is useful in synthesis because the alkylated products on hydrolysis and decarboxylation or, better, by direct decarbalkoxylation under neutral conditions with an alkali metal salt (for example, lithium chloride) in a dipolar aprotic solvent such as dimethylformamide (Krapcho, 1982) afford carboxylic acids (esters) or ketones. From alkylated malonic or cyanoacetic esters, substituted acetic acids or esters are obtained, and alkylated acetoacetic esters give methyl ketones.

1.3 γ-Alkylation of 1,3-dicarbonyl compounds; dianions in synthesis

Alkylation of a 1,3-diketone or a β-keto-ester at a 'flanking' methyl or methylene group instead of at the doubly activated methylene or methine does not usually take place to any significant extent under 'ordinary' conditions. It can be accomplished selectively and in good yield, however, by way of the corresponding *dianion*, itself prepared from the diketone and two equivalents of a suitable strong base such as sodamide or lithium diisopropylamide, by reaction with one equivalent of alkylating agent (Harris and Harris, 1969). Thus, 2,4-pentanedione is converted into 2,4-nonanedione in 82 per cent yield (1.13) by reaction at the more reactive less resonance-stabilised carbanion and 1,6-diphenyl-1,3-pentanedione is obtained in 77 per cent yield by reaction of the dianion of benzylacetone with benzyl chloride. Keto acids and triketones can also be obtained by reaction of the dianions with carbon dioxide or with esters.

$$CH_3COCH_2COCH_3 \xrightarrow[\text{liq. } NH_3]{\text{2 equivs } KNH_2} CH_3CO\bar{C}HCO\bar{C}H_2 \leftrightarrow CH_3\overset{O^-}{\overset{|}{C}}{=}CH{-}\overset{O^-}{\overset{|}{C}}{=}CH_2 \xrightarrow[(2)\ H_3O^+]{(1)\ C_4H_9Br} CH_3COCH_2COCH_2C_4H_9 \ (82\%) \qquad (1.13)$$

With unsymmetrical diketones which could apparently give rise to two different dianions, it is found in practice that in most cases only one is formed, and a single product results on alkylation. Thus, with 2,4-

hexanedione alkylation at the methyl group greatly predominates over that at the methylene group, and 2-acetylcyclohexanone and 2-acetylcyclopentanone are both alkylated exclusively at the methyl group. In general, the ease of alkylation follows the order $C_6H_5CH_2 > CH_3 > CH_2$.

The reaction can be applied equally well to β-keto aldehydes and β-keto esters, and, in the latter case, provides a useful route to 'mixed' Claisen ester condensation products. Reactions with β-keto aldehydes are generally effected by treating the prepared monosodium salt of the aldehyde with alkali amide, to prevent self-condensation of the aldehyde. With β-keto esters the dianions are conveniently prepared by reaction with two equivalents of lithium diisopropylamide, and give γ-alkylated products in high yield with a wide range of alkylating agents (Huckin and Weiler, 1974). The dianion generated in this way from ethyl acetoacetate, for example, has been used in the synthesis of a number of natural products. An example is seen in the synthesis of the ten-membered lactone (±)-diplodialide A (6) where, in fact, the reaction is used twice; once to prepare the β-keto-ester (3) by alkylation of the dianion of ethyl acetoacetate with the bromide (2), and again to introduce the double bond of diplodialide A by reaction of the dianion of the β-keto lactone (4) with phenylselenyl bromide to give the γ-phenylselenide (5). Elimination by way of the derived selenoxide (cf. p. 121) then led to diplodialide A (Ishida and Wada, 1979).

$CO_2C_2H_5$ (1) 1 equiv. NaH, THF hexamethylphosphoramide (2) 1 equiv. C_4H_9Li → OC_2H_5

OTHP Br (2)

OTHP OC_2H_5 (3) (78%)

several steps

(4)

(1) 2 equivs. $LiN(iso\text{-}C_3H_7)_2$
(2) C_6H_5SeBr

SeC_6H_5 (5)

(1) H_2O_2, CH_2Cl_2 0 °C
(2) →25 °C

(6)

(1.14)

Dianions are now being widely employed in synthesis and their application is not confined to γ-alkylation of β-dicarbonyl compounds. Dianions derived from β-keto sulphoxides are alkylated at the γ-carbon atom, and the auxiliary sulphoxide group can subsequently be removed after alkylation by pyrolysis or reductive-cleavage to give $\alpha\beta$-unsaturated or α-alkylated ketones (Grieco and Pogonowski, 1974) (1.15).

$C_6H_5S(\rightarrow O)$ ... 2NaH; THF, hexamethylphosphoramide → dianion; (1) C_4H_9Br (2) H_3O^+; boil CCl_4 (78%); Zn, CH_3CO_2H (1.15)

Synthetically useful results have been obtained recently with dianions derived from nitroalkanes. Primary nitro-alkanes, RCH_2NO_2, can be deprotonated twice in the α-position to give dianions (7) which, in contrast to simple nitronates (the *mono*anions) give *C*-alkylated products in good yield. With the monoanions alkylation takes place mainly on oxygen (Seebach *et al.*, 1977).

$$C_2H_5CH_2NO_2 \xrightarrow{\text{2 equivs } C_4H_9Li} C_2H_5\bar{C}{=}NO_2^- \, 2Li^+ \quad (7)$$

$$\xrightarrow{(1)\ C_6H_5CH_2Br \quad (2)\ H_3O^+} C_2H_5CH(CH_2C_6H_5)NO_2 \quad (53\%) \qquad (1.16)$$

If there is only one α-hydrogen atom, as in 2-nitropropane, the $\alpha\beta$-doubly deprotonated species (8) is formed and on alkylation affords the β-alkyl derivative, again by reaction at the less stabilised carbanion. The same is true when another activating group such as an aryl group or a carbethoxy group is present in the β-position. Ethyl β-nitropropionate, for example, by reaction with two equivalents of butyl-lithium, forms a dianion which is readily alkylated with primary or secondary alkyl halides at the position α to the ester group.

2 equivs C_4H_9Li; THF, −80 °C, hexamethylphosphoramide → (8) $2Li^+$; (1) $C_6H_5CH_2Br$, −78 °C → −30 °C; (2) H_3O^+

same conditions → $2Li^+$; (1) $C_6H_{13}I$; (2) H_3O^+ (76%)

$C_2H_5O_2C$ ... NO_2 same conditions → C_2H_5O ... NO_2^- $2Li^+$ (1.17)

(1) CH_3I; (2) $LiN(iso\text{-}C_3H_7)_2$; (3) $C_6H_5CH_2Br$ → (78%)

Similarly the γδ-unsaturated nitro compound (9) undergoes double deprotonation to give a dark red solution of the αβ-dianion (10). With electrophiles, for example methyl iodide, this gives mainly the δ-substitution product (11*b*). This can be converted into an αβ-unsaturated aldehyde by reduction of the nitro group with Ti(III) chloride. In this sequence the nitro compound (9) is synthetically equivalent to the crotonaldehyde anion (12) (Seebach, Henning and Lehr, 1978).

In a different kind of application, β-hydroxy carboxylic esters have been converted stereoselectively into α-alkyl derivatives without protection of the hydroxyl group (1.19) (Fräter, 1979).

NO_2 (9) $\xrightarrow[\text{THF, }-80\,^\circ\text{C, HMPA}]{2\,C_4H_9Li}$ NO_2 $2Li^+$ (10) $\xrightarrow{CH_3I}$ NO_2 with CH_3 (11a) + H_3C … NO_2 (11b) $\xrightarrow{TiCl_3,\ H_3O^+}$ H_3C … CHO (1.18)

[– CHO] (12)

HO … $CO_2C_2H_5$ $\xrightarrow[(2)\ C_4H_9Br]{(1)\ 2\text{ equivs }LiN(iso\text{-}C_3H_7)_2\text{, THF, HMPA}}$ OH … $CO_2C_2H_5$, C_4H_9 (97) + OH … $CO_2C_2H_5$, C_4H_9 (3) (1.19)

1.4. Alkylation of ketones

Alkylation of monofunctional carbonyl compounds, aldehydes, ketones and esters, is more difficult than that of the 1,3-dicarbonyl compounds discussed above. As can be seen from Table 1.1, a methyl or methylene group which is activated by only one carbonyl, ester or cyano group requires a stronger base than sodium ethoxide or methoxide to convert it into the enolate anion in high enough concentration to be useful for subsequent alkylation. Alkali metal salts of tertiary alcohols, such as t-butanol or t-amyl alcohol, in solution or suspension in the corresponding alcohol or in an inert solvent, have been used with success, but suffer from the disadvantage that they are not sufficiently basic to convert the ketone completely into the enolate anion, thus allowing the possibility of an aldol condensation between the anion and unchanged carbonyl compound. An alternative procedure is to use a much stronger base which will convert the compound completely into the anion. Typical bases of this type are sodium and potassium amide, sodium hydride, and sodium and potassium triphenylmethyl, in such solvents as ether, benzene, dimethoxyethane or dimethylformamide. The alkali metal amides are often used in solution in liquid ammonia. Although these bases can convert ketones essentially quantitatively into their enolate anions, aldol condensation may again be a difficulty with sodium hydride or sodamide in inert solvents, because of the insolubility of the reagents. Formation of the anion takes place only slowly in the heterogeneous reaction medium and both the ketone and the enolate ion are present at some point. This difficulty does not arise with the lithium dialkylamides, such as lithium diisopropylamide or lithium 2,2,6,6-tetramethylpiperidide or the alkali metal salts of bis(trimethylsilyl)amine (p. 4), which are soluble in non-polar solvents, and these bases have now largely replaced the more traditional reagents for the generation of enolate anions.

(1.20)

Examples illustrating the base-catalysed intermolecular alkylation of a ketone and an ester are given in (1.20). Intramolecular alkylations also take place readily in appropriate cases, and reactions of this kind have been widely used in the synthesis of cyclic compounds. *Cis*-9-methyl-1-decalone, for example, is smoothly obtained from the ketone (13) in presence of sodium t-amylate(pentanoate). In these alkylation reactions the alkylating agent generally approaches the enolate from the less hindered side and in a direction orthogonal to the plane of the enolate anion.

(1.21)

A common side reaction in the direct alkylation of ketones is the formation of di- and poly-alkylated products through interaction of the original enolate with the monoalkylated compound (1.22).

$$\mathrm{RCH{=}C(O^-)R'} \xrightarrow{\mathrm{CH_3I}} \mathrm{RCH(CH_3)COR'} \underset{}{\overset{\mathrm{RCH{=}C(O^-)R'}}{\rightleftharpoons}} \mathrm{RCH_2COR'} + \mathrm{RC(CH_3){=}C(O^-)R'} \xrightarrow{\mathrm{CH_3I}} \mathrm{RC(CH_3)_2COR'} \qquad (1.22)$$

This difficulty can be avoided to some extent by adding a solution of the enolate in a polar co-ordinating solvent such as dimethoxyethane to a large excess of the alkylating agent. The enolate may therefore be rapidly consumed before equilibration with the alkylated ketone can take place. Nevertheless, formation of polysubstituted products is a serious problem in the direct alkylation of ketones and often results in decreased yields of the desired monoalkyl compound.

Alkylation of symmetrical ketones and of ketones which can enolise in one direction only can, of course, give only one mono-*C*-alkylated product. With unsymmetrical ketones, however, two different monoalkylated products may be formed by way of the two structurally isomeric enolates (1.23). But if one of the isomeric enolate anions is stabilised by conjugation with another group such as cyano, nitro or ethoxycarbonyl, then for all practical purposes only this stabilised anion is formed and alkylation takes place at the position activated by both groups. Even an α-phenyl or a conjugated carbon–carbon double bond provides sufficient stabilisation of the resulting anion to direct substitution into the adjacent position (1.23).

$$\mathrm{RCH_2COCH_2R'} \overset{\text{base}}{\rightleftharpoons} \mathrm{RCH{=}\underset{}{C}(O^-){-}CH_2R'} + \mathrm{RCH_2C(O^-){=}CHR'}$$

$$\downarrow \mathrm{AlkX}$$

$$\mathrm{RCH(Alk.)COCH_2R'} + \mathrm{RCH_2COCH(Alk.)R'} \qquad (1.23)$$

$$\mathrm{C_6H_5CH_2COCH_3} \xrightarrow[\text{dimethoxy-ethane}]{\mathrm{(C_6H_5)_3CK}} \mathrm{C_6H_5CH{=}C(O^-)CH_3} + \mathrm{C_6H_5CH_2C(O^-){=}CH_2}$$

$$\downarrow \mathrm{CH_3I}$$

$$\underset{(93\%\ \text{of product})}{\mathrm{C_6H_5CH(CH_3)COCH_3}} + \underset{(<1\%\ \text{of product})}{\mathrm{C_6H_5CH_2COCH_2CH_3}}$$

Sometimes, specific lithium enolates of unsymmetrical carbonyl compounds are formed because of chelation of the lithium with some suitably placed substituent. Thus, reaction of the cyclopentanone (14) with lithium diisopropylamide followed by alkylation gave largely the product formed by alkylation at the more hindered α-position, adjacent to the phenyl substituent. This is believed to be due to preferential formation of the C-2 lithio enolate because of chelation of the lithium atom in the enolate with the phenyl substituent (Posner and Lenz, 1979). Again, lithiation of the mixed ester (15) took place α to the MEM ester group, presumably as a result of intramolecular chelation of lithium with the ethereal oxygen (Cox, Heaton and Horbury, 1980).

(14) $\xrightarrow[\text{(2) } CH_3I]{\text{(1) LiN(iso-}C_3H_7)_2\text{, THF}}$ 2-CH_3, 3-C_6H_5, 3-CH_3 cyclopentanone + diastereomer (1.24)

$MEMO_2C$–(15)–CO_2CH_3 $\xrightarrow[\text{(2) } C_6H_5CH_2Cl]{\text{(1) LiN(iso-}C_3H_7)_2\text{, THF}}$ $MEMO_2C$–$CH(CH_2C_6H_5)$–…–CO_2CH_3 (57%)

$MEM = CH_3OCH_2OCH_2$

Alkylation of unsymmetrical ketones bearing only α-alkyl substituents, however, generally leads to mixtures containing both α-alkylated products. The relative amounts of the two products depend on the structure of the ketone and may also be influenced by experimental factors such as the nature of the cation and the solvent (see Table 1.2). In the presence of the free ketone or of another proton source such as a protic solvent, equilibration of the two enolate ions can take place. Therefore if the enolate is prepared by slow addition of the base to the ketone, or if an excess of the ketone remains after the addition of base is complete, the equilibrium mixture of enolate anions is obtained containing preponderantly the more substituted enolate. Slow addition of the ketone to an

(1.25)

TABLE 1.2 *Composition of mixtures of enolate anions generated from the ketone and a triphenylmethyl metal derivative in dimethoxyethane*

(House, 1967)

Ketone, cation and reaction conditions	Enolate anion composition %	
2-methylcyclopentanone (CH_3)	enolate (O^-, CH_3), more substituted	enolate (O^-, CH_3), less substituted
K^+ (kinetic control)	55	45
K^+ (equil. control)	78	22
Li^+ (kinetic control)	28	72
Li^+ (equil. control)	94	6
trans-decalone (H, H, O)	enolate (H, H, O^-)	enolate (H, H, O^-)
Li^+ (kinetic control)	13	87
Li^+ (equil. control)	53	47
$C_4H_9CH_2COCH_3$	$C_4H_9CH{=}C(O^-){-}CH_3$	$C_4H_9CH_2C(O^-){=}CH_2$
K^+ (kinetic control)	46	54
K^+ (equil. control)	58	42
Li^+ (kinetic control)	30	70
Li^+ (equil. control)	87	13

excess of a strong base in an aprotic solvent, on the other hand, leads to the kinetic mixture of enolates; under these conditions the ketone is completely converted into the anion and equilibration does not occur.

The composition of mixtures of enolates formed under kinetic conditions differs from that of mixtures formed under equilibrium conditions. In general, enolate mixtures formed under kinetic conditions contain more of the less highly substitued enolate than the equilibrium mixture, reflecting the fact that the less hindered α-protons are removed more rapidly by the base. However, whichever method is used, mixtures of both structurally isomeric enolates are generally obtained, and mixtures of products result on alkylation. Di- and tri-alkylation products may also be formed (see p. 13) and it is not always easy to isolate the pure monoalkylated compound from the resulting complex mixtures. This is a serious problem in synthesis since in many cases it results in the loss of valuable starting materials.

A number of methods have been used to improve selectivity in the alkylation of unsymmetrical ketones and to reduce the amount of polyalkylation. One widely used procedure is to introduce temporarily an activating group at one of the α-positions to stabilise the corresponding enolate anion; this group is removed later after the alkylation has been effected. Common activating groups used for this purpose are the ethoxycarbonyl, ethoxyoxalyl and formyl groups. Thus, to prepare 2-methylcyclohexanone from cyclohexanone the best procedure is to go through the 2-ethoxycarbonyl derivative, which is easily obtained from the ketone by reaction with ethyl carbonate, or by condensation with diethyl oxalate followed by decarbonylation. Conversion into the enolate ion with a base such as sodium ethoxide takes place exclusively at the doubly activated position. Methylation with methyl iodide, and removal of the β-ketoester group with acid or base gives 2-methylcyclohexanone, free from polyalkylated products.

$(CO_2C_2H_5)_2$, $NaOC_2H_5$, C_2H_5OH; $COCO_2C_2H_5$; heat, borax; $CO_2C_2H_5$; (1) $NaOC_2H_5$, C_2H_5OH (2) CH_3I; CH_3, $CO_2C_2H_5$; boil, HCl or LiCl, DME; CH_3 (1.26)

Another technique is to block one of the α-positions by introduction of a removable substituent which *prevents* formation of the corresponding enolate. A widely used method is acylation with ethyl formate and transformation of the resulting formyl or hydroxymethylene substituent into a group that is stable to base, such as an enamine, an enol ether or an enol thioether. An example of this procedure is shown below in the preparation of 9-methyl-1-decalone from *trans*-1-decalone. Direct alkylation of this compound gives mainly the 2-alkyl derivative (1.27). A useful alternative procedure is alkylation of the dianion prepared from the formyl derivative with potassium amide (p. 9).

Specific enolate anions may also be obtained from unsymmetrical ketones by way of the structurally specific enol acetates or, better, the trimethylsilyl enol ethers. Reaction of the latter with one equivalent of methyl-lithium in dimethoxyethane affords the corresponding lithium

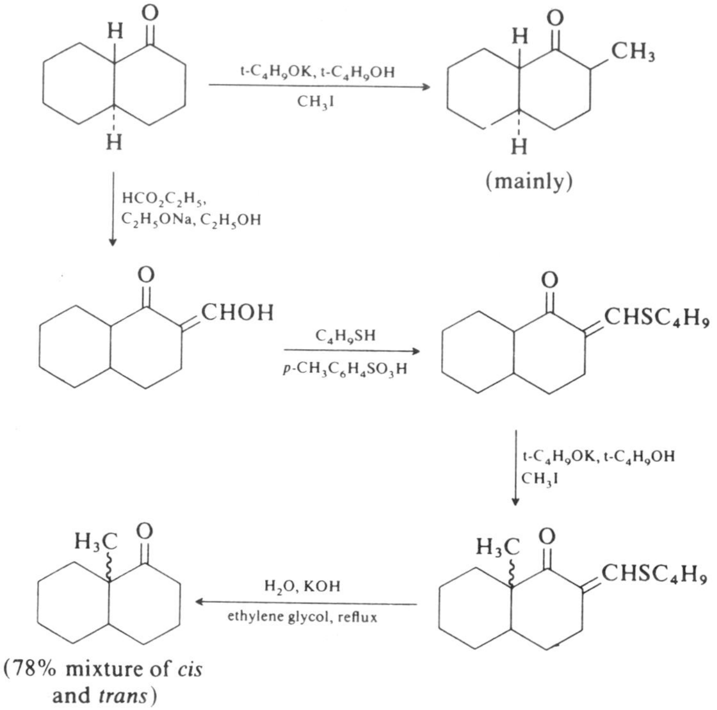

H O

H

$t\text{-}C_4H_9OK$, $t\text{-}C_4H_9OH$

CH_3I

H O CH_3

H

(mainly)

$HCO_2C_2H_5$,

C_2H_5ONa, C_2H_5OH

O CHOH

C_4H_9SH

p-$CH_3C_6H_4SO_3H$

O $CHSC_4H_9$

(1.27)

$t\text{-}C_4H_9OK$, $t\text{-}C_4H_9OH$

CH_3I

H_3C O

H_2O, KOH

ethylene glycol, reflux

H_3C O $CHSC_4H_9$

(78% mixture of *cis* and *trans*)

enolate along with inert tetramethylsilane. Equilibration of the enolate does not take place as long as care is taken to ensure the absence of proton donors such as an alcohol or an excess of ketone. Reaction with an alkyl halide then affords, largely, a specific monoalkylated ketone. It is rarely possible to obtain completely selective alkylation, because as soon as some monoalkyl ketone is formed in the reaction mixture it can bring about equilibration of the original enolate (cf. p. 15) but this difficulty is lessened by using the covalent *lithium* enolates which give maximum stabilisation of the enolate consistent with a resonable rate of alkylation.

This procedure is limited in its scope by the fact that methyl-lithium cannot be employed in the presence of certain other functional groups in the molecule, and, further, in its reactions with trimethylsilyl ethers, it gives rise to the relatively unreactive lithium enolates. It has recently been found that for the generation of enolate ions from trimethylsilyl ethers benzyltrimethylammonium fluoride can be used with advantage. The fluoride ion serves well to cleave the silyl ethers (p. 319), and the ammonium enolates produced are more reactive than the lithium analogues. Even relatively unreactive alkylating agents such as 1-iodobutane give reasonable yields of specifically alkylated products.

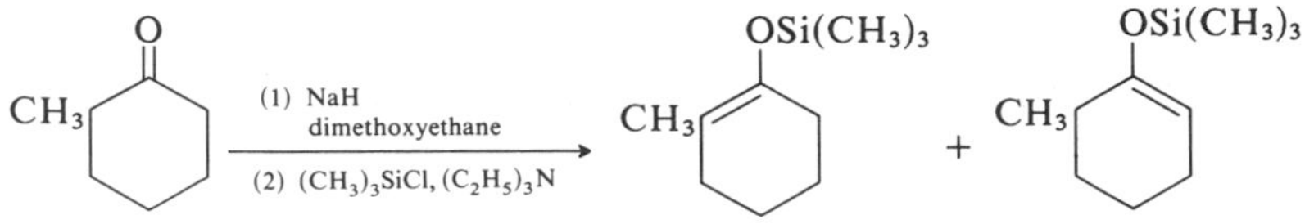

(1) NaH, dimethoxyethane; (2) $(CH_3)_3SiCl$, $(C_2H_5)_3N$

(78% of mixture) (22% of mixture)

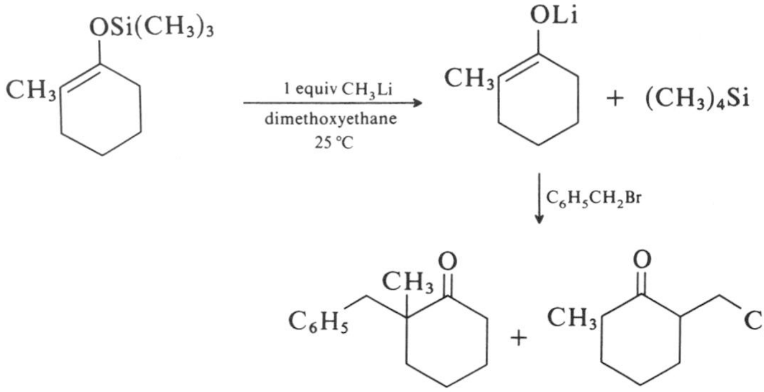

1 equiv CH_3Li, dimethoxyethane, 25 °C

$+ (CH_3)_4Si$ (1.28)

$C_6H_5CH_2Br$

84% 7%

Esters, epoxides and ketone functional groups are unaffected under the reaction conditions (Kuwajima, Nakamura and Shimizu, 1982).

The success of this approach to specific enolates is dependent on the availability of the pure silyl ethers. The more highly substituted isomers usually predominate in the mixture produced by reaction of the enolates, prepared under equilibrium conditions, with trimethylsilyl chloride (1.28) and in favourable cases they may be purified by distillation or by gas-liquid chromatography. The less highly substituted silyl ethers are obtained from the enolate prepared from the ketone under kinetic conditions with lithium diisopropylamide.

In addition to their use for the preparation of specific lithium enolates, trimethylsilyl enol ethers are also excellent substrates for the *acid*-catalysed alkylation of carbonyl compounds. In the presence of Lewis acids ($TiCl_4$, $SnCl_4$, ZnX_2) they react readily with *tertiary* alkyl halides to give the alkylated product in high yield (Reetz, 1982). This procedure thus complements the more common base-catalysed alkylation of enolates which fails with tertiary halides. It is supposed that the Lewis acid promotes ionisation of the halide, RX, to form the cation, R^+, which is

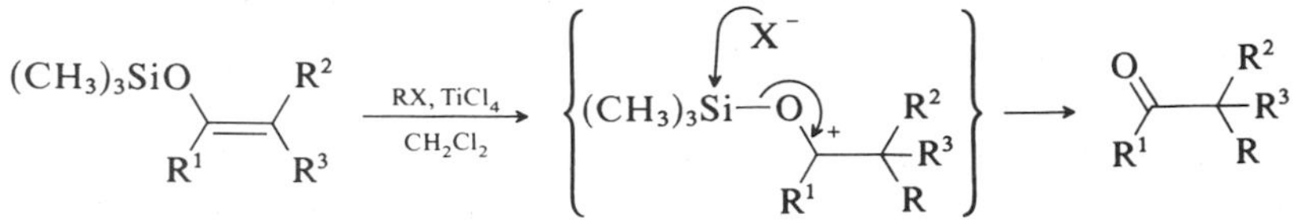

$(CH_3)_3SiO$, R^1, R^2, R^3 — RX, $TiCl_4$, CH_2Cl_2 → $\{(CH_3)_3Si$—O, X^-, R^1, R^2, R^3, R$\}$ → O, R^1, R^2, R^3, R

(16) (1.29)

trapped by the silyl enol ether to give the addition product (16) which is rapidly desilylated.

Thus, reaction of the thermodynamic trimethylsilyl enol ether (17) of 2-methylcyclohexanone with t-butyl chloride in the presence of titanium tetrachloride gave the product (18), containing two adjacent quaternary carbon atoms, in a remarkable 48 per cent yield. The aldehyde (19) similarly gave the α-t-butyl derivative (20) (Chan, Paterson and Pinsonnault, 1977).

$OSi(CH_3)_3$ (17) — (1) t-C_4H_9Cl, $TiCl_4$, CH_2Cl_2, −23 °C; (2) aq. Na_2CO_3 → (18) (48%) (1.30)

(19) CHO → $OSi(CH_3)_3$ — (1) t-C_4H_9Cl, $TiCl_4$, CH_2Cl_2, −23 °C; (2) aq. Na_2CO_3 → CHO

Alkylation can also be effected with reactive secondary halides, such as substituted benzyl halides, and with chloromethylphenyl sulphide, a primary halide. Thus, reaction of the benzyl chloride (21) with the trimethylsilyl enol ether derived from mesityl oxide in the presence of zinc bromide afforded a short and efficient route to the sesquiterpene (±)-ar-turmerone (Paterson, 1979). Reaction of chloromethylphenyl sulphide with the trimethylsilyl enol ethers of lactones in the presence of zinc bromide followed by pyrolytic elimination of the sulphoxide of the initial sulphide (22), provides a good route to the α-methylene lactone unit common in many cytotoxic sesquiterpenes. Desulphurisation of the sulphide with Raney nickel, instead of oxidation and elimination,

O — (1) LiN(iso-$C_3H_7)_2$, THF, −78 °C; (2) $(CH_3)_3SiCl$ → $(CH_3)_3SiO$ — Cl (21), $ZrBr_2$, CH_2Cl_2 → O (80%) (1.31)

affords the α-methyl derivatives (Paterson and Fleming, 1979). Esters and ketones behave similarly.

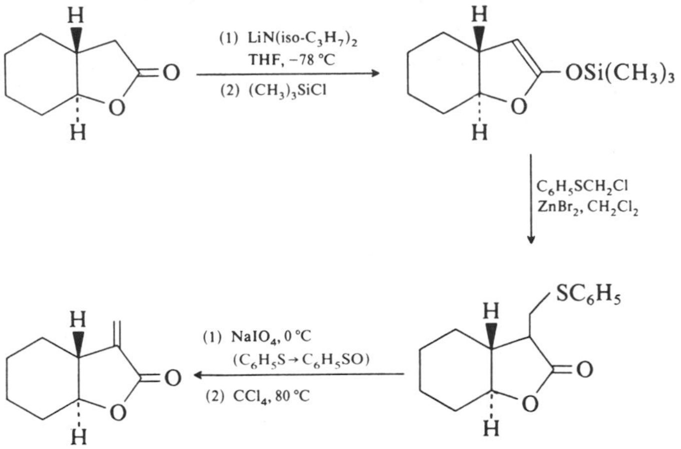

(1) $LiN(iso\text{-}C_3H_7)_2$, THF, −78 °C; (2) $(CH_3)_3SiCl$ → $OSi(CH_3)_3$

$C_6H_5SCH_2Cl$, $ZnBr_2$, CH_2Cl_2 → SC_6H_5 (22)

(1) $NaIO_4$, 0 °C ($C_6H_5S \rightarrow C_6H_5SO$); (2) CCl_4, 80 °C

(1.32)

Specific enolate ions of unsymmetrical ketones can also be obtained by reduction of the $\alpha\beta$-unsaturated ketones with lithium in liquid ammonia (cf. p. 441). Alkylation of the intermediate enolate affords an α-alkyl derivative of the corresponding *saturated* ketone (1.33) which may not be the same as that obtained by direct base-catalysed alkylation of the saturated ketone itself. For example, direct base-catalysed alkylation of 2-decalone generally leads to 3-alkyl derivatives, but by proceeding

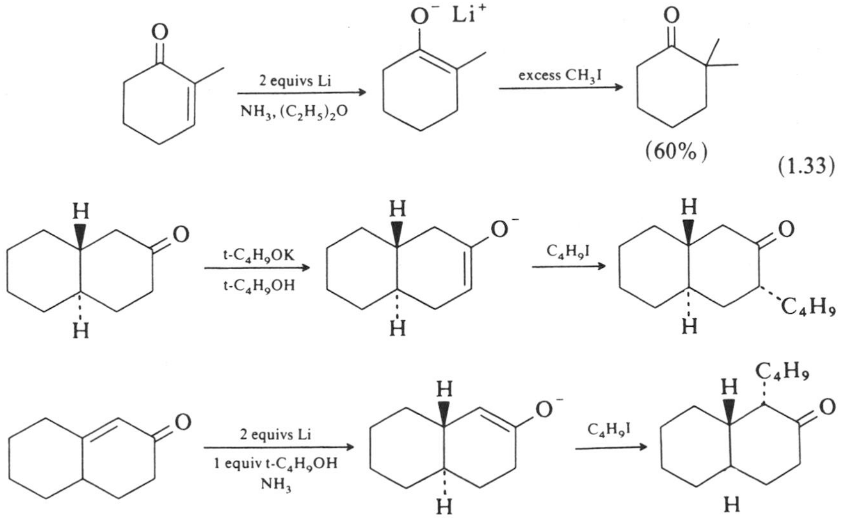

2 equivs Li, NH_3, $(C_2H_5)_2O$ → $O^-\ Li^+$ → excess CH_3I → (60%)

(1.33)

$t\text{-}C_4H_9OK$, $t\text{-}C_4H_9OH$ → O^- → C_4H_9I → C_4H_9

2 equivs Li, 1 equiv $t\text{-}C_4H_9OH$, NH_3 → O^- → C_4H_9I → C_4H_9

from the 1(9)enone (21) the 1-alkyl derivative is obtained. The success of this procedure depends on the fact that in liquid ammonia the alkylation step is faster than equilibration of the initially formed enolate with its isomer. Again it is essential to use the *lithium* enolates; with sodium and potassium salts equilibration takes place and mixtures of alkylated products are obtained. In practice, alkylations are best effected with methyl iodide and primary halides, or activated halides such as allyl or benzyl compounds. With secondary halides reactions are slower, leading to a loss in selectivity (Caine, 1976).

Alkylation of $\alpha\beta$-unsaturated ketones *without reduction of the double bond* can be arranged to give either the α, β or α' alkyl derivative. In general, treatment of an $\alpha\beta$-unsaturated ketone with a strong non-nucleophilic base such as lithium diisopropylamide under aprotic conditions results in the controlled formation of the α'-enolate anion (as 22). If alkylation of the latter is fast compared with enolate ion equilibration, the corresponding α'-alkyl derivative is obtained in good yield. Thus, 5,5-dimethyl-2-cyclohexenone on reaction with lithium diisopropylamide in tetrahydrofuran followed by methyl iodide gave the α'-methyl derivative (23) in 83 per cent yield (Lee *et al.*, 1973).

$\xrightarrow[\text{THF, 0 °C}]{\text{LiN(iso-}C_3H_7)_2}$ (22) $\xrightarrow{CH_3I}$ (23) (83%) (1.34)

In a different approach α'-alkylcyclohexenones have been obtained in high yield by reaction of the silyl enol ethers of the cyclohexenone epoxides with alkyl- or phenyl-cuprate reagents. 6-t-Butyl-2-cyclohexenone, for example, was obtained in quantitative yield by reaction of lithium t-butylcyanocuprate with the trimethylsilyl ether of cyclohexenone epoxide (Marino and Jaén, 1982). With Grignard reagents or alkyl-lithiums, however, the α-alkyl compound is formed in an S_N2 reaction (Wender, Erhardt and Letendre, 1981).

$OSi(CH_3)_3$ $\xrightarrow[\text{ether, −78 °C}]{(t\text{-}C_4H_9CuCN)Li}$ $t\text{-}C_4H_9$, $OSi(CH_3)_3$, OH $\xrightarrow{HCl,\ H_2O}$ $t\text{-}C_4H_9$ (1.35)

In contrast, if an $\alpha\beta$-unsaturated ketone is treated with a base under conditions which promote enolate ion equilibration, for example in the presence of a protic solvent, then not the α' but the more stable dienolate anion (24) is produced. Alkylation of (24) occurs preferentially at the α-carbon atom to give first the mono-alkyl-$\beta\gamma$-unsaturated ketone. The α-proton in this compound is readily removed by interaction with the base or the original enolate (24), since it is activated both by the carbonyl group and the carbon–carbon double bond. In the presence of excess of alkylating agent the resulting anion is again alkylated at the α-position and the α,α-dialkyl-$\beta\gamma$-unsaturated ketone is produced. If the alkylating agent is restricted, however, so that further alkylation does not occur, the thermodynaically more stable α-alkyl $\alpha\beta$-unsaturated ketone gradually accumulates (1.36).

$$\overset{\alpha'}{RCH_2}CO\overset{\alpha}{C}H{=}\overset{\beta}{C}H\overset{\gamma}{C}H_2R^1 \underset{}{\overset{\text{base (slow)}}{\rightleftharpoons}} RCH_2\overset{O^-}{\overset{|}{C}}{=}CH{-}CH{=}CHR^1 \;(24)$$

$$(24) \xrightarrow{R^2X} RCH_2{-}\overset{O^-}{\overset{|}{C}}{=}\underset{R^2}{\underset{|}{C}}{-}CH{=}CHR^1 \overset{\text{base (fast)}}{\rightleftharpoons} RCH_2CO\underset{R^2}{\underset{|}{C}}HCH{=}CHR^1$$

$$RCH_2{-}\overset{O^-}{\overset{|}{C}}{=}\underset{R^2}{\underset{|}{C}}{-}CH{=}CHR^1 \xrightarrow{H_3O^+} RCH_2CO\underset{R^2}{\underset{|}{C}}{=}CHCH_2R^1$$

$$RCH_2{-}\overset{O^-}{\overset{|}{C}}{=}\underset{R^2}{\underset{|}{C}}{-}CH{=}CHR^1 \xrightarrow{R^2X} RCH_2COC(R^2)(R^2){-}CH{=}CHR^1$$

In accordance with this scheme it is found that dialkylation is diminished by slow addition of the alkylating agent or by use of a less reactive alkylating agent (for example, methyl chloride instead of methyl iodide in 1.37), thus allowing isomerisation of the $\beta\gamma$-unsaturated ketone to the $\alpha\beta$-isomer to take place before alkylation.

Steroid enone (H_3C at ring junction) $\xrightarrow[\text{(2) } CH_3I \text{ added rapidly}]{\text{(1) } t\text{-}C_4H_9OK,\ t\text{-}C_4H_9OH}$ 4,4-dimethyl (H_3C, CH_3) $\beta\gamma$-unsaturated ketone (65%) + 4-CH_3 $\alpha\beta$-unsaturated ketone (3%)

(1.37)

Steroid enone (H_3C at ring junction) $\xrightarrow[\text{(2) } CH_3Cl \text{ added over 20 h}]{\text{(1) } t\text{-}C_4H_9OK,\ t\text{-}C_4H_9OH}$ 4-CH_3 $\alpha\beta$-unsaturated ketone (40%) + 4,4-dimethyl (H_3C, CH_3) $\beta\gamma$-unsaturated ketone (12%)

A disadvantage of the above procedure is that it generally gives mixtures of products, particularly in experiments aimed at the monoalkylated compound. A much better route to the latter proceeds through the metalloenamines derived from cyclohexylimines or *N,N*-dimethylhydrazones (Stork and Benaim, 1971) (1.38). Dialkylation does not occur because transfer of a proton from the monoalkylated compound to the metalloenamine is slow.

CH_3 $C_6H_{11}N$ — slightly less than 1 equiv of $LiN(iso\text{-}C_3H_7)_2$, THF, 8 h → CH_3 Li^+ $C_6H_{11}\bar{N}$

CH_3I (1.38)

CH_3 O CH_3 (90%) ← CH_3COOH, H_2O, reflux — CH_3 $C_6H_{11}N$ CH_3

$\alpha\beta$-Unsaturated esters on treatment with lithium diisopropylamide followed by an alkylating agent similarly give α-alkyl derivatives of the corresponding $\beta\gamma$-unsaturated esters by a deconjugative alkylation. With $\beta\gamma$-unsaturated esters, alkylation of the lithium enolates affords the α-alkyl ester, with complete retention of double-bond geometry and position (Kende and Toder, 1982) (1.39).

$CO_2C_2H_5$ — (1) $LiN(iso\text{-}C_3H_7)_2$, THF, −78 °C (2) CH_3I → $CO_2C_2H_5$ (90%)

(1.39)

$CO_2C_2H_5$ → $CO_2C_2H_5$ (84%)

$\alpha\beta$-Unsaturated ketones are converted into β-alkyl derivatives of the corresponding *saturated* ketone by conjugate addition of Grignard reagents or organocopper reagents (see p. 81). Reaction of the intermediate metal enolates with phenylselenyl bromide affords the α-phenylseleno-ketone from which the β-alkyl-$\alpha\beta$-unsaturated ketone is smoothly obtained by oxidation followed by selenoxide elimination (Reich, Renga and Reich, 1974) (1.40).

$$C_6H_5COCH{=}CHCH_3 \xrightarrow{(CH_3)_2CuLi} C_6H_5C(OM){=}CHCH(CH_3)_2 \xrightarrow{C_6H_5SeBr} C_6H_5COCH(SeC_6H_5)CH(CH_3)_2 \xrightarrow{H_2O_2,\ CH_2Cl_2} C_6H_5COCH(Se(\rightarrow O)C_6H_5)CH(CH_3)_2 \xrightarrow[CH_2Cl_2]{\text{reflux}} C_6H_5COCH{=}C(CH_3)_2\ (88\%) \quad (1.40)$$

The stereochemistry of the product obtained in the alkylation of cyclic ketones is important in synthesis but is not always easy to predict. If the product still contains a hydrogen atom attached to the carbon which has been alkylated, equilibration under the alkaline conditions of the reaction generally leads to formation of the more stable isomer, irrespective of the initial direction of attack on the enolate anion. Thus, in the methylation of cholestanone, the product is the 2-α-methyl compound in which the methyl group has the more stable equatorial conformation.

If the alkylated product has no hydrogen atom at the alkylated position, so that equilibration is not possible, then the product is usually that formed by orthogonal attack on the less hindered side of the enolate anion. This is particularly the case with relatively rigid ketones which cannot easily adopt different conformations. Hence, methylation of norcamphor (25) with sodium triphenylmethyl and methyl iodide leads almost entirely to the *exo* methyl derivative, even though equilibration studies show that the *endo* isomer is (slightly) more stable. Further alkylation of this methyl derivative with methylpentenyl chloride then

$$\text{(25)} \xrightarrow[\text{ether}]{NaC(C_6H_5)_3} \text{enolate } O^- \xrightarrow{CH_3I} \text{exo-}CH_3,\ H \xrightarrow[(2)\ (CH_3)_2C{=}CH(CH_2)_2Cl]{(1)\ NaNH_2,\ THF} -(CH_2)_2CH{=}C(CH_3)_2,\ CH_3 \quad (1.41)$$

gives the *endo*-methyl-*exo*-methylpentenyl derivative, again by attack of the alkylating agent from the less hindered *exo* side of the enolate ion (1.41). Reversing the order of the alkylations leads to the epimeric product, showing that the sterochemistry is controlled by the direction of attack on the enolate ion and not by the size of alkyl group. If the enolate anion is flexible, both conformational factors and steric hindrance in the anion may influence the steric course of the alkylation, and it is often difficult to predict what the stereochemical result in a particular case will be. Mixtures of isomers often result particularly in the alkylation of open-chain enolates.

1.5. The enamine and related reactions

The enamine reaction, introduced by Stork and his co-workers and widely used in synthesis (Stork *et al.*, 1963; Cook, 1968; Hickmott, 1982; Whitesell and Whitesell, 1983), provides a valuable alternative method for the selective alkylation and acylation of aldehydes and ketones.

Enamines are $\alpha\beta$-unsaturated amines and are simply obtained by reaction of an aldehyde or ketone with a secondary amine in the presence of a dehydrating agent such as potassium carbonate or, better, by heating in benzene solution in the presence of a catalytic amount of toluene-*p*-sulphonic acid, with azeotropic removal of the water formed (1.42). The amines found most generally useful in forming enamines are pyrrolidine, morpholine and piperidine in descending order of reactivity. All of the steps of the reaction are reversible and enamines are readily hydrolysed by water to reform the carbonyl compound. All reactions of enamines must therefore be conducted under anhydrous conditions, but once reaction has been effected the modified carbonyl compound is easily liberated from the product by addition of dilute acid to the reaction mixture.

$$\mathrm{RCH_2COR^1 + HNR^2R^3} \rightleftharpoons \mathrm{RCH_2C(R^1)(OH)NR^2R^3}$$

$$\rightleftharpoons \mathrm{RCH_2C(R^1){=}\overset{+}{N}R^2R^3\ OH^-} \rightleftharpoons \mathrm{RCH{=}C(R^1){-}NR^2R^3} \qquad (1.42)$$

The usefulness of enamines in synthesis is due to the fact that there is some negative charge on the β-carbon atom (i.e. the α-carbon atom of the original carbonyl compound) which can therefore act as a nucleophile in reactions with alkyl and acyl halides and with electrophilic

$$\text{>C=C-}\ddot{\text{N}}\text{<} \leftrightarrow \text{>}\bar{\text{C}}\text{-C=}\overset{+}{\text{N}}\text{<} \qquad (1.43)$$

alkenes. Reaction with alkyl halides, for example, leads irreversibly to *C*-alkylated and *N*-alkylated products. Subsequent hydrolysis of the *C*-alkylated iminium salt gives the alkylated ketone; the *N*-alkylated product is usually water soluble and is unaffected by the hydrolysis.

$$RCH{=}C(R^1){-}\ddot{N}R^2R^3 \leftrightarrow R\bar{C}H{-}C(R^1){=}\overset{+}{N}R^2R^3$$

$$\xrightarrow{CH_3I,\ C_6H_6,\ \text{reflux}} RCH{=}C(R^1){-}\overset{+}{N}(CH_3)R^2R^3\ I^- + RCH(CH_3){-}C(R^1){=}\overset{+}{N}R^2R^3\ I^- \qquad (1.44)$$

$$\xrightarrow{\text{hydrolysis}} RCH(CH_3){-}C(R^1){=}O$$

Enamine alkylation has a number of advantages over direct base-catalysed alkylation of aldehydes and ketones. Since no base or other catalyst is required, there is less tendency for wasteful self-condensation reactions of the carbonyl compound, and even aldehydes can be alkylated and acylated in good yield. A valuable feature of the enamine reaction is that it is regioselective. In the alkylation of an unsymmetrical ketone the product of reaction at the *less* substituted α-carbon atom is formed in preponderant amount, in contrast to direct base-catalysed alkylation of unsymmetrical ketones which usually gives a mixture of products. For example, reaction of the pyrrolidine enamine of 2-methylcyclohexanone with methyl iodide gives 2,6-dimethylcyclohexanone almost exclusively. This selectivity derives from the fact that the enamine from an unsymmetrical ketone consists mainly of the more reactive isomer in which the double bond is directed toward the less substituted carbon. In the 'more-substituted' enamine there is decreased interaction between the nitrogen

$$\text{2-methylcyclohexanone} + \text{pyrrolidine} \xrightarrow[C_6H_6]{p\text{-}CH_3C_6H_4SO_3H} \text{(85\% of mixture)} + \text{(15\% of mixture)} \qquad (1.45)$$

lone pair and the π-system of the double bond because of steric interference between the α-substituent (CH_3 in the example) and the α-methylene group of the amine (cf. Doering, *et al.*, 1985).

Alkylation of enamines with alkyl halides generally proceeds in only poor yield because the main reaction is *N*-alkylation rather than *C*-alkylation. Good yields of alkylated products are obtained using reactive halides such as benzyl or allyl halides, and it is believed that in these cases there is migration of the substituent group from nitrogen to carbon. This may take place in some cases by an intramolecular pathway, resulting in rearrangement of allyl substituents or, in other cases, by dissociation of the *N*-alkyl derivative followed by irreversible *C*-alkylation. Fortunately, a modification of the original procedure provides a way round this difficulty. It is found that many *imines* formed from aliphatic primary amines and enolisable aldehydes and ketones are readily deprotonated on treatment with ethyl-magnesium bromide (Stork and Dowd, 1963, 1974) or with lithium diisopropylamide at a low temperature (Wittig and Frommeld, 1964). The metal salts so formed give high yields of monoalkylated carbonyl compounds on reaction with primary and secondary alkyl halides. With imines derived from methyl ketones, and reaction at low temperature, alkylation takes place at the methyl group; with other dialkyl ketones regioselective alkylation at either α position can be realised by judicious choice of experimental conditons (1.46) (Smith *et al.*, 1983). Aldehydes and ketones also react readily with the metalloenamines to form aldols or $\alpha\beta$-unsaturated carbonyl compounds.

CHO + t-$C_4H_9NH_2$ —(benzene, reflux)→ NC_4H_9-t

—(C_2H_5MgBr, THF, reflux)→ $\bar{N}C_4H_9$-t, $\overset{+}{M}gBr$

—((1) $C_6H_5CH_2Cl$; (2) H_3O^+)→ CHO, C_6H_5 (80%)

(1.46)

O → NC_4H_9 —((1) LiN(iso-C_3H_7)$_2$, DME, −60 °C; (2) CH_3I; (3) H_3O^+)→ O (98%)

A useful alternative route to lithio-enamines proceeds from allylamines. The corresponding Schiff bases prepared from aryl aldehydes are rapidly converted into the N-alkenylimines on treatment with potassium t-butoxide, whence the metalloenamine is obtained by addition of t-butyl-lithium (1.47) (Wender and Schaus, 1978).

NH_2 $\xrightarrow{C_6H_5CHO}$ N C_6H_5 $\xrightarrow[THF]{t\text{-}C_4H_9OK}$ N C_6H_5

t-C_4H_9Li, THF, −78 °C

CHO (94%) $\xleftarrow[(2)\ H_3O^+]{(1)\ C_2H_5I,\ -78\ °C}$ N C_6H_5 t-C_4H_9

(1.47)

A valuable application of metalloenamines is in the preparation of α-alkyl-$\alpha\beta$-unsaturated ketones as shown on p. 24. These are difficult to prepare from the $\alpha\beta$-unsaturated ketones themselves because of the facile formation of the α,α-dialkyl-$\beta\gamma$-unsaturated ketones.

The metalloenamine reaction forms the basis of a highly efficient method for the enantionselective alkylation of ketones, using the optically active amine (27) to make the imines (Meyers *et al.*, 1981: Meyers, Williams, White and Erickson, 1981). Alkylation of ketone enolates themselves does not lend itself to enantioselective control because of the symmetry of the enolate π-system. But with the imine (28) prepared from cyclohexanone and (27), optically active 2-ethylcyclohexanone was obtained with high optical purity. The methoxy group in (27) plays an important part in the asymmetric induction, for in its absence optical yields are much lower. It is suggested that reaction takes place by way of a chelated lithium compound of the form (29) in which the five-membered chelate ring imparts some rigidity and conformational control in the molecule. Alkylation is then directed specifically to one face of the double bond (the top face in the example) by the lithium atom which

C_6H_5 H H_2N OCH_3 (27)

C_6H_5 H N OCH_3 (28) $\xrightarrow[(2)\ C_2H_5I\ (3)\ H_3O^+]{(1)\ LiN(iso\text{-}C_3H_7)_2,\ THF,\ -20\ °C}$ O C_2H_5 H

(82% yield; 94% enantiomeric excess)

(1.48)

is co-ordinated also to the halogen of the alkylating agent. This is not the whole story, however, and other factors including the experimental conditions may also be involved.

The reaction can also be used with acyclic ketones although an ambiguity is introduced by the possibility of geometrical isomerism about the carbon–carbon double bond in the lithiated imine. Thus, the imine (30) from 5-nonanone, metallated at −30 °C with lithium diisopropylamide gave the kinetic lithio-enamine (31), but this was converted into the thermodynamically more stable (32) on refluxing the solution. Alkylation of (32) with methyl iodide and hydrolysis gave the α methylated ketone (33) of 94 per cent optical purity. Methylation of (31) gave the same product but in only 31 per cent optical purity. This limitation appears to be general. High optical yields of alkylated open chain ketones can apparently only be obtained by the metalloenamine route if the thermodynamically favoured lithioenamine is used in the alkylation step.

Alkylation of enamines of aldehydes and ketones can also be effected with electrophilic alkenes, such as $\alpha\beta$-unsaturated ketones, esters or nitriles to give high yields of monoalkylated carbonyl compounds, and the sequence provides a useful alternative to base-catalysed Michael addition. In these reactions N-alkylation is reversible and good yields of C-alkylated products are usually obtained. Reaction again takes place at the less substituted α-carbon atom (1.51).

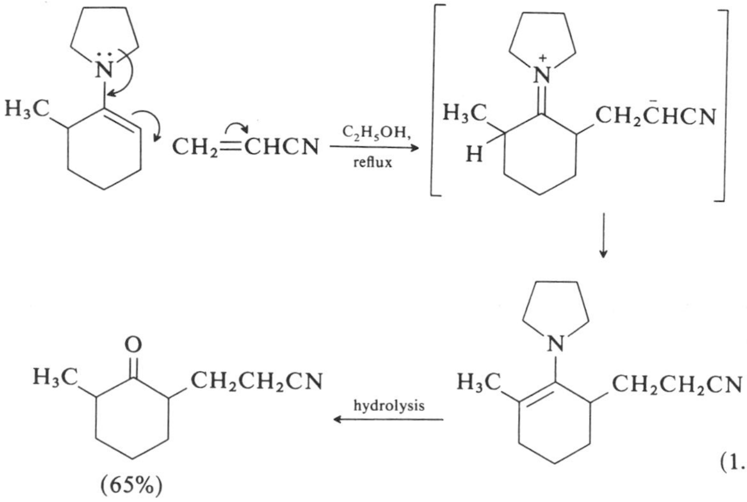

H_3C ... $CH_2{=}CHCN$ $\xrightarrow[\text{reflux}]{C_2H_5OH,}$ [H_3C, H ... $CH_2\bar{C}HCN$] → H_3C ... CH_2CH_2CN $\xrightarrow{\text{hydrolysis}}$ H_3C ... CH_2CH_2CN (65%)

(1.51)

$$C_2H_5CH{=}CH{-}N\langle\;\rangle + CH_2{=}CHCO_2CH_3 \xrightarrow[(2)\ H_3O^+]{(1)\ CH_3CN,\ \text{reflux}} C_2H_5CH(CHO)CH_2CH_2CO_2CH_3$$

(67%)

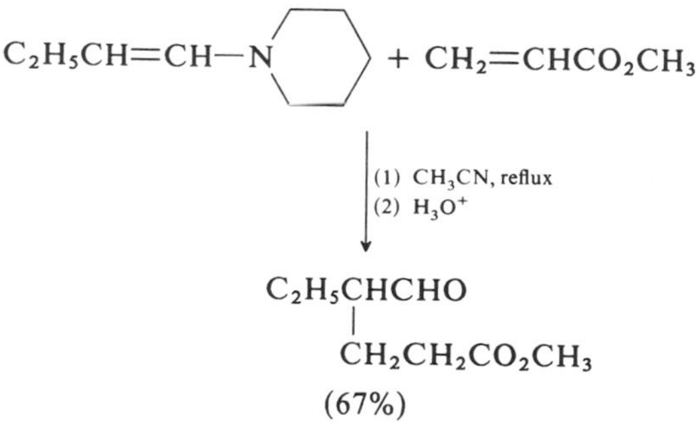

Enamines also react readily with acid chlorides or anhydrides to give products which, on hydrolysis, afford β-diketones or β-ketoesters. Reaction at the nitrogen atom is again reversible, and good yields of C-acylated products are obtained. The morpholine enamine of cyclohexanone, for example, and heptanoyl chloride give 2-heptanoylcyclohexanone in 75 per cent yield. In these reactions triethylamine is often added to neutralise the hydrogen chloride formed which would otherwise combine with the enamine; alternatively, two equivalents of enamine may be used (1.52)

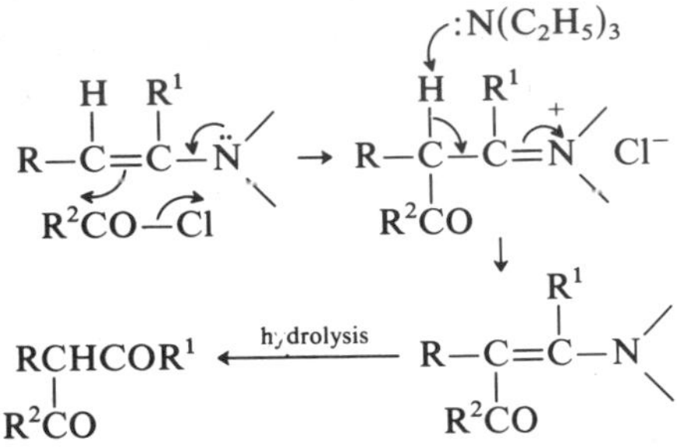

$$\mathrm{R{-}C(H){=}C(R^1){-}\ddot{N}<} + \mathrm{R^2CO{-}Cl} \rightarrow \mathrm{R{-}C(H)(COR^2){-}C(R^1){=}\overset{+}{N}<\ Cl^-} \xrightarrow{:N(C_2H_5)_3} \mathrm{R{-}C(COR^2){=}C(R^1){-}N<} \xrightarrow{\text{hydrolysis}} \mathrm{RCH(COR^2)COR^1} \quad (1.52)$$

Endocyclic enamines such as pyrrolines and Δ^2-tetrahydropyridines are useful for the synthesis of more complex nitrogen heterocyclic compounds, as in the alkaloids (Wenkert, 1968; Stevens, 1977). Thus, reaction

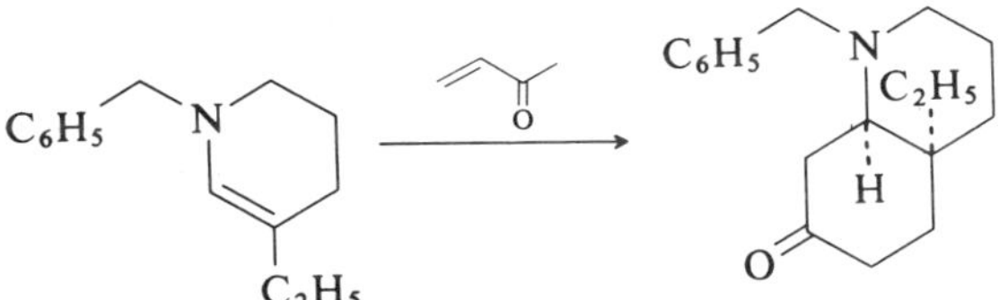

(34) → (35)

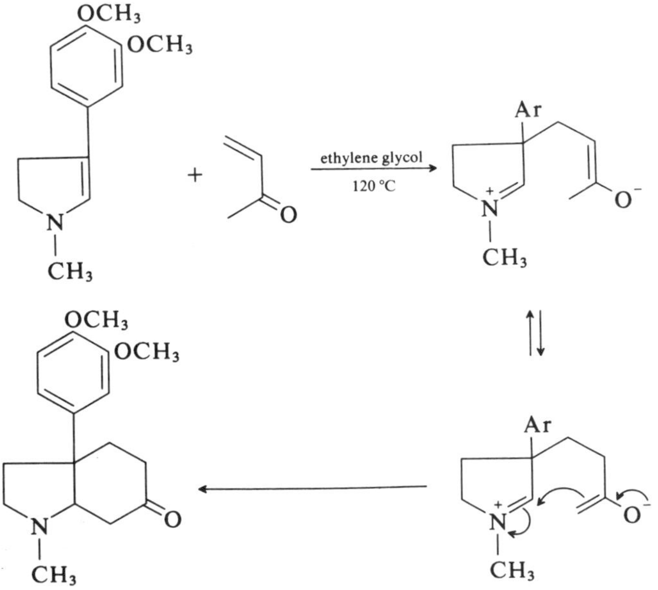

ethylene glycol, 120 °C

Mesembrine (56%)

of the Δ^2-tetrahydropyridine (34) with methyl vinyl ketone gave the *cis* fused hydroquinolone (35), subsequently converted into the alkaloid aspidospermine, and the alkaloid mesembrine was obtained as shown.

A useful alternative to the metalloenamine route to alkyl derivatives proceeds not from an imine but from the dimethylhydrazone of an aldehyde or ketone (Corey and Enders, 1976). These compounds, on reaction with lithium diisopropylamide or *n*-butyl-lithium, are converted into lithium derivatives which can be alkylated with alkyl halides, epoxides and carbonyl compounds. At the end of the sequence the dimethylhydrazine grouping is removed by oxidation with sodium periodate or other means, liberating the alkylated aldehyde or ketone. Reactions take place easily under mild conditions and with high positional and steroselectivity. Generally alkylation occurs at the less substituted α-position of an unsymmetrical ketone, and with cyclohexanone derivatives, *axial* methylation is highly favoured. Thus, methyl pentyl ketone was converted into ethyl pentyl ketone in 95 per cent yield, and 2-methyl-cyclohexanone gave trans-2,6-dimethylcyclohexanone (1.54). With epoxides, γ-hydroxycarbonyl compounds and thence, by oxidation, 1,4-dicarbonyl compounds are obtained in high yield, and reaction with aldehydes and ketones leads to β-hydroxycarbonyl compounds formed, in effect, by a 'directed' aldol condensation (cf. p. 48). One advantage which lithiated dimethylhydrazones have over the metalloenamines is

NN(CH$_3$)$_2$ —[LiN(iso-C$_3$H$_7$)$_2$ / THF, 0 °C]→ N(CH$_3$)$_2$, N, Li$^+$

↓ CH$_3$I

NN(CH$_3$)$_2$ —[NaIO$_4$ / CH$_3$OH, H$_2$O]→ O

NN(CH$_3$)$_2$, CH$_3$ —[(1) LiN(iso-C$_3$H$_7$)$_2$ (2) CH$_3$I]→ NN(CH$_3$)$_2$, CH$_3$, CH$_3$ (1.54)

—[NaIO$_4$ / CH$_3$OH, H$_2$O]→ O, CH$_3$, CH$_3$

(95%; 97% *trans*)

that they are readily converted into organocuprates which themselves take part in a range of carbon–carbon bond-forming reactions (p. 78). Thus, the keto-aldehyde (36) was obtained in 70 per cent yield by reaction of the cuprate derived from acetaldehyde dimethylhydrazone with cyclohexenone. The lithium compound itself reacts at the carbonyl group.

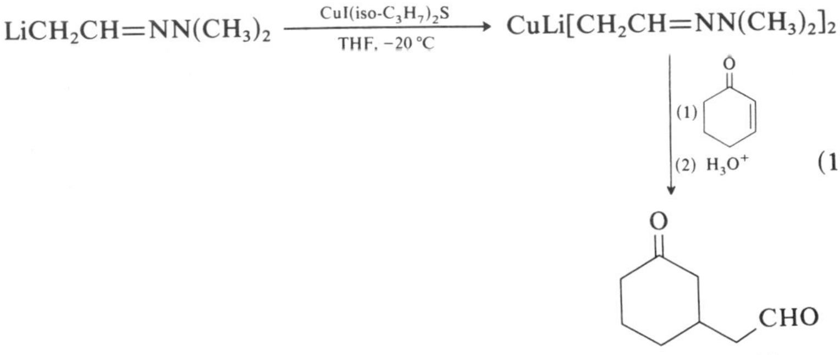

$$LiCH_2CH{=}NN(CH_3)_2 \xrightarrow[\text{THF, }-20\,^{\circ}C]{CuI(iso\text{-}C_3H_7)_2S} CuLi[CH_2CH{=}NN(CH_3)_2]_2 \xrightarrow[(2)\ H_3O^+]{(1)\ \text{cyclohexenone}} \quad (1.55)$$

(36) (70%)

Alkylation of lithiated dimethylhydrazones forms the basis of another efficient method for the asymmetric alkylation of aldehydes and ketones, using the optically active hydrazines (*S*)-1-amino-2-(methoxymethyl)-pyrrolidine (SAMP) (37) and its enantiomer (RAMP) as chiral auxiliary groups. Deprotonation of the optically active hydrazones, alkylation and removal of the chiral auxiliary hydrazine under mild conditions (ozonolysis or acid hydrolysis of the *N*-methyl quaternary salt) affords the alkylated aldehyde or ketone with, generally, greater than 95 per cent

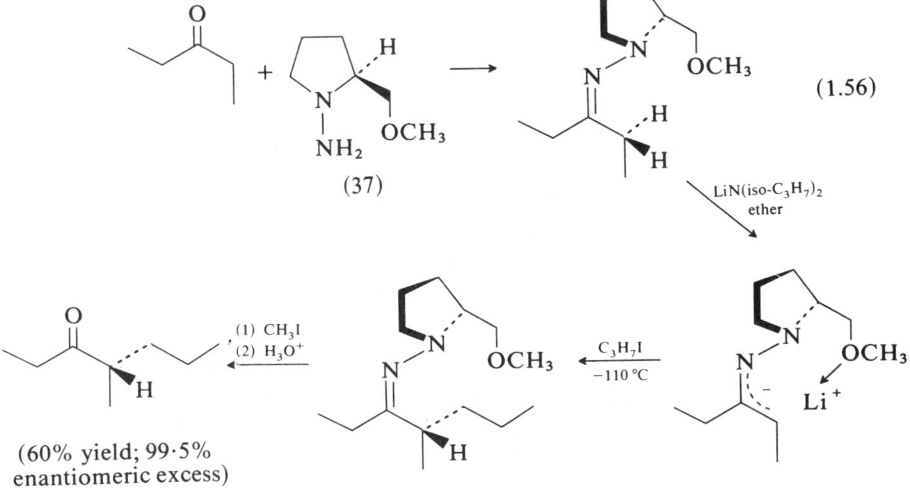

(1.56)

(37)

$LiN(iso\text{-}C_3H_7)_2$, ether

C_3H_7I, $-110\,^{\circ}C$

(1) CH_3I (2) H_3O^+

(60% yield; 99·5% enantiomeric excess)

optical purity (Enders *et al.*, 1984). This procedure has already been exploited in the asymmetric synthesis of several natural products. Thus, (+)-(*S*)-4-methyl-3-heptanone, the principal alarm pheromone of the leaf-cutting ant *Atta texana* was synthesised from 3-pentanone in greater than 99 per cent optical purity as shown in (1.56).

Another method for the regioselective alkylation of unsymmetrical ketones proceeds from the oximes (Kofron and Yeh, 1976). Lithium derivatives of oximes are formed preferentially at the α-carbon atom which is *syn* to the oxygen, presumably as a result of chelate formation. Reaction with an alkyl halide then occurs only on that side to give a specific α-alkyl derivative. Directed aldol condensations are also possible. Thus, dilithiocyclohexanone oxime and acetone gave only 2-isopropylidenecyclohexanone in 48 per cent yield, after cleavage of the first-formed oxime.

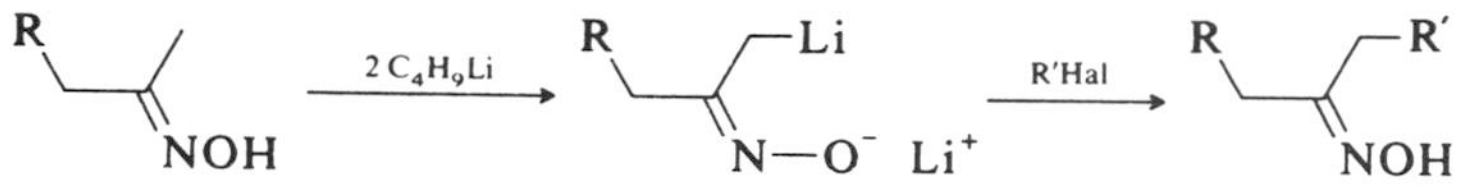

R, NOH —($2\,C_4H_9Li$)→ R, Li, $N{-}O^-\ Li^+$ —(R'Hal)→ R, R', NOH (1.57)

Base-catalysed alkylation can also be used for the preparation of α-alkylcarboxylic esters but the products, of course, are racemic. A highly effective method for the asymmetric synthesis of α-alkylcarboxylic acids by way of an optically active oxazoline is described on p. 72. Another procedure makes use of optically active amides derived from prolinol (38) (Evans and Takacs, 1980) or the two optically active 2-oxazolidones (39) and (40), readily available from (*S*)-valine and (1*S*, 2*R*)-norephedrine respectively (Evans, Ennis and Mathre, 1982). With lithium diisopropylamide, amides derived from (39) and (40) and carboxylic

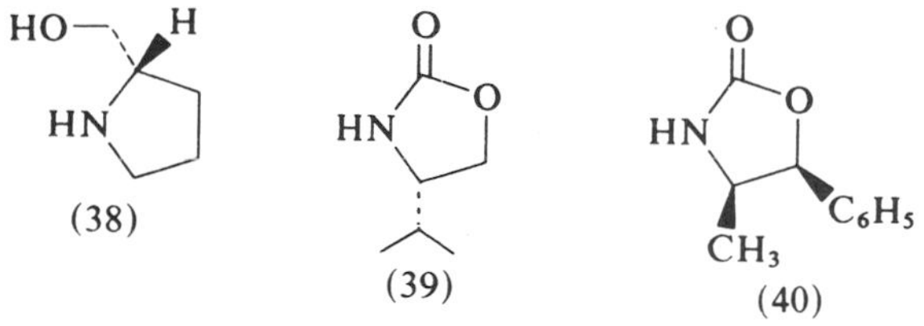

(38) (39) (40) (1.58)

acids are cleanly converted into their respective *Z*-metal enolates with >100:1 stereoselection. Alkylation with reactive alkylating agents (methyl iodide, allyl and benzyl halides) affords alkylated products with very high diastereoselection and thence by careful hydrolysis, the corresponding carboxylic acids. Thus the amide (41) derived from (40) gave the alkylated product (42) with greater than 99:1 diastereoselection and thence, by reaction with the lithium salt of benzyl alcohol and hydrogenolysis of the resulting benzyl ester, almost optically pure

(*S*)-α-benzylpropanoic acid (1.59). A similar sequence of reactions using the propionamide derived from (39) gave optically pure (*R*)-α-benzylpropanoic acid.

(41) → LiN(iso-C_3H_7)$_2$, THF, −78 °C → ; $C_6H_5CH_2Br$ → (42) (75%; >99:1) → $C_6H_5CH_2OLi$, $C_6H_5CH_2OH$ → ; H_2, Pd → (1.59)

Some intriguing examples of the α-alkylation of α-amino acids without loss of chirality and without the use of an auxiliary chiral reagent have been effected. Thus, proline gave optically pure α-methylproline by the sequence illustrated in (1.60) (Seebach *et al.*, 1983). Reaction of proline

t-C_4H_9CHO, H^+ → (43) → LiN(iso-C_3H_7)$_2$, THF, −78 °C → (44) → CH_3I → (45) → H_3O^+ → (1.60) with pivaldehyde gave the single stereoisomer (43) which was deprotonated with lithium diisopropylamide to the chiral non-racemic enolate (44). Attack on the enolate with asymmetric induction by the acetal centre takes place entirely on the same side of the bicyclic system as the t-butyl group to give (45), with the methyl and t-butyl groups on the exo side of the bicyclic system, which is readily hydrolised to optically pure α-methylproline. Related sequences were employed to prepare α-alkyl derivatives of cysteine (Seebach and Weber, 1983) and threonine (Seebach and Aebi, 1983).

α-Alkylation of amines is a valuable synthetic transformation. The amino group itself is not sufficiently 'activating' to allow conversion of an α-methyl or methylene group into an alkali metal salt, but certain derivatives of secondary amines (46), where X is a group which confers some acidity on the α-hydrogen atoms of the amine and at the same time confers some stability on the lithium salt, can be converted into lithium salts with a strong base such as lithium diisopropylamide. The lithium salts react readily with alkyl bromides and iodides, and with aldehydes and acid chlorides. Removal of the group X then affords the α-alkylated amine.

$$R{-}NH(CH_2R^1) \rightarrow \underset{(46)}{R{-}N(X){-}CH_2R^1} \longrightarrow R{-}N(X){-}CH(Li)R^1 \xrightarrow{R^2X} R{-}N(X){-}CH(R^1)R^2 \rightarrow R{-}NH{-}CHR^1R^2 \qquad (1.61)$$

Successful results have been obtained with certain amides (Beak and Reitz, 1978) and nitroso derivatives (Seebach and Enders, 1975), of

$$(H_3C)_2NH \xrightarrow[\text{toluene, cat. }(NH_4)_2SO_4\text{, reflux}]{(H_3C)_2N{-}CH{=}NC_4H_9\text{-t}} \underset{(47)}{(H_3C)_2N{-}CH{=}NC_4H_9\text{-t}} \xrightarrow{t\text{-}C_4H_9Li,\ -78\,^{\circ}C} CH_3N(CH_2Li){-}CH{=}NC_4H_9$$

$$\xrightarrow{(1)\ C_6H_5CH_2Br;\ (2)\ KOH,\ CH_3OH,\ H_2O} \underset{(48)}{C_6H_5CH_2CH_2NHCH_3}$$

$$\xrightarrow{(1)\ C_6H_5CHO;\ (2)\ KOH,\ CH_3OH,\ H_2O} \underset{(49)}{C_6H_5CH(OH)CH_2NHCH_3} \qquad (1.62)$$

secondary amines, but the most promising route to date uses the readily available α-formamidines (as 47). Dimethylamine, for example, was readily converted into N-methylphenylethylamine (48) and the hydroxy derivative (49), and pyrrolidine gave 2-n-butylpyrrolidine in 70 per cent yield (Meyers *et al.*, 1984).

By using an optically active amine to prepare the amidine instead of t-butylamine, asymmetric alkylation of secondary amines has been achieved (Meyers and Fuentes, 1983). The best results so far have been obtained using the optically active amine (50). With this auxiliary agent, 1,2,3,4-tetrahydroisoquinoline has been converted into a series of *S*-1-alkyl derivatives with greater than 90 per cent optical purity.

$$\xrightarrow[(2)\ RX]{(1)\ LiN(iso\text{-}C_3H_7)_2} \qquad \xrightarrow[CH_3CO_2H]{H_2NNH_2} \qquad (1.63)$$

NR* ... H R NR* ... NH H R

$(CH_3)_3SiO$, C_6H_5, H, H, H_2N, $OSi(CH_3)_3$

(50)

Primary amines also have been converted into α-alkyl derivatives by a sequence involving formation of the corresponding azoxy derivative, alkylation with an alkyl-lithium and subsequent cleavage of the α-alkylated azo compound (Barton *et al.*, 1979). Another procedure uses *N*-nitroso-*N*-(1-methoxyethyl) derivatives of the amine (Saavedra, 1983).

1.6. Alkylation of α-thio and α-seleno carbanions

Because of its electronegativity and the possibility of delocalisation of the unshared pair on carbon in the anion $-\overset{(-)}{C}H-S-R$, sulphur can stabilise a neighbouring carbanion, and carbanions derived from sulphur compounds are frequently used in synthesis (Block, 1978). The sulphur atom may be present in a sulphide, a sulphoxide or a sulphone (Magnus, 1977) and may be removed from the product after reaction, if desired, by reductive cleavage, by hydrolysis to a carbonyl compound in the case of vinyl sulphides, or by elimination to give an alkene (see p. 120). Controlled rearrangement of the alkylated product before removal of the sulphur extends the scope of the synthetic sequence in many cases.

Selenium analogues can often be used in place of the sulphur compounds but because of the greater expense of selenium reagents and their toxicity the sulphur reagents would generally be employed unless some particular advantage lay in the use of the selenium derivatives (Clive, 1978; Reich, 1979; Liotta, 1984).

Because of the activating effect of the SO and SO_2 groups, sulphoxides and sulphones are much more easily converted into α-carbanions than sulphides.

Alkyl sulphides are apparently not acidic enough to be converted easily into anions under experimentally useful conditions; the presence of another activating group, such as a second sulphur atom or a double bond, is generally desirable for convenient reaction. Thus, allyl sulphides (allyl *phenyl* sulphides are usually employed) are readily converted by butyl-lithium into carbanions which are alkylated with active alkylating agents such as methyl iodide or allyl or benzyl halides, mainly at the position α to the sulphur atom, although in some cases γ-alkylation is also observed. Reduction of the products with lithium and ethylamine removes the sulphur substituent to give the α-alkylated alkene. The sequence provides an excellent method for coupling allyl groups; squalene, for example, was synthesised from farnesyl bromide and farnesyl phenyl sulphide in 70 per cent yield. Coupling of two different allyl groups is easily effected, and intramolecular reactions take place easily as in (1.64). A practical advantage is that no dialkylation products

C_6H_5S — C_4H_9Li, THF, −78 °C → C_6H_5S $^-$ Li^+ ; CH_2Cl ; SC_6H_5 ; Li, $C_2H_5NH_2$

(59% overall) (1.64)

HO ; C_6H_5S — (1) C_4H_9Li, THF, −78 °C (2) Li, $C_2H_5NH_2$

are formed because the monoalkylated compound is not sufficiently acidic to form an anion under the reaction conditions. Alkylation of the allyl vinyl sulphide (51) may be combined with thio-Claisen rearrangement to give, after hydrolysis of the product, a $\gamma\delta$-unsaturated aldehyde (1.65). In this sequence the anion (52) is synthetically equivalent to the anion $\overset{(-)}{C}H{=}CHCH_2CH_2CH{=}O$. [2,3]-Sigmatropic rearrangement of sulphur ylids has also been exploited to make carbon–carbon bonds and employed in complex syntheses (Vedejs, 1984) (cf. p. 140).

(51) $\xrightarrow[\text{THF, }-78\,^\circ C]{s\text{-}C_4H_9Li}$ (52) $\xrightarrow{C_5H_{11}Br}$ C_5H_{11}-substituted allyl vinyl sulphide $\xrightarrow{\text{heat in aqueous dimethoxyethane}}$ [$C_5H_{11}CH{=}CHCH_2CH_2CH{=}S$] $\longrightarrow$ $C_5H_{11}CH{=}CHCH_2CH_2CH{=}O$ (1.65)

Alkyl vinyl sulphides form α-lithio derivatives at the vinyl group on treatment with s-butyl-lithium. These react readily with alkyl halides to give α-alkyl derivatives which on hydrolysis afford ketones (1.66). This is a useful alternative to the dithiane route to ketones. Alkyl vinyl selenides can be used in the same way.

$$C_2H_5S{-}CH{=}CH_2 \xrightarrow[\text{THF, }-78\,^\circ C]{s\text{-}C_4H_9Li} C_2H_5S{-}C(Li){=}CH_2 \xrightarrow{C_8H_{17}Br} C_2H_5S{-}C(C_8H_{17}){=}CH_2 \xrightarrow[CH_3CN]{Hg^{2+},\ H_2O} O{=}C(C_8H_{17}){-}CH_3\ (90\%) \quad (1.66)$$

α-(Phenylthio)- and α-(phenylseleno)-carbonyl compounds (aldehydes, ketones, esters, lactones) are also useful in synthesis (Trost, 1978; Clive, 1978). They are easily formed by sulphenylation or selenenylation of the enolates (cf. p. 9) and are alkylated at the carbon bearing the heteroatom. The alkylated products can be reductively cleaved with lithium and ethylamine to give the α-alkyl carbonyl compound, or oxidised to the sulphoxide or selenoxide which, on elimination, affords the α-alkyl-$\alpha\beta$-unsaturated carbonyl derivative. The α-methylene lactone moiety, frequently found in some classes of natural products, is readily constructed as shown in (1.67). The specific formation of the exocyclic methylene group comes about because of the stereospecific alkylaton of the lactone enolate with diphenyldiselenide, which establishes the required *anti* relationship between the α-phenylseleno substituent and the adjacent methine proton; thence *syn* elimination of selenoxide can only lead to the exocyclic alkene.

OLi C₄H₉ $\xrightarrow[\text{THF}]{C_6H_5SSC_6H_5}$ O C₄H₉ SC₆H₅ $\xrightarrow[(2)\ CH_3I]{(1)\ Li,\ NH_3}$ O C₄H₉ CH₃ (84%)

$\xrightarrow[(2)\ CH_3I]{(1)\ LiN(iso\text{-}C_3H_7)_2,\ THF}$ CH₃ (1.67)

(1) $LiN(iso\text{-}C_3H_7)_2$ THF
(2) $C_6H_5SeSeC_6H_5$

CH₂ $\xleftarrow[(2)\ warm]{(1)\ H_2O_2,\ THF\ 0\ °C}$ SeC₆H₅ CH₃

1.7. Umpolung (dipole inversion)

A general class of reactions of importance in synthesis, since it widens the range of transformations possible, consists of processes which reverse temporarily the characteristic reactivity, nucleophilic or electrophilic, of an atom or group. One such process results in the transformation of the normally electrophilic carbon of a carbonyl group into a nucleophilic carbon. This inversion of the normal polarisation of a functional group has come to be known as 'umpolung' (Seebach, 1979*a*, 1979*b*). It is commonly encountered in the laboratory in the benzoin condensation which results in effect, from addition of the benzoyl anion, $C_6H_5-\overset{(-)}{C}=O$, to the carbonyl group of benzaldehyde, the intermediate (53) behaving as a 'benzoyl anion equivalent'. The process is important in nature in reactions catalysed by thiamine coenzyme in which

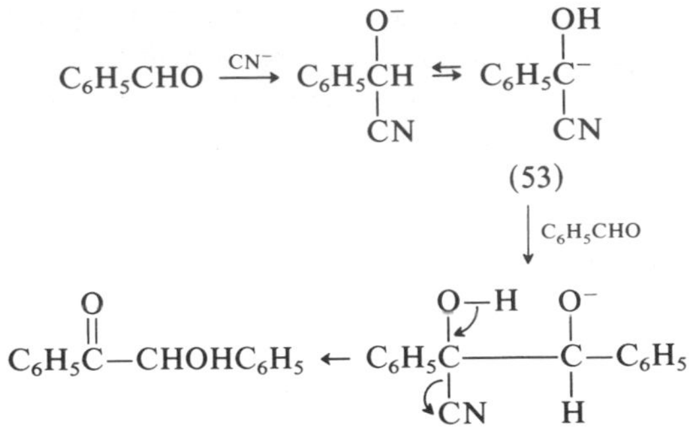

$$C_6H_5CHO \xrightarrow{CN^-} C_6H_5\underset{CN}{\overset{O^-}{C}}H \rightleftarrows C_6H_5\underset{CN}{\overset{OH}{C}}{}^- \; (53) \xrightarrow{C_6H_5CHO} C_6H_5\underset{CN}{\overset{O-H}{C}}-\underset{H}{\overset{O^-}{C}}-C_6H_5 \rightarrow C_6H_5\overset{O}{\overset{\|}{C}}-CHOHC_6H_5 \quad (1.68)$$

acyl anion equivalents, $[RC{=}O]^-$, are generated with the aid of the thiazole ring of the thiamine molecule. For example, in the transketolase reaction the intermediate ($54 \equiv X-\overset{(-)}{C}{=}O$) goes to the ketol (55) by reaction with an aldehyde. This sequence has been emulated in the laboratory for the preparation of α-hydroxyketones from aldehydes.

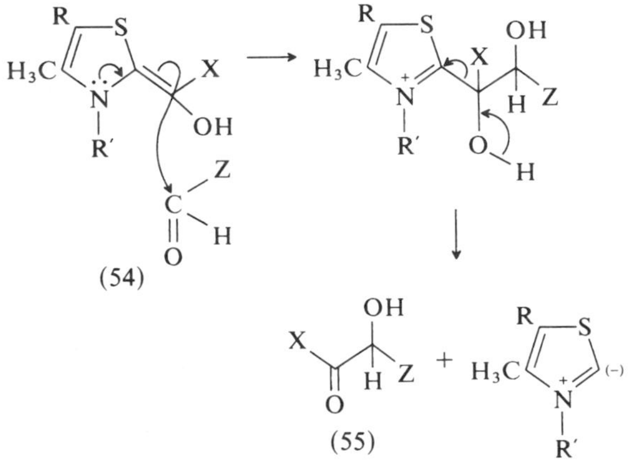

(1.69)

(54) (55)

Thiazolium salts have also been used to catalyse the conjugate addition of acyl anions to $\alpha\beta$-unsaturated ketones and esters (cf. Albright, 1983). A good intramolecular example, which formed the key step in a synthesis of hirsutic acid, is seen in (1.70).

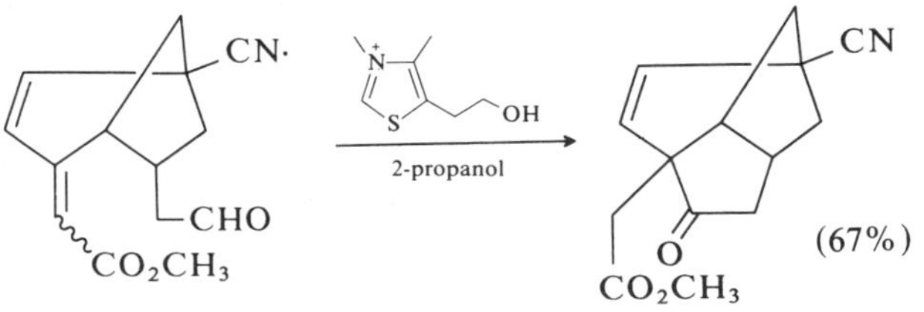

(1.70)

2-propanol

(67%)

A number of other methods for the generation of acyl anion equivalents from aldehydes have been developed for use in the laboratory (Lever, 1976). One proceeds from cyanohydrins, as in the benzoin condensation. Aldehyde cyanohydrins, protected as the acetals with ether vinyl ether, or as their trimethylsilyl ethers (p. 341), are readily transformed into anions by treatment with lithium diisopropylamide. Reaction with an alkyl halide gives the protected cyanohydrin of a ketone from which the parent compound is easily liberated by successive treatment with dilute sulphuric acid and dilute sodium hydroxide. Intramolecular reactions again take place readily. With aldehydes and ketones α-hydroxyketones are produced.

$$CH_3CH(OSi(CH_3)_3)CN \xrightarrow[(2)\ \text{cyclopentyl-Br}]{(1)\ LiN(iso\text{-}C_3H_7)_2} CH_3C(OSi(CH_3)_3)(CN)-C_5H_9 \xrightarrow{H_3O^+} CH_3C(OH)(CN)-C_5H_9 \xrightarrow{NaOH} CH_3COC_5H_9\ (85\%)$$

Other methods for effecting umpolung of carbonyl reactivity employ sulphur-containing reagents (Gröbel and Seebach, 1977). In one widely used route to acyl anions from aldehydes the key step is the conversion of the aldehyde carbonyl group into a 1,3-dithiane, which, because of the stabilising effect of the two electronegative sulphur atoms, is easily converted into the corresponding carbanion at C-2, i.e. at the carbon of the original carbonyl group, by reaction with butyl-lithium. The resulting lithium compounds are stable in solution at low temperatures, and undergo the whole range of reactions shown by other organolithium compounds. After reaction, the 1,3-dithiane can be reconverted into the carbonyl group by hydrolysis with acid in the presence of mercuric ion.

$$HS(CH_2)_3SH \xrightarrow[HCl,\ CHCl_3]{C_6H_5CHO} \text{2-}C_6H_5\text{-2-H-1,3-dithiane} \xrightarrow[THF,\ -10\,^\circ C]{C_4H_9Li} \text{2-}C_6H_5\text{-2-Li-1,3-dithiane} \equiv [C_6H_5\bar{C}O]$$

$$\xrightarrow{C_6H_5CH_2Br} \text{2-}CH_2C_6H_5\text{-2-}C_6H_5\text{-1,3-dithiane} \xrightarrow{H_3O^+,\ Hg^{2+}} C_6H_5COCH_2C_6H_5 \quad (1.72)$$

$$\text{1,3-dithiane (56)} \xrightarrow[(2)\ (CH_3)_2C{=}CHCH_2Br]{(1)\ C_4H_9Li} \text{2-}((CH_3)_2C{=}CHCH_2)\text{-1,3-dithiane} \xrightarrow{H_3O^+,\ Hg^{2+}} (CH_3)_2C{=}CHCH_2CHO$$

Thus, primary and secondary alkyl halides (iodides are best) react readily to form 2-alkyl-1,3-dithianes which, on hydrolysis afford aldehydes or ketones. The sequence provides a method for converting an aldehyde into a ketone, or, using the commercially available parent 1,3-dithiane (56), an alkyl halide into the homologous aldehyde. In the latter sequence the dithiane (56) is operationally equivalent to the formyl anion $[HC{=}O]^-$. Two alkyl groups can be introduced by two successive reactions without isolation of intermediates, and this sequence has been applied in a convenient synthesis of three- to seven-membered cyclic ketones. A variety of other reactions has been effected. Epoxides readily form thio-acetals of β-hydroxy aldehydes or ketones, aldehydes and ketones give derivatives of α-hydroxy aldehydes or ketones, and acid chlorides and esters give thio-acetals of 1,2-dicarbonyl compounds.

A limitation on the use of dithianes for carbonyl umpolung is that although the anions react readily with alkylating agents they do not add to Michael acceptors, reacting, for example, with $\alpha\beta$-unsaturated aldehydes and ketones at the carbonyl group. A more versatile reagent is ethyl ethylthiomethyl sulphoxide (57), the anion of which undergoes both alkylation with halides and conjugate addition to enones, allowing the formation of a wide range of aldehydes and ketones and 1,4-dicarbonyl compounds.

C_2H_5S, $C_2H_5S \rightarrow O$ (57) — $LiN(iso\text{-}C_3H_7)_2$, THF, 20 °C → Li^+ — C_2H_5I → ; $LiN(iso\text{-}C_3H_7)_2$; (1) C_3H_7Br, (2) H_3O^+, Hg^{2+} → (90%) ; (1) [enone], (2) hydrolysis ← (1.73)

Metallated enol ethers also serve as versatile and efficient acyl anion equivalents. The reagents are prepared by action of t-butyl-lithium on an enol ether in tetrahydrofuran at low temperature. Most work so far has been done with the parent compound, methoxyvinyl-lithium, but a few substituted enol ethers have also been used. The crotonyl anion equivalent (58) for example is obtained by metallation of 1-methoxybutadiene. Reaction of the lithium compounds with electrophiles (alkyl halides, aldehydes and ketones) gives, initially, vinyl ethers which may be further elaborated or converted by mild hydrolysis with acid into

$$CH_2{=}C(H)OCH_3 \xrightarrow{t\text{-}C_4H_9Li} CH_2{=}C(Li)OCH_3 \equiv [CH_3\overset{O}{\overset{\|}{C}}{}^-] \xrightarrow{E^+} CH_3\overset{O}{\overset{\|}{C}}{-}E \qquad (1.74)$$

$$CH_2{=}CH{-}CH{=}CH{-}OCH_3 \xrightarrow{t\text{-}C_4H_9Li} CH_2{=}CH{-}CH{=}C(Li){-}OCH_3 \;(58) \equiv [CH_2{=}CH{-}CH_2{-}\overset{O}{\overset{\|}{C}}{}^-]$$

the corrresponding carbonyl compounds. Reaction with alkyl halides leads to ketones, and aldehydes and ketones give α-ketols. (1.75). $\alpha\beta$-Unsaturated compounds react at the carbonyl group; conjugate addition can be achieved, if desired, by first converting the lithio derivatives into the corresponding cuprates (see p. 81).

$$CH_2{=}CH{-}OCH_3 \xrightarrow[\text{THF, }-65\,^\circ C]{t\text{-}C_4H_9Li} CH_2{=}C(Li){-}OCH_3 \xrightarrow{(CH_3)_2C{=}CHCH_2Br} (CH_3)_2C{=}CHCH_2C(OCH_3){=}CH_2$$

$$CH_2{=}C(Li){-}OCH_3 \xrightarrow{(1)\ CH_3CH_2CH_2CHO;\ (2)\ H_3O^+} CH_3CH_2CH_2CH(OH)COCH_3 \ (63\%)$$

$$(CH_3)_2C{=}CHCH_2C(OCH_3){=}CH_2 \xrightarrow{\text{dil. HCl, }25\,^\circ C} (CH_3)_2C{=}CHCH_2COCH_3 \ (74\%) \qquad (1.75)$$

Other routes to acyl anion equivalents from t-butylhydrazones (Adlington *et al.*, 1983), from carboxylic acids via the derived amidrazones (Baldwin and Bottaro, 1981) and from dithio esters (Meyers, Tait and Commins, 1978) have been described recently and may offer advantages in particular cases.

Another group of versatile reagents are the alkylidene dithianes or ketene thioacetals (59). A number of methods are used to prepare these compounds (cf. Hase and Koskimies, 1982). The valuable derivatives (60), for example, are obtained from commercially available 1,3-dithiane by way of the trimethylsilyl derivative (61) by reaction with aldehydes or ketones (p. 135). They undergo a variety of useful reactions. Alkyl groups attached to the double bond are readily metallated by butyllithium or lithium diisopropylamide to give the corresponding allyl

$$(R^1S)(R^2S)C{=}C(R^3)(R^4) \ (59) \qquad \text{1,3-dithian-2-ylidene}{=}C(R)(R^1) \ (60) \qquad (1.76)$$

anions. Reaction of these with various electrophiles takes place predominantly at the carbon adjacent to the sulphur, giving products which are hydrolysed to $\alpha\beta$-unsaturated ketones (1.77). In this sequence the allyl anions (as 62) serve as masked $\alpha\beta$-unsaturated acyl anions.

$Si(CH_3)_3$ Li (61) CHO C_4H_9Li, THF, −20 °C Li^+ (62) CH_3I CH_3 H_3O^+, Hg^{2+} (90%) (1.77)

The dianion (63), obtained from 2-propenethiol with butyl-lithium, is a synthetic equivalent of the homoenolate ion (64). It reacts with alkyl halides and carbonyl compounds mainly at the γ-carbon atom to give thioenolates which are hydrolysed to carbonyl compounds formed, in effect, by electrophilic attack on (64). This sequence complements the more familiar nucleophilic attack at the β-carbon atom of acrylic aldehyde and other $\alpha\beta$-unsaturated aldehydes.

SH 2 equivs C_4H_9Li THF, 0 °C (63) $2Li^+$ ≡ (−) CHO (64) $C_6H_{13}Br$ (1.78) C_6H_{13} H SCH_3 hydrolysis C_6H_{13} CHO

More generally, β-acylcarbanion equivalents are obtained by deprotonation of allyl ethers. The anions thus obtained, by proper choice of experimental conditions, react with alkyl halides at the γ-carbon atom giving enol ethers which are hydrolysed to carbonyl compounds (1.79).

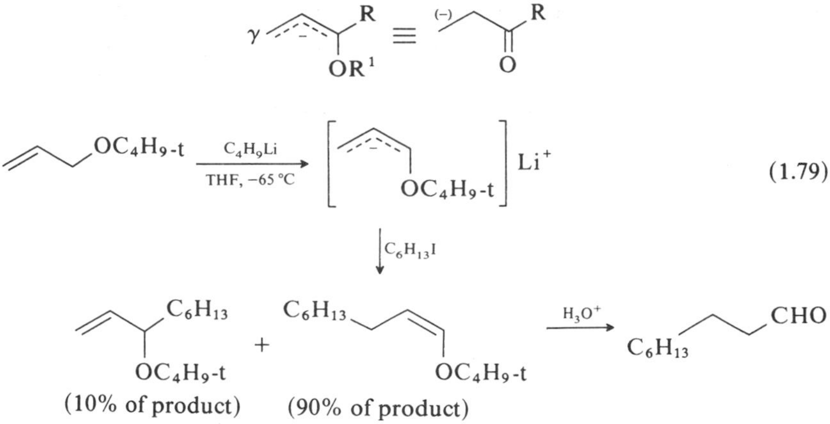

$$\gamma\ \text{(allyl anion, R, OR}^1\text{)} \equiv \overset{(-)}{\ }\text{CH}_2\text{CH}_2\text{C(=O)R}$$

$$CH_2{=}CHCH_2OC_4H_9\text{-t} \xrightarrow[\text{THF, } -65\,°C]{C_4H_9Li} \left[\text{allyl anion}-OC_4H_9\text{-t}\right]Li^+ \tag{1.79}$$

$\downarrow C_6H_{13}I$

C_6H_{13}, OC_4H_9-t (10% of product) + C_6H_{13}, OC_4H_9-t (90% of product) $\xrightarrow{H_3O^+}$ C_6H_{13} CHO

Aliphatic nitro compounds are being used increasingly in organic synthesis (Seebach, 1979). They also serve as good acyl anion equivalents, for after electrophilic alkylation at the α-carbon atom they are easily converted into carbonyl compounds by the Nef reaction or by reductive hydrolysis with $TiCl_3$ (see p. 487). Thus, in a synthesis of jasmone, Michael addition of nitroethane to the $\alpha\beta$-unsaturated ketone (65) was followed by conversion of the product into (66), formed, in effect, by Michael addition of CH_3CO^- to (65) (Dubs and Stüssi, 1978). Nitromethane has been used similarly as an operational equivalent of the formyl anion $H\overset{(-)}{C}{=}O$.

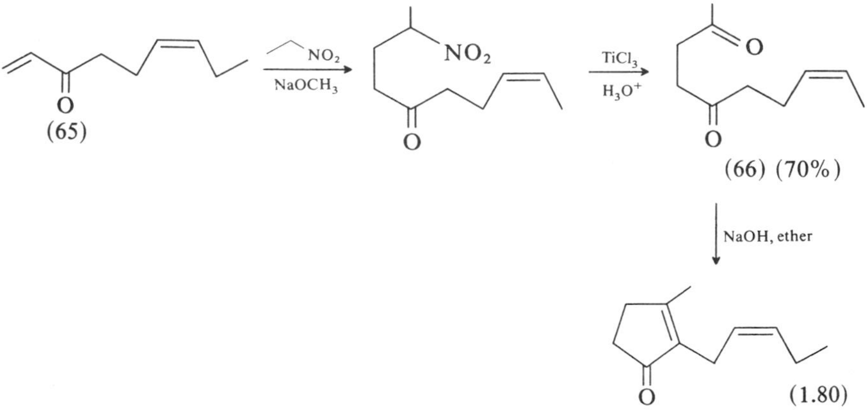

(65) $\xrightarrow[NaOCH_3]{CH_3CH_2NO_2}$ NO_2 $\xrightarrow[H_3O^+]{TiCl_3}$ (66) (70%) $\xrightarrow{\text{NaOH, ether}}$

(1.80)

Nitro compounds can also be used as agents for umpolung at C-1 of amines. Nucleophilic substitution at C-1 of amines is readily effected by way of the corresponding imminium ions as in the Mannich reaction. Electrophilic substitution can be brought about using the *N*-nitroso compounds or certain amides (see p. 37). Another way is to start with the corresponding nitro compound as in the example in (1.81); here the dianion of nitroethane is synthetically equivalent to the α-amino-anion $CH_3\overset{(-)}{C}HNH_2$.

$$CH_3{-}C(O^-){=}CH_2 \quad CH_2{=}\overset{+}{N}(CH_3)_2 \rightarrow CH_3COCH_2CH_2N(CH_3)_2$$

$$CH_3CH_2NO_2 \xrightarrow[\text{THF, hexamethylphosphoramide, } -60\,^{\circ}C]{\text{2 equivs } C_4H_9Li} \left[CH_3C(Li){=}\overset{+}{N}(O^-)_2\right] Li^+ \equiv [-\bar{C}H{-}NH_2]$$

$$\xrightarrow[(2)\ H_3O^+]{(1)\ C_6H_5CH_2Br} CH_3CH(CH_2C_6H_5)NO_2 \longrightarrow CH_3CH(CH_2C_6H_5)NH_2 \qquad (1.81)$$

1.8. The aldol reaction

The aldol reaction between an aldehyde and another aldehyde or a ketone is a good method for making carbon–carbon bonds (Nielson and Houlihan, 1968), but hitherto it has not been used much in synthesis, for two main reasons. The first is that, as ordinarily effected the intermolecular reaction between two different aldehydes or an aldehyde and a ketone very often gives a mixture of products, or the wrong product. The second reason is that in a reaction such as that depicted in (1.82)

$$R{-}CHO + H_3C{-}CH_2{-}CO{-}R^1 \xrightarrow{\text{base}} R{-}CH(OH){-}CH(CH_3){-}CO{-}R^1 \ (\textit{syn (erythro)}) + R{-}CH(OH){-}CH(CH_3){-}CO{-}R^1 \ (\textit{anti (threo)}) \qquad (1.82)$$

between an aldehyde and an ethyl ketone, say, to give an α-alkyl-β-hydroxycarbonyl compound, a mixture of the *syn* (*erythro*) and *anti* (*threo*) isomers is ordinarily formed and has to be separated into its components. To be able to employ the aldol reaction in synthesis these difficulties have to be faced. We must be able to 'direct' the course of the reaction between unlike components so that only the product required is obtained, or at least is formed predominantly, and we must be able to control the stereochemical course of the reaction so that the *syn* or *anti* isomer is obtained at will. Both of these difficulties have now been largely overcome as a result of intensive study of the aldol reaction occasioned by the presence of the α-alkyl-β-hydroxycarbonyl functional group in the structures of many naturally occurring macrolides and ionophores.

A number of methods have been developed to bring about 'directed' aldol reaction between two different carbonyl compounds to give a particular aldol product. Most of them proceed from the preformed enolate or enol ether of one of the components (Mukaiyama, 1982). The success of these procedures hinges on the fact that the reactions are carried out in such a way that the newly formed aldol is trapped as a metal chelate complex and is thus protected from wasteful side reactions, particularly from dissociation into the free carbonyl components which could undergo non-selective condensation. Thus, aldol reaction of acetone with acetaldehyde, ordinarily inefficient because the acetaldehyde reacts more readily with itself, is easily effected using preformed acetone enolate (1.83). A number of metal counterions have been used,

$$CH_3COCH_3 \longrightarrow CH_2{=}C(OM){-}CH_3 \xrightarrow{CH_3CHO} \text{[metal chelate]} \xrightarrow{H_3O^+} CH_3CHOHCH_2COCH_3 \qquad (1.83)$$

but the best results have been obtained with lithium and boron enolates. Lithium enolates have been widely used. Lithium efficiently traps the aldol by chelate formation in aprotic solvents such as ether and tetrahydrofuran, and lithium enolates themselves are readily obtained from carbonyl compounds (see p. 12). Thus, 2-pentanone, on deprotonation with lithium diisopropylamide and reaction of the enolate with acetaldehyde gave the 'mixed' aldol (67) in 90 per cent yield.

$$\text{2-pentanone} \xrightarrow[\text{ether, } -70\,^\circ\text{C, hexamethylphosphoramide}]{LiN(iso\text{-}C_3H_7)_2} \text{lithium enolate (OLi)} \xrightarrow[(2)\ H_3O^+]{(1)\ CH_3CHO} (67) \qquad (1.84)$$

Even better results have been obtained using vinyloxyboranes as the enolate component. Vinyloxyboranes are readily made from ketones by reaction with a dialkylboryl trifluoromethanesulphonate (triflate) in the presence of a tertiary base and, furthermore, either enolate can be prepared from an unsymmetrical ketone by varying the dialkylboryl triflate. Dibutylboryl triflate gives the kinetic (less substituted) vinyloxyborane, and with 9-borabicyclo[3,3,1]-9-nonanyl triflate (9-BBN triflate, see p. 291) the thermodynamic (more substituted) vinyloxyborane is obtained. These vinyloxyboranes react readily with aldehydes and ketones to give aldols in excellent yield (1.85). Thus, 4-methyl-2-pentanone with dibutylboryl triflate gave the vinyloxyborane (68) and thence the aldols (69); with 9-BBN triflate the vinyloxyborane (70) was formed from which the isomeric aldols (71) were obtained.

$(C_4H_9)_2BSO_3CF_3$, $C_2H_5N(iso\text{-}C_3H_7)_2$ → $OB(C_4H_9)_2$ (68) → RCHO, ether, −78 °C → $(C_4H_9)_2B{-}O$ → H_2O_2, H_2O → (69)

9-BBN.SO_3CF_3, 2,6-lutidine → OB (70) → (1) RCHO (2) H_2O_2, H_2O → (71)

(1.85)

Acid-catalysed 'directed' aldol reactions make use of enol ethers instead of metal enolates. Enol ethers, particularly silyl enol ethers, of both aldehydes and ketones react smoothly with aldehydes and ketones in the presence of Lewis acids to give aldols in good yield. Enol ethers from unsymmetrical ketones react specifically at the carbon atom at the terminus of the double bond. Various catalysts have been used, but the best results have been obtained with titanium tetrachloride. Reactions are believed to proceed as in (1.86). The undesirable retro-aldol reaction of the product is prevented by the formation of the stable titanium chelate complex which yields the aldol on hydrolysis. However, this method suffers from the disadvantage compared with the lithium enolate and vinyloxyborane procedures that it is not stereoselective and does not allow control of stereochemistry in reactions which give rise to aldols containing several chiral centres.

Another important feature of the aldol reaction has to do with the stereochemistry of the product in a reaction such as that depicted in

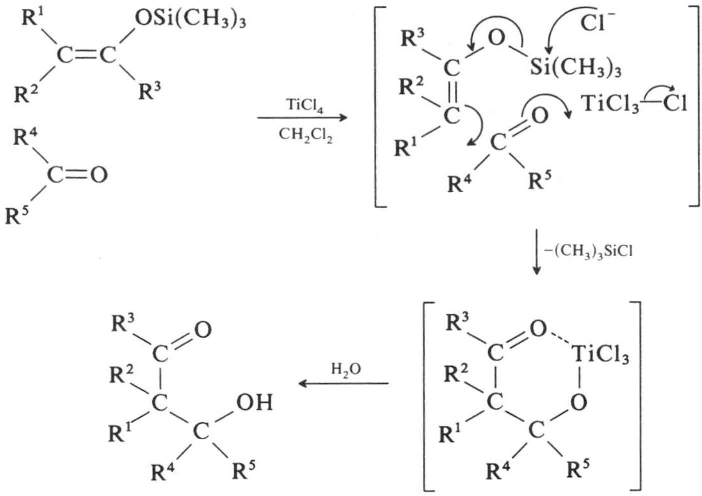

(1.86)

(1.87) where the β-hydroxy carbonyl compound also has a substituent at the α-position. Here there are four possible stereoisomeric products

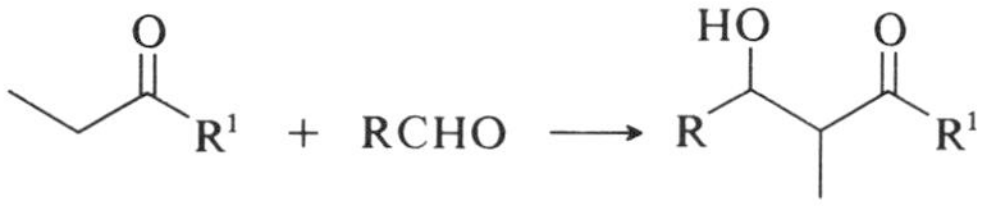

(1.87)

(1.88), two *syn* or *erythro* products and two *anti* or *threo* products. Consequently there are two stereochemical aspects of the reaction which have to be considered, namely internal stereochemical control or *diastereoselection* (that is whether (±)-(72) or (±)-(73) will be the

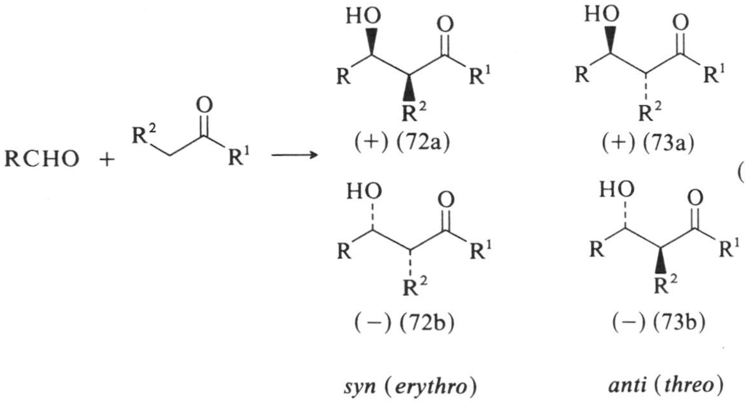

(1.88)

(+) (72a) (+) (73a)

(−) (72b) (−) (73b)

syn (*erythro*) *anti* (*threo*)

favoured product) and absolute stereochemical control or *enantioselection* (that is, for the given diastereomer (73), say, whether the product will be the (+) or (−) epimer). Both aspects have been widely studied recently and a high degree of control has been realised in many cases (Evans, Nelson and Taber, 1982). The system of nomenclature used here to designate the diastereomers is that proposed by Masamune (Masamune, Kaiho and Garvey, 1982). The main carbon chain is drawn in the extended zig-zag form and that isomer in which the two substituents at C-2 and C-3 are disposed in the same direction, away from or towards the observer, is designated *syn* and the other *anti.* Some authors would call these the *erythro* and *threo* isomers, respectively.

To achieve good diastereoselection the lithium and boron enolates have been most widely used. The stereochemical course of the reactions with these enolates depends on whether the reaction is run under thermodynamic or kinetic conditions and on the geometry of the enolate. For the kinetic reaction enolate geometry is important; it is found in general that the *Z*-enolates (74) give mainly the 2,3-*syn* aldols while the *E*-enolates (75) lead to the 2,3-*anti* aldols (Heathcock, 1981). For reactions with the lithium enolates the size of the substituents R^1 and R^2 also

$$\mathrm{RCHO} + \underset{(74)}{R^1CH{=}C(OLi)R^2} \longrightarrow \mathrm{R{-}CH(OH){-}CH(R^1){-}CO{-}R^2}\ (syn)$$

$$\mathrm{RCHO} + \underset{(75)}{R^1CH{=}C(OLi)R^2} \longrightarrow \mathrm{R{-}CH(OH){-}CH(R^1){-}CO{-}R^2}\ (anti) \qquad (1.89)$$

has to be taken into account. The larger the size of R^2, the greater the selectivity. For example, the *Z*-enolate from 2,2-dimethyl-3-pentanone gave the *syn* aldol (76) almost exclusively on reaction with benzaldehyde in ether at −70 °C; under the same conditions 3-pentanone gave a mixture containing 30 per cent of the *anti* isomer. In general, reactions with *E*-enolates are less stereoselective than those with *Z*-enolates.

$$\mathrm{CH_3CH_2COC_4H_9\text{-}t} \xrightarrow[\text{THF, }-70\,^\circ\text{C}]{\mathrm{LiN(iso\text{-}C_3H_7)_2}} \mathrm{CH_3CH{=}C(OLi)C_4H_9\text{-}t} \xrightarrow{\mathrm{C_6H_5CHO}} \underset{\substack{(76)\\(>98\%\ syn)}}{\mathrm{C_6H_5{-}CH(OH){-}CH(CH_3){-}CO{-}R}} \qquad (1.90)$$

In contrast, reactions effected under equilibrating conditions give mainly the 2,3-*anti* aldol, irrespective of the geometry of the enolate. The aldol condensation is easily reversible and *syn–anti* equilibration is often observed when an ethereal solution of the aldol is allowed to stand for a period of time. Thus equilibration can sometimes be used to achieve *anti* stereoselection.

Excellent stereoselection has also been obtained in the reactions of vinyloxyboranes with aldehydes (cf. Masamune, 1981; Evans, *et al.* 1981; Masamune and Choy, 1982). *Z*-Vinyloxyboranes are readily obtained from ketones by reaction with a dialkyboron trifluoromethanesulphonate (triflate), and they react with aldehydes under mild conditions to give the *syn* aldols with high selectivity. Thus, reaction of 3-pentanone with dibutylboryl triflate affords the corresponding *Z*- and *E*-enolates in a ratio of >99:1; subsequent reaction with benzaldehyde gave the *syn* and *anti* aldols in a ratio >97:3. The same condensation effected by way of the lithium enolates gave a ratio of only 80:20.

$(C_4H_9)_2BOTf$; $C_2H_5N(iso\text{-}C_3H_7)_2$, ether, −78 °C → $OB(C_4H_9)_2$ enolate (>99:1); C_6H_5CHO → C_6H_5, HO, O, CH_3 (>97% *syn*)

(1.91)

E-Boron enolates, necessary for the preparation of 2,3-*anti* aldols, are not so readily obtained from ketones but they can be prepared from thio esters. Thus, treatment of (*S*)-t-butyl propanethioate with dicyclopentylboryl triflate gave the corresponding *E*-enolate almost exclusively; reaction with a variety of aldehydes then led mainly to the *anti* 2-methyl-3-hydroxy thio esters.

O, SC_4H_9-t; $(C_5H_9)_2BOTf$; $C_2H_5N(iso\text{-}C_3H_7)_2$, ether, 0 °C → $OB(C_5H_9)_2$, SC_4H_9-t; CHO →

HO, O, SC_4H_9-t, CH_3 + HO, O, SC_4H_9-t, CH_3 (9:1)

(1.92)

These reactions are believed to proceed by way of a chair-like six-membered cyclic transition state in which the ligated metal atom is bonded to the oxygen atoms of the aldehyde and the enolate. For the reaction of a Z-enolate with aldehyde R^1CHO the transition state could be represented as (77) or (78). However, the latter is disfavoured by 1,3-diaxial interaction between substituents R^1 and R^3 and reaction thus takes place mainly via transition state (77) to give the 2,3-*syn* aldol. Similarly, reaction with the E-enolate proceeds preferentially through transition state (80) leading to the 2,3-*anti* aldol. The superiority of the boron enolates over other metal enolates in these reactions is ascribed to the shortness of the boron–carbon and boron–oxygen bonds, resulting in a more compressed transition state in which steric effects are enhanced.

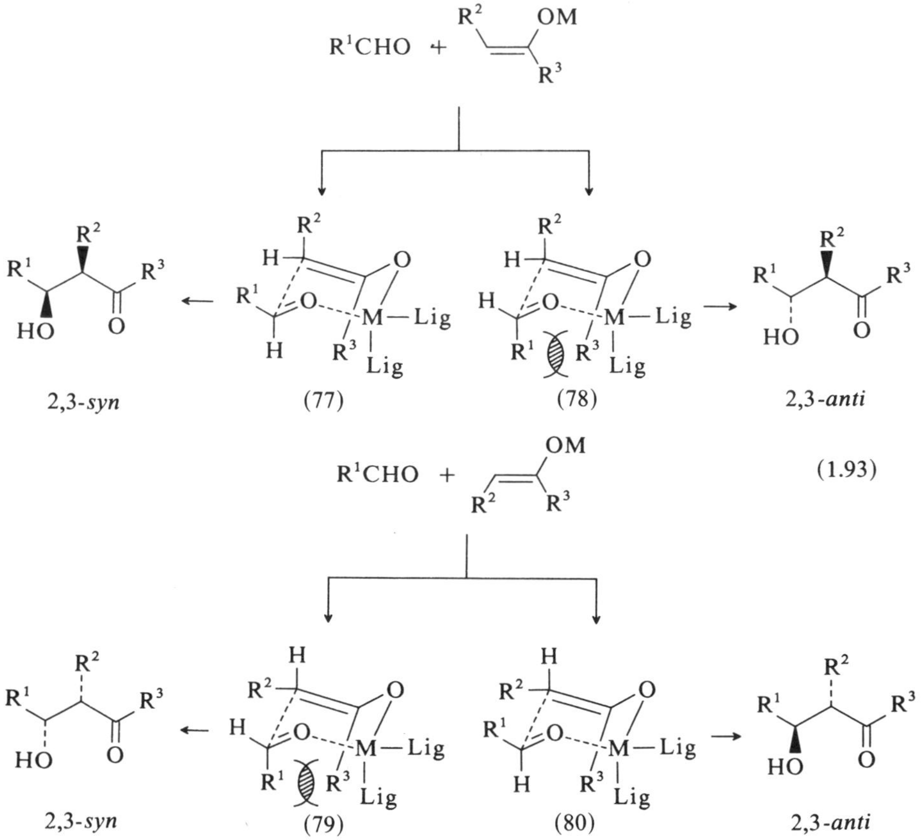

R^1CHO + (Z-enolate: R^2, OM, R^3)

2,3-*syn* ← (77) ; (78) → 2,3-*anti*

R^1CHO + (E-enolate: R^2, OM, R^3) (1.93)

2,3-*syn* ← (79) ; (80) → 2,3-*anti*

The second important stereochemical feature of the aldol reaction, which comes into play when one of the components is optically active, concerns the absolute stereochemistry of the aldol products. In this area highly enantioselective reactions have been effected recently using boron enolates carrying chiral auxiliary groups. For example, reaction of the dicyclopentylboron enolate derived from the optically active ketone (81) with isobutyraldehyde gave the *syn* and *anti* adducts in the ratio 91:9; and of the two possible *syn* diastereomers the isomer (82) was obtained exclusively. Baeyer–Villiger oxidation of the crystallised product and hydrolysis of the ester produced then gave the optically pure acid (83)

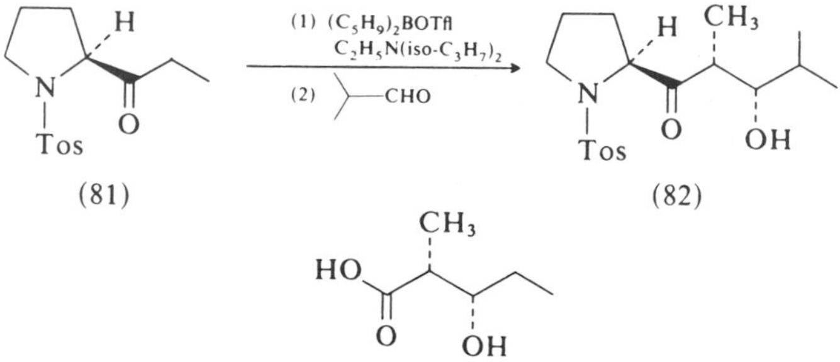

(1) $(C_5H_9)_2BOTf$, $C_2H_5N(iso\text{-}C_3H_7)_2$; (2) $(CH_3)_2CHCHO$

(81) (82) (83) (1.94)

(Evans, Nelson *et al.*, 1981). Excellent results have also been obtained with imides derived from the optically active 2-oxazolidines (84, R = H) and (85, R = H). The derived boron enolates undergo highly enantioselective condensations with aldehydes. Thus, the propionamide (84, R = COC_2H_5), as its *Z*-boron enolate, reacted with aldehydes to give the *syn* aldols (86) almost exclusively. Hydrolysis afforded enantiomerically pure 3-hydroxy-2-methyl carboxylic acids (87). The same sequence applied to the imide (85, R = COC_2H_5) gave the corresponding epimeric acids (Evans, Bartoli and Shih, 1981; see also Evans and Bartoli, 1982).

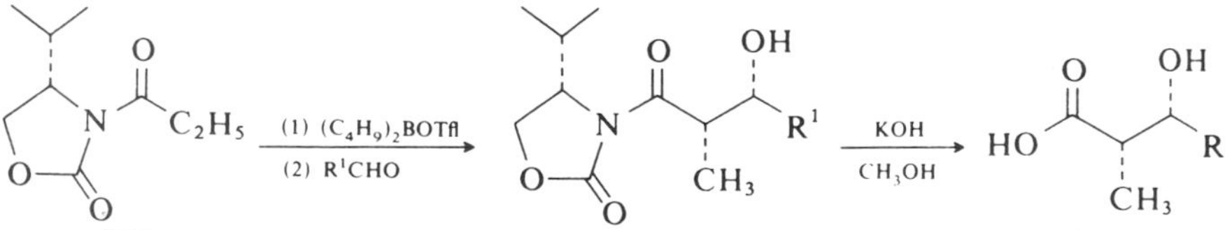

(1) $(C_4H_9)_2BOTf$; (2) R^1CHO — KOH, CH_3OH

(84) (86) (87) (1.95)

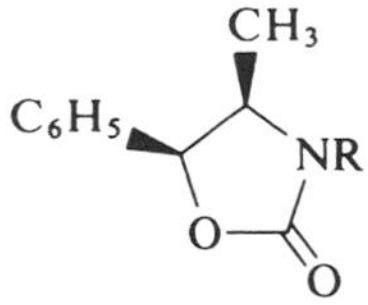

(85)

The situation becomes a little more complex if the aldehyde contains a chiral centre, as in 2-phenylpropanal. Now the aldol produced contains three chiral centres and there are eight possible stereoisomers, two different racemates derived from the *Z*-enolate and two others from the *E*-enolate (1.96). The relative stereochemistry of the substituents at C-2

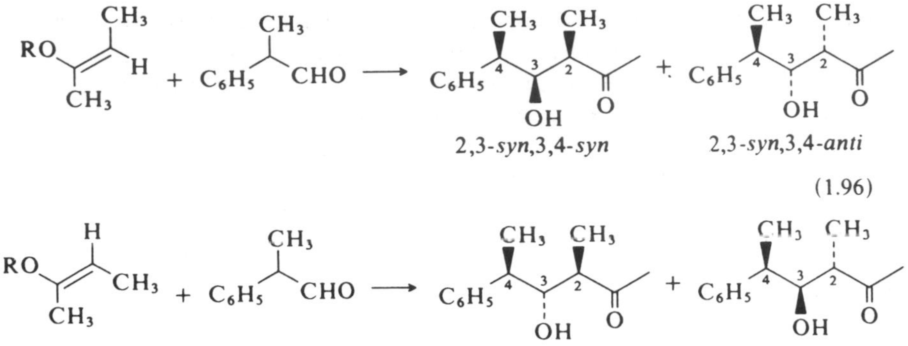

2,3-*syn*,3,4-*syn* 2,3-*syn*,3,4-*anti*

(1.96)

2,3-*anti*,3,4-*anti* 2,3-*anti*,3,4-*syn*

and C-3 is controlled by the geometry of the enolate (p. 52). The C-3,C-4-stereochemistry, on the other hand, depends on the direction of approach of the enolate and the carbonyl group to each other. This is illustrated in (1.97) for the reaction of 2-phenylpropanal with the *Z*-enolate of t-butyl ethyl ketone. In a chiral aldehyde the two faces of

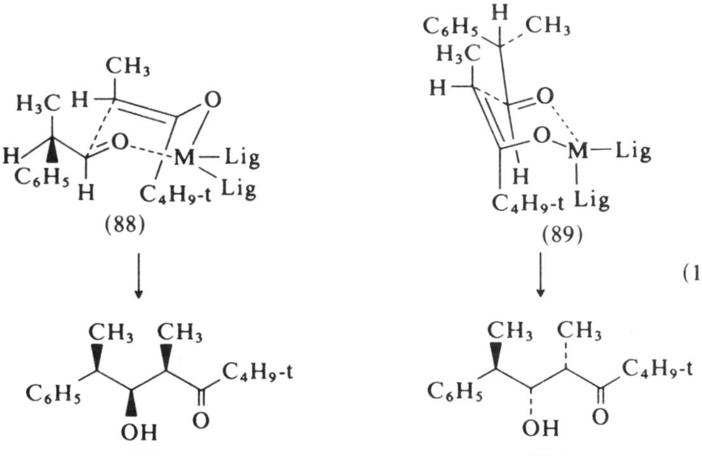

(88) (89)

(1.97)

(90) 2,3-*syn*,3,4-*syn*

(91) 2,3-*syn*,3,4-*anti*

the carbonyl group are not equivalent and when an achiral reagent, such as the enolate anion, approaches the aldehyde it shows some preference for one face over the other with the result that unequal amounts of the two diastereomers are formed. This is designated *diastereofacial selectivity.* Thus, in the present reaction the 2,3-*syn*-3,4-*syn* compound (90), formed by approach of the enolate to the α (*re*) face of the aldehyde (as in 88) is produced in larger amount than the 2,3-*syn*-3,4-*anti* isomer, which arises by β-attack (89) (both isomers have the 2,3-*syn* configuration because they are derived from the *Z*-enolate (1.98)). Similarly, reaction of an achiral aldehyde with a chiral enolate leads to some degree of diastereofacial selectivity and the two diastereoisomeric products are not produced in equal amounts. Generally the degree of selectivity shown in these reactions is not high, not high enough to be of practical value in a multistep synthesis, but it may be increased by a proper manner of proceeding. One way in which the selectivity has been increased to a more practically useful level is by the use of *double stereodifferentiation.*

C_6H_5 CHO + C_4H_9-t OLi ⟶

CH_3 CH_3 C_6H_5 4 3 2 C_4H_9-t OH O + CH_3 CH_3 C_6H_5 4 3 2 C_4H_9-t OH O

2,3-*syn*,3,4-*syn* 2,3-*syn*,3,4-*anti*

ratio 86 14

(1.98)

In this, a chiral aldehyde is made to react with a chiral enolate - hence the name given to the procedure; the diastereofacial selectivity of *both* components is exploited. If one of the components is optically active then we may hope to achieve not merely a stereoselective synthesis but an enantioselective one, i.e. produce an optically active aldol. Some very highly efficient enantioselective aldol reactions of this kind have now been realised. In practice, it is usually the enolate which is made the optically active component by introduction of temporary or 'auxiliary' chiral groups (Masamune *et al.*, 1980; Masamune and Choy, 1982; Masamune *et al.*, 1985; Heathcock, 1981). The procedure depends on the well-known fact that in a kinetically controlled reaction which leads to the formation of new chiral centres, certain pairs of optically active substrates react to form one diastereomer predominantly, whereas other combinations result in inferior stereoselection. It is best illustrated by an example.

For double-stereodifferentiation experiments to be effective the two reaction partners must show some diastereofacial selectivity in their reactions with achiral partners, and this is evaluated separately. Thus, in the reaction of the (*S*)-aldehyde (92) with the achiral enolate (93) the two diastereomers (94) and (95) are formed in the ratio 2.7:1. The β-orientation of the C-3 hydroxyl substituent in the main product (94) shows that it has been formed through a transition state like (88) and that in aldehyde (92) there is some preference for attack on the α-(*re*)

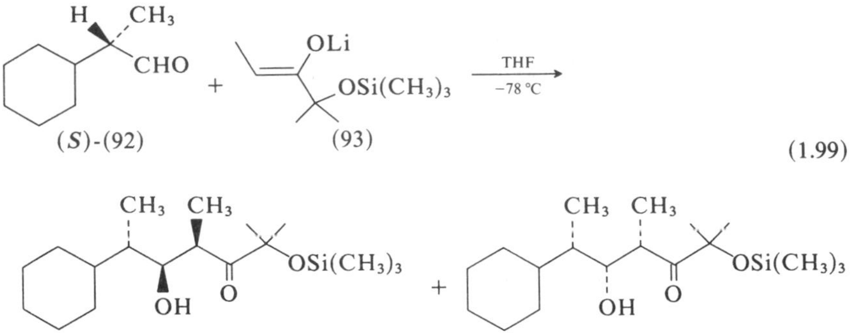

H CH_3 CHO + OLi $OSi(CH_3)_3$ $\xrightarrow[-78\,^\circ C]{THF}$

(*S*)-(92) (93) (1.99)

CH_3 CH_3 OH O $OSi(CH_3)_3$ + CH_3 CH_3 OH O $OSi(CH_3)_3$

(94) (2.7 parts) (95) (1 part)

face of the carbonyl group. Similarly, reaction of benzaldehyde with the optically active (*S*)-enolate (96) produces the two aldols (97) and (98) in the ratio 3.5:1, reaction again favouring the product formed by α-attack on the carbonyl group of the aldehyde. It might be expected, therefore, that reaction of the (*S*)-aldehyde (92) with the (*S*)-enolate (96) would lead to enhanced stereoselection, since both reagents favour α-attack at the carbonyl group (they are designated a 'matched pair').

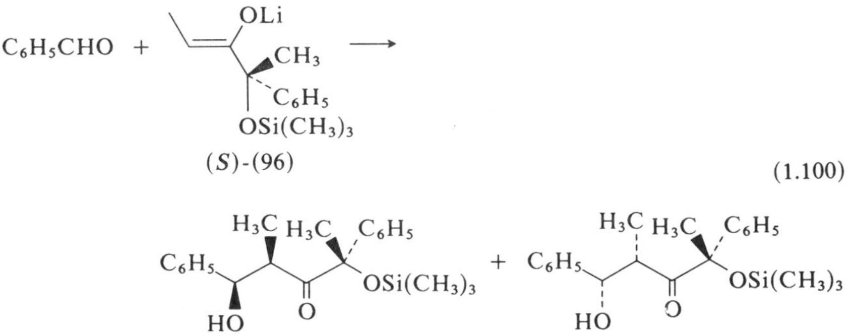

C_6H_5CHO + OLi CH_3 C_6H_5 $OSi(CH_3)_3$ $\longrightarrow$

(*S*)-(96) (1.100)

C_6H_5 H_3C H_3C C_6H_5 HO O $OSi(CH_3)_3$ + C_6H_5 H_3C H_3C C_6H_5 HO O $OSi(CH_3)_3$

(97) (3.5 parts) (98) (1 part)

This is indeed found to be the case; the diastereoselection is now increased to 8:1 (1.101). On the other hand, reaction of the (*S*)-aldehyde (α-selective) with the (*R*)-enolate (β-selective), in which the two reagents are acting in opposition (a mismatched pair) leads to reduced stereoselection. Even better results have been obtained with the cyclohexyl analogue of (96) (cyclohexyl for phenyl) which showed an increase in stereoselection in the reaction with benzaldehyde to 15:1. By far the most efficient reagents found so far, however, are the (*R*)- and (*S*)-boron enolates (as 99, R = 9BBN, n-C_4H_9, or cyclopentyl) which are readily obtained from (*R*)- and (*S*)-mandelic acid. Thus, reaction of the (*S*)-dicyclopentyl compound (99, R = C_5H_9) with propanal gave the 2,3-*syn* isomers (100)

(S)-92+(S)-96 ⟶ (matched pair)

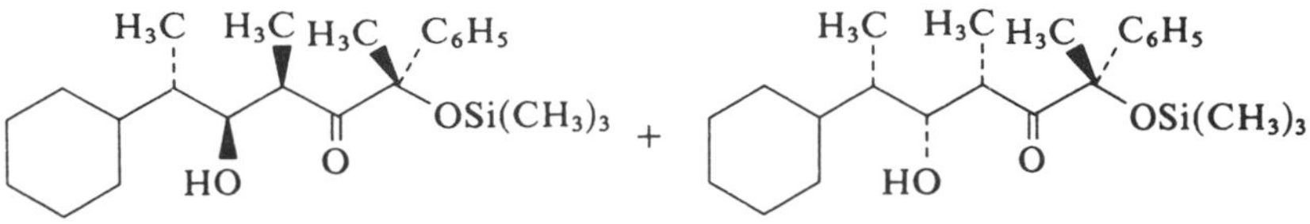

H_3C H_3C H_3C C_6H_5 $OSi(CH_3)_3$ HO O + H_3C H_3C H_3C C_6H_5 $OSi(CH_3)_3$ HO O

ratio 8 1

(*S*)-92+(*R*)-96 ⟶ *ratio* 1:1.5
(mismatched pair)

(1.101)

and (101) in a ratio of >100:1. Removal of the silyl protecting group and cleavage of the resulting α-hydroxy ketone with sodium metaperiodate gave the optically active 3-hydroxy-2-methylcarboxylic acid (102) of >98 per cent optical purity. A similar sequence of reactions using the (R)-boron enolate gave the epimeric acid. This sequence provides a route to enantiomerically pure 3-hydroxy-2-methylcarboxylic acids, a structural unit found in many natural products derived from propionate. Thus the insect pheromone serricornine (103) was obtained from (102).

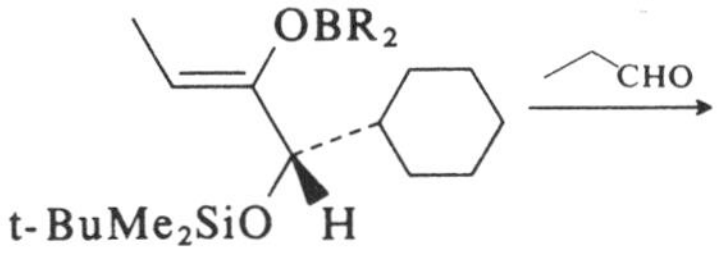

OBR_2 t-$BuMe_2SiO$ H
(*S*)-(99)

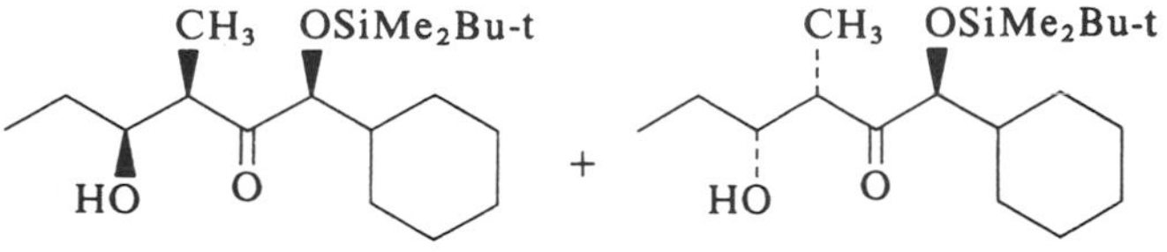

⟶ (CHO reagent)

CH_3 $OSiMe_2Bu$-t HO O + CH_3 $OSiMe_2Bu$-t HO O

(100) (101)

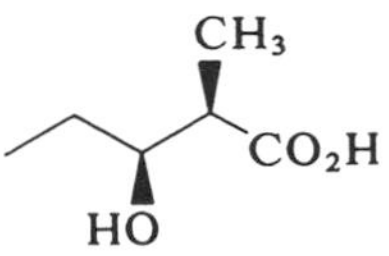

CH_3 CO_2H HO
(102)

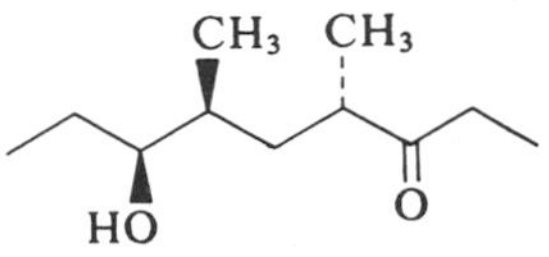

CH_3 CH_3 HO O
(103)

(1.102)

The diastereofacial selectivity of these boron enolates is so high that it completely outweighs any stereochemical preference the aldehyde may have to the extent that either the 3,4-*anti* or 3,4-*syn*-aldol adduct can be obtained as the predominant product irrespective of the aldehyde. Thus, reaction of the (S)-boron enolate (99, R = 9—BBN) with the laevorotatory aldehyde (104) gave a 40:1 ratio of the 3,4-*anti* (105) and 3,4-*syn* (106) isomers; use of the (*R*)-boron enolate reversed the ratio to 1:15. Further manipulation of (105) gave optically pure Prelog–Djerassi lactone (107) a key intermediate in the synthesis of several natural products.

CH_3 CH_3 CH_3O_2C CHO (−)-(104) → (OB, X)

(105) CH_3 CH_3 CH_3 CH_3O_2C 4 3 2 R HO O + (106) CH_3 CH_3 CH_3 CH_3O_2C R HO O

(1.103)

	(105)	(106)
(*S*)-enolate, X = cyclohexyl–CH(H)OSiMe$_2$Bu-t	40	1
(*R*)-enolate, X = cyclohexyl–CH(H)OSiMe$_2$Bu-t	1	15

t-BuMe$_2$SiO, O, OH, CO_2CH_3, CH_3 CH_3 CH_3 → (1) HF-CH_3CN (2) $NaIO_4$ → HO, O, O, O, CH_3, CH_3 CH_3 (107)

The aldol reaction with *Z*-boron enolates provides a route for the enantioselective synthesis of 2,3-*syn*-3-hydroxy-2-methylcarboxylic acids (102) but attempts to reach the corresponding optically active 2,3-*anti* acids from the *E*-boron enolates have been frustrated by difficulties in preparing the *E*-enolates. Enolate precursors EtCOR which have a bulky chiral auxiliary group R, necessary for the reaction sequence, tend to form the *Z*-enolates.

Other metal enolates besides the boron derivatives have been employed. A lithium derivative has been used in double stereodifferentiation experiments (Heathcock *et al.*, 1981) and high diastereofacial selectivity has been obtained in Lewis acid-catalysed reactions of chiral aldehydes with enol silyl ethers (Heathcock and Flippin, 1983). A *syn* selective reaction that is independent of the geometry of the enolate uses zirconium enolates and this has been exploited in a highly enantioselective aldol reaction using optically active amide enolates (Evans and McGee, 1981). Similarly, tin(II) enolates, conveniently prepared from ketones by reaction with $Sn(OTf)_2$ in the presence of a tertiary amine react readily with ketones under very mild conditions to give selectively the *syn* aldol (see Mukaiyama, 1983). Tin(IV) enolates have also been used (Shenoi and Stille, 1982).

The addition of allylmetal compounds to aldehydes is operationally equivalent to the aldol addition and is highly selective (1.104). The initial adducts can be converted into aldols by oxidative cleavage of the double bond (cf. Hoffmann, 1982). A number of allylmetal compounds have been used but synthetically the most useful appear to be the boron and chromium derivatives. Thus, the 2-butenyl boronates (108) and (110) on reaction with various aldehydes gave the homoallylic alcohols (109) and (111) with diastereoselectivities of more than 95 per cent (Hoffmann and Zeiss, 1981).

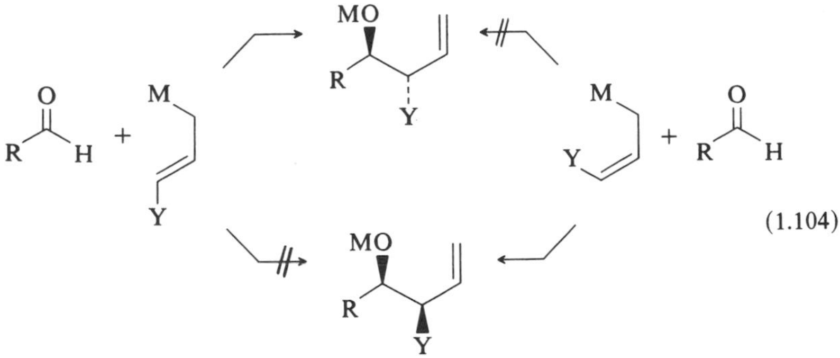

(1.104)

The reactions are believed to take place by way of the cyclic six-membered transition states (112) and (113) in which the group R of the aldehyde occupies an equatorial position. The alternative orientation for R is disfavoured by 1,3-interaction with the oxygen atoms on boron, one of which is necessarily axial.

Allyl chromium compounds are useful for the preparation of *anti*-3-hydroxy-2-methylcarboxylic acids. Reaction of 1-bromo-2-butene with aldehydes in the presence of Cr(II) ion leads to *anti*(*threo*) condensation products with high selectivity, irrespective of the configuration of the

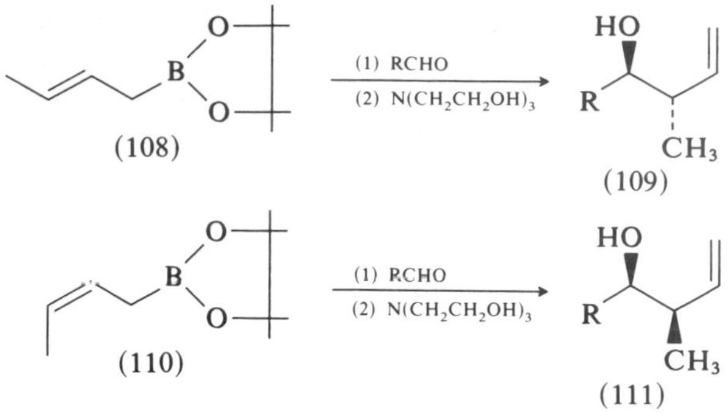

(108) → (1) RCHO, (2) $N(CH_2CH_2OH)_3$ → (109)

(110) → (1) RCHO, (2) $N(CH_2CH_2OH)_3$ → (111)

(1.105)

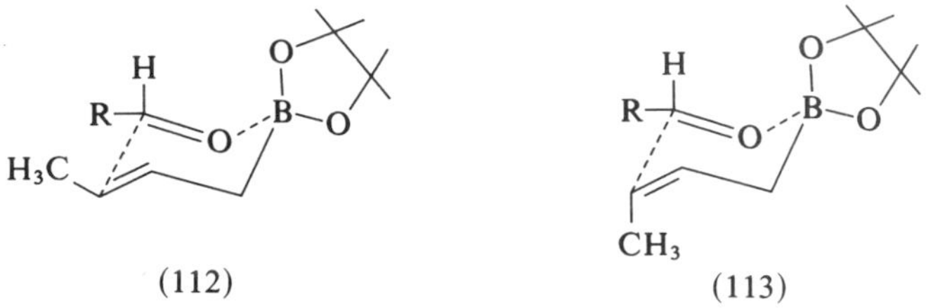

(112) (113)

(1.106)

double bond in the butene. Oxidative cleavage of the double bond in the resulting homoallylic alcohol provides the *anti*-3-hydroxy-2-methylcarboxylic acids. For example, reaction with benzaldehyde gave the *anti* homoallylic alcohol (114) exclusively in 96 per cent yield and thence, by ozonisation, the carboxylic acid (115) (Buse and Heathcock, 1978; Hiyama, Kimura and Nozaki, 1981).

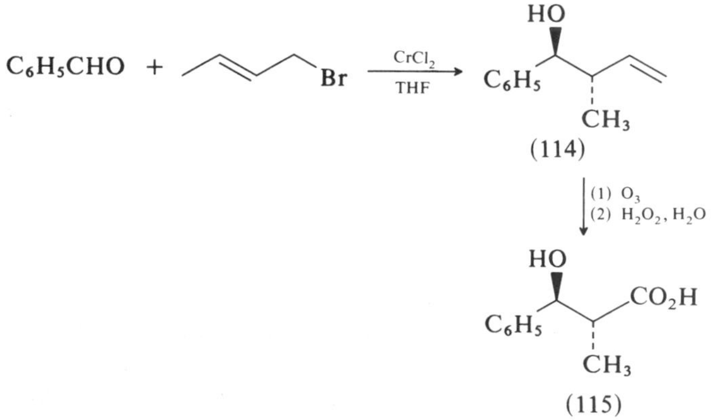

C_6H_5CHO + (crotyl bromide) $\xrightarrow[THF]{CrCl_2}$ (114) $\xrightarrow{(1)\ O_3;\ (2)\ H_2O_2,\ H_2O}$ (115)

(1.107)

The reaction (1.108) formed an important step in the synthesis of the antibiotic rifamycin-S. A cyclic transition state containing chromium is proposed for these reactions. Isomerisation of the double bond in the allyl group probably takes place in this complex.

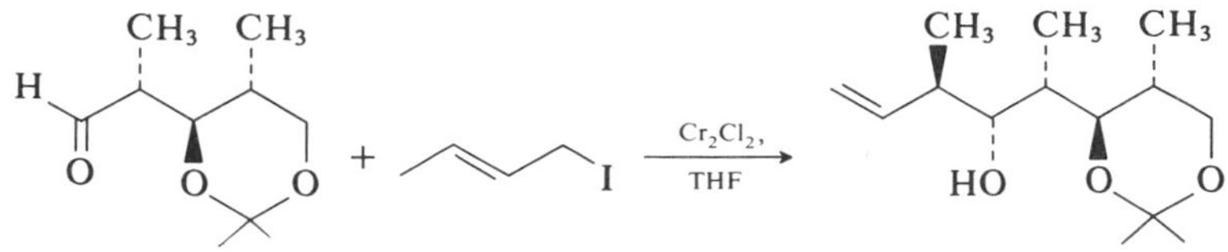

(1.108)

The 'nitro-aldol' reaction between the enolate of a nitro compound and an aldehyde frequently gives low yields as ordinarily effected, except in the case of nitromethane. Good yields can be obtained, however, by using the silyl nitronates. These react readily with a wide range of aldehydes in the presence of fluoride ion, which triggers the release of nitronate ion from the silyl ether (Colvin, Beck and Seebach, 1981). The main product is the *anti* isomer but this can be converted largely into the *syn* compound by careful acidification of the *bis*-lithium salt after desilylation (1.109) (Seebach *et al.*, 1982). Reduction of the products with hydrogen and Raney nickel affords the corresponding α-amino alcohols.

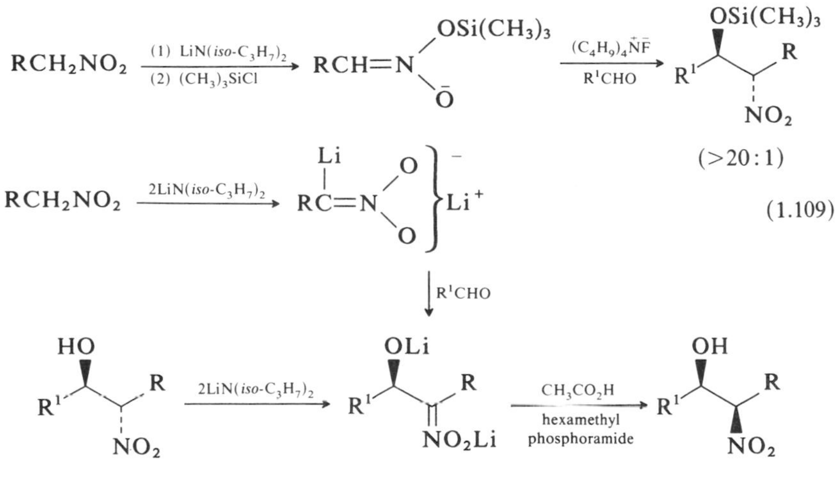

(>20:1)

(1.109)

1.9. Allylic alkylation of alkenes

Alkylation of a methyl or methylene group adjacent to a carbon–carbon double bond is not so easy as alkylation at the α-position to a carbonyl group but it can sometimes be effected, and conditions arranged to give replacement of hydrogen by an electrophilic or nucleophilic substituent. Metallation of alkenes with very strong bases generates allylic metallo derivatives which behave as typical organometallic compounds. Thus, reaction of limonene with butyl-lithium in the presence of tetramethylethylenediamine results in selective metallation at C-10, giving an allyl-lithium species which undergoes the whole range of reactions shown by other organo-lithium compounds. Reaction with 1-bromo-3-methyl-2-butene, for example, gave β-bisabolene (1.110). However, reactions are not always so selective and, in any case, the strongly basic conditions required in this method severely limit its application.

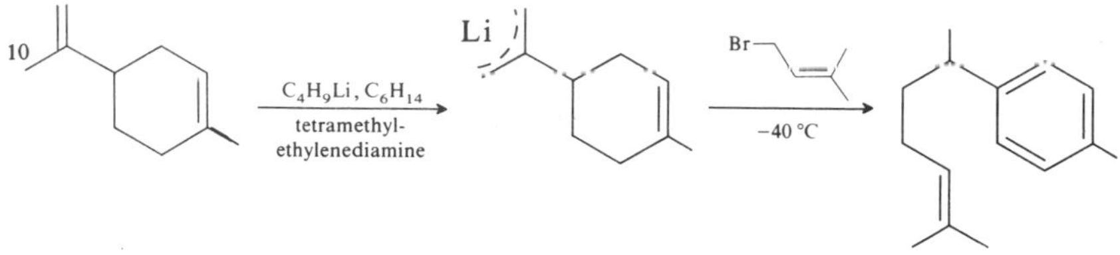

10 —[C_4H_9Li, C_6H_{14} / tetramethyl-ethylenediamine]→ Li —[Br, −40 °C]→

(1.110)

An alternative mild procedure involves activation of the allylic position of the alkene by formation of a π-allylpalladium complex, followed by reaction of this ambident electrophile with a suitable anion (Trost, 1980, 1981; Hegedus, 1984).

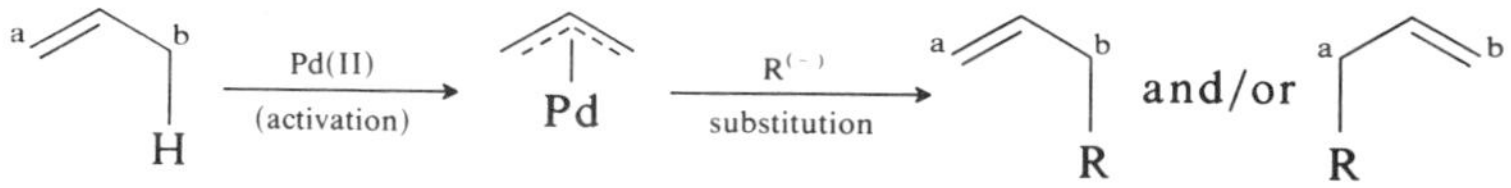

a, b, H —[Pd(II) (activation)]→ Pd —[$R^{(-)}$ substitution]→ a, b, R and/or a, b, R (1.111)

π-Allylpalladium complexes can be obtained directly from alkenes by reaction with a stoichiometric amount of a Pd(II) salt, for example palladium(II) chloride, in methanol or acetic acid. The reaction is believed to involve a palladium hydride species, and the following pathway has been suggested. Loss of hydrogen from the alkene takes place preferentially from the position adjacent to the more substituted

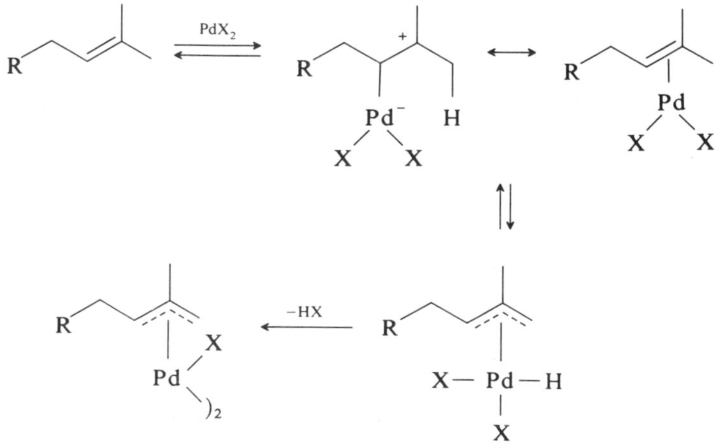

R ⇌[PdX_2] R, +, Pd^-, H, X, X ⟷ R, Pd, X, X

⇅ (1.112)

R, X, Pd, $)_2$ ←[−HX] R, X— Pd —H, X

end of the double bond, and in the sequence $CH_3 > CH_2 \gg CH$. Depending on the structure of the alkene a mixture of complexes may be obtained from which the individual components can sometimes be separated by fractional crystallisation or chromatography.

$PdCl_2$, NaCl; $CuCl_2$, CH_3CO_2H, CH_3CO_2Na → Pd, Cl,)$_2$ (4) + Pd, Cl,)$_2$ (1) (1.113)

Better is a *catalytic* process in which the allylic position is 'activated' by a leaving group, X. Reaction of these allyl compounds with palladium(0) complexes such as $Pd(PPh_3)_4$ brings about ionisation to form the same kind of π-allyl palladium complexes as in the stoichiometric reaction. The cationic Pd(II) allyl complexes (116) are attacked by a

X + PdL_4 ⇌ X^-, $\overset{+}{Pd}(II)$, L, L (116) —Na^+Nu^-→ Nu, Pd^0, L, L ↓ (allyl–X) Nu + X^-, $\overset{+}{Pd}$, L, L (1.114)

variety of nucleophiles, including amines and soft carbon nucleophiles with pK_as in the range 10–17, such as the anion of diethyl malonate. After nucleophilic attack a Pd(0) complex is produced which interacts with the allylic-X compound to regenerate the π-allyl palladium complex (116) and continue the cycle. The leaving group X may be a group such as a halide or ether but the best results so far have been obtained with allyl acetates, although it has recently been claimed that the corresponding phosphates are better still (Tanigawa *et al.*, 1982). Allylic lactones and epoxides have also been used (Trost and Klein, 1979; Trost and Molander, 1981; Tsuji, 1982).

Organometallic reagents are generally thought of as nucleophilic but it is to be noted that these π-allyl palladium complexes are electrophilic and react with nucleophiles. It has been thought that suitable carbon nucleophiles were restricted to relatively stabilised carbanions such as those derived from diethyl malonate and other 1,3-dicarbonyl compounds but conditions which allow the use of enolate anions derived from ketones have recently been discovered, thus greatly widening the scope of the method (Fiaud and Malleron, 1981). Harder anions, such as those in methyl-lithium, methyl-magnesium iodide or lithium dimethylcuprate, do not give coupled products, possibly due to preferential attack on the palladium rather than the allyl group. With an unsymmetrical π-allyl complex, reaction can take place at either end of the allyl system. Other things being equal attack at the less substituted (less sterically hindered) position predominates, but in many cases mixtures are obtained.

Some examples are given in equation (1.115) showing how the sequence can be used to convert an allylic methyl group into an ethyl group and to effect preferential substitution of an allylic acetate group in the presence of a normally more reactive halide.

$PdCl_2$, CH_3CO_2Na / $CuCl_2$, CH_3CO_2H → (Pd–Cl π-allyl dimer)$_2$

$NaCH(SO_2C_6H_5)(CO_2CH_3)$, $(C_6H_5)_3P$, THF → $CH(SO_2C_6H_5)(CO_2CH_3)$ product

(1) NaOH (2) 180 °C (3) Li, $C_2H_5NH_2$

Br…$OCOCH_3$ —$Pd[P(C_6H_5)_3]$, THF→ Br…PdL_2 —$NaCH(CO_2C_2H_5)_2$→ Br…$CH(CO_2C_2H_5)_2$

$Pd[P(C_6H_5)_3]_4$, THF, reflux, $NaCH(CO_2CH_3)_2$ → $HOOC$…$CH(CO_2CH_3)_2$

(1.115)

Intramolecular reactions take place readily, and have been used to prepare five- and six-membered rings and a variety of medium-ring compounds. Thus, a key step in the synthesis of the twelve-membered ring lactone (±)-recifeilide (118) involved the palladium-catalysed cyclisation of the allylic acetate (117).

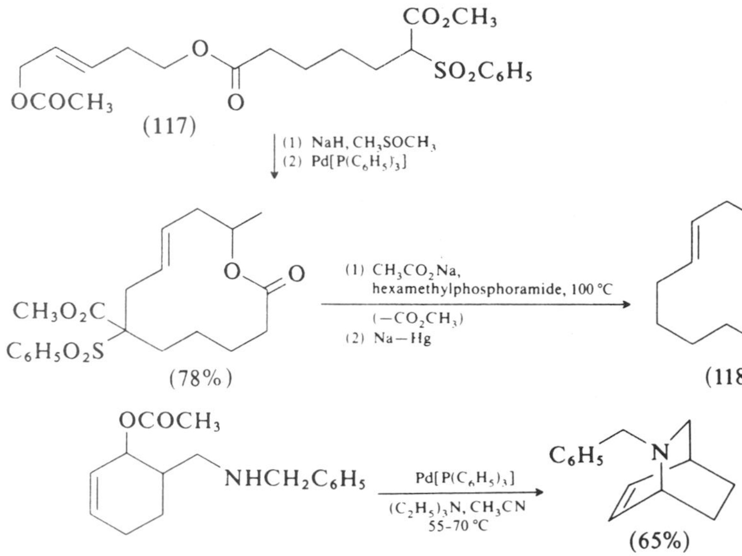

CO_2CH_3 $SO_2C_6H_5$ $OCOCH_3$ O (117)

(1) NaH, CH_3SOCH_3
(2) $Pd[P(C_6H_5)_3]$

CH_3O_2C $C_6H_5O_2S$ (78%)

(1) CH_3CO_2Na, hexamethylphosphoramide, 100 °C ($-CO_2CH_3$)
(2) Na–Hg

(118)

(1.116)

$OCOCH_3$ $NHCH_2C_6H_5$ $\xrightarrow[(C_2H_5)_3N,\ CH_3CN,\ 55\text{-}70\ °C]{Pd[P(C_6H_5)_3]}$ C_6H_5 N (65%)

These palladium-catalysed displacements are highly stereoselective and take place with net retention of configuration, at least in reactions with carbon nucleophiles (1.117). Retention of configuration is, in fact, the result of a double inversion. Displacement of acetate by the palladium to form the π-allylpalladium complex takes place with inversion; subsequent attack by the nucleophile on the side opposite the bulky palladium leads again to inversion with resulting retention of configuration in the product (1.118). Thus a key step in the synthesis of the insect

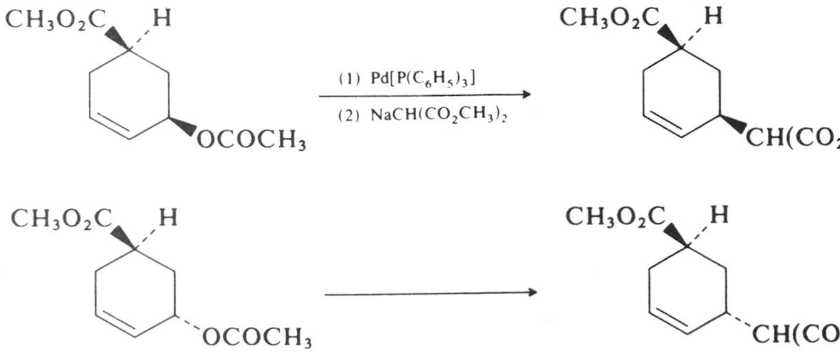

CH_3O_2C H $OCOCH_3$ $\xrightarrow[(2)\ NaCH(CO_2CH_3)_2]{(1)\ Pd[P(C_6H_5)_3]}$ CH_3O_2C H $CH(CO_2CH_3)_2$

CH_3O_2C H $OCOCH_3$ $\longrightarrow$ CH_3O_2C H $CH(CO_2CH_3)_2$ (1.117)

moulting hormone, ecdysone (121), involved the controlled conversion of the allylic acetate (119) into (120) with retention of configuration.

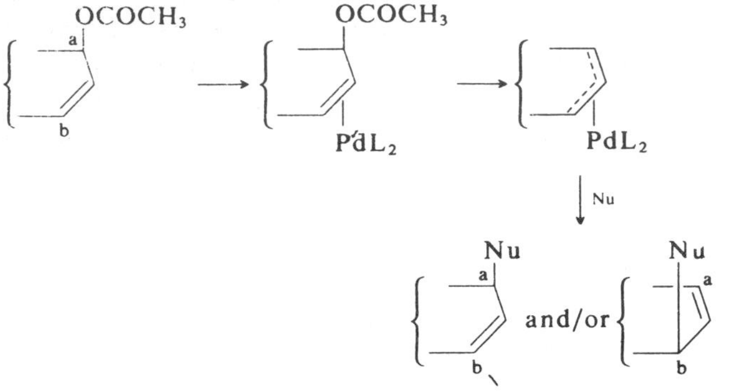

$OCOCH_3$ a b → $OCOCH_3$ PdL_2 → PdL_2 ↓ Nu (1.118)

Nu a b and/or Nu a b

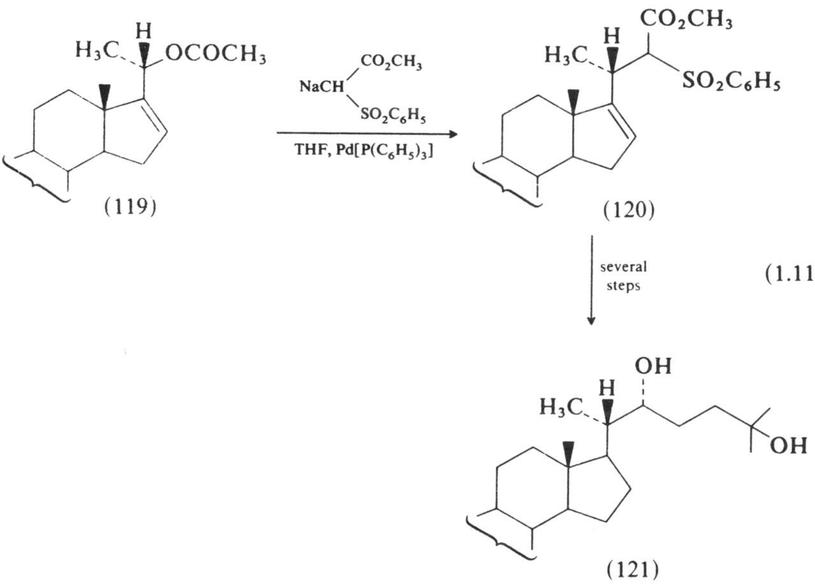

(119) H_3C, H, $OCOCH_3$ → $NaCH(CO_2CH_3)(SO_2C_6H_5)$, THF, $Pd[P(C_6H_5)_3]$ → (120) H_3C, H, CO_2CH_3, $SO_2C_6H_5$

↓ several steps (1.119)

(121) H_3C, H, OH, OH

Attempts have been made to achieve asymmetric allylic alkylations by using chiral phosphine ligands in the palladium catalyst instead of PPh_3, but optical yields obtained are not yet of practical synthetic value.

Another very useful palladium-catalysed reaction involves organopalladium species formed from organic bromides and iodides and pal-

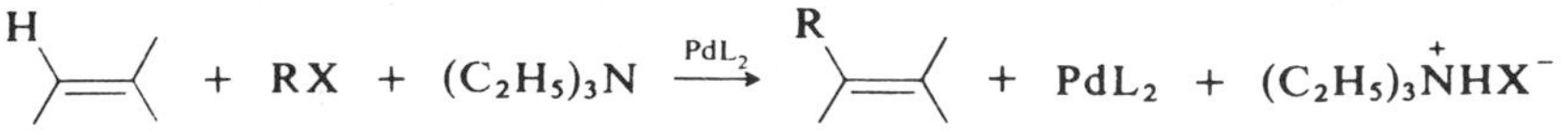

$$\text{H–C=C} + RX + (C_2H_5)_3N \xrightarrow{PdL_2} \text{R–C=C} + PdL_2 + (C_2H_5)_3\overset{+}{N}HX^- \quad (1.120)$$

ladium acetate in the presence of triphenylphosphine. These species react readily with alkenes in the presence of a base such as triethylamine with replacement of a *vinylic* hydrogen atom, providing a good route to substituted alkenes which cannot readily be obtained directly by any other route (Heck, 1979, 1982). The organic halide must not contain a β-hydrogen atom, in which case it simply undergoes elimination to form an alkene. Effectively, therefore, the reaction is limited to aryl, benzyl and vinyl bromides and iodides. The mechanism of the reaction is not known in detail, but the stereochemistry of the products obtained is best explained by *syn* addition of the organopalladium compound, $RPdL_2X$, to the alkene followed by *syn* elimination of palladium hydride. Thus, bromobenzene and (*Z*)-1-phenyl-1-propene gave (*Z*)-1,2-diphenyl-1-propene in 73 per cent yield, whereas the *E*-alkene gave (*E*)-1,2-diphenyl-1-propene.

C_6H_5Br + C_6H_5 CH_3 $\xrightarrow[P(C_6H_5)_3,\ (C_2H_5)_3N]{Pd(OCOCH_3)_2}$ CH_3 C_6H_5 C_6H_5 (73%) (1.121)

Reaction of aryl bromides with acrylic acid derivatives gives β-aryl $\alpha\beta$-unsaturated acids, and if vinylic bromides are used conjugated dienoic acids are obtained. In general, substitution at double bonds bearing an electron-withdrawing group on one carbon takes place mainly at the other. In other cases mixtures may be obtained unless one carbon of the double bond is sterically hindered, when again reaction takes place at the other carbon.

Br + CO_2CH_3 $\xrightarrow[2\%\ P(C_6H_5)_3,\ (C_2H_5)_3N]{1\%\ Pd(OCOCH_3)_2}$ CO_2CH_3 (75%) (1.122)

Intramolecular versions of the reaction are useful for making heterocyclic compounds. For example indoles are formed from *ortho*-halo-*N*-allylanilines (1.123). Indoles are also formed from *ortho*-allylanilines by reaction with palladium chloride in acetonitrile and the sequence can be made catalytic in palladium by using benzoquinone to re-oxidise Pd(0) to Pd(II). Cyclisation is believed to proceed by palladium-assisted nucleophilic attack on the double bond of the allyl group. Cyclisation

Br N H $\xrightarrow[(C_2H_5)_3N,\ CH_3CN,\ 110\,°C]{Pd(OCOCH_3)_2}$ N H (87%) (1.123)

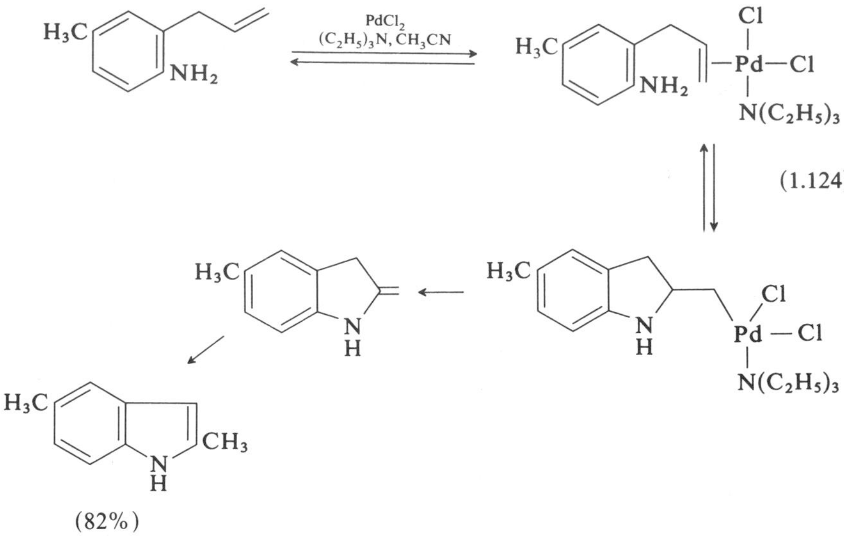

(1.124)

(82%)

of aliphatic amino alkenes to alicyclic nitrogen heterocycles requires conversion of the amine into its *p*-toluenesulphonyl derivative. This prevents harmful co-ordination of the more basic aliphatic amino group with the palladium catalyst. *ortho*-Allylphenols are similarly easily converted into benzofurans and unsaturated open chain alcohols with a suitable disposition of the hydroxyl group and the double bond form di- or tetra-hydrofuran derivatives. Acetylenic alcohols and amines have also been used to make derivatives of furan and pyrrole (Utimoto, 1983).

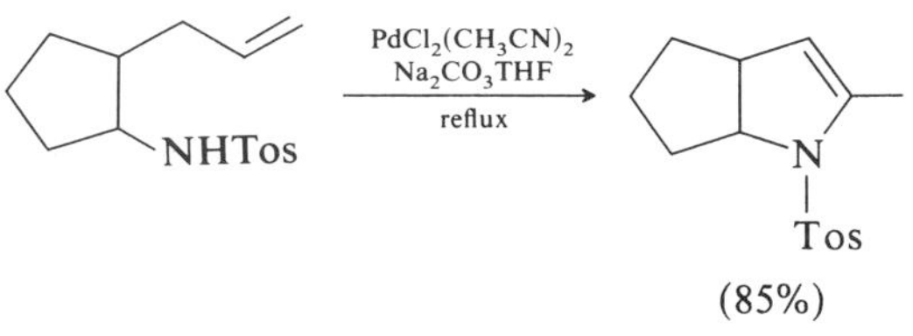

$PdCl_2(CH_3CN)_2$, Na_2CO_3THF, reflux (1.125)

(85%)

1.10. The dihydro-1,3-oxazine synthesis of aldehydes and ketones

A versatile method for the synthesis of aldehydes and ketones from alkyl halides, described by Meyers and his co-workers (1973), is illustrated in (1.126) in general terms. The starting material is a dihydro-1,3-oxazine derivative (available commercially or by a simple preparation from 2-methyl-2,4-pentandiol and a nitrile), the 'activated' alkyl group of which is readily metallated by treatment with butyl-lithium at −80 °C. The resulting anions react rapidly with a wide variety of alkyl halides (bromides and iodides) to give high yields of substituted di-hydro-

CH_2X $\xrightarrow{C_4H_9Li,\ THF,\ -80\ °C}$ CHLi(X) $\xrightarrow{RHal}$ CHR(X) $\xrightarrow{NaBH_4,\ THF,\ H_2O,\ C_2H_5OH}$ CHR(X)

RCHCHO(X) $\xleftarrow{H_2O,\ oxalic\ acid}$

(1.126)

oxazines which are smoothly reduced in quantitative yield with sodium borohydride to the tetrahydro derivatives. These compounds are readily cleaved in dilute acid to give the aldehyde. By this method halides are converted into aldehydes with two more carbon atoms, and the sequence thus complements the synthesis from lithio-1,3-dithiane described above (p. 43) which gives an aldehyde with one more carbon. A useful feature of the reaction is that, by reduction with borodeuteride instead of borohydride, C-1 labelled aldehydes can be obtained. By appropriate choice of starting materials a variety of different kinds of aldehydes and ketones can be synthesised. Thus, successive alkylation with two molecules of alkyl halide leads to $\alpha\alpha$-disubstituted aldehydes, and with one molecule of an $\alpha\omega$-dihalide, alicyclic aldehydes can be obtained (Politzer and Meyers, 1971). Aldehydes and ketones afford precursors of $\alpha\beta$-unsaturated aldehydes, and with epoxides γ-hydroxyaldehydes are formed. Hydrolysis at the dihydro-oxazine stage leads to carboxylic acids instead of aldehydes.

A limitation of the dihydro-oxazine synthesis of aldehydes is that, in general, it is suitable only for the preparation of monosubstituted acetaldehydes. Another difficulty is that dihydro-1,3-oxazines are not always readily accessible, and an alternative procedure using 2-alkyl-4,4-dimethyl-2-oxazolines, which are easily prepared by heating carboxylic acids with 2-amino-2-methyl-1-propanol, is often more convenient (Meyers and Mihelich, 1976). The lithio derivatives, prepared with butyllithium, react with alkyl halides giving 2-alkyloxazolines which may be hydrolysed to homologated acetic acids. Dialkylation leads to α,α-dialkylacetic acids. By reaction with aldehydes and ketones, followed by hydrolysis, $\alpha\beta$-unsaturated acids are formed, and epoxides give γ-butyrolactones.

$$\text{HOCH}_2\text{C(CH}_3)_2\text{NH}_2 \xrightarrow[100\text{–}150\,°\text{C}]{\text{RCO}_2\text{H}} \text{2-R-4,4-dimethyloxazoline}$$

$$\text{2-CH}_3\text{-oxazoline} \xrightarrow[(2)\ \text{C}_6\text{H}_5\text{CH}_2\text{Cl}]{(1)\ \text{C}_4\text{H}_9\text{Li}} \text{oxazoline-CH}_2\text{CH}_2\text{C}_6\text{H}_5 \xrightarrow[(2)\ \text{CH}_3\text{I}]{(1)\ \text{C}_4\text{H}_9\text{Li}} \text{oxazoline-CH(CH}_3)\text{CH}_2\text{C}_6\text{H}_5 \xrightarrow{\text{H}_3\text{O}^+} (\text{C}_6\text{H}_5\text{H}_2\text{C})(\text{H}_3\text{C})\text{CHCO}_2\text{H}\ (89\%) \quad (1.127)$$

An interesting development is the use of chiral oxazolines (as 122) to effect asymmetric synthesis of α-alkylcarboxylic acids (1.128). High optical yields are obtained, and either enantiomer of the alkylated acids can be prepared by selecting the order of introduction of the substituent groups (Meyers, 1978). The high asymmetric induction in the alkylation

$$(122) \xrightarrow[\text{THF},\ -80\,°\text{C}]{\text{LiN}(iso\text{-C}_3\text{H}_7)_2} (6\%) + (123)\ (94\%) \xrightarrow{\text{RX}} \text{transition state (R–X--Li}\leftarrow\text{OCH}_3) \xrightarrow{-\text{LiX}} \text{H(CH}_3)\text{(R)C-oxazoline} \xrightarrow{\text{H}_3\text{O}^+} \text{H(CH}_3)\text{(R)C–CO}_2\text{H}$$

(S)-acid if $R > CH_3$
for $R = C_4H_9$, 75% optically pure (1.128)

step is ascribed to the formation of the rigid chelated asymmetric lithium compound (123). The methoxy group is essential, for in its absence the optical yields fall drastically. Attack on the lithium compound by the alkylating agent takes place from the underside of the molecule as drawn in (123) i.e. on the side opposite the bulky phenyl substituent. If the phenyl substituent is replaced by a smaller group optical yields again fall.

By reaction with aldehydes instead of alkyl halides, high yields of *threo*(*anti*)-3-hydroxy-2-methylcarboxylic acids are obtained. High optical yields of 3-alkylcarboxylic acids have also been obtained by conjugate addition of organolithium compounds to vinyloxazolines (e.g. 124). This is illustrated in (1.129) for the synthesis of (+)-ar-turmerone (125) the chief constituent of the essential oil from rhizomes of *curcuma longa linn* (Meyers and Smith, 1979).

(124) — (1) $p\text{CH}_3\text{C}_6\text{H}_4\text{Li}$, THF, −78 °C; (2) CH_3CO_2H → (99% optically pure)

↓ H_3O^+, C_2H_5OH

several steps

(125) (93% enantiomeric excess) (1.129)

The oxazoline ring is, in effect, a masked carboxyl group, and conversion into an oxazoline is a useful method for protecting carboxyl groups against attack by Grignard reagents or lithium aluminium hydride.

1.11. Coupling of organonickel and organocopper complexes

Formation of carbon–carbon bonds by straightforward reaction of organic halides with organometallic compounds, as in the Wurtz reaction or the reaction of Grignard reagents with halides, is of only limited use in synthesis. Yields are often poor, owing to intervening side reactions such as α- and β-eliminations, and in 'mixed' reactions, as with RHal and R′M, substantial amounts of RR and R′R′ are formed besides the desired RR′ because of fast halogen-metal interchange. In the presence of certain transition metal complexes, however, notably those of nickel (Kumada, 1980) and copper (see p. 87) greatly improved yields of coupled products can be obtained from Grignard reagents and organohalides, and these catalysed reactions provide a highly efficient method for the formation of carbon-carbon bonds. The reaction with monohalo-alkenes takes place stereospecifically with complete retention of the geometrical configuration of the double bond. For example, (*Z*)- and (*E*)-β-bromostyrene are converted exclusively into *cis*- and *trans*-1-phenylpropene by reaction with methylmagnesium iodide in ether in the presence of $NiCl_2(PPh_3)_2$.

Another method for the selective combination of unlike groups uses the π-allylnickel(I) bromides, represented as in (1.130), which are

R=H, CH_3, $CO_2C_2H_5$, etc. (1.130)

obtained by reaction of allylic bromides with excess of nickel carbonyl in dry benzene (Semmelhack, 1972; Baker, 1973). In polar co-ordinating solvents such as dimethylformamide or *N*-methylpyrrolidine these complexes react smoothly with a wide variety of organic halides (iodides are best) to give the coupled product in high yield (1.131). Vinyl and aryl halides react just as well as alkyl halides, and, helpfully, hydroxyl and carbonyl groups do not interfere with the reaction. Carbonyl groups do react with the complexes, but more slowly and at higher temperatures.

$$2R'Hal. + (\text{R-allyl-NiBr})_2 \longrightarrow 2R'-CH_2-C(R)=CH_2 \quad (1.131)$$

R' = 4-hydroxycyclohexyl, R = CH_3 ⟶ 88%
R' = C_6H_5 R = CH_3 ⟶ 98%

The complex from $\alpha\alpha$-dimethylallyl bromide generally undergoes preferential coupling at the primary rather than at the tertiary position. In cases where substituents at C-1 and C-3 of the allyl group allow the possibility of geometrically isomeric products both isomers are usually produced (1.132).

$$CH_3I + (\text{(CH}_3)_2\text{-allyl-NiBr})_2 \longrightarrow CH_3-CH_2-CH=C(CH_3)-CH_3 \quad (90\%)$$

(1.132)

$$\text{C}_6\text{H}_{11}\text{I} + (\text{CH}_3,\text{R-allyl-NiBr})_2 \longrightarrow \text{C}_6\text{H}_{11}-CH_2-CH=C(CH_3)R$$

(40% *cis*, 60% *trans*)
R = $(CH_3)_2C=CHCH_2CH_2-$

Exactly how the new carbon–carbon bond is formed in these reactions is still uncertain, but it is thought that a complex formed by the route shown in (1.133) may be involved.

$$\text{(allyl)Ni(Br)(L)} + \text{RX} \rightleftharpoons \text{(allyl)Ni(Br)(L)} \leftarrow \text{R—X} \longrightarrow \text{(allyl)Ni(Br)(X)(R)} + \text{L}$$

L = e.g. dimethylformamide (1.133)

There is evidence that in some cases at least it proceeds by a free radical chain mechanism. Thus, reaction of π-(2-methoxyallyl)nickel bromide with (S)-(+)-2-iodo-octane gave 4-methyl-2-decanone which was completely racemic. Furthermore, the reaction was totally inhibited by addition of a small amount of *meta*-dinitrobenzene, a potent radical anion scavenger. A possible chain-initiation step involves electron transfer from the nickel complex to the halide to produce the corresponding radical anion.

$$[\text{(allyl)NiBr}] + \text{RX} \rightarrow \text{RX}^{\cdot -} + [\text{(allyl)NiBr}]^{\cdot +}$$

Allyl halides undergo a more complex reaction with π-allylnickel(I) complexes due to halogen–metal exchange and cannot be used in coupling reactions of the type described above, except where the allyl group in the halide and in the complex is the same. In other cases, mixtures of products are obtained.

The reaction has been usefully extended to the synthesis of cyclic 1,5-dienes, which are very easily obtained in remarkably high yield by reaction of $\alpha\omega$-bisallyl bromides with nickel carbonyl (1.134). The products consist very largely of the *trans, trans* cyclic dienes, irrespective of whether the starting material contained *cis* or *trans* double bonds.

$$(CH_2)_n(CH{=}CHCH_2Br)_2 \xrightarrow[\text{dimethylformamide}]{Ni(CO)_4} (CH_2)_n\begin{matrix} CH{=}CHCH_2 \\ | \\ CH{=}CHCH_2 \end{matrix} \quad (1.134)$$

$n = 6$ (59%)
$n = 8$ (70%)
$n = 12$ (84%)

With the bisallyl bromide where $n = 2$, however, the main reaction was the formation of a *six-* and not an eight-membered ring; 4-vinyl-cyclohexene was the predominant product. Similarly, for $n = 4$, the product consisted entirely of *cis-* and *trans-*1,2-divinylcyclohexane; no cyclodeca-1,5-diene was detected. Evidently, in these cases, six-membered ring formation is so much favoured over eight- and ten-membered ring formation that the usual strong preference for the joining of primary carbon atoms is overcome.

This cyclisation procedure, because it leads to cyclic 1,5-dienes, makes available a wide variety of cyclic structures which are not readily obtainable by any other method, and has already been used in the synthesis of a number of natural products (Billington, 1985). The important sesquiterpene hydrocarbon humulene, for example, was readily synthesised using this method and the ester (126) was converted into the macrolide (127).

CH_2Br O O $Ni(CO)_4$ *N*-methyl-pyrrolidone O O (1.135)

CH_2Br (70%)

(126) (127)

Ni(0)-complexes such as bis(1,5-cyclo-octadiene)nickel and tetrakis(triphenylphosphine)nickel have been used to catalyse the coupling of vinyl and aryl halides, providing a useful route to symmetrical 1,3-dienes and biaryl derivatives (Semmelhack *et al.*, 1981). Conditions are mild and compatible with most common functional groups. Thus, 4-bromobenzaldehyde gave 4,4′-diformylbiphenyl in 79 per cent yield with a catalytic amount of the 1,5-cyclo-octadiene complex, and methyl 2-bromoacrylate gave butadiene-2,3-dicarboxylic ester in 99 per cent yield. Intramolecular reactions take place readily as illustrated in the final step of the synthesis of (128), the dimethyl ether of the naturally

CH_3O I O I CH_3O $Ni[P(C_6H_5)_3]_4$ dimethyl-formamide 50 °C CH_3O O CH_3O (1.136)

(128) (52%)

occurring bridged biphenyl alnusone. Intermolecular reactions are limited to the formation of symmetrical coupled products. Unsymmetrical products have been obtained by catalysed condensation of alkenyl-diisobutylalanes with iodoalkenes (Negishi, 1982) (1.137).

$$(E)\text{-}C_5H_{11}CH{=}CHAl(iso\text{-}C_4H_9)_2 + (E)\text{-}ICH{=}CHC_4H_9 \xrightarrow[\text{ether-hexane, 25 °C}]{5\%\ Ni[P(C_6H_5)_3]_4} C_5H_{11}CH{=}CHCH{=}CHC_4H_9 \quad (70\%;\ 95\%\ E,E) \tag{1.137}$$

A number of other organic reactions leading to the formation of new carbon–carbon bonds are promoted by copper and copper ions. Some of these may involve the transient formation of organocopper species. A very useful reaction is the oxidative coupling of terminal alkynes, which is easily effected by agitation of the alkyne with an aqueous mixture of copper(I) chloride in an atmosphere of air or oxygen (the Glaser reaction) or by reaction of the ethynyl compound with copper(II) acetate in pyridine solution. Good yields of coupled products can be obtained from substrates containing a variety of functional groups, and the reaction has been applied to the synthesis of cyclic compounds from $\alpha\omega$-diacetylenes, and as a key step in the synthesis of a variety of natural products, for example, the fungal polyene, corticrocin (128) (1.138). The mechanism of the reaction is not entirely clear, but it seems to be generally held that it proceeds by an initial ionisation step, facilitated probably by copper(I) ion complex formation, followed by a one-electron transfer involving copper(II) and subsequent dimerisation of the radical formed. There is no doubt, however, that the reaction provides one of the easiest methods of forming carbon–carbon bonds and ranks with the Ullmann, Kolbe and acyloin reactions for molecular duplication.

$$HO_2CCH{=}CH(CH_2)_2C{\equiv}CH \xrightarrow{Cu_2Cl_2,\ O_2,\ H_2O,\ NH_4Cl} HO_2CCH{=}CH(CH_2)_2C{\equiv}CC{\equiv}C(CH_2)_2CH{=}CHCO_2H\ (100\%) \xrightarrow{20\%\ aq.\ KOH} \underset{(128)}{HO_2C(CH{=}CH)_6CO_2H} \tag{1.138}$$

$$HC{\equiv}C(CH_2)_2OCO(CH_2)_8C{\equiv}CH \xrightarrow[\text{pyridine}]{Cu(OCOCH_3)_2} \text{cyclic } (CH_2)_2(C{\equiv}C)_2(CH_2)_8 \text{ with } O\text{—}CO \text{ bridge} \quad (88\%)$$

The direct coupling of terminal alkynes discussed above is obviously not so well suited for the synthesis of unsymmetrical products by coupling of two different alkynes. Happily, this difficulty has been overcome by the discovery that, in presence of a base and a catalytic amount of copper(I) ion, terminal alkynes react rapidly with 1-bromo-alkynes with elimination of hydrogen bromide, to form the unsymmetrical diyne in high yield. The reaction is believed to take the course shown in (1.139). The second step regenerates the copper(I) ion which is best kept at low concentration to avoid self-coupling of the bromoalkyne. The addition of a base facilitates the reaction by removing the liberated acid and assisting in the solution of the copper(I) derivative. This valuable reaction has already been used to synthesise a variety of unsymmetrical diynes, including a number of naturally occurring polyacetylenic compounds.

$$\begin{aligned} RC{\equiv}CH + Cu^+ &\rightarrow RC{\equiv}CCu + H^+ \quad \text{(fast)} \\ RC{\equiv}CCu + BrC{\equiv}CR' &\rightarrow R(C{\equiv}C)_2R' + CuBr \end{aligned} \tag{1.139}$$

1.12. Reactions of lithium organocuprates: copper-catalysed reactions of Grignard reagents

A more general procedure for coupling unlike organohalides which is not restricted to the coupling of allylic with non-allylic halides as in the reactions of π-allylnickel halides discussed above, makes use of the now widely employed lithium organocuprates, which are generally represented as LiR_2Cu (Posner, 1975*b*). These reagents are more stable and more reactive than the well-known organocopper(I) reagents (see p. 85). They are prepared *in situ* and not isolated, most conveniently by reaction of two equivalents of an organolithium compound with cuprous iodide in ether. Aryl, alkenyl and primary alkyl cuprates are readily obtained by this procedure. Secondary and tertiary alkyl cuprates are best obtained from the corresponding lithium compound and an ether-soluble derivative of copper(I) iodide such as the complex with

$$\begin{aligned} &CH_3Li + CuI \xrightarrow{\text{ether}} CH_3Cu \xrightarrow{CH_3Li} Li(CH_3)_2Cu \\ &2(CH_3)_3CLi + CuI(C_4H_9)_3P \xrightarrow{\text{ether}} Li[(CH_3)_3C]_2CuP(C_4H_9)_3 \end{aligned} \tag{1.140}$$

tributylphosphine or dimethyl sulphide. The composition of the reagent solutions is not well defined. The state of aggregation of the complexes is uncertain, but it seems likely that, in many, the organic ligands are bonded to tetrahedral clusters of four metal atoms. Recent spectroscopic studies of lithium dimethylcuprate suggest that in ether solution it exists as a dimer $[LiCu(CH_3)_2]_2$. These reagents undergo a variety of synthetically useful reactions (Carruthers, 1982). They are excellent reagents for the specific replacement of iodine or bromine in organic halides by alkyl, alkenyl or aryl groups. Reactions take place readily at or below room temperature to give high yields of substitution products. Primary alkyl tosylates also react very well. Ketonic carbonyl groups react only slowly under the conditions employed, so that selective reaction in the presence of an unprotected ketonic carbonyl group is possible, but aldehydes undergo ordinary carbonyl addition unless the temperature is kept below about −90 °C. Some typical reactions are shown in (1.141). Secondary alkyl bromides and iodides generally do not give very good yields of product on reaction with the ordinary cuprates LiR_2Cu, but this difficulty can be overcome by reaction with the so-called 'higher order' cuprates, $Li_2R_2Cu(CN)$, formed *in situ* by reaction of two molecular equivalents of an organo-lithium compound with cuprous cyanide. These reagents also react very readily with primary alkyl bromides which sometimes give only poor yields with ordinary cuprates (Lipshutz *et al.*, 1983). It should be noted, however, that many of these displacement reactions can be effected just as well, or even better in certain cases, with the appropriate Grignard reagent and a catalytic amount of a copper(I) salt (see p. 87).

OTos $\xrightarrow[\text{ether, } -75\,°C]{(C_4H_9)_2CuLi}$ $C_{10}H_{22}$ (98%)

H₃C OH H ... Br Br $\xrightarrow[\text{ether, } -15\,°C]{(CH_3)_2CuLi}$ H₃C OH H ... H₃C CH₃ (60%)

(1.141)

Br⁀Cl $\xrightarrow[\text{THF, } -78\,°C]{(s\text{-}C_4H_9)_2Cu(CN)Li_2}$ ...Cl (89%)

I $\xrightarrow[\text{THF, } 0\,°C]{(CH_2{=}CH)_2Cu(CN)Li_2}$ (90%)

Reaction of lithium diphenylcuprate with (−)-(*R*)-2-bromobutane takes place with predominant inversion of configuration (1.142) and this and other features of the reaction of organocuprates with secondary alkyl halides suggests that they proceed by S_N2 displacement at carbon.

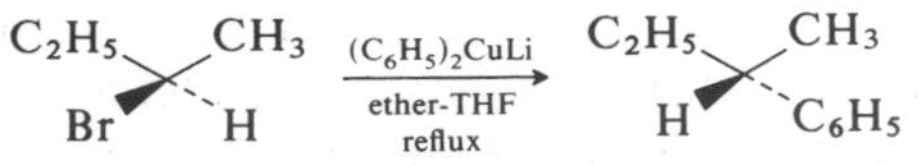

$$\text{(1.142)}$$

However, the corresponding iodide gives a racemic product on reaction with lithium diphenylcuprate, and there is evidence that reaction of cuprates with iodides at any rate takes place by a one-electron transfer process and not by S_N2 displacement (Ashby *et al.*, 1982).

In contrast, alkenyl halides react with organocuprates to give the substituted alkene with retention of configuration of the double bond. Likewise, alkenylcuprates react with retention of the geometry of the double bond (1.143). Since alkenyl free radicals are easily interconverted these results suggest that alkenyl radicals are not involved in these reactions, but little is yet known of the detailed course of the reactions.

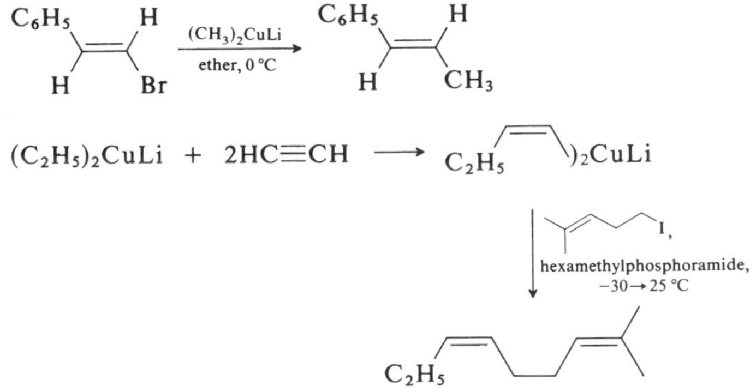

$$(C_2H_5)_2CuLi + 2HC{\equiv}CH \longrightarrow (C_2H_5CH{=}CH)_2CuLi$$

hexamethylphosphoramide, −30→25 °C

(63%; 95% *Z*)

(1.143)

As illustrated in the second example in (1.143) *Z*-alkenylcuprates are readily obtained by *syn* addition of organocuprates to acetylenes. Alkylation of these cuprates provides a good route to *Z*-1,2-disubstituted alkenes, and by reaction with alkenyl halides, best in the presence of zinc bromide and catalytic $Pd(PPh_3)_4$, conjugated dienes are obtained with high stereoselectivity. Aryl halides give styrene derivatives (1.144) (Jabri, Alexakis and Normant, 1981).

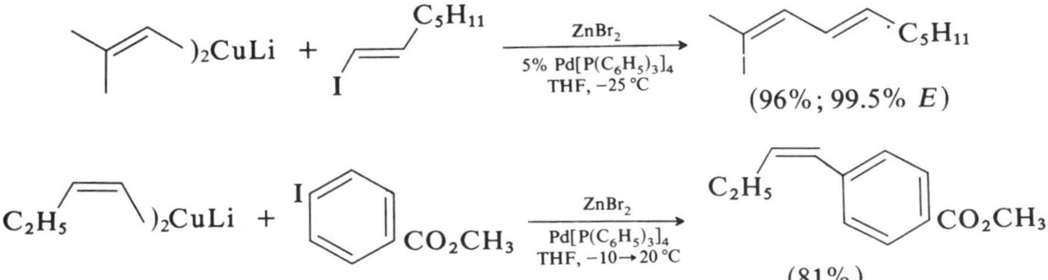

$ZnBr_2$, 5% $Pd[P(C_6H_5)_3]_4$, THF, −25 °C

(96%; 99.5% *E*)

$ZnBr_2$, $Pd[P(C_6H_5)_3]_4$, THF, −10→20 °C

(81%)

(1.144)

Acid chlorides also react readily to give ketones, and epoxides form alcohols by attack at the less substituted carbon. Better yields are obtained with the 'higher order' cuprates $Li_2R_2Cu(CN)$. With $\alpha\beta$-unsaturated epoxides allylic alcohols are formed by reaction at the allylic position (1.145).

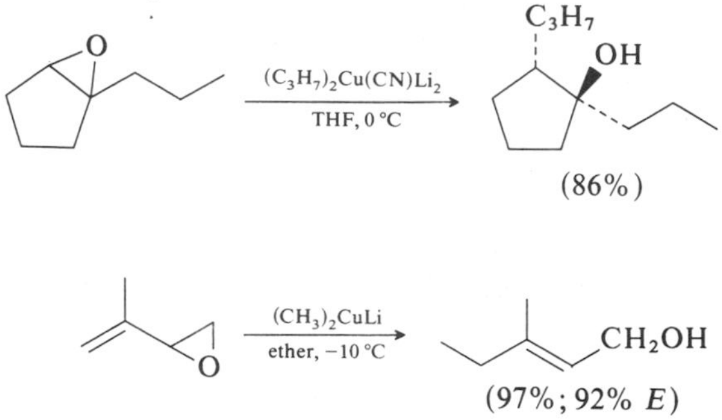

O

$(C_3H_7)_2Cu(CN)Li_2$
THF, 0 °C

C_3H_7
OH

(86%)

(1.145)

$(CH_3)_2CuLi$
ether, −10 °C

O

CH_2OH

(97%; 92% *E*)

One of the most useful reactions of organocuprates is their conjugate addition to $\alpha\beta$-unsaturated ketones to give β-substituted derivatives of the corresponding saturated ketone (1.146) (Posner, 1975*a*); non-conjugated ketonic carbonyl groups do not react under the conditions employed. Conjugate addition of Grignard reagents to $\alpha\beta$-unsaturated ketones is a well-known reaction, but higher yields are often obtained with the cuprates, particularly in reactions with higher order cuprates $Li_2R_2Cu(CN)$. Reaction with $\alpha\beta$- and $\beta\beta$-disubstituted enones is often not very successful because of steric crowding of the double bond, but in such cases very much better yields are obtained by effecting the reaction in the presence of a Lewis acid, particularly boron trifluoride etherate

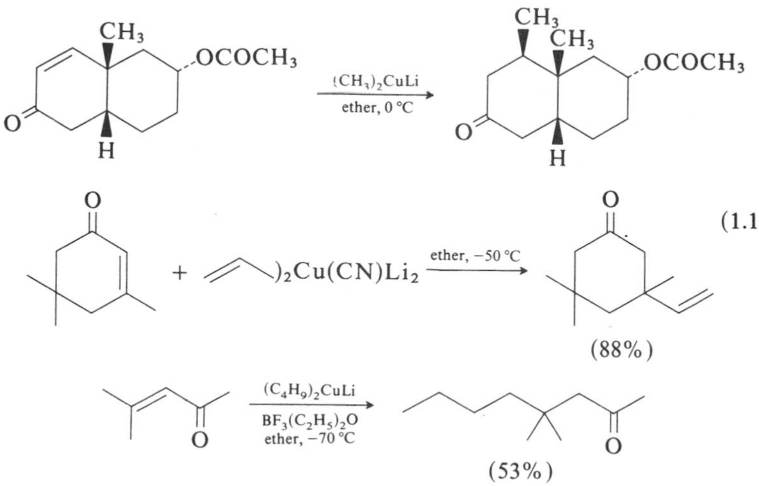

CH_3
$OCOCH_3$
O
H

$(CH_3)_2CuLi$
ether, 0 °C

CH_3
CH_3
$OCOCH_3$
O
H

(1.146)

O
+ $)_2Cu(CN)Li_2$
ether, −50 °C
O

(88%)

$(C_4H_9)_2CuLi$
$BF_3(C_2H_5)_2O$
ether, −70 °C

O

O

(53%)

(1.146) (Yamamoto *et al.*, 1982). The factors controlling the stereochemistry of addition of organocuprates to cyclic conjugated enones are not completely understood. Mixtures of geometrical isomers are often produced, but generally one isomer predominates, formed by approach of the reagent in a direction orthogonal to the plane of the enone system. Both steric and stereoelectronic factors play a part and the solvent may also have an effect.

The mechanism of the transfer of organo groups from organocuprates to the β-position of conjugated ketones is uncertain. One suggestion, supported by a considerable amount of evidence (House, 1976) is that reaction proceeds by initial transfer of one electron from the organocopper(I) species to the ketone to give an anion radical followed by coupling and intramolecular transfer of the organo group from the metal to the β-carbon atom (1.147). In reactions with enones the R groups of the organocuprate, R_2CuLi, are transferred with retention of configuration, apparently excluding the intermediacy of the free radical $R^{\cdot}$, and in line with the formation of an intermediate complex (129) in which the group R migrates intramolecularly from the cuprate to the enone.

$$R_4Cu_2Li_2 + CH_3CH{=}CHCOCH_3 \xrightarrow{\text{electron transfer}} CH_3\dot{C}HCH{=}C(O^-)CH_3 + [R_4Cu_2Li_2]^{\cdot +}$$

$$\xrightarrow{\text{coupling}} \underset{(129)}{[R_4Cu_2Li_2\text{-}CH(CH_3)CH{=}C(O^-)CH_3]} \longrightarrow CH_3CH(R)CH{=}C(O^-)CH_3 + [R_3Cu_2Li_2]^+ \qquad (1.147)$$

$$\longrightarrow CH_3CH(R)CH_2COCH_3 + Li^+ + RCu + \tfrac{1}{2}R_4Cu_2Li_2$$

Whatever the exact mechanism of the conjugate addition reaction, it seems clear that enolate anions are formed as intermediates and they can be alkylated by reaction with alkyl halides. Both inter- and intramolecular examples are known. Thus, 3-butyl-2-methylcyclohexanone was obtained from cyclohexenone in 84 per cent yield by successive reaction with lithium dibutylcuprate and methyl iodide, and the *cis*-decalone (130), an intermediate in the synthesis of the sesquiterpene valerane, was obtained as shown in (1.148).

Conjugate addition of 'ordinary' organocuprates to $\alpha\beta$-unsaturated aldehydes is not a synthetically useful reaction because of the simultaneous formation of products formed by reaction at the carbonyl group. However, conjugate addition can be effected with the complex

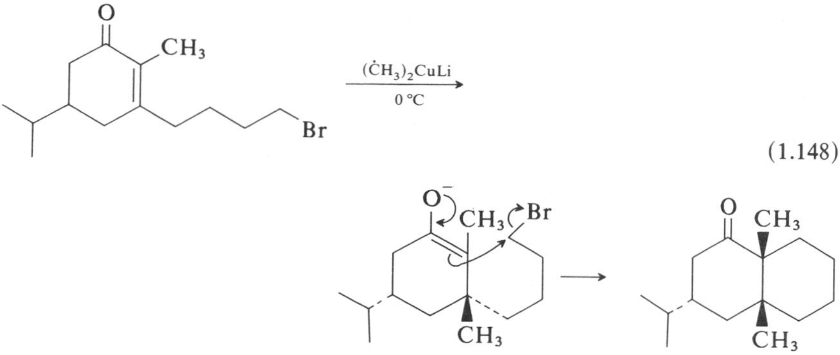

$$\xrightarrow[0\,°C]{(CH_3)_2CuLi}$$

(1.148)

(130)

$Me_5Cu_3Li_2$, formed by addition of the appropriate amount of methyllithium to a suspension of copper(I) iodide in ether. It converts $\alpha\beta$-unsaturated aldehydes efficiently into β-methyl aldehydes and, unlike Me_2CuLi, gives negligible attack on the carbonyl group, even when conjugate addition results in formation of a quaternary carbon atom. Thus, the cyclohexylidene derivative (131) gave the β-methyl aldehyde (132) in 90 per cent yield, a synthetically useful transformation since decarbonylation affords the *gem*-dimethyl group found in many natural products.

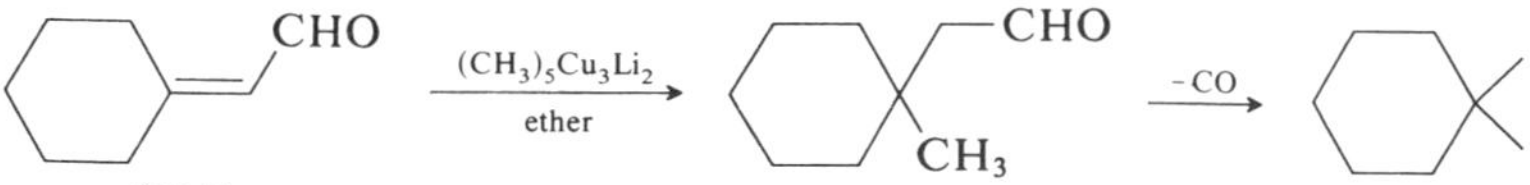

$$\xrightarrow[ether]{(CH_3)_5Cu_3Li_2} \qquad \xrightarrow{-CO}$$

(131)

(132)
(90%; 99%
1,4-addition)

(1.149)

Conjugate addition is also a feature of the reaction of lithium organocuprates with $\alpha\beta$-acetylenic esters to give $\beta\beta$-disubstituted acrylic esters. By conducting the reaction at −78 °C high yields of *cis*-addition compounds are obtained; that is the reaction yields an acrylic ester in which the substituents in the acetylenic precursor are *cis* to each other (1.150). $\alpha\beta$-Olefinic esters do not react with organocuprates under the usual mild conditions, but conjugate addition can be effected in the presence of boron trifluoride etherate, or by copper(I)-catalysed reaction with Grignard reagents (see p. 87). Addition of alkylcopper–boron trifluoride complexes to $\alpha\beta$-unsaturated esters in which the acid is esterified with an appropriate optically active alcohol has been employed in an asymmetric synthesis of β-substituted carboxylic acids.

(*S*)-(−)-Citronellic acid, for example, was obtained in 98.5 per cent optical purity as shown in (1.150) (Oppolzer *et al.*, 1983).

C_7H_{15}—≡—CO_2CH_3 $\xrightarrow[\text{THF, −78 °C}]{(CH_3)_2CuLi}$ C_7H_{15}, CO_2CH_3, CH_3, Cu $\xrightarrow{H_3O^+}$ C_7H_{15}, CO_2CH_3, CH_3, H

O, O, O, H, $C(CH_3)_3$, CH_3 $\xrightarrow[\text{CuI BF}_3\ -78\ °C]{\text{Li}}$ O, R*O, H, CH_3 (1.150)

↓ hydrolysis

H_3C, H, HO_2C

Allylic halides and acetates also react readily with organocuprates. Reaction can occur at either end of the allylic system to give either unrearranged or rearranged products. The course of the reactions is dependent on the structure of the allylic compound and to some extent on the experimental conditions. Thus the bromide (133) and an equimolecular amount of di-n-butylcuprate gave mainly the rearranged compound (134), but with the acetate (135) and lithium dimethylcuprate the predominant product was formed by direct replacement of the acetate group. Allylic lactones give unsaturated acids.

A disadvantage of the straightforward reactions of organocuprates described above is that in many cases an excess of reagent has to be employed, and in the conjugate addition to enones at any rate only one of the two organo groups in the cuprate takes part in the reaction; the other is effectively wasted. One of the advantages of the 'higher order' organocuprates, $R_2Cu(CN)Li_2$, is that only a small excess of the reagent is required. A number of other 'mixed' reagents, R_rR_tCuLi, have been developed in which the group R_r is tightly bound to copper and only R_t is transferred. Two reagents which have been used have 1-pentynyl

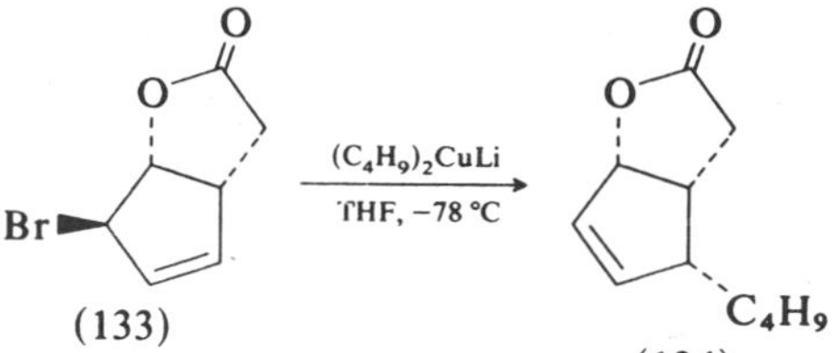

(133) $\xrightarrow[\text{THF, } -78\,^{\circ}\text{C}]{(C_4H_9)_2CuLi}$ (134)

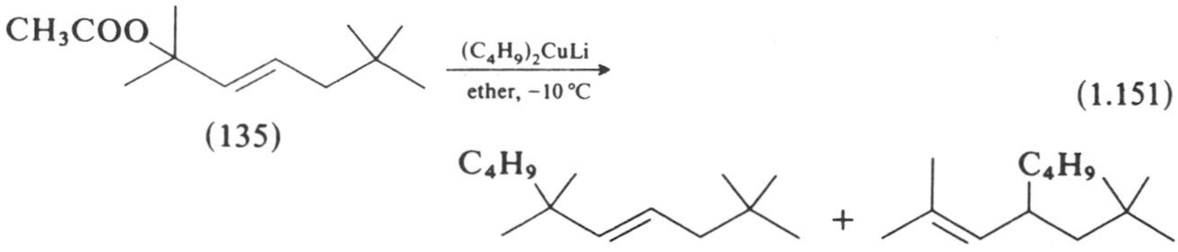

CH_3COO (135) $\xrightarrow[\text{ether, } -10\,^{\circ}\text{C}]{(C_4H_9)_2CuLi}$ (1.151)

C_4H_9 … 83% of product + C_4H_9 … 17% of product

$\xrightarrow[\text{THF, } -30\,^{\circ}\text{C}]{(C_4H_9)_2CuLi}$ C_4H_9 … CO_2H (93%)

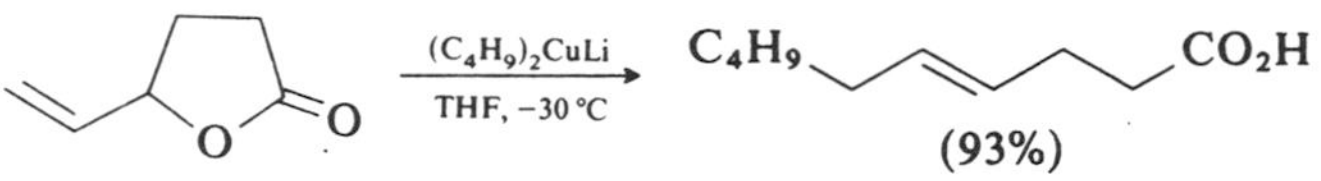

($-C{\equiv}C-C_3H_7$) and t-butylethynyl ($-C{\equiv}C-CMe_3$) as the residual bound group R_r. They react readily with cyclohexenones transferring only the other group R_t to give the β-substituted cyclohexanones. The so-called hetero cuprates (RCuCN)Li and (RCuSPh)Li, formed by treating copper(I) cyanide or copper phenylthioate with one equivalent of an alkyl-lithium, are also very effective. They react readily with alkyl halides and αβ-unsaturated ketones to give products formed by transfer of R in high yield (Lipshutz, Wilhelm and Floyd, 1981).

Organocopper(I) reagents: copper catalysed reaction of Grignard reagents

Organocopper(I) reagents, RCu, have been less extensively used in synthesis than the organocuprates largely because they are less stable and, in general, since they have to be employed at low temperatures, less reactive. They are prepared, generally, from an organolithium compound and an equimolecular amount of a copper(I) salt. Interest in organocopper(I) reagents has revived recently, due to the discovery that the reactive species produced from *Grignard reagents* and an equimolecular amount of copper(I) bromide, variously represented as $RCu.MgBr_2$ (Normant *et al.*, 1974) or [RCuBr]MgBr (Westmijze *et al.*, 1976) add to terminal alkynes to give synthetically useful 1-alkenyl-copper(I) compounds (1.152). In these reactions the R group of the

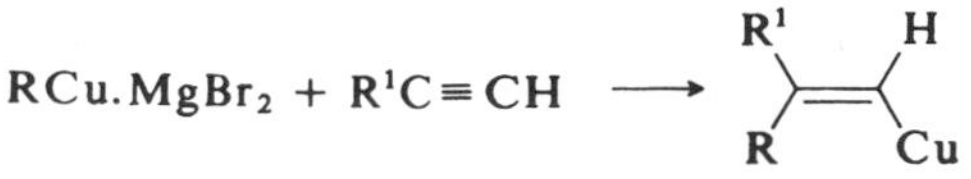

$$RCu.MgBr_2 + R^1C{\equiv}CH \longrightarrow (R^1)(R)C{=}C(H)(Cu) \qquad (1.152)$$

Grignard reagent and the copper add to the same side of the triple bond, and the copper bonds to the terminal carbon atom.

The alkenylcopper(I) reagents prepared in this way undergo a range of useful reactions leading to substituted alkenes and 1,3-dienes. Hydrolysis affords terminal alkenes and reaction with alkyl halides gives trisubstituted alkenes. Thus, reaction of the reagent prepared from hexylmagnesium bromide and copper(I) bromide with phenylacetylene, followed by hydrolysis, gave 2-phenyl-1-octene; with methyl iodide (*E*) 3 phenyl 2-nonene was obtained. Other transformations, illustrated in (1.153) give homoallylic alcohols and 1-halogenoalkenes, in each case with retention of the stereochemistry of the double bond. Oxidative dimerisation leads to 1,3-butadienes, again with high stereoselectivity.

$$C_6H_{13}MgBr + CuBr \xrightarrow[-20\,°C]{THF} C_6H_{13}CuBr.MgBr$$

$\downarrow$ $C_6H_5{-}{\equiv}$, THF, −20 °C

$(C_6H_{13})(C_6H_5)C{=}CH(CH_3) \xleftarrow{CH_3I} (C_6H_{13})(C_6H_5)C{=}CH(Cu) \xrightarrow{aq.\ NH_4Cl} (C_6H_{13})(C_6H_5)C{=}CH_2$

$$C_4H_9CuBr.MgBr \xrightarrow[-40\,°C]{CH{\equiv}CH,\ ether} (H)(C_4H_9)C{=}C(H)(Cu) \qquad (1.153)$$

$\xrightarrow{I_2}$ $C_4H_9CH{=}CHI$; $\xrightarrow{CO_2}$ $C_4H_9CH{=}CHCO_2H$; $\xrightarrow{\text{ethylene oxide}}$ $C_4H_9CH{=}CHCH_2CH_2OH$

$$iso\text{-}C_3H_7Cu.MgBr_2 \xrightarrow{HC{\equiv}CH} [iso\text{-}C_3H_7CH{=}CHCuBr]MgBr$$

$\downarrow$ O_2, THF, 0 °C

$iso\text{-}C_3H_7CH{=}CH{-}CH{=}CH{-}C_3H_7\text{-}iso$

A number of reactions of Grignard reagents which are *catalysed* by copper(I) salts, as opposed to those using stoichiometric amounts of copper salts discussed above, are of value in synthesis. Some of these reactions, in fact, proceed more easily and in better yield than the corresponding stoichiometric reactions using organocopper(I) or cuprate reagents, both in substitution and addition (cf. Normant, 1978; Erdik, 1984).

The reaction of Grignard reagents with $\alpha\beta$-unsaturated ketones and esters in the presence of a catalytic amount of copper(I) salts gives good yields of the corresponding β-substituted saturated carbonyl compound (1.154). Although the reaction was discovered more than 40 years ago it is still not completely understood. It has been suggested that a cuprate reagent is formed transiently and reacts competitively with the initial Grignard reagent.

O O
MgCl
cat. $CuCl.(CH_3)_2S$

(1.154)

CO_2CH_3 $\xrightarrow[\text{ether, } Cu_2I_2]{CH_3MgI}$ CO_2CH_3

Straightforward reaction of Grignard reagents, RMgBr, with alkyl halides, R′Hal, to give coupled products, R–R′, is generally unsatisfactory (cf. p. 73), but it can be effected in good yield in the presence of catalytic amounts of copper(I) salts (Erdik, 1984) or of lithium tetrachlorocuprate, Li_2CuCl_4, itself easily prepared from lithium chloride and copper(II) chloride. Primary halides react best, but a wide variety of different Grignard reagents has been used. The catalytically active species is believed to be an organocopper(I) complex produced by rapid metathesis between copper(I) and copper(II) halides and the Grignard reagents. With the tetrachlorocuprate $\alpha\omega$-dibromides are alkylated only once at -78 °C, providing a route to a variety of aliphatic, alicyclic and aromatic compounds otherwise relatively inaccessible. 6-Methyl-1-heptyl bromide, for example, is readily prepared from isopropylmagnesium bromide and 1,5-dibromopentane, and the difficultly accessible odd-numbered alkyl chlorides, such as tridecyl and pentadecyl chlorides are easily obtained in excellent yield from 1-bromo-3-chloropropane and the appropriate Grignard reagent.

1.13. Synthetic applications of carbenes and carbenoids

A carbene is a neutral intermediate containing bivalent carbon, in which a carbon atom is covalently bonded to two other groups and has two valency electrons distributed between two non-bonding orbitals. If the two electrons are spin-paired the carbene is a singlet; if the spins of the electrons are parallel it is a triplet.

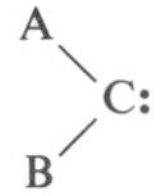

$$\begin{matrix} A \\ & \diagdown \\ & & C: \\ & \diagup \\ B \end{matrix}$$

A singlet carbene is believed to have a bent sp^2 hybrid structure in which the paired electrons occupy the vacant sp^2-orbital. A triplet carbene may be either a bent sp^2 hybrid with an electron in each unoccupied orbital, or a linear *sp* hybrid with one electron in each of the unoccupied *p*-orbitals. Structures intermediate between the last two are also possible (1.155). The results of experimental observations and molecular orbital

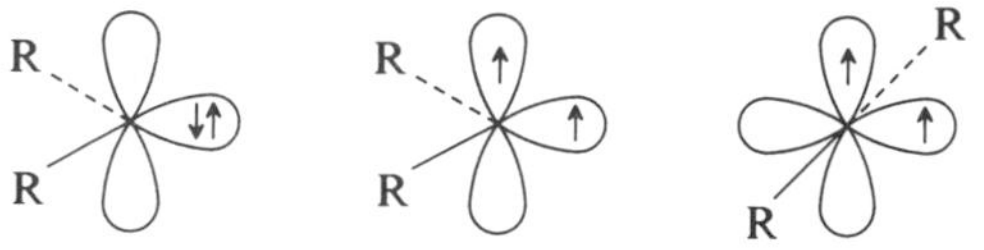

(1.155)

lowest singlet triplet triplet

calculations indicate that many carbenes have a non-linear triplet ground state. Exceptions are the dihalogenocarbenes and carbenes with oxygen, nitrogen and sulphur atoms attached to the bivalent carbon, which are probably singlets. The singlet and triplet states of a carbene do not necessarily show the same chemical behaviour. For example, addition of singlet carbenes to olefinic double bonds to form cyclopropane derivatives is much more stereoselective than addition of triplet carbenes.

A variety of methods is available for the generation of carbenes, but for synthetic purposes they are usually obtained by thermal or photolytic decomposition of diazoalkanes, or by α-elimination of hydrogen halide from a haloform or of halogen from a *gem*-dihalide by action of base or a metal. In many of these latter reactions it is doubtful whether a 'free' carbene is actually formed. It seems more likely that in these reactions the carbene is complexed with a metal or held in a solvent cage with a salt, or that the reactive intermediate is, in fact, an organometallic compound and not a carbene. Such organometallic or complexed intermediates which, while not 'free' carbenes, give rise to products expected of carbenes are usually called carbenoids (1.156). Carbenes produced

$$RCHN_2 \xrightarrow{h\nu} [RCH:] + N_2$$

$$N_2CHCO_2C_2H_5 \xrightarrow{\Delta} [:CHCO_2C_2H_5] + N_2$$

$$CHCl_3 \xrightarrow{B^-} BH + :CCl_3^- \longrightarrow :CCl_2 + Cl^- + BH \qquad (1.156)$$

$$R_2CBr_2 + R'Li \longrightarrow R_2CBrLi + R'Br \xrightarrow{?} [R_2C:] + LiBr$$

by photolysis of diazoalkanes are highly energetic species and indiscriminate in their action, and photolysis is not, therefore, a good method for generating alkylcarbenes for synthesis. Thermal decomposition of diazoalkanes often produces a less energetic, and more selective, carbene, particularly in the presence of copper powder or copper salts. Copper–carbene complexes are probably involved in these reactions. Another convenient and widely used route to alkylcarbenes is the thermal or photolytic decomposition of the lithium or sodium salts of toluene-*p*-sulphonylhydrazones. The diazoalkane is first formed and decomposes under the reaction conditions (1.157).

$$\triangleright\!-CH{=}NNHSO_2C_6H_4CH_3 \xrightarrow[\text{diglyme, 150 °C}]{CH_3ONa} \triangleright\!-CHN_2 \longrightarrow \triangleright\!-CH\!:$$

(1.157)

Ketocarbenes and alkoxycarbonylcarbenes are usually produced by heating or photolysing diazoketones and diazoesters. Dihalogenocarbenes are obtained by the action of base on a haloform (1.156) or from the halomethylmercury compounds PhHgCXYZ (X, Y, Z = halogens). These are easily prepared by reaction of a phenylmercury(II) halide with a haloform and potassium t-butoxide and some are commercially available. When heated in benzene solution in the presence of an alkene they form dihalogenocyclopropanes by addition of transiently formed dihalogenocarbene to the double bond (1.158)

$$\text{trans-2-butene} \xrightarrow[\text{benzene, 80°C; [:CBr}_2]]{C_6H_5HgCBr_3} \text{1,1-dibromo-}trans\text{-2,3-dimethylcyclopropane (Br, Br; H, CH}_3\text{; H}_3\text{C, H)}$$

(1.158)

$$C_6H_5HgCBr_3 \longrightarrow C_6H_5Hg\cdots\overset{Br}{\cdots}CBr_2 \longrightarrow C_6H_5HgBr + [:CBr_2]$$

Carbenes, in general, are very reactive electrophilic species. Their activity depends to some extent on the method of preparation, on the nature of the substituent groups R and R′ in R—$\ddot{C}$—R′, and also on the presence or absence of certain metals or metallic salts (see p. 92). Carbenes undergo a variety of reactions, including insertion into C—H bonds, addition to carbon–carbon double and triple bonds, and skeletal rearrangements, some of which are useful in synthesis. Insertion into C—H bonds is not of general synthetic value because mixtures are often produced. Methylene itself attacks primary, secondary and tertiary C—H bonds indiscriminately, although other carbenes may be more selective. With alkylcarbenes intramolecular insertion into C—H bonds is the preferred course of reaction. The major product is usually an alkene, formed by insertion at the β-C—H bond; insertion at the γ-C—H bond gives a cyclopropane as a second product. In general, no intermolecular reactions are observed when intramolecular insertion is possible (1.159).

$$(CH_3)_2CHCHN_2 \rightarrow [(CH_3)_2CHCH:] \downarrow CH_3-\underset{CH_3}{C}=CH_2 + CH_3-\overset{H}{C}(-CH_2-CH_2-) \quad (1.159)$$

(75%) (25%)

Only in favourable cases where the possibilities of different reactions are limited by geometric factors are insertion reactions of carbenes synthetically useful. For example, camphor toluene-*p*-sulphonylhydrazone, when heated with sodium methoxide in diglyme, gives tricyclene by intramolecular insertion of the derived carbene. A similar reaction was a key step in a synthesis of α-santalene (1.160).

Br ... =O —(1) H_2NNH_2, C_2H_5OH (2) HgO, CH_3OH, reflux→ [Br ... :] → Br ... (68%)

(1.160)

Probably the most useful application of carbenes in synthesis is in the formation of three-membered rings by addition to carbon–carbon double and triple bonds. This is a common reaction of all carbenes which do not undergo intramolecular insertion. Particularly useful synthetically are the halocarbenes, which are readily obtained from a variety of precursors (p. 89). Generated in presence of an alkene they give rise to halocyclopropanes which are valuable intermediates for the preparation of cyclopropanes, allenes, and ring-expanded products (Parham and Schweizer, 1963, Krapcho, 1978) (1.161). Addition of halocarbenes to alkenes is a stereospecific *cis* reaction, but this is not necessarily the case with other carbenes. Intramolecular additions are also possible. The toluene-*p*-sulphonylhydrazones of αβ-unsaturated aldehydes and ketones, for example, on reaction with sodium methoxide in aprotic media at 160–220 °C yield alkyl substituted cyclopropenes, presumably by way of the corresponding alkenylcarbene (1.161). *gem*-Dihalocyclopropenes are obtained by reaction of dihalocarbenes with disubstituted acetylenes. They have been used to prepare cyclopropenones.

$$(CH_3)_2C{=}C(CH_3)CH{=}NNHTos \xrightarrow[\text{diglyme, 160–220 °C}]{CH_3ONa,} [(CH_3)_2C{=}C(CH_3)\ddot{C}H] \longrightarrow \text{1,3,3-trimethylcyclopropene } (72\%)$$

$$\textit{cis-}CH_3CH{=}CHCH_3 \xrightarrow[t\text{-}C_4H_9OK\ [:CBr_2]]{CHBr_3} \text{cis-1,1-dibromo-2,3-dimethylcyclopropane } (79\%) \xrightarrow[\text{ether}]{LiAlH_4} \text{cis-1,2-dimethylcyclopropane} \qquad (1.161)$$

$$\text{6,6-dibromobicyclo[3.1.0]hexane} \xrightarrow[H_2O]{AgNO_3} \text{2-bromocyclohex-2-en-1-ol}$$

Skell attributed the stereospecificity of the reactions with dihalogenocarbenes to the fact that these carbenes have a singlet ground state which allows concerted addition of the carbene to the alkene, since the two new σ-bonds of the cyclopropane can be formed without changing the spin of any of the electrons involved. Addition of triplet carbenes, on the other hand, would be stepwise, going through a triplet diradical intermediate, with the possibility of rotation about one of the bonds before spin inversion and closure to the cyclopropane could take place. It is now believed that the addition of dihalogenocarbenes is one of a general class of non-linear, cheletropic reactions which are symmetry-allowed and concerted. A necessary feature of such reactions is that both electrons must be in the same orbital, as they are in singlet carbenes.

Addition of carbenes to aromatic systems leads to ring-expanded products. Methylene itself, formed by photolysis of diazomethane, adds to benzene to form cycloheptatriene in 32 per cent yield; a small amount of toluene is also formed by an insertion reaction. Better yields of cycloheptatriene are obtained in the presence of copper salts (p. 92). One of the oldest known carbene reactions is the addition of ethoxycarbonylcarbene, from diazoacetic ester, to benzene to form a mixture of cycloheptatrienylcarboxylic esters. Under the conditions of the reaction the intermediate norcaradiene undergoes a Cope rearrangement to give the ring expanded product. In the reaction with polycyclic aromatic hydrocarbons the norcaradiene can often be isolated. Dichlorocarbene also adds to some aromatic compounds. It is insufficiently reactive to attack benzene, but it adds readily to more reactive aromatic compounds. In most cases ring expansion proceeds spontaneously.

The composition of the mixture of products obtained from reactions of carbenes is profoundly altered by the presence of certain transition metals, notably copper and its salts. Under these conditions the intermediates obtained, for example, by decomposition of diazo compounds are more selective than 'free' carbenes. Insertion reactions are suppressed and higher yields of addition products are obtained in reactions with alkenes and aromatic compounds. Thus, benzene reacts readily with diazomethane in the presence of cuprous chloride to form cycloheptatriene in 85 per cent yield. The reaction is general for aromatic systems, substituted benzenes giving a mixture of the corresponding substituted cycloheptatrienes. Related reactions of cyclic and acyclic alkenes produce cyclopropanes in good yield and with complete retention of configuration. Intramolecular addition to carbon–carbon double bonds also takes place readily in presence of copper(I) salts, and this was exploited in a neat synthesis of the hydrocarbon sesquicarene (136) from *cis,trans*-farnesal (1.162). Unsaturated diazoketones and diazoacetic esters likewise form

CH=NNH$_2$ —(MnO_2, CH_2Cl_2, 0 °C)→ [N$_2$CH] —(Cu_2I_2)→ (136)

(1.162)

intramolecular addition products in the presence of copper(I) salts and reactions of this type have been used to prepare bridged bicyclic and 'cage' ketones (1.163). In the absence of copper the main reaction of diazoketones is the Wolff rearrangement (p. 96).

It is very unlikely that 'free' carbenes are involved in any of these catalysed reactions, and cyclopropane formation is believed to involve a copper–carbene–alkene complex similar to that invoked in the Simmons–Smith reaction (1.170). The view that the reactive intermediate is some kind of carbene–copper complex gains strong support from kinetic studies and from the observation that catalysed decomposition of ethyl

$$CH_2{=}CH(CH_2)_3COCl \xrightarrow{CH_2N_2} CH_2{=}CH(CH_2)_3COCHN_2 \xrightarrow{Cu,\ 80\ °C} \text{bicyclic cyclopropyl ketone}\ (50\%) \quad (1.163)$$

$$N_2CHCO_2CH_2CH{=}CHCH_3 \xrightarrow[\text{cyclohexane, boil}]{Cu_2O} \text{bicyclic lactone}\ (66\%)$$

diazoacetate with optically active copper complexes in presence of styrene gave a mixture of cyclopropane derivatives which was optically active (Cowell and Ledwith, 1970).

Intramolecular insertion reactions of keto carbenes at an unactivated C—H bond have not been so widely employed in synthesis as the addition reactions described above but they sometimes allow transformations which would be otherwise difficult to achieve (Burke and Grieco, 1979). Geometrically rigid structures favour intramolecular insertions, but this is not necessary. For example, diazocamphor was catalytically converted into cyclocamphanone (137) in high yield, but the open-chain diazo compound (138) equally readily gave the cyclopentanone derivative (139).

$$\text{diazocamphor} \xrightarrow[\text{reflux}]{\text{Cu, benzene}} (137)$$

$$(138) \xrightarrow[CH_2Cl_2]{Rh_2(OCOCH_3)_4} (139)\ (60\%) \quad (1.164)$$

An insertion reaction of an intermediate alkylidenecarbene (Stang, 1982) is a feature of the Dreiding synthesis of cyclopentenones (cf. Karpf, Huguet and Dreiding, 1982). In this useful reaction thermal cyclisation of an alkynyl alkyl ketone with at least one hydrogen atom at a β'-position leads specifically to a substituted 2-cyclopentenone. The reaction leads

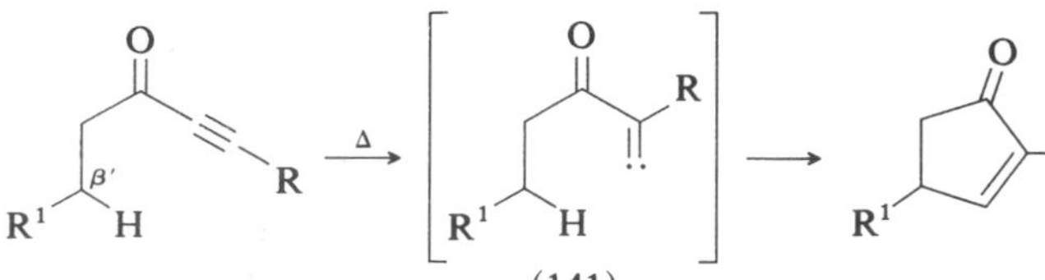

(140) (141) (1.165)

to formation of a new carbon–carbon bond at a non-activated β'-carbon atom of (140) with a [1,2]shift of the acetylenic substituent. It is explained by the formation of an intermediate alkylidene-carbene which inserts into the β'-carbon–hydrogen bond. A range of cyclic, polycyclic, spiro and propellane skeletons has been synthesised by this route. If there is more than one β'-position then, of course, a mixture of products is formed (1.166).

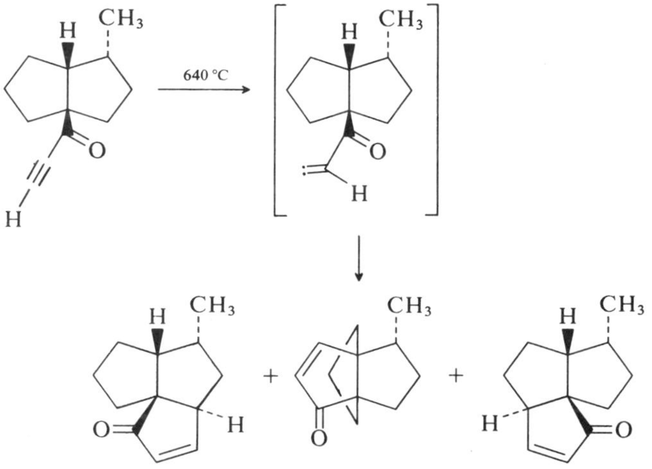

640 °C

(1.166)

A useful reaction which may be mentioned here although it does not involve carbenes, is the acid-catalysed cyclisation of diazoketones by electrophilic attack on suitably situated carbon–carbon double bonds or aromatic rings (Smith and Dieter, 1981). Thus, the unsaturated diazoketone (142) readily gave the fused cyclopentenone (143) and the bridged tricyclic compound (145) was obtained from (144) in 80 per cent yield, with generation of a quaternary carbon centre. Reaction is believed to proceed by initial protonation of the diazoketone (or complexation if

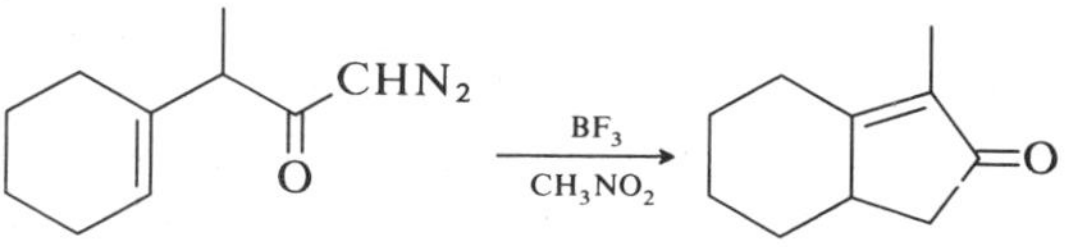

$\xrightarrow[CH_3NO_2]{BF_3}$

(142) (143)

$$\text{(144)} \xrightarrow{CF_3CO_2H} \text{(145)} \qquad (1.167)$$

a Lewis acid is used) followed by nucleophilic displacement of nitrogen from the resultant diazonium ion by the double bond or aromatic ring.

$$RCOCHN_2 \xrightarrow{H^+} RCOCH_2\overset{+}{N}{\equiv}N \xrightarrow{X^-} RCOCH_2X + N_2$$

Fused cyclopentenes can also be made by intramolecular addition of ketocarbenes to 1,3-dienes followed by thermal rearrangement of the vinyl-cyclopropanes formed initially (Hudlicky *et al.*, 1981).

$$\xrightarrow[\text{benzene}]{Cu(acac)_2} \quad \xrightarrow[\text{in vac.}]{500\,^{\circ}C} \quad (65\%) \qquad (1.168)$$

Related to these copper-catalysed reactions of diazoalkanes and diazoketones is the valuable Simmons–Smith reaction (Simmons, Cairns, Vladuchick and Hoiness, 1973), which is widely used for the synthesis of cyclopropane derivatives from alkenes by reaction with methylene iodide and zinc–copper or, better, zinc–silver couple. This is a versatile reaction and has been applied with success to a wide variety of alkenes. Many functional groups are unaffected, making possible the formation of a variety of cyclopropane derivatives. Dihydrosterculic acid, for example, was obtained from methyl oleate in 51 per cent yield (1.169).

$$CH_3(CH_2)_7CH{=}CH(CH_2)_7CO_2CH_3 \xrightarrow[\text{(2) } H_2O,\ NaOH]{\text{(1) } CH_2I_2,\ Zn\text{-}Cu,\ \text{ether}} CH_3(CH_2)_7\text{-}\triangle\text{-}(CH_2)_7CO_2H \ (51\%) \qquad (1.169)$$

The reaction is stereospecific and takes place by *cis* addition of methylene to the less hindered side of the double bond. The reactive intermediate is believed to be an iodomethylenezinc iodide complex which reacts with the alkene in a bimolecular process to give a cyclopropane and zinc iodide (1.170). A valuable feature of the reaction in synthesis is the

$$\text{>C=C<} + CH_2(ZnI)I \longrightarrow [\text{transition state}] \longrightarrow \text{cyclopropane} + ZnI_2 \quad (1.170)$$

stereochemical control exerted on the developing cyclopropane ring by a suitably situated hydroxyl group in the alkene. With allylic and homoallylic alcohols or ethers, the rate of the reaction is greatly increased and in five- and six-membered cyclic allylic alcohols the product in which the cyclopropane ring is *cis* to the hydroxyl group is formed stereospecifically. A methoxycarbonyl group in the α-position to the double bond has a similar directing effect (1.171). These effects are ascribed to co-ordination of oxygen to the zinc, followed by transfer of methylene to the nearer face of the adjacent double bond. In eight- and nine-membered cyclic allylic alcohols, reaction with methylene iodide and zinc-copper couple takes place stereospecifically to give products in which the cyclopropane ring is *trans* to the hydroxyl group. Presumably, in these reactions the conformation of the ring in the intermediate is such that it is the *trans* face of the double bond which is nearer to the co-ordinated reagent.

$$\xrightarrow[\text{ether}]{CH_2I_2,\ Zn\text{-}Cu} \quad (1.171)$$

Carbenes undergo a number of skeletal rearrangements, some of which are useful in synthesis. The most important of these is the Wolff rearrangement of diazoketones to ketenes, which is brought about by heat, light or by action of some metallic catalysts. This reaction is the key step in the well-known Arndt–Eistert method for converting a carboxylic acid into its next higher homologue (Bachman and Struve, 1942) (1.172). With

$$RCOCl \xrightarrow{CH_2N_2} RCOCHN_2 \longrightarrow [R\text{–}C(=O)\text{–}\ddot{C}H] \longrightarrow RCH{=}C{=}O \xrightarrow{R'OH} RCH_2CO_2R' \quad (1.172)$$

cyclic diazoketones the rearrangement leads to ring contraction, and this reaction has been widely used to prepare derivatives of strained small-ring compounds such as bicyclo[2,1,1]-hexane and benzocyclobutene (Meinwald and Meinwald, 1966) (1.173).

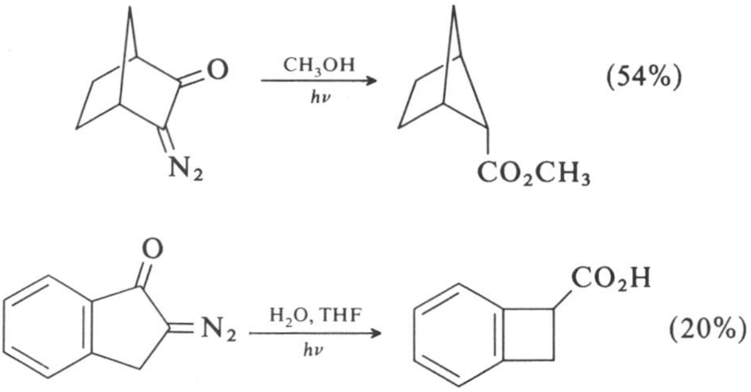

$$\xrightarrow[h\nu]{CH_3OH} \quad (54\%)$$

(1.173)

$$\xrightarrow[h\nu]{H_2O,\ THF} \quad (20\%)$$

There is no clear evidence that a carbene is involved in any of these rearrangements. A concerted migration and expulsion of nitrogen is usually a valid alternative (1.174).

$$-\underset{R}{\overset{|}{C}}-CH-\overset{+}{N_2} \longrightarrow \;>C{=}CHR+N_2 \qquad (1.174)$$

1.14. Formation of carbon–carbon bonds by addition of free radicals to alkenes

Free radical reactions are now being used increasingly in synthesis both for the formation of carbon–carbon bonds and for the manipulation of functional groups. Their application in the functionalisation of unactivated carbon–hydrogen bonds (p. 263), in the deoxygenation of alcohols (ROH → RH) (p. 454) and in the conversion of carboxylic acids into the corresponding nor-hydrocarbons and nor-bromides (p. 104) is referred to elsewhere in the book. Here we discuss the formation of carbon–carbon bonds by addition of free radicals to alkenes.

It has been known for many years that alkyl free radicals add to the double bond of alkenes with the formation of a new carbon–carbon bond. This reaction, of course, forms the basis of an important industrial method for making polymers, but it is only recently that it has come into use in systematic organic synthesis. Free-radical reactions of this kind have a number of advantages over the more familiar ionic reactions. For example, free radicals do not undergo the molecular rearrangements or unwanted elimination reactions sometimes encountered in ionic reactions involving carbonium ions or carbanions with a leaving group on the adjacent carbon. Further, the rate of reaction of alkyl radicals with many functional groups (hydroxyl, ester, carbonyl, halogen) is slow compared with their rates of addition to carbon–carbon double bonds so that it is often possible to bring about the addition without the need for protection and deprotection of functional groups which often acts as a brake in syntheses involving polar reagents.

To be most effective in synthesis, radicals need to act by a chain mechanism and to be continually trapped and regenerated, as shown in (1.175). The feasibility of this sequence hangs on a delicate balance of

Initiation	$RX + In^{\bullet} \rightarrow$	$R^{\bullet} + In\text{-}X$	
Addition	$R^{\bullet} + CH_2{=}CHR' \rightarrow$	$RCH_2{-}\dot{C}HR'$	(1.175)
Chain transfer	$RCH_2\dot{C}HR' + RX$	$RCH_2CH(X)R' + R^{\bullet}$	
or	$RCH_2\dot{C}HR' + InH$	$RCH_2CH_2R' + In^{\bullet}$	

reaction rates which themselves depend on the structures of the radical and the alkene involved in the reaction (Tedder and Walton, 1980). For a successful outcome the initial radical $R^{\bullet}$ must add to the alkene faster than it abstracts hydrogen from a hydrogen donor, but the adduct radical $RCH_2\dot{C}HR'$ has to react faster with the hydrogen donor (In–H for a successful chain reaction) than with the alkene, which would simply lead to polymerisation. Both addition and chain transfer steps must compete favourably with chain-terminating radical combination (Giese, 1983; Beckwith, 1981). The rate of addition of radicals to alkenes depends on the nature of the substituent groups on the double bond and for alkyl radicals is greater in the presence of electron-withdrawing substituents. The orientation of addition is strongly influenced by steric factors, and addition of alkyl radicals to monosubstituted alkenes takes place very largely at the unsubstituted carbon of the double bond (1.176).

$$C_6H_{11}^{\bullet} + CH_2{=}CHZ \longrightarrow C_6H_{11}CH_2{-}\dot{C}HZ$$

k rel.	Z	
	$=H$	1
	$=C_6H_5$	65
	$=CO_2CH_3$	450
	$=CHO$	2300

$C_6H_{11}^{\bullet}$ adding to $CH_2{=}CHCO_2CH_3$: 99.8 (at CH_2), 0.2 (at CH)

$CF_3^{\bullet}$ adding to $CH_2{=}CHCF_3$: 98 (at CH_2), 2 (at CH) (1.176)

$C_6H_{11}^{\bullet}$ adding to citraconic anhydride (O=C–O–C=O ring with H_3C on the double bond): 3.0 (at the methyl-bearing carbon), 97.0 (at the unsubstituted carbon)

$CF_3^{\bullet}$ adding to $CH_2{=}CHCH_3$: 92 (at CH_2), 8 (at CH)

The alkyl radicals used in these reactions may be generated in a number of ways. The initiator may be light of a suitable wavelength or another organic free radical generated, for example, from benzoyl peroxide or azobisisobutyronitrile. They are commonly obtained also by reduction of an alkyl or alkenyl halide or an alkyl selenide with a trialkyltin hydride, or by reaction of an organomercurial with a sodium borohydride. In the latter reaction an alkylmercuric hydride is formed and undergoes homolytic fission to give first the alkylmercuric radical and thence the akyl radical.

$$RCH_2HgOAc \xrightarrow{NaBH_4} RCH_2HgH \rightarrow RCH_2Hg^{\cdot} + H^{\cdot} \rightarrow RCH_2^{\cdot} + Hg \quad (1.177)$$

Alkyl radicals are nucleophilic and add readily to electron-deficient alkenes such as derivatives of acrylonitrile and acrylic ester, and reactions of this kind have been employed recently in a number of useful synthetic transformations (Giese and Hueck, 1979, 1981). Thus, the organomercurial (146), formed by methoxymercuration of cyclohexene, on reduction with sodium trimethoxyborohydride in the presence of acrylonitrile gave the coupled product (148), as a mixture of stereoisomers, by way of the radical (147), and the *C*-glycopyranoside (149) was obtained from tetraacetylglucosyl bromide in 72 per cent yield. The high stereoselectivity in the latter reaction is ascribed to stabilisation of the radical (150) by

(1) $Hg(OCOCH_3)_2$, CH_3OH; (2) NaCl → OCH_3, HgCl (146) → $NaBH(OCH_3)_3$, CH_2Cl_2 → OCH_3 (147) → CN → OCH_3, CN → $NaBH(OCH_3)_2$ → OCH_3, CN (148) (77%)

OAc, AcO, O, AcO, AcO, Br → $(C_4H_9)_3SnH$, $CH_2{=}CHCN$, ether, $h\nu$ → OAc, AcO, O, AcO, AcO, CH_2CH_2CN (149) (72%)

AcO, AcO, O, AcO, AcO (150)

$$R{-}Br \xrightarrow[-(C_4H_9)_3SnBr]{(C_4H_9)_3Sn^{\cdot}} R^{\cdot} \xrightarrow{CH_2{=}CHCN} R\dot{C}H\text{-}CH_2CN \xrightarrow{(C_4H_9)_3SnH} RCH_2CH_2CN + (C_4H_9)_3Sn^{\cdot} \quad (1.178)$$

interaction of the half-occupied orbital with the non-bonding electron pair on the ring oxygen atom (Giese and Dupuis, 1983; Dupuis *et al.*, 1984). In reactions such as this where the intermediate radical is formed from an alkyl halide and a trialkyltin hydride, the trialkyltin radical is continuously regenerated and continues the chain (1.178). These free-radical addition reactions are strongly influenced by steric factors and the yields obtained depend on the degree of substitution of the carbon atom attacked. Methyl crotonate, for example, reacts with cyclohexyl radicals much more slowly ($\times 10^{-2}$) than methyl acrylate.

Intramolecular radical cyclisations are now being increasingly used in synthesis. Thus, the *cis*-tetrahydroindane derivative (152) was readily obtained from the bromo-diene (151), and the α-acylamino radical (153) gave the indolizidine derivative (154). The latter reaction illustrates the general point that kinetically controlled cyclisation of 5-hexenyl radicals leads mainly to five- rather than six-membered rings (Julia, 1971, 1974). There is a geometric reason for this; the interconnecting chain is not long enough to allow the radical to approach the other carbon atom of the double bond along the correct trajectory. Formation of the five-membered ring is thus favoured even although this requires generation of a somewhat less stable primary radical.

Br; CO_2CH_3 (151) —$(C_4H_9)_3SnH$, azobisisobutyronitrile, benzene, 65 °C→ [CO_2CH_3] → H; CO_2CH_3 (152)

SC_6H_5; N; O → [N; O] (153) → H_3C, H, H; N; O (154) (1.179)

Br → → $(C_4H_9)_3SnH$ → (>95%)

→ $(C_4H_9)_3SnH$ → (<5%)

The example in (1.180) illustrates one of the advantages of radical reactions in synthesis. The bicyclic ketone is formed, in effect, by 'Michael addition' of the β-acetoxyalkyl radical (156) to the double bond. The corresponding β-acetoxy carbanion, in contrast, would simply undergo elimination of acetoxy anion to generate the alkene (155); such a pathway, with expulsion of the acetoxy radical, is not open to (156). Here the β-acetoxy radical (156) serves as an operational equivalent of the inaccessible β-acetoxy carbanion (Danishefsky, Chackalamannil and Uang, 1982). Radicals containing other leaving groups on the β-carbon atom can be used in the same way as synthetic equivalents of the corresponding carbanions.

(155) $\xrightarrow[CH_3CO_2H]{Hg(OCOCH_3)_2}$ $HgOCOCH_3$, $OCOCH_3$ $\xrightarrow{NaBH(OCH_3)_3}$ [(156) $OCOCH_3$] $\longrightarrow$ H, H, $OCOCH_3$ (58%) (1.180)

$OCOCH_3$ $\equiv$ (−) $OCOCH_3$

Stork has pointed out that cyclisation by addition of a vinyl radical rather than an alkyl radical to a double bond has the advantage that the product contains a double bond in a defined position which may be used in further chemical transformations (Stork and Baine, 1982) and he has exploited this in the synthesis of functionalised polycyclic compounds. Thus, the vinyl bromide (157) was smoothly cyclised to the indane derivative (158) in 70 per cent yield without the need to protect the hydroxyl and cyano groups (Stork and Mook, 1983). Another noteworthy feature of this reaction is the high yield obtained even although the cyclisation leads to the formation of a quaternary carbon atom. Another

(157) $\xrightarrow[\text{azobisisobutyronitrile, benzene, reflux}]{(C_4H_9)_3SnH}$ (158) (70%) (1.181)

route to vinyl radicals is by addition of alkyl radicals to alkynes. An application of this procedure which serves as a model for the synthesis of the C/D ring system of cardiac aglycones is shown in (1.182). Here

(159) $\xrightarrow{(C_4H_9)_3SnH}$ [ ] → (160) → (161) (1.182)

the initial alkyl radical attacks the triple bond to form an intermediate vinyl radical which reacts further with the double bond of the cyclohexene ring to give (160), converted by further manipulation into the $\alpha\beta$-unsaturated lactone (161). Again cyclisation leads to a high yield of product containing a quaternary carbon atom (Stork, 1983; Stork *et al.*, 1983).

In another promising application useful stereo-control of addition to the double bond of allylic alcohols has been achieved as illustrated in (1.183) (Stork and Kahn, 1985).

A practical limitation of the intermolecular reactions described earlier is that the alkene has to be used in excess and, in many cases yields of coupled product are reduced by competing hydrogen abstraction by the

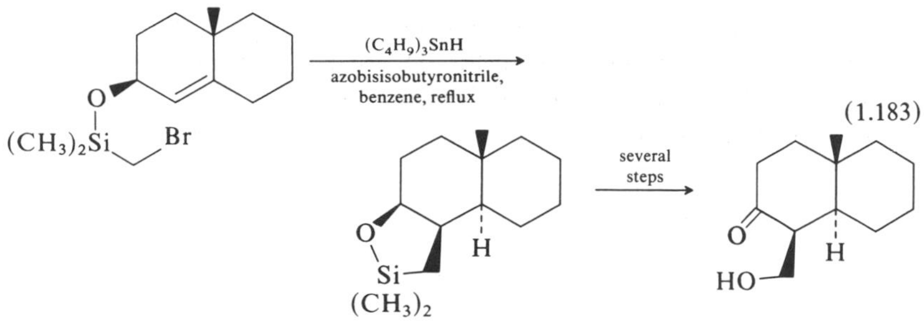

(1.183)

initial radical. A limited but valuable way round these difficulties lies in the use of allyltributyl stannane as the radical trap to effect allylation of alkyl radicals. Alkyl radicals add to the γ-carbon of the allyl stannane and the resulting β-stannyl radical then fragments to give the allylated product and the stannyl radical which continues the chain. Thus, cyclohexyl bromide was converted into allylcyclohexane in 88 per cent yield, and the bromide (162) gave the allyl derivative (163) in 76 per cent yield (Keck and Yates, 1982).

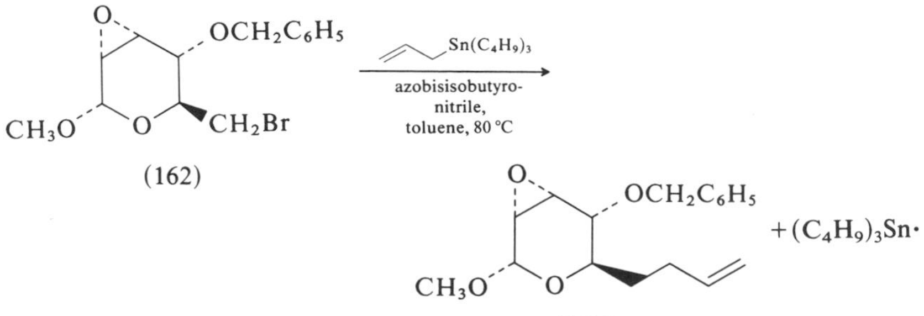

(1.184)

In a modification of this procedure β-substituted acrylic esters have been generated directly by a radical addition–elimination sequence using the tributylstannyl derivative (164) (Baldwin, Kclly and Ziegler, 1984). The tributylstannyl radical eliminated continues the cycle.

$$(C_4H_9)_3Sn\cdot + RBr \longrightarrow R\cdot + (C_4H_9)_3SnBr$$

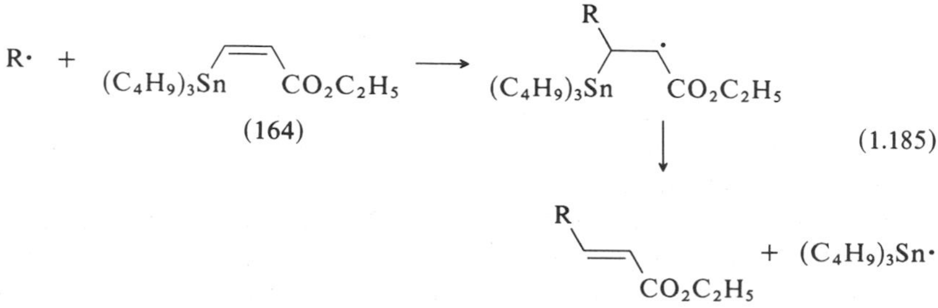

(1.185)

Another route to derivatives of acrylic ester proceeds by addition of alkyl radicals to the ester (166) with expulsion of the t-butylthiyl radical which continues the chain. In these experiments the alkyl radicals were derived from carboxylic acids by radical fragmentation of the thiohydroxamic acid (165). Even tertiary radicals reacted to give (167) in this scheme (Barton and Crich, 1984*b*).

$\cdot$S–C_4H_9-t SSC_4H_9-t R–$CO_2^{\cdot}$ –CO_2

(165)

$CO_2C_2H_5$ R + $\cdot$S–C_4H_9-t ← $CO_2C_2H_5$ $\cdot$R S–C_4H_9-t

(167) (166)

(1.186)

1.15. Some photocyclisation reactions

Few areas of organic chemistry have been more productive of new synthetic reactions in recent years than organic photochemistry. In general, absorption of light by an organic molecule can produce three types of activated molecule not accessible by normal thermal means, namely the electronically excited singlet and triplet states, and, often, a vibrationally 'hot' ground state. Each of these excited states may undergo different chemical reactions in proceeding back to the ground state. The triplet excited state, which generally has a relatively long lifetime, is frequently encountered in photochemical reactions. Because of the high energy of the excited states photochemical reactions often lead to strained structures which would be difficult to obtain by thermal reactions.

Many photochemical transformations are of value in synthesis, and some are referred to elsewhere (see pp. 269, 392). Particularly useful are photoelectrocyclic reactions of the type (1.187), which lead to the interconversion of acyclic conjugated trienes and 1,3-cyclohexadienes. The

(1.187)

formation of an acyclic hexatriene from a 1,3-cyclohexadiene was first noted in the case of the light-induced formation of calciferol from ergosterol via precalciferol, but many other examples have since been recorded (Barton, 1959). The reaction is reversible and under appropriate conditions certain acyclic 1,3,5-trienes are converted into cyclohexadienes. Thus, irradiation of *trans,cis,trans*-2,4,6-octatriene in ether affords a stationary state containing 10 per cent of *trans*-1,2-dimethyl-3,5-cyclohexadiene. By far the most useful application of this reaction in synthesis is in the conversion of stilbene derivatives, in which two of the double bonds of the 'triene' are contained in benzene rings, into 4*a*,5*a*-dihydrophenanthrenes and thence into phenanthrenes (1.188). Stilbene

(1.188)

undergoes a rapid *Z*–*E* isomerisation under the influence of ultra-violet light, and *Z*-stilbene, upon further irradiation, cyclises to give the dihydrophenanthrene. The dihydrophenanthrene has not been isolated but there is convincing evidence for its formation. In presence of mild oxidising agents such as oxygen or iodine it is readily converted into phenanthrene, and the sequence of reactions has been widely used to prepare a variety of phenanthrene derivatives containing alkyl, chloro, bromo, methoxy, phenyl, and carboxyl substituents from the appropriately substituted stilbene derivatives. Polycyclic aromatic hydrocarbons can also be obtained by photocyclisation of the appropriate stilbene analogues (Blackburn and Timmons, 1969; Laarhoven, 1983*a*). Thus 1-α-styrylnaphthalene is converted into chrysene and 2-α-styrylnaphthalene gives benzo[*c*]phenanthrene. In the latter case, cyclisation takes place at the 1-position of the naphthalene nucleus only. No benz[*a*]anthracene, formed by cyclisation at the 3-position was obtained. The convenience of this synthetic method compensates for the moderate yields obtained in some cases. The reaction has also been very useful for the preparation of helicenes and heterohelicenes. Phenanthrenes do not always result, however. Stilbenes which have one or two electron-withdrawing substituents on the double bond give a 9,10-dihydrophenanthrene on irradiation and not a phenanthrene, and 9,10-dihydrophenanthrenes are also obtained from 2-vinylbiphenyl derivatives even in the presence of oxygen (Laarhoven, 1983*b*).

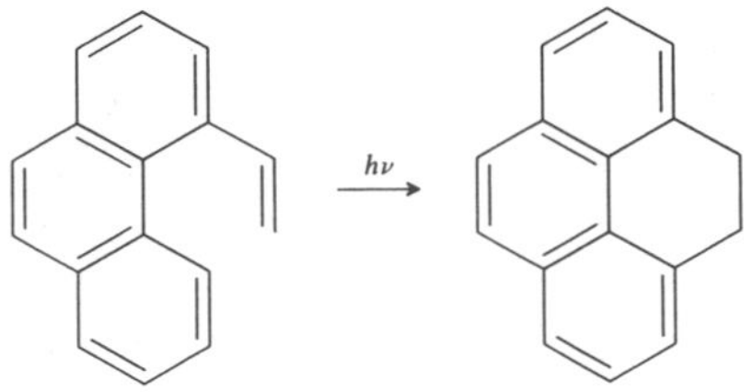

$\xrightarrow{h\nu}$ (1.189)

A wide range of heterocyclic compounds has also been prepared by this procedure. Thus, 2-α-styrylpyridine gave 1-azaphenthrene and 2-α-styrylthiophen cyclised to naphtho[2,1-b]thiophen. *N*-Benzoylenamines derived from cyclohexanones give *trans*-octahydrophenanthridones, and a reaction of this kind was used in a synthesis of ($\pm$)-crinan, the basic ring skeleton of the *Lycoris* alkaloids (1.190).

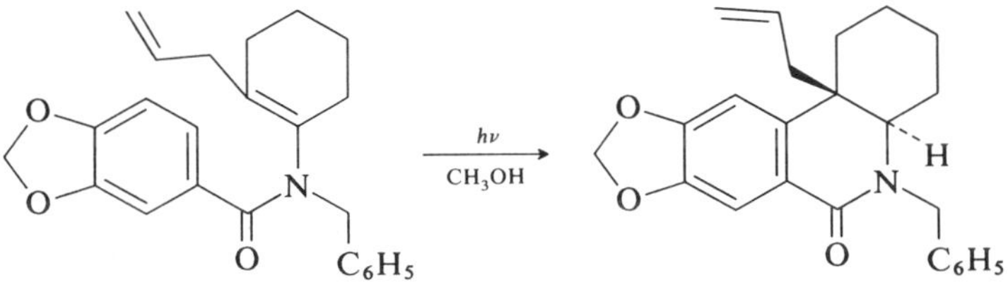

$\xrightarrow[CH_3OH]{h\nu}$

(1.190)

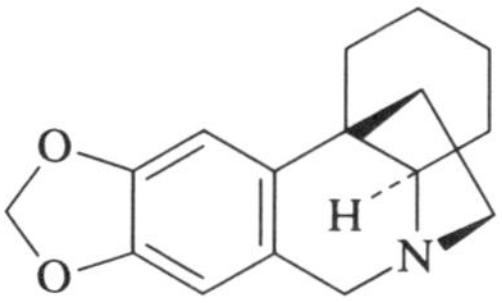

Crinan

The double bond in stilbene can be replaced by a heteroatom with a lone pair of electrons and still allow cyclisation. Thus, diphenylamine derivatives are converted into carbazoles and under non-oxidative conditions the cyclohexenone derivative (168) gave the *trans*-hexahydrocarbazole (169) in 90 per cent yield (cf. Schultz, 1983).

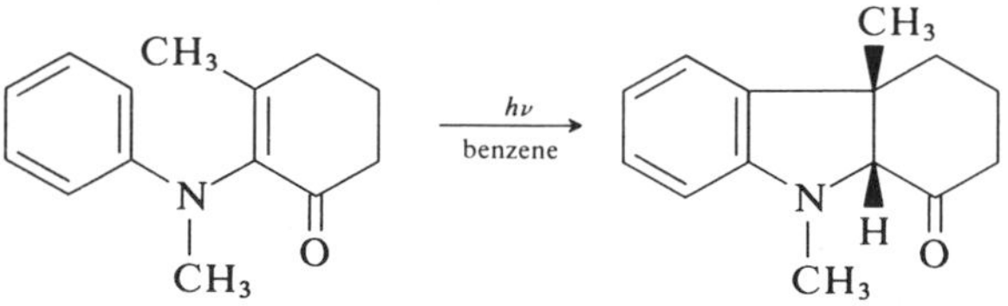

(168) $\xrightarrow[\text{benzene}]{h\nu}$ (169) (1.191)

An alternative procedure for the cyclisation of stilbenes makes use of the ready photolysis of the carbon–iodine bond in iodoaromatic compounds. It had been found earlier that photolysis of iodoaromatic compounds in aromatic solvents, particularly benzene, was useful for the preparation of biphenyl and polyphenyl derivatives. In an intramolecular version of the reaction 2-iodostilbene gave phenanthrene in 90 per cent yield by photolysis in hexane solution. This procedure is particularly useful for the preparation of nitrophenanthrenes which cannot be obtained by irradiation of nitrostilbenes. It seems likely that these reactions proceed by a radical mechanism and not by way of a dihydrophenanthrene-type of intermediate.

Another reaction of practical value in synthesis is the cyclo-addition of alkenes to double bonds to form four-membered rings. Most simple alkenes absorb in the far ultra-violet and in the absence of sensitisers undergo mainly fragmentations and *trans–cis* isomerisation, but conjugated alkenes which absorb at longer wavelengths form cyclo-addition compounds readily (Dilling, 1966; Eaton, 1968). Thus butadiene on irradiation in dilute solution with light from a high pressure mercury arc forms cyclobutene and bicyclo[1,1,0]butane. Many substituted acyclic 1,3-butadienes behave similarly and conjugated cyclic dienes also are often converted into cyclobutenes by direct irradiation. Thus, *trans,trans*-2,4-hexadiene gives *cis*-3,4-dimethylcyclobutene and the anhydride (170) forms the tricyclic compound (171), a reaction which was exploited in the preparation of 'Dewar-benzene' (van Tamelen, Pappas and Kirk, 1971). These cyclisations, and the reverse ring-opening reactions, are stereospecific. The stereochemical course of the reactions is controlled by the symmetries of the engaged orbitals in reactants and products – in general terms by the number of electrons involved and the conditions, thermal or photochemical, of the experiment (Woodward and Hoffmann, 1970; Gilchrist and Storr, 1979).

$h\nu$, ether

(170) (171) (1.192)

In the presence of a sensitiser reaction may follow a different course. Thus butadiene itself on irradiation with a sensitiser dimerises to form, mainly, *trans*-1,2-divinylcyclobutane, and many other acyclic dienes behave in a similar way.

$\alpha\beta$-Unsaturated carbonyl compounds also readily undergo photochemical cyclo-addition to alkenes. Since they absorb at sufficiently long wavelengths sensiters are not required in these reactions. In general, reactions are brought about by irradiation with light of wavelength greater than 300 nm, often by conducting the reactions in pyrex vessels. In this way the destructive effect of short wavelength irradiation is avoided. Reaction takes place through a triplet excited state of the enone formed by intersystem crossing from the initial $n \rightarrow \pi^*$ excited singlet.

Both inter- and intra-molecular reactions have been used in synthesis (Sammes, 1970; de Mayo, 1971; Oppolzer, 1982). Thus, the first step in Corey's synthesis of caryophyllene involved addition of isobutene to cyclohexenone to give the *trans*-cyclobutane derivative (172), and the remarkable intramolecular cyclo-addition of (173) to give (174) with three contiguous quaternary chiral centres, formed an important step in a recent synthesis of the sesquiterpene (±)-isocomene. Addition of (Z)- and (E)-2-butene to cyclohexenone gave the same mixture of addition products in each case, suggesting that the reactions proceed in a stepwise manner through radical intermediates

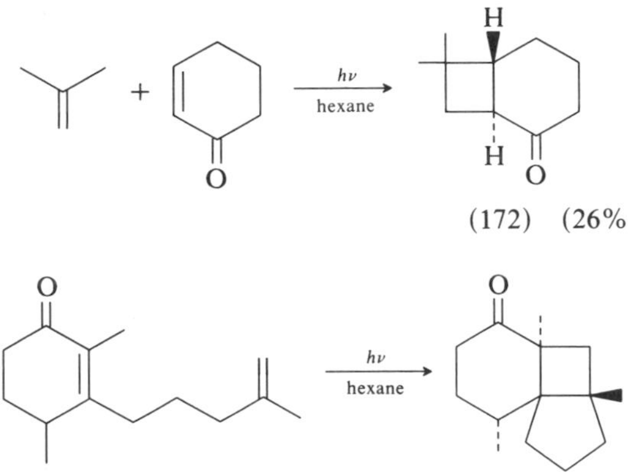

hν, hexane

(172) (26%)

(1.193)

hν, hexane

(173)

(174) (77%)

Acetylenes also add to $\alpha\beta$-unsaturated ketones on irradiation to form cyclobutenes.

The synthetic usefulness of these cyclo-addition reactions extends beyond the immediate formation of cyclobutane derivatives. Rearrangements encouraged by the relief of strain in the cyclobutane ring can be used to build up complex ring systems as in the synthesis of α-caryophyllene alcohol (1.194). In another application (de Mayo, 1971) photoaddition of an alkene to the enolised form of a 1,3-diketone results in the formation of a β-hydroxyketone, retroaldolisation of which, with ring opening, provides a 1,5-diketone. Thus, irradiation of a solution of acetylacetone in cyclohexene affords the 1,5-diketone (176) by spontaneous retroaldolisation of the intermediate β-hydroxyketone (175). With enol acetates of cyclic 1,3-diketones the sequence results in ring expansion by two carbon atoms and reactions of this kind have been

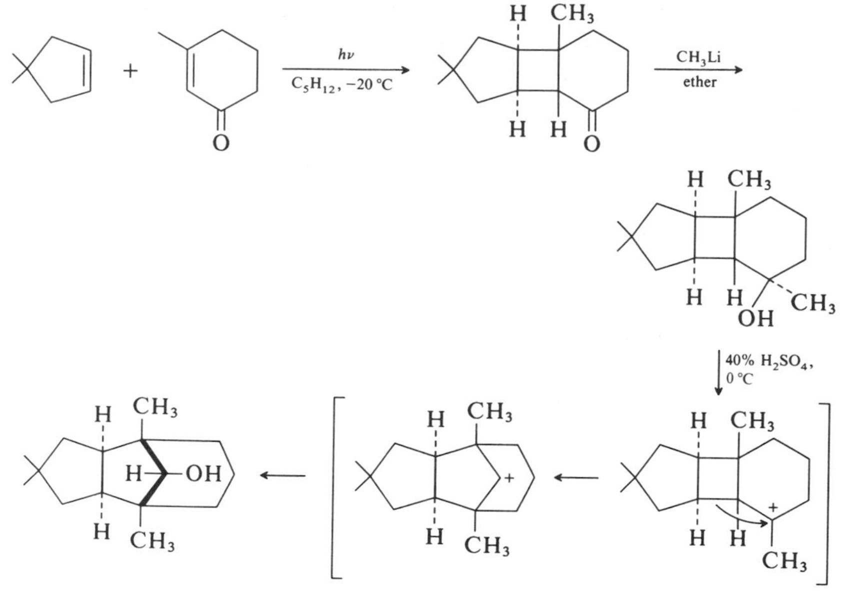

$h\nu$, C_5H_{12}, $-20\,°C$; CH_3Li, ether; 40% H_2SO_4, 0 °C

(1.194)

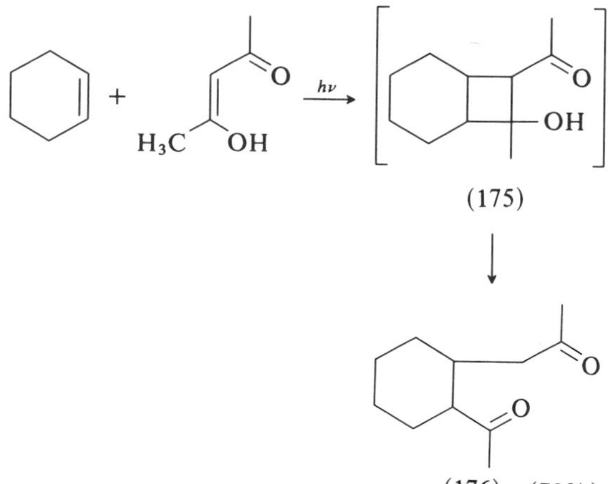

$h\nu$

(175)

(176) (78%)

(1.195)

applied in the synthesis of a number of polycyclic natural products (Oppolzer, 1982). Thus, irradiation of the enol acetate (177) gave the tricyclic product (178) in quantitative yield. Hydrolysis and retroaldolisation then formed the bicyclo[6,3,0]undecandione (179). More recently it has been found that alkenes add readily to the double bond of allylic *alcohols* on irradiation in the presence of copper(I) trifluoromethanesulphonate (triflate). Intramolecular reactions with 4-butenylallylic alcohols

(177) $\xrightarrow[\text{hexane}]{h\nu}$ (178) $\xrightarrow[C_2H_5OH]{KOH}$ → (179) (78%) (1.196)

give bicyclo[3,2,0]heptan-2-ols. The derived ketones fragment cleanly at 500 °C, providing a novel route to cyclopent-2-en-1-ones (1.197) (Salomon *et al.*, 1982). This procedure has preparative advantages for in some cases where an $\alpha\beta$-unsaturated ketone did not react with an alkene or gave mixtures of isomers the catalysed reaction with the corresponding allylic alcohol proceeded readily.

$\xrightarrow[\text{ether, CuOTfl}]{h\nu}$ (91%) ↓ Jones oxidation; $\xleftarrow{500\ °C}$ (78%) (1.197)

It seems that in these reactions the copper is co-ordinated to the hydroxyl group and the two double bonds. For successful cyclisation the double bonds must be so disposed that co-ordination with the copper is possible, as in (180).

(180)

2 Formation of carbon–carbon double bonds

2.1. β-Elimination reactions

The formation of carbon–carbon double bonds is important in synthesis, not only for the obvious reason that the compound being synthesised may contain a double bond but also because formation of the double bond followed by reduction may be the most convenient route to a new carbon–carbon single bond.

One of the most commonly used methods for forming carbon–carbon double bonds is by β-elimination reactions of the type (2.1) where X = e.g. OH, OCOR, halogen, OSO_2Ar, $\overset{+}{N}R_3$, $\overset{+}{S}R_2$. Included among these reactions

$$\underset{\mathrm{H}}{-\overset{|}{\underset{|}{\mathrm{C}}}}-\underset{\mathrm{X}}{\overset{|}{\underset{|}{\mathrm{C}}}}- \longrightarrow -\overset{|}{\mathrm{C}}=\overset{|}{\mathrm{C}}- + \mathrm{HX} \qquad (2.1)$$

are acid-catalysed dehydrations of alcohols, solvolytic and base-induced eliminations from alkyl halides and sulphonates, and the well-known Hofmann elimination from quaternary ammonium salts. They proceed by both E1 and E2 mechanisms and synthetically useful reactions are found in each category. Some examples are given in (2.2). These reactions,

$$\text{HO-(bicyclic alcohol)} \xrightarrow{PCl_5} \text{(bicyclic alkene)} \qquad (2.2)$$

$$CH_3CHBrCH_2CH_3 \xrightarrow[C_2H_5OH]{C_2H_5ONa} \underset{(81\%\ \text{of product})}{CH_3CH{=}CHCH_3} + \underset{(19\%\ \text{of product})}{CH_3CH_2CH{=}CH_2}$$

$$\underset{^{+}S(CH_3)_2}{CH_3\underset{|}{C}HCH_2CH_3} \xrightarrow[C_2H_5OH]{C_2H_5ONa} \underset{(26\%\ \text{of product})}{CH_3CH{=}CHCH_3} + \underset{(74\%\ \text{of product})}{CH_3CH_2CH{=}CH_2}$$

$$\underset{^{+}N(CH_3)_3I^-}{CH_3\underset{|}{C}H(CH_2)_2CH_3} \xrightarrow[130\ ^\circ C]{H_2O,\ KOH} \underset{(98\%\ \text{of product})}{CH_2{=}CH(CH_2)_2CH_3} + \underset{(2\%\ \text{of product})}{CH_3CH{=}CHCH_2CH_3}$$

although often used, leave much to be desired as synthetic procedures. One disadvantage is that in many cases elimination can take place in more than one way, so that mixtures of products are obtained. Mixtures of geometrical isomers may also be formed (see below). The direction of elimination in unsymmetrical compounds is governed largely by the nature of the leaving group, but may be influenced to some extent by the experimental conditions. It is found in general that acid-catalysed dehydration of alcohols, and other E1 eliminations, as well as eliminations from alkyl halides and arylsulphonates with base, gives rise to the more highly substituted alkene as the principal product (the Saytzeff rule), whereas base-induced eliminations from quaternary ammonium salts and from sulphonium salts gives predominantly the less substituted alkene (the Hofmann rule). Exceptions to both rules are observed, however. If there is a conjugating substituent such as CO or C_6H_5 at one β-carbon atom, elimination will take place towards that carbon to give the conjugated alkene, irrespective of the method used. For example, (*trans*-2-phenylcyclohexyl)trimethylammonium hydroxide on Hofmann elimination gives 1-phenylcyclohexene exclusively; none of the 'expected' 3-phenylcyclohexene is detected. Another exception is found in the elimination of hydrogen chloride from 2-chloro-2,4,4-trimethylpentane, which gives mainly the terminal alkene; presumably in this case the intermediate leading to the expected Saytzeff product is destabilised by steric interaction between a methyl group and the t-butyl substituent (2.3). An additional disadvantage of acid-catalysed dehydration of

$$CH_3{-}C(CH_3)_2{-}CH_2{-}C(CH_3)_2{-}Cl \xrightarrow{\text{base}} CH_3{-}C(CH_3)_2{-}CH_2{-}C(CH_3){=}CH_2 \ (81\%) + CH_3{-}C(CH_3)_2{-}CH{=}C(CH_3){-}CH_3 \ (19\%) \quad (2.3)$$

alcohols, and other E1 eliminations, is that since they proceed through an intermediate carbonium ion, elimination is frequently accompanied by rearrangement of the carbon skeleton. Thus, if the α-hydroxy compound in the first equation of (2.2) is replaced by the β-isomer, dehydration is accompanied by ring contraction to give an isopropylidenecyclopentane derivative. Numerous other examples are known in the terpene series, as in the conversion of camphenilol into santene (2.4).

The Hofmann reaction with quaternary ammonium salts and base-induced eliminations from alkyl halides and arylsulphonates are generally

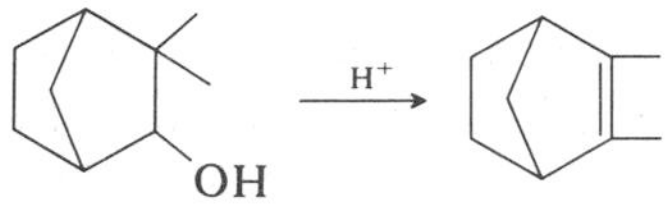

$\xrightarrow{H^+}$ (2.4)

OH

anti elimination processes, that is to say the hydrogen atom and the leaving group depart from opposite sides of the incipient double bond although examples involving *syn* elimination are known. This has been established in a number of ways. For example, reaction of ethanolic potassium hydroxide with *meso*-stilbene dibromide gave *cis*-bromostilbene whereas the (±)-dibromide gave the *trans*-stilbene. Again, the quaternary ammonium salts derived from *threo*- and *erythro*-1,2-diphenylpropylamine were found to undergo stereospecific elimination on treatment with sodium ethoxide in ethanol as shown in (2.5). In agreement with the relatively rigid requirements of the transition state, the *erythro* isomer, with eclipsed phenyl groups, reacted more slowly than the *threo*.

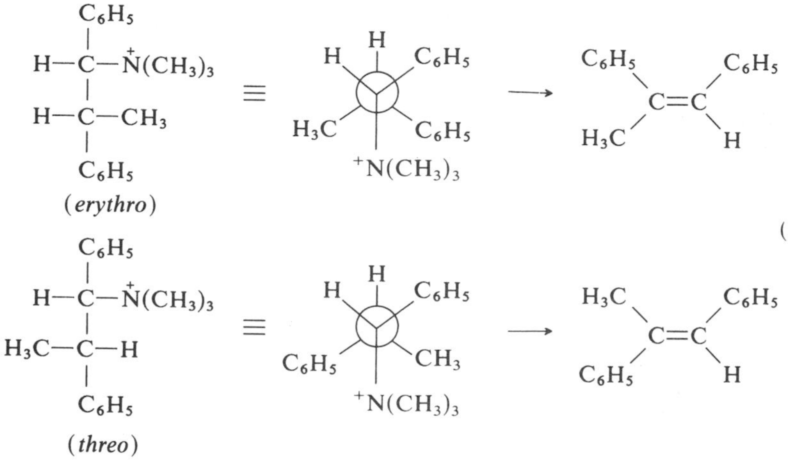

C_6H_5; H—C—$\overset{+}{N}(CH_3)_3$; H—C—CH_3; C_6H_5 (*erythro*) ≡ H, H, C_6H_5, H_3C, C_6H_5, $^+N(CH_3)_3$ ⟶ C_6H_5, C_6H_5, C=C, H_3C, H

(2.5)

C_6H_5; H—C—$\overset{+}{N}(CH_3)_3$; H_3C—C—H; C_6H_5 (*threo*) ≡ H, H, C_6H_5, C_6H_5, CH_3, $^+N(CH_3)_3$ ⟶ H_3C, C_6H_5, C=C, C_6H_5, H

For stereoelectronic reasons elimination takes place most readily when the hydrogen atom and the leaving group are in the anti-periplanar arrangement (1), although a number of '*syn*' eliminations have been observed in which hydrogen and the leaving group are eclipsed (2) (2.6).

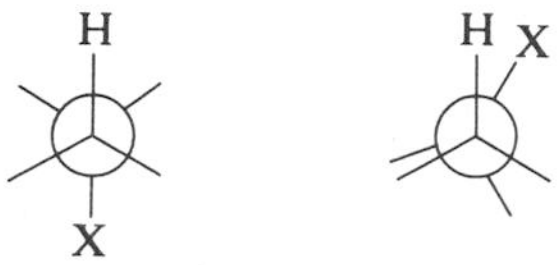

H, X (1) H X (2) (2.6)

In open-chain compounds the molecule can usually adopt a conformation in which H and X are anti-periplanar, but in cyclic systems this is not always so, and this may have an important bearing on the direction of elimination in cyclic compounds. In cyclohexyl derivatives antiperiplanarity of the leaving groups requires that they be diaxial even if this is the less stable conformation, and on this basis we can understand why menthyl chloride (3) on treatment with ethanolic sodium ethoxide gives only 2-menthene (2.7), while neomenthyl chloride (4) gives a mixture of 2- and 3-menthene in which the 'Saytzeff product' predominates. The elimination from menthyl chloride is much slower than that from neomenthyl chloride because the molecule has to adopt an unfavourable conformation with axial substituents before elimination can take place.

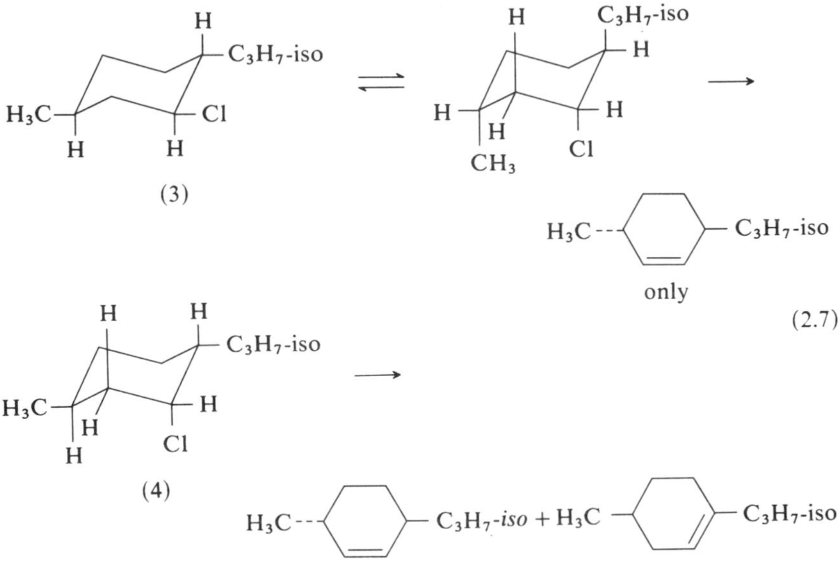

(3)

only

(2.7)

(4)

(25% of product) (75% of product)

Syn eliminations are apparently more common than was formerly appreciated, particularly in reactions involving quaternary ammonium salts, where *syn* and *anti* eliminations often take place at the same time (Sicher, 1972). They are also observed in reactions of some bridged bicyclic compounds where *anti*-elimination is disfavoured by steric or conformational factors (2.8), and in compounds where a strongly electron-attracting substituent on the β-carbon atom favours elimination in that direction, outweighing other effects (cf. Schlosser, 1972).

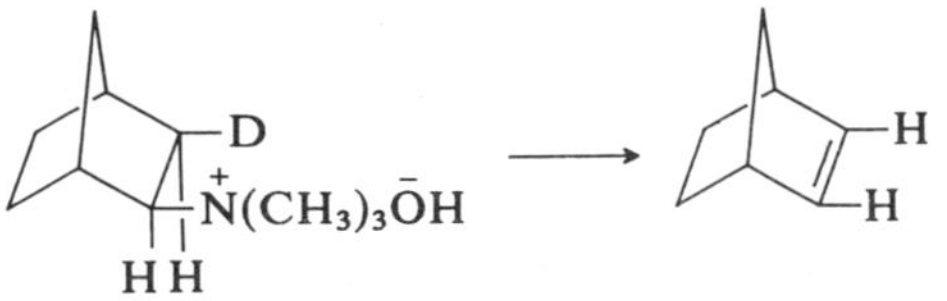

(2.8)

In most cases, however, *anti* elimination is preferred.

In spite of the disadvantages acid-catalysed dehydration of alcohols and base-induced eliminations from halides and arylsulphonates are widely used in the preparation of alkenes. The bases used in the latter reactions include alkali metal hydroxides and alkoxides, as well as organic bases such as pyridine and triethylamine. Superior results have been obtained using 1,5-diazabicyclo[3,4,0]non-5-ene (6) and the related 1,5-diazabicyclo-[5,4,0]undec-5-ene. Thus, the chloroester (5) could not be dehydrochlorinated with pyridine or quinoline but with the base (6) at 90 °C the enyne was obtained in 85 per cent yield (2.9). Other methods are discussed by Schlosser (1972).

$$CH{\equiv}CCH_2CH(Cl)(CH_2)_8CO_2CH_3\ (5) + (6) \xrightarrow{90\ °C} CH{\equiv}CCH{=}CH(CH_2)_8CO_2CH_3 \qquad (2.9)$$

2.2. Pyrolytic *syn* eliminations

Another important group of alkene-forming reactions, some of which are useful in synthesis, are pyrolytic eliminations. Included in this group are the pyrolyses of carboxylic esters and xanthates, which provide valuable alternative methods for dehydration of alcohols without rearrangement, and pyrolysis of amine oxides. For the majority of cases these reactions are believed to take place in a concerted manner by way of a cyclic transition state and, in contrast to the eliminations discussed above, they are necessarily *syn* eliminations; i.e. the hydrogen atom and the leaving group depart from the same side of the incipient double bond (2.10).

The *syn* character of these pyrolytic eliminations has been demonstrated in a number of ways. Thus, pyrolysis of the *erythro* and *threo* isomers of 1-acetoxy-2-deutero-1,2-diphenylethane gave in each case *trans*-stilbene, but the stilbene from the *erythro* compound retained nearly all its deuterium, whereas the stilbene from the *threo* compound had lost most of its deuterium. Either the hydrogen or the deuterium could be *syn* to the acetoxy group, but the preferred conformations (shown) are those in which the phenyl groups are as far removed from each other as possible (2.11). Similarly, the oxide of *cis*-1-phenyl-2-dimethylaminocyclohexane gave 3-phenylcyclohexene, elimination involving the

$$\text{ester} \longrightarrow -\overset{|}{C}=\overset{|}{C}- + \text{HOCOR}$$

$$\text{xanthate} \longrightarrow \overset{|}{C}=\overset{|}{C} + \text{HSCOSR} \quad (\text{COS}+\text{RSH}) \tag{2.10}$$

$$\text{amine oxide } (\overset{+}{N}R_2, \bar{O}) \longrightarrow -\overset{|}{C}=\overset{|}{C}- + \text{HONR}_2$$

syn and not the *anti* hydrogen even although the latter is activated by the phenyl group.

Pyrolysis of esters to give an alkene and a carboxylic acid is usually effected at a temperature of about 300–500 °C and may be carried out by simply heating the ester if its boiling point is high enough, or by passing the vapour through a heated tube. Yields are usually good, and the absence of solvents and other reactants simplifies the isolation of the product. In practice acetates are nearly always employed, but esters of other carboxylic acids can be used. The reaction provides an excellent method for preparing pure terminal alkenes from primary acetates and, because it does not involve either acidic or basic reagents, it is especially

(*erythro*) $\xrightarrow{\Delta}$ $C_6H_5(H)C=C(^2H)C_6H_5$

(*threo*) $\xrightarrow{\Delta}$ $C_6H_5(H)C=C(H)C_6H_5$ (2.11) useful for preparing highly reactive or thermodynamically disfavoured dienes and trienes. An example is the preparation of 4,5-dimethylenecyclohexene, which is obtained from 4,5-diacetoxymethylcyclohexene without extensive rearrangement to *o*-xylene (2.12). With secondary and tertiary acetates, where elimination can take place in more than one direction, mixtures of products are usually obtained.

$$CH_3CH_2CH_2CH_2OCOCH_3 \xrightarrow[N_2]{500\,^\circ C} CH_3CH_2CH{=}CH_2 \quad (100\%)$$

$$\text{4,5-bis}(CH_2OCOCH_3)\text{cyclohexene} \xrightarrow[N_2]{500\,^\circ C} \text{4,5-bis}(=CH_2)\text{cyclohexene} \quad (47\%) \qquad (2.12)$$

The high temperature required for elimination is a disadvantage in these reactions, and in some cases the alkene formed may not be stable under the reaction conditions. Thus, pyrolysis of 1-cyclopropylethyl acetate affords mainly cyclopentene by rearrangement of the initially formed 1-vinylcyclopropane. In some cases also rearrangement of the ester may take place before elimination, leading to mixtures of products. This is especially liable to occur with allylic esters as in the example (2.13).

$$CH_3CH(OCOCH_3){-}CH{=}CH(CH_2)_2CH_3 \rightleftharpoons CH_3CH{=}CH{-}CH(OCOCH_3)(CH_2)_2CH_3 \qquad (2.13)$$

$$\xrightarrow{\Delta} CH_2{=}CHCH{=}CH(CH_2)_2CH_3 + CH_3CH{=}CHCH{=}CHCH_2CH_3$$

Pyrolysis of amine oxides (the Cope reaction) (Cope and Trumbull, 1960) and of methyl xanthates (the Chugaev reaction) (Nace, 1962) takes place at much lower temperatures (100–200 °C) than that of carboxylic esters, with the result that further decomposition of sensitive alkenes can often be avoided in these reactions. On the other hand, separation of the alkene from the other products of the reaction may be more troublesome, particularly in the Chugaev reaction, where sulphur-containing impurities can be troublesome. Pyrolysis of amine oxides generally takes place under particularly mild conditions, and this often allows the generation of a new carbon–carbon double bond without migration into conjugation with other unsaturated systems in the molecule, as in the synthesis of 1,4-pentadiene (2.14). If an allyl or benzyl group is attached to the nitrogen atom, rearrangement to give an *O*-substituted hydroxylamine may compete with elimination.

$$\text{t-}C_4H_9-\underset{\displaystyle H}{\overset{\displaystyle CH_3}{\overset{|}{\underset{|}{C}}}}-OCSSCH_3 \xrightarrow{170\,^\circ C} \text{t-}C_4H_9CH{=}CH_2 \quad (71\%)$$

$$CH_2{=}CH(CH_2)_3\underset{\displaystyle O}{\underset{\downarrow}{N}}(CH_3)_2 \xrightarrow{140\,^\circ C} CH_2{=}CHCH_2CH{=}CH_2 \quad (61\%) + (CH_3)_2NOH \qquad (2.14)$$

$$CH_2{=}CHCH_2\underset{\displaystyle O}{\underset{\downarrow}{N}}(C_2H_5)_2 \xrightarrow{150\,^\circ C} (C_2II_5)_2NOCII_2CII{=}CII_2 \quad (59\%) + CH_2{=}CH_2 + CH_2{=}CH-CH_2-\underset{\displaystyle OH}{\underset{|}{N}}C_2H_5$$

Like the base-induced eliminations discussed above, pyrolytic *syn* eliminations have the disadvantage that with derivatives of unsymmetrical secondary and tertiary alcohols and with unsymmetrical amine oxides, mixtures of products are formed. With open-chain compounds, the three methods give closely similar results. If there is a conjugating substituent in the β-position elimination takes place to give the conjugated olefin, but otherwise the composition of the product is determined mainly by the number of hydrogen atoms on each β-carbon. For example, pyrolysis of s-butyl acetate affords a mixture containing 57 per cent 1-butene and 43 per cent 2-butene, in close agreement with the 3 : 2 distribution predicted on the basis of the number of β-hydrogen atoms. Of the 2-butenes the E-isomer is formed in larger amount, presumably because there is less steric interaction between the two methyl substituents in the transition state which leads to the E-alkene (2.15). It is found in general with

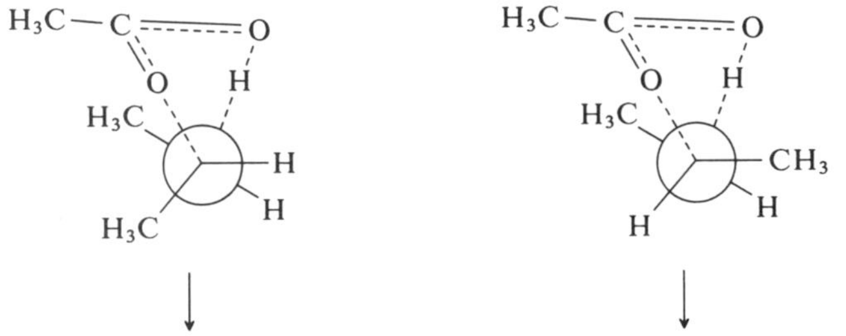

H₃C—C=O, O, H, H₃C, H, H, H₃C → (Z)-2-butene (15%)

H₃C—C=O, O, H, H₃C, CH₃, H, H → (E)-2-butene (28%) (2.15)

aliphatic esters, xanthates or amine oxides which could form either a Z- or E-alkene on pyrolysis that the more stable E-isomer is formed in larger amount (2.16). In the third example in (2.16), the necessity for *syn* elimination ensures the formation of Z-alkene from the *threo* isomer and of the E-alkene from the *erythro* isomer of the amine oxide.

With alicyclic compounds some restrictions are imposed by the conformation of the leaving groups and the necessity to form the cyclic intermediate. Thus, the cyclohexyl acetate (7) in which the leaving group is

$$CH_3CH_2CHCH_2CH_2CH_3 \text{ (with } OCSSCH_3 \text{ on C-3)} \xrightarrow{250\,°C} CH_3CH_2CH{=}CHCH_2CH_3 \; (17\%\ Z)\ (33\%\ E) + CH_3CH{=}CH(CH_2)_2CH_3 \; (16\%\ Z)\ (34\%\ E) \quad (2.16)$$

$$CH_3CH_2CHCH_3 \text{ (with } N(CH_3)_2 \rightarrow O \text{ on C-2)} \longrightarrow CH_3CH{=}CHCH_3 \; (12\%\ Z)\ (21\%\ E) + CH_3CH_2CH{=}CH_2 \; (67\%)$$

$$C_6H_5CH(CH_3){-}CH(N(CH_3)_2{\rightarrow}O)CH_3 \longrightarrow C_6H_5C(CH_3){=}CHCH_3 \; (93\% \text{ of product}) + 7\% \text{ unconjugated}$$

threo ⟶ 94% *E*, 0.1% *Z*
erythro ⟶ 4% *E*, 89% *Z*

axial, in contrast to its isomer (8) does not form a double bond in the direction of the ethoxycarbonyl group, even though it would be conjugated, because the necessary cyclic transition state is not sterically possible from this conformation. With the xanthate (9), however, in which the leaving group is equatorial, the six-membered cyclic transition state can lead equally well to abstraction of a hydrogen atom from either β-carbon, and an equimolecular mixture of two products results (2.17).

(7) cyclohexane with H, $CO_2C_2H_5$; H, $OCOCH_3$ $\xrightarrow{435\,°C}$ cyclohexene $CO_2C_2H_5$

(8) cyclohexane with H, $CO_2C_2H_5$; $OCOCH_3$, H $\xrightarrow{435\,°C}$ cyclohexene $CO_2C_2H_5$ (conjugated) (2.17)

(9) cyclohexane with CH_3, H; $OCSCH_3$ (=S), H $\xrightarrow{\Delta}$ methylcyclohexene CH_3 + methylcyclohexene CH_3

(49% of product) (51% of product)

Pyrolysis of alicyclic amine oxides does not necessarily lead to product mixtures similar in composition to those obtained from the corresponding acetates and xanthates. A notable difference is found in the pyrolysis of 1-methylcyclohexyl derivatives. The acetates and xanthates afford mixtures containing 1-methylcyclohexene and methylenecyclohexane in a ratio of 3 : 1, whereas pyrolysis of the oxide of 1-dimethylamino-1-methylcyclohexane gives methylenecyclohexane almost exclusively. The reason for this is thought to be that in the oxide the coplanar five-membered cyclic transition state only allows abstraction of hydrogen from the methyl substituent, whereas with the more flexible six-membered transition state of the ester pyrolysis, hydrogen abstraction is also possible from the ring. With larger more flexible rings, where the requisite planar transition state is more easily attained, the cycloalkene is the preponderant product from amine oxides as well.

As a general rule, formation of compounds with a double bond exocyclic to a ring is not favoured in pyrolysis of esters, presumably because of their relative instability with respect to other isomers. For instance, 1-cyclohexylethyl acetate on pyrolysis gives mainly vinylcyclohexane.

Sulphoxides with a β-hydrogen atom readily undergo *syn* elimination on pyrolysis to form alkenes. These reactions also take place by way of a concerted cyclic pathway. They are highly stereoselective, the *erythro* sulphoxide (10), for example, leading predominantly to *trans*-methylstilbene, while the corresponding *threo* isomer gives mainly *cis*-methylstilbene (2.18).

CH_3 $\overset{\bar{O}}{}\ \overset{+}{S}C_6H_5$, C_6H_5, H, H, C_6H_5 (10) ≡ H, $\overset{+}{S}C_6H_5$, C_6H_5, H, H_3C, C_6H_5 $\xrightarrow[\text{dioxan}]{80\ ^\circ\text{C}}$ C_6H_5, H, H_3C, C_6H_5 (alkene) + C_6H_5SOH (2.18)

Since sulphoxides are readily obtained by oxidation of sulphides, the reaction provides another useful method for making carbon–carbon double bonds.

Pyrolysis of sulphoxides is a feature of a useful new method for introducing unsaturation at the α-position of aldehydes, ketones and esters. Reaction of the corresponding enolates with dimethyl or diphenyl disulphide affords an α-methylthio derivative (as 11) which on oxidation and elimination of sulphenic acid in boiling toluene is converted into the $\alpha\beta$-unsaturated carbonyl compound. Yields are usually good. The queen substance of honey bees was synthesised from methyl 9-oxodecanoate as shown in (2.19) after initial protection of the ketone function as the ketal. The *E*-isomer usually predominates in reactions

CO_2CH_3 —— (1) CH_2OH–CH_2OH, H^+, benzene; (2) $LiN(iso\text{-}C_3H_7)_2$ ——→ OCH_3

↓ CH_3SSCH_3

(1) oxidation $NaIO_4$; (2) 120 °C; (3) H_3O^+ ←—— CO_2CH_3, SCH_3 (11)

CO_2CH_3

(2.19)

leading to disubstituted ethylenes, but tri- and tetra-substituted compounds may be obtained as mixtures of isomers unless, of course the *threo* and *erythro* sulphoxides (as 10) are separated before pyrolysis. For the preparation of $\alpha\beta$-unsaturated carbonyl compounds, this procedure has a number of advantages over the usual sequence of bromination–dehydrobromination. It takes place under comparatively mild conditions, it is specific, since only enolate anions react at the first stage, and it can be carried out in the presence of other functional groups, such as carbon–carbon double bonds, which might be affected by bromination.

Even better results are obtained by using selenoxides. Alkyl phenyl selenoxides with a β-hydrogen atom undergo *syn* elimination to form alkenes under milder conditions than sulphoxides - at room temperature or below - and this mild reaction has been exploited for the preparation of a variety of different kinds of unsaturated compounds (Reich, 1979; Clive, 1978; Liotta, 1984). The selenoxides are readily obtained from the corresponding selenides by oxidation with hydrogen peroxide or other reagents, and they generally undergo elimination under the reaction conditions to give the alkene directly. The selenides themselves can be prepared from alkyl halides or tosylates by reaction with phenyl selenide anion or directly from the alcohol with the relatively odourless *N*-phenylselenophthalimide (Grieco *et al.*, 1981) or by reaction with *o*-nitrophenyl selenocyanate (Grieco, Gilman and Nishizawa, 1976). Better yields of alkenes are sometimes obtained by elimination from *o*-nitrophenyl selenoxides than from the phenylselenoxides themselves. Elimination following alkylation of the selenides or selenoxides provides a route to substituted alkenes and, in conjunction with the conjugate addition of organocuprates (p. 81) forms a step in a useful method for β-alkylation of $\alpha\beta$-unsaturated ketones. It can also be used to make α-alkyl $\alpha\beta$-unsaturated ketones by way of the α-selenoketone, as in the last example in (2.20). 1,2-Disubstituted alkenes prepared by selenoxide elimination are generally obtained largely as the E-isomers, following *syn* elimination from a cyclic transition state (12) in which the two substituent groups are staggered.

$$C_6H_5CH_2Br \xrightarrow[C_2H_5OH]{C_6H_5SeNa} C_6H_5CH_2SeC_6H_5 \xrightarrow[(2)\ C_2H_5Br]{(1)\ LiN(iso\text{-}C_3H_7)_2,\ THF,\ -78\ ^\circ C} C_6H_5CH(C_2H_5)SeC_6H_5$$

$$\xrightarrow{H_2O_2,\ THF} \left[C_6H_5CH(CH_3CH{-}H)\text{-}\overset{+}{Se}(C_6H_5){-}O^- \right] \ (12) \longrightarrow (E)\text{-}C_6H_5CH{=}CHCH_3 \ (80\%)$$

$$\text{cyclopent-2-enone} \xrightarrow{(C_6H_5)_2CuLi} \text{3-}C_6H_5\text{-cyclopentenolate (OM)} \xrightarrow{C_6H_5SeBr} \text{2-}SeC_6H_5\text{-3-}C_6H_5\text{-cyclopentanone} \xrightarrow{H_2O_2} \text{3-}C_6H_5\text{-cyclopent-2-enone}$$

$$\text{(}OCH_3\text{, ketal, } SeC_6H_5\text{ ketone)} \xrightarrow[(3)\ H_2O_2]{(1)\ LiN(iso\text{-}C_3H_7)_2;\ (2)\ (CH_3)_2C{=}CHCH_2Br} \text{(}\alpha\text{-prenyl enone)} \ (65\%) \qquad (2.20)$$

Selenoxide elimination also forms a step in a good method for converting epoxides into allylic alcohols (see p. 350) and for introducing unsaturation at the α-position of carbonyl compounds. In the latter conversion it is superior to the sulphoxide elimination described on p. 120 because of the milder conditions employed. The unsaturated compound is usually obtained directly on oxidation of the selenide, without isolation of the selenoxide. Thus, the pollen attractant of honey bees was obtained from methyl linoleate as shown in (2.21) without interference by the other double bonds in the molecule.

Another useful application is in the synthesis of the α-methylene lactone structural unit found in cytotoxic sesquiterpenes (2.22).

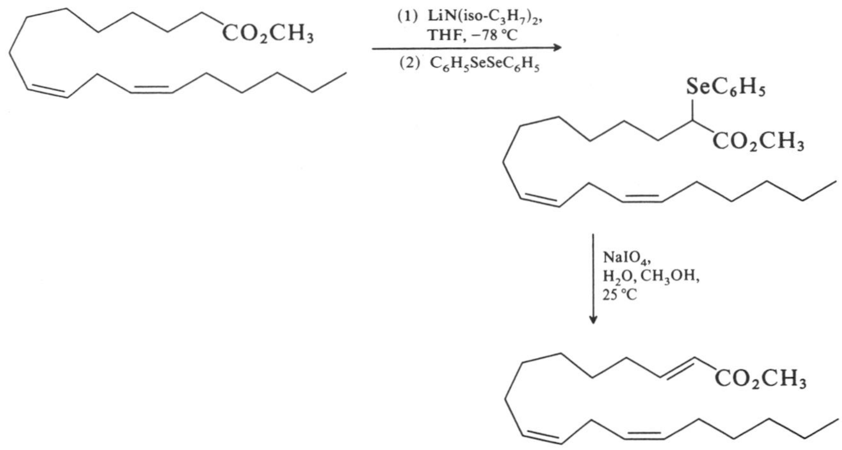

CO_2CH_3 —— (1) $LiN(iso\text{-}C_3H_7)_2$, THF, −78 °C; (2) $C_6H_5SeSeC_6H_5$ → SeC_6H_5 / CO_2CH_3

NaIO₄, H_2O, CH_3OH, 25 °C ↓

CO_2CH_3

(80%) (2.21)

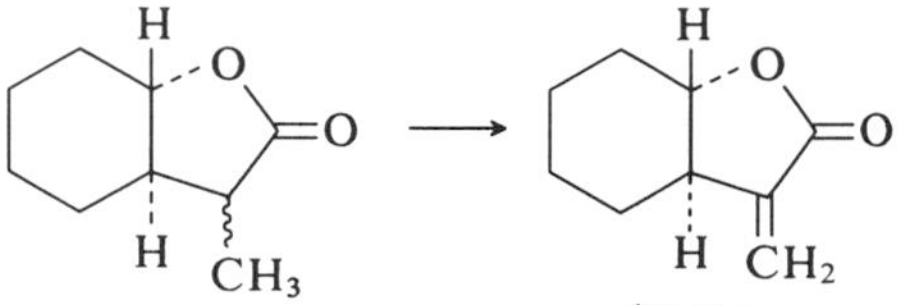

H, O, =O, H, CH_3 → H, O, =O, H, CH_2 (2.22)

(96%)

Alkenes are also obtained by elimination of C_6H_5SeOH from β-hydroxy selenides, best by the action of methanesulphonyl chloride and triethylamine, although acidic conditions can also be used. The reaction is highly stereospecific and proceeds by *anti* (*trans*) elimination, probably by way of the episelenonium ion (13). High yields are usually obtained

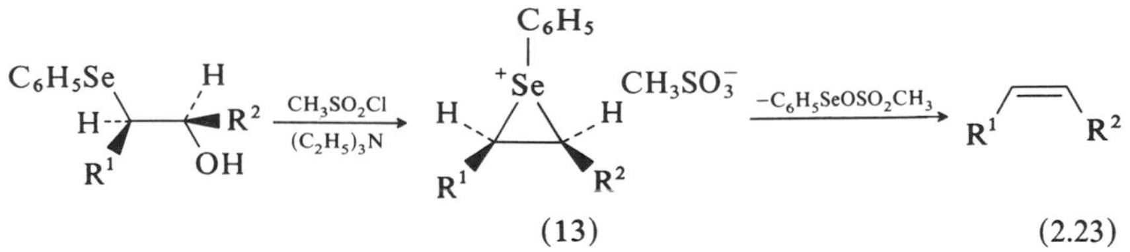

C_6H_5Se, H, H, R^2, R^1, OH —— CH_3SO_2Cl, $(C_2H_5)_3N$ → C_6H_5, ^+Se, $CH_3SO_3^-$, H, H, R^1, R^2 —— $-C_6H_5SeOSO_2CH_3$ → R^1, R^2

(13) (2.23)

and the reaction can be used for the preparation of di-, tri- and tetra-substituted alkenes. In certain cases it provides a useful alternative to the Wittig reaction when the phosphonium salt cannot be readily obtained. The β-hydroxy selenides can be prepared by a number of methods (Clive, 1978) for example from α-selenoketones by reduction or by reaction with a Grignard reagent, or from α-lithioselenides by reaction with an aldehyde or ketone.

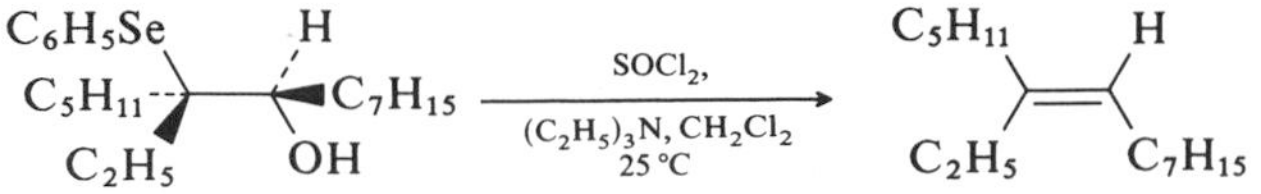

C_6H_5Se, H, C_5H_{11}, C_7H_{15}, C_2H_5, OH —— $SOCl_2$, $(C_2H_5)_3N$, CH_2Cl_2, 25 °C → C_5H_{11}, H, C_2H_5, C_7H_{15} (2.24)

2.3. Sulphoxide–sulphenate rearrangement; synthesis of allyl alcohols

Allylic sulphoxides, on gentle warming, are partly converted in a reversible reaction into rearranged allyl sulphenate esters (14). The reactions are [2,3]-sigmatropic rearrangements and take place through a five-membered cyclic transition state. The equilibrium is usually much in favour of the sulphoxide, but if the mixture is treated with a 'thiophile' (trimethyl phosphite and diethylamine are often used) the sulphenate is irreversibly converted into the allylic alcohol and, even although the sulphenate is present in low equilibrium concentration with the sulphoxide, its removal by reaction with the thiophile results in conversion of the sulphoxide into the rearranged allylic alcohol in high yield. Combined with alkylation of the sulphoxides the reaction provides a versatile synthesis of di- and tri-substituted allylic alcohols (Evans and Andrews, 1974). Equally good results have been obtained using selenoxides (Reich, 1975).

$$H_2C{=}CH{-}CH_2{-}\overset{+}{S}(O^-){-}R \rightleftharpoons [H_2C{\cdots}O{-}SR\ \text{cyclic transition state}] \rightleftharpoons H_2C(OSR){-}CH{=}CH_2\ (14) \quad (2.25)$$

$$(14) \xrightarrow{\text{'thiophile', methanol, 25 °C}} CH_2{=}CHCH_2OH$$

As indicated in (2.26), the rearrangement step is highly stereoselective, leading, in the acyclic series, mainly to the *trans* allylic alcohol. The factors controlling this are not clearly understood.

$$C_6H_5\overset{+}{S}(O^-)CH_2CH{=}CH_2 \xrightarrow[\text{(2) } CH_3I]{\text{(1) } LiN(iso\text{-}C_3H_7)_2,\ THF,\ -60\ °C} C_6H_5\overset{+}{S}(O^-)CH(CH_3)CH{=}CH_2 \rightleftharpoons C_6H_5\overset{+}{S}(O^-)CH(CH_3)CH{=}CH_2$$

$$\rightleftharpoons H_3C(H)C{=}CHCH_2OSC_6H_5 \xrightarrow{(CH_3O)_3P,\ CH_3OH,\ 25\ °C} H_3C(H)C{=}CHCH_2OH\ (74\%) \quad (2.26)$$

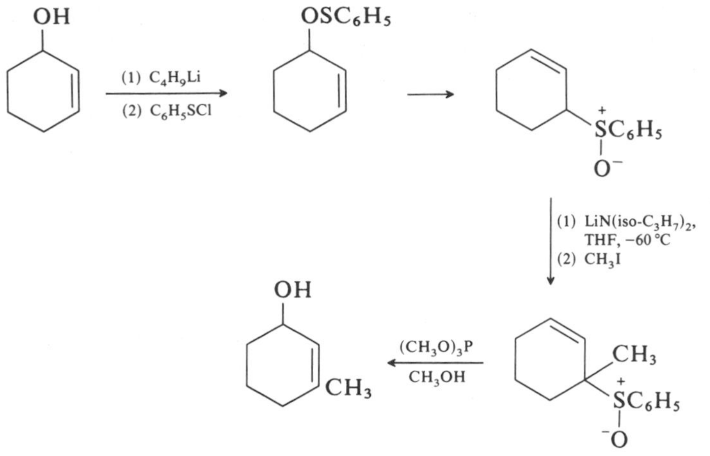

OH OSC$_6$H$_5$

(1) C_4H_9Li
(2) C_6H_5SCl

$\overset{+}{S}C_6H_5$
O^-

(1) $LiN(iso\text{-}C_3H_7)_2$, THF, −60 °C
(2) CH_3I

OH

$(CH_3O)_3P$
CH_3OH

CH$_3$ CH$_3$ $\overset{+}{S}C_6H_5$ ^-O

(2.26)

In this series of reactions the anions derived from the sulphoxides, such as (15), are equivalent to the vinyl anions (16) and the sequence provides, in fact, a method for effecting β-alkylation of allylic alcohols. This is exemplified in the second example in (2.26) in which the sulphoxide is obtained by rearrangement of the sulphenate ester prepared from the unsubstituted allyl alcohol.

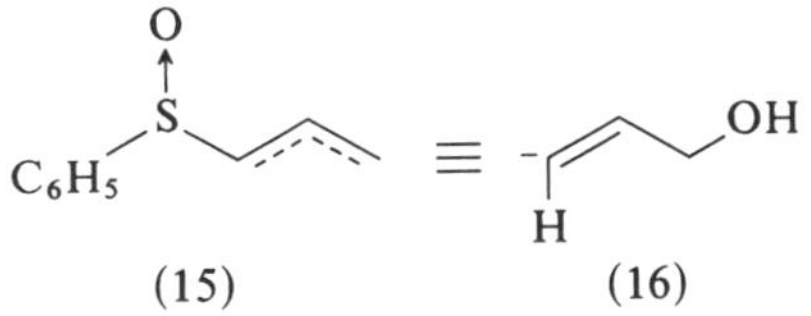

O S C_6H_5 ≡ OH H

(15) (16)

(2.27)

2.4. The Wittig and related reactions

The reaction between a phosphorane or phosphonium ylid, and an aldehyde or ketone to form a phosphine oxide and an alkene is known as the Wittig reaction after the German chemist, Georg Wittig, who first showed the value of this procedure in the synthesis of alkenes (Maercker, 1965; Johnson, 1966; Gosney and Rowley, 1979). The reaction is easy

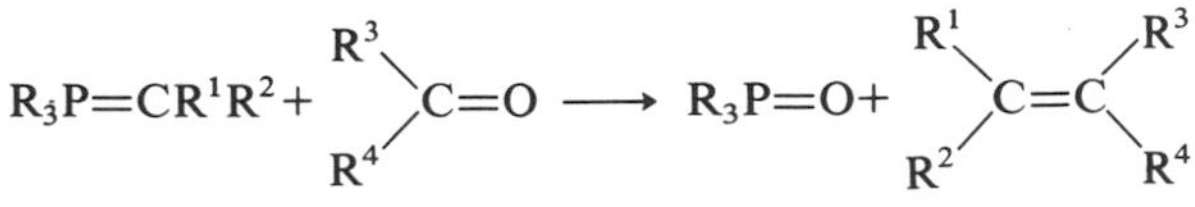

$$R_3P{=}CR^1R^2 + R^3R^4C{=}O \longrightarrow R_3P{=}O + R^1R^2C{=}CR^3R^4 \quad (2.28)$$

to carry out and proceeds under mild conditions. A valuable feature of the Wittig procedure is that, in contrast to the elimination and pyrolytic reactions discussed above, it gives rise to alkenes in which the position of the double bond is unambiguous, as illustrated in the examples (2.29). The reaction generally leads to high yields of di- and tri-substituted alkenes from aldehydes and ketones but, because of steric effects, yields of tetra-substituted alkenes from ketones are often poor. Sluggish reactions can sometimes be forced by addition of hexamethylphosphoramide or a catalytic amount of a crown ether to the reaction medium, or by conducting the reaction at a higher temperature than usual in refluxing benzene or toluene with sodium t-amylate as base.

$$(C_6H_5)_3P{=}CH_2 + C_7H_{15}COCH_3 \xrightarrow{\text{ether}} C_7H_{15}C(CH_3){=}CH_2$$

$$(C_6H_5)_3P{=}CHCH_3 + CH_3COCO_2C_2H_5 \xrightarrow[\text{reflux}]{\text{ether}} CH_3CH{=}C(CH_3)CO_2C_2H_5 \qquad (2.29)$$

Phosphoranes are resonance-stabilised structures in which there is some overlap between the carbon *p* orbital and one of the *d* orbitals of phosphorus as shown in (2.30). Reaction with a carbonyl compound

$$R^1R^2C{=}PR_3 \longleftrightarrow R^1R^2\bar{C}{-}\overset{+}{P}R_3 \qquad (2.30)$$

phosphorane form ylid form

takes place by attack of the carbanionoid carbon of the ylid form on the electrophilic carbon of the carbonyl group with the formation of a betaine which collapses to the products by way of a four-membered cyclic transition state (2.31), the driving force being provided by formation of the very strong phosphorus–oxygen bonds. Depending on the reactants,

$$R_3\overset{+}{P}{-}\bar{C}R^1R^2 + O{=}CR^3R^4 \rightleftharpoons R_3\overset{+}{P}{-}CR^1R^2{-}CR^3R^4{-}O^-$$

$$\downarrow \qquad (2.31)$$

$$R_3P{=}O + R^1R^2C{=}CR^3R^4 \longleftarrow \underset{\text{(four-membered ring } R_3P{-}CR^1R^2{-}CR^3R^4{-}O)}{}$$

either the first or the second step may be rate-determining. It has never been observed that the last step is the slowest, and it is uncertain whether the four-membered ring compound is a true intermediate or a transition state. Evidence for the formation of a betaine in the first stage of the reaction is provided by the isolation of compounds of this type in certain cases.

The reactivity of the phosphorane depends on the nature of the groups R, R^1 and R^2. In practice, R is nearly always phenyl. Alkylidene trialkylphosphoranes, in which the formal positive charge on the phosphorus is lessened by the inductive effect of the alkyl groups, are more reactive than alkylidenetriphenylphosphoranes in the initial addition to a carbonyl group to form a betaine, but by the same token decomposition of the betaine becomes more difficult, and alkylidene trialkylphosphoranes are superior to the triphenyl compounds only in certain special cases. In the alkylidene part of the phosphorane, if R^1 or R^2 is an electron-withdrawing group (e.g. CO or CO_2R) the negative charge in the ylid becomes delocalised into R^1 or R^2 and the nucleophilic character, and reactivity towards carbonyl groups, is decreased. Reagents of this type are much more stable and less reactive than those in which R^1 and R^2 are alkyl groups, and with them the rate-determining step in reactions with carbonyl groups is the initial addition to form the betaine. The more electrophilic the carbonyl group the more readily does the reaction proceed. The preparatively important carbalkoxymethylenetriphenylphosphoranes, for example, react fairly readily with aldehydes, but give only poor yields in reaction with the less reactive carbonyl group of ketones (2.32).

$$(C_6H_5)_3P{=}CHCO_2C_2H_5 \xrightarrow[\text{benzene, reflux}]{\text{crotonaldehyde,}} CH_3CH{=}CHCH{=}CHCO_2C_2H_5\ (80\%)$$

$$(C_6H_5)_3P{=}CHCO_2C_2H_5 \xrightarrow[\text{reflux}]{\text{cyclohexanone, benzene}} \text{(cyclohexylidene)}{=}CHCO_2C_2H_5\ (25\%) \qquad (2.32)$$

In the majority of Wittig reagents R^1 and R^2 are alkyl groups which have little effect on the carbanionoid character of the molecule. These reagents are markedly nucleophilic and react readily with carbonyl and other polar groups. Addition of the ylid to carbonyl groups takes place rapidly and decomposition of the betaine now becomes the rate-determining step of the reaction. Since the polarity of the carbonyl group is of little consequence when the second step is rate determining, aldehydes and ketones usually react equally well with these reagents.

Alkylidenephosphoranes can be prepared by a number of methods, but in practice they are usually obtained by action of base on alkyltriphenylphosphonium salts, which are themselves readily available from an alkyl halide and triphenylphosphine (2.33). The phosphonium

$$(C_6H_5)_3P + R^1R^2CHX \longrightarrow (C_6H_5)_3\overset{+}{P}{-}CHR^1R^2\ X^-$$
$$\text{base} \downarrow\uparrow \qquad (2.33)$$
$$(C_6H_5)_3P{=}CR^1R^2 + HB$$

salt can usually be isolated and crystallised but the phosphorane is generally prepared in solution and used without isolation. Formation of the phosphorane is reversible, and the strength of base necessary and the reaction conditions depend entirely on the nature of the ylid. A common procedure is to add the stoichiometric amount of an ethereal solution of butyl-lithium to a solution or suspension of the phosphonium salt in ether, benzene or tetrahydrofuran, followed, after an appropriate interval, by the carbonyl compound. A suspension of sodium hydride in ether and lithium and sodium alkoxides, in solution in the corresponding alcohol or in dimethylformamide, are also very commonly used. The latter procedure is simpler than that using organolithium compounds, and has the added advantage that it can be used to prepare phosphoranes from compounds containing functional groups, such as the ethoxycarbonyl group, which would react with organolithium compounds. By this method, also, good yields of products can be obtained from unstable ylids by generating the ylid in presence of the carbonyl compound with which it is to react.

Reactions involving non-stabilised ylids must be conducted under anhydrous conditions and in an inert atmosphere, because these ylids react both with oxygen and with water. Water effects hydrolysis, with formation of a phosphine oxide and a hydrocarbon; the most electronegative group is cleaved. Benzylidenetriphenylphosphorane, for example, reacts with water to give triphenylphosphine oxide and toluene. With oxygen, reaction leads in the first place to triphenylphosphine oxide and a carbonyl compound which undergoes a Wittig reaction with unoxidised ylid to form a symmetrical alkene. This is in fact quite a useful route to symmetrical alkenes. The reaction is conveniently effected by passing oxygen through a solution of the phosphorane or by reaction of lithium ethoxide with the phosphonium periodate in ethanol (2.34) (Bestmann, Armsen and Wagner, 1969).

$$(C_6H_5)_3P{=}CHC_6H_5 \xrightarrow[\text{or } NaIO_4]{O_2} (C_6H_5)_3PO + C_6H_5CHO$$
$$\downarrow (C_6H_5)_3P{=}CHC_6H_5 \qquad (2.34)$$
$$C_6H_5CH{=}CHC_6H_5 \quad (55\%)$$

Wittig reactions with stabilised phosphoranes sometimes proceed only slowly. A valuable alternative makes use of phosphonate esters, themselves readily obtained from an alkyl halide and triethyl phosphite via an Arbuzov rearrangement. Reaction of the phosphonates with a suitable base gives the corresponding carbanions (as 17), which are more nucleophilic than the related phosphoranes, since the negative charge is no longer attenuated by delocalisation into *d* orbitals of the adjacent positively charged phosphorus atom. They react readily with the carbonyl group of aldehydes and ketones to form an alkene and a water-soluble phosphate ester (Wadsworth, 1977; Walker, 1979). Thus, the anion (17) from ethyl bromoacetate and triethyl phosphite reacts rapidly with cyclohexanone at room temperature to give ethyl cyclohexylideneacetate in 70 per cent yield compared with a 25 per cent yield obtained in the reaction with the triphenylphosphorane. The mechanism of the reaction is analogous to that of the Wittig reaction, proceeding by way of *erythro* and *threo* intermediate oxyanions which undergo *syn* elimination of phosphate, possibly by way of a four-membered cyclic intermediate.

$$(C_2H_5O)_3P + BrCH_2CO_2C_2H_5 \longrightarrow \left[(C_2H_5O)_2\overset{\overset{O-C_2H_5\ Br^-}{|}}{P^+}-CH_2CO_2C_2H_5\right]$$

$$\downarrow$$

$$\underset{(17)}{(C_2H_5O)_2\overset{\overset{O\ Na^+}{\|}}{P}\bar{C}HCO_2C_2H_5} \xleftarrow[\text{dimethoxy-ethane}]{\text{NaH}} (C_2H_5O)_2\overset{\overset{O}{\|}}{P}CH_2CO_2C_2H_5 \qquad (2.35)$$

$$\downarrow \text{cyclohexanone}$$

$$\underset{(70\%)}{\text{cyclohexylidene}=CHCO_2C_2H_5} + (C_2H_5O)_2\overset{\overset{O}{\|}}{P}-ONa$$

Where it is applicable this procedure, sometimes called the Wadsworth-Emmons reaction, is generally superior to the Wittig reaction with resonance-stabilised phosphoranes and it is widely employed in the preparation of $\alpha\beta$-unsaturated esters and ketones and other conjugated systems. It gives better yields than the Wittig reaction, the phosphonate esters are readily available and it has the useful practical advantage that the phosphate formed as by-product is water soluble and easily removed from the reaction mixture. It is unsuitable for the preparation of alkenes from non-stabilised reagents; in these cases reaction stops at the betaine stage and no alkene is produced.

$\alpha\beta$-Unsaturated ketones can be made from β-keto phosphonates and carbonyl compounds (2.36). β-Keto phosphonates are themselves obtained by reaction of the lithium salt of dimethyl methylphosphonate with an ester, as in the example, or by reaction with an aldehyde followed by oxidation of the initial hydroxyphosphonate with pyridinium chlorochromate.

THPO ... $CO_2C_2H_5$ —($LiCH_2P(O)(OCH_3)_2$, THF, −70 °C)→ THPO ... $P(O)(OCH_3)_2$

—((1) NaH (2) $>$—CHO (3) H_3O^+)→ HO ... + $(CH_3O)_2P(O)ONa$

(THP = tetrahydropyranyl) (78%) (2.36)

Intramolecular reactions take place readily under the appropriate conditions to give cyclic compounds (Becker, 1980) and the reaction has been used to make macrocyclic lactones related to naturally occurring antibiotics of this class [2.37] (Stork and Nakamura, 1979; Nicolaou, Seitz and Pavia, 1981).

$P(OCH_3)_2$ —(NaH, dimethoxyethane)→ =O, CH_3 (70%)

(2.37)

$P(OC_2H_5)_2$, CHO —($LiOC_3H_7$-iso, THF)→ (66%)

Because of the mild conditions under which they take place, their versatility and the unambiguous position of the double bonds formed, the Wittig and the Wadsworth–Emmons reactions make almost ideal alkene syntheses. Their main disadvantage is that, as originally described they are not subject to steric control and where the structure of the alkene allows it a mixture of *Z*- and *E*-isomers is often produced. In the reaction of phosphonate anions and resonance-stabilised ylids with aldehydes the *E*-alkene generally predominates. Non-stabilised ylids, on the other hand, usually give more of the *Z*-alkene. High yields of *Z*-1,2-disubstituted alkenes have also been obtained using sodium hexamethyldisilazide, $NaN[Si(CH_3)_3]_2$, as base (Corey and Lansbury, 1983). Benzylidenetriphenylphosphorane and benzaldehyde, for example, give a product containing 25 per cent *Z*- and 75 per cent *E*-stilbene, and even higher proportions of *E*-products have been obtained in some other reactions. This preference for the *E*-isomer is a result of the fact that, with stabilised ylids, formation of the intermediate betaines is reversible, allowing interconversion to the more stable *threo* form which collapses to the *E*-alkene. With the non-stabilised ylid,

$$R_3\overset{+}{P}-\overset{-}{C}HR^1 + R^2CHO \rightleftharpoons \left[\begin{array}{l} R_3\overset{+}{P}-C(H)(R^1) \\ \quad | \\ {}^{-}O-C(H)(R^2) \end{array}\right] (\textit{erythro}) \longrightarrow R^2(H)C{=}C(H)R^1 \; (Z)$$

$$R_3\overset{+}{P}-\overset{-}{C}HR^1 + R^2CHO \rightleftharpoons \left[\begin{array}{l} R_3\overset{+}{P}-C(H)(R^1) \\ \quad | \\ {}^{-}O-C(R^2)(H) \end{array}\right] (\textit{threo}) \longrightarrow R^2(H)C{=}C(H)R^1 \; (E) \qquad (2.38)$$

ethylidenetriphenylphosphorane, however, benzaldehyde gives a mixture of *Z*- and *E*-β-methylstyrene containing more of the *E*-isomer (87 per cent). Here formation of the betaine is irreversible and conversion into the alkene proceeds mainly from the kinetically favoured *erythro* betaine. Recent work has shown that the steric course of the reactions can be substantially altered by varying the reaction conditions, and by proper choice of experimental conditions a high degree of steric control can be achieved, at least for reactions leading to disubstituted alkenes. Thus, reaction of stabilised ylids with aldehydes in the presence of protic solvents or lithium salts gives increased amounts of *Z*-alkenes. With non-stabilised ylids, salt-free conditions and non-polar solvents give high selectivity for *Z*-alkene. These variations in the ratio of *Z*- and *E*-products can be ascribed to the influence of the solvents and additives on the relative stabilities and rates of decomposition of the *threo* and *erythro* betaines (Schlosser, 1970; Reucroft and Sammes, 1971).

Two other variations of the Wittig reaction which overcome the stereochemical limitation to some extent and which can be used for the stereoselective synthesis of *Z*- and *E*-1,2-disubstituted alkenes, are based on the readily available phosphine oxides (18) and phosphonobis-*N,N*-dialkylamides (20). Reactions employing anions derived from phosphine

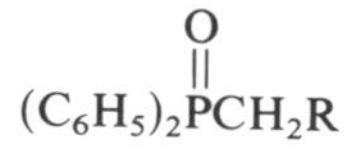

$$(C_6H_5)_2\overset{O}{\overset{\|}{P}}CH_2R \qquad (C_6H_5)_2\overset{O}{\overset{\|}{P}}CH(R)CH(OH)R^1 \qquad RCH_2\overset{O}{\overset{\|}{P}}[N(CH_3)_2]_2$$

(18) 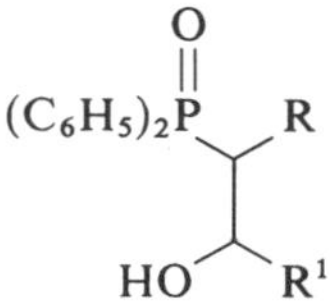 (19) 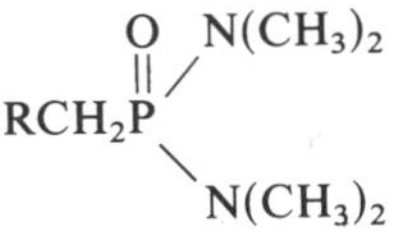 (20)

oxides (the Horner–Wittig reaction) are now finding increasing application (Lythgoe *et al.*, 1976; Clough and Pattenden, 1981; Buss and Warren, 1981). The oxides are readily obtained by quaternisation of triphenylphosphine and hydrolysis of the phosphonium salt, or by reaction of lithiodiphenylphosphide with an alkyl halide or tosylate and oxidation of the resulting phosphine with hydrogen peroxide. The derived lithio derivatives react with aldehydes and ketones to give *β*-hydroxyphosphine oxides (as 19) which smoothly eliminate water-soluble $(C_6H_5)_2\overset{O}{\overset{\|}{P}}ONa$ on treatment with sodium hydride to form the corresponding alkene. The elimination step is stereospecific, *erythro* hydroxyphosphine oxide giving the *Z*-alkene and *threo* hydroxyphosphine oxide the *E*-alkene by preferential *syn* elimination, probably by way of a four-membered cyclic transition state similar to that invoked in the Wittig reaction. Separation of the *erythro* and *threo* isomers of the hydroxyphosphine oxide before elimination thus provides a route to the pure *Z*- and *E*- isomers of the alkene. In practice, reaction of lithio derivatives of alkyldiphenylphosphine oxides with aldehydes generally leads predominantly to the *erythro*-alcohols. Purification by chromatography or crystallisation and elimination then affords pure *Z*-alkene in high yield. The *E*-alkene is best obtained by reduction of the ketone formed by acylation of the lithio diphenylphosphine oxide with an ester or a lactone, followed by elimination from the *threo*-alcohol thus produced (2.39). A component (21) of the pheromone of the Mediterranean fruit fly was made this way.

Lithio derivatives of phosphonobis-*N,N*-dialkylamides (20) similarly react with aldehydes and ketones to give *β*-hydroxyphosphonamides which decompose on heating in benzene solution to form alkenes in high

$(C_6H_5)_2P(O)$ — (1) C_4H_9Li, THF, $-78\,°C$ (2) R^1CHO → NaH, dimethylformamide

erythro : *threo* (9:1)

(1) C_4H_9Li, THF, $-78\,°C$ (2) $R^1CO_2C_2H_5$ → $NaBH_4$ → NaH, dimethylformamide

(1) C_4H_9Li, THF, $-78\,°C$ (2) $CH_2(CH_2)_4OCO$ → (81%) → $NaBH_4$ → (85%) → NaH, dimethylformamide → (90%) (21)

(2.39)

yield. These eliminations also are stereospecific, providing another route for the formation of *Z*- and *E*-alkenes (Corey and Kwiatkowsky, 1968).

The value of the Wittig reaction is clearly shown by the fact that is has already been used in the synthesis of very many alkenes, including a considerable number of natural products (Maercker, 1965). It is a versatile synthesis and can be used for the preparation of mono-, di-, tri-, and tetra-substituted ethylenes, and also of cyclic compounds. The carbonyl components may contain a wide variety of other functional groups such as hydroxyl, ether, ester, halogen and terminal acetylene, which do not interfere with the reaction. In compounds that contain both ester and carbonyl groups the latter react preferentially as long as the Wittig reagent is not present in excess. The mild conditions of the reaction make it an ideal method for the synthesis of sensitive alkenes such as carotenoids and other polyunsaturated compounds, as in (2.40).

$$OHCCH{=}CH(C{\equiv}C)_2CH{=}CHCHO + 2(C_6H_5)_3P{=}CH(CH{=}CH)_2CH_3$$

$$\downarrow$$

$$CH_3(CH{=}CH)_4(C{\equiv}C)_2(CH{=}CH)_4CH_3 \qquad (2.40)$$

$$2C_6H_5CH{=}CHCHO + (C_6H_5)_3P{=}CHCH{=}CHCH{=}P(C_6H_5)_3$$

$$\downarrow$$

$$C_6H_5(CH{=}CH)_5C_6H_5$$

One especially useful application of the Wittig reaction is in the formation of exocyclic double bonds. The Wittig reaction is the method of choice for converting a cyclic ketone into an exocyclic alkene. The Grignard method gives the endocyclic isomer almost exclusively. Thus, cyclohexanone and methylenetriphenylphosphorane give methylenecyclohexane, and the same reaction has been used to prepare a variety of methylene steroids. Inhoffen's synthesis of vitamins D_2 and D_3 involved the Wittig reaction in three of the steps of which the first is shown in (2.41).

$$\text{(bicyclic ketone, R, H)} + (C_6H_5)_3P{=}CHCH{=}CH_2 \longrightarrow \text{(bicyclic diene, R, H, }{=}CH{-}CH{=}CH_2\text{)}$$

$$\text{cyclohexanone} + (C_6H_5)_3P{=}CHOCH_3 \longrightarrow \text{cyclohexylidene}{=}CHOCH_3 \xrightarrow{H_3O^+} \text{cyclohexane(H)(CHO)} \qquad (2.41)$$

A valuable group of Wittig reagents is derived from α-haloethers. They react with aldehydes and ketones to form vinyl ethers which on acid hydrolysis are converted into aldehydes containing one more carbon atom. Thus cyclohexanone is converted into formylcyclohexane (2.41).

The simplest Wittig reagent, methylenetriphenylphosphorane (22) does not react easily with unreactive substrates such as some epoxides or hindered ketones. A useful reactive alternative is the doubly deprotonated lithio derivative (23) which is readily prepared from (22) by reaction with one equivalent of t-butyl-lithium (Corey, Kang and Kyler, 1985). For example, fenchone, which is unaffected by methylenetriphenylphosphorane itself at temperatures up to 50 °C, reacts with the new reagent to give the exomethylene derivative (24) in 87 per cent yield. With epoxides, a γ-oxido ylid is first formed which reacts further with an added aldehyde to give a homoallylic alcohol. Cyclopentene oxide, for example, which does not react with methylenetriphenylphosphorane, gave the *trans* homoallylic alcohol (25) in 65 per cent yield.

$$(C_6H_5)_3P{=}CH_2 \xrightarrow[\text{t-}C_4H_9Li]{\text{1 equiv}} (C_6H_5)_3P{=}CHLi \xleftarrow[\text{s-}C_4H_9Li]{\text{2 equivs}} (C_6H_5)_3\overset{+}{P}{-}CH_3\ \bar{X}$$

(22) (23)

$\xrightarrow{(C_6H_5)_3P=CHLi}$ adduct $\xrightarrow{\text{t-}C_4H_9OH}$ CH_2

(24) (87%)

(2.42)

O $\xrightarrow{(C_6H_5)_3P=CHLi}$ O^- $P(C_6H_5)_3$ $\xrightarrow{C_6H_5CHO}$ OH C_6H_5

(25) (65%)

There is a silicon version of the Wittig reaction, known as the Peterson reaction (Peterson, 1968; Ager, 1984). It entails the elimination of trimethylsilanol, $(CH_3)_3SiOH$, from a β-hydroxyalkyltrimethylsilane, and has the practical advantage over the Wittig reaction that the by-product of the reaction, hexamethyldisiloxane, is volatile and much easier to remove from the reaction product than triphenylphosphine oxide. Further, the steric course of the reaction is more easily controlled, at least for the generation of disubstituted alkenes. Both the *Z*-and *E*-forms of an alkene can be separately obtained from a single stereomer of the hydroxysilane depending on whether the elimination is effected under basic or acidic conditions (2.43). As ordinarily effected the Peterson

$(CH_3)_3Si$ H C_3H_7 OH C_3H_7 H

threo (26)

KH, THF (*syn.* elimination) → H C_3H_7 / C_3H_7 H (96%; 95% *E*-)

$BF_3{\cdot}(C_2H_5)_2O$, CH_2Cl_2 (*anti* elimination) → H H / C_3H_7 C_3H_7 (99%; 94% *Z*)

(2.43)

reaction would give a mixture of *Z*- and *E*-alkenes where this is possible, due to the fact that the β-hydroxysilane is generally obtained as a mixture of the *threo* and *erythro* forms. However, the actual elimination is highly stereoselective, and with a pure diastereomer of the hydroxysilane elimination can be controlled to give either *Z*- or *E*-alkene. Under basic conditions *syn* elimination takes place, probably by way of a cyclic four-membered transition state like that in the Wittig reaction, so that, for example, the *threo* hydroxysilane (26) gives mainly *E*-4-octene. Under acidic conditions the elimination is *anti* and the *Z*-octene is formed. In the same way the *erythro* hydroxysilane gives the *Z*-alkene with base and the *E*-alkene with acid.

The reaction can also be used for the preparation of $\alpha\beta$-unsaturated esters, and it is an attractive alternative to the phosphonate modification of the Wittig reaction for this, since the silicon reagents are, in general, more reactive than phosphonate anions. Even easily enolisable ketones, which often give poor yields in the Wittig reaction, react readily with the silicon reagents.

$$(CH_3)_3SiCH_2CO_2C_2H_5 \xrightarrow[\text{THF, }-78\,°C]{(C_6H_{11})_2NLi} (CH_3)_3SiCHLiCO_2C_2H_5$$

$$\downarrow \text{cyclohexanone, THF, }-78\,°C$$

$$\text{cyclohexylidene}{=}CHCO_2C_2H_5 \quad (95\%) \qquad (2.44)$$

The β-hydroxysilanes required for the Peterson elimination are prepared by reaction of α-silylcarbanions with aldehydes and ketones, or from β-ketosilanes by reduction or by reaction with a Grignard reagent or an organolithium compound. α-Silylcarbanions are themselves obtained from α-chlorosilanes, by addition of alkyl-lithium reagents to vinylsilanes or by metallation of weakly acidic silanes with butyl-lithium (2.45).

$$(CH_3)_3SiCH_2Cl \xrightarrow{Mg} (CH_3)_3SiCH_2MgCl$$

$$(CH_3)_3SiCH{=}CH_2 \xrightarrow{RLi} (CH_3)_3SiCH(Li)CH_2R \qquad (2.45)$$

$$(CH_3)_3SiCH_2SCH_3 \xrightarrow{C_4H_9Li} (CH_3)_3SiCH(Li)SCH_3$$

Diastereomeric β-hydroxysilanes can also be obtained from $\alpha\beta$-epoxysilanes by reaction with lithium organocuprates. Reaction occurs selectively at the carbon bearing the silicon atom to give the *erythro* hydroxysilane from the *Z*-epoxide and the *threo* hydroxysilane from the *E*-epoxide. The series of reactions therefore furnishes a method for the stereoselective synthesis of alkenes from 1-alkenylsilanes (2.46).

$$(CH_3)_3Si(H)C{=}C(H)C_3H_7 \xrightarrow{m\text{-}ClC_6H_4CO_3H} \text{epoxide} \xrightarrow[\text{ether, }-78\,°C]{(C_3H_7)_2CuLi} (CH_3)_3Si(H)(C_3H_7)C{-}C(H)(OH)(C_3H_7)\ \textit{erythro}$$

$$(CH_3)_3Si(H)C{=}C(C_3H_7)H \xrightarrow{m\text{-}ClC_6H_4CO_3H} \text{epoxide} \xrightarrow{(C_3H_7)_2CuLi} (CH_3)_3Si(H)(C_3H_7)C{-}C(H)(OH)(C_3H_7)\ \textit{threo} \qquad (2.46)$$

Sulphur ylids also have been finding increasing application in synthesis (Trost and Melvin, 1975). Sulphur ylids are formally zwitterions in which a carbanion is stabilised by interaction with an adjacent sulphonium

$$>\overset{+}{S}-\overset{-}{C}< \longleftrightarrow >S{=}C< \qquad (2.47)$$

centre. They are usually prepared by proton abstraction from a sulphonium salt with a suitable base or, less commonly, by reaction of a sulphide with a carbene formed, for example, by thermal or photolytic decomposition of a diazo compound. The latter reactions take place by conversion of the diazo compound into an electrophilic singlet carbene which attacks a non-bonding electron pair of the sulphur atom to give the ylid (2.48) (Ando, 1977; Kosarych and Cohen, 1982).

$$(CH_3)_2\overset{+}{S}-CH_3\overset{-}{I} \xrightarrow[\text{dimethyl sulphoxide}]{NaH} (CH_3)_2\overset{+}{S}-\overset{-}{C}H_2$$

$$(C_6H_5)_2S + N_2C(CO_2CH_3)_2 \xrightarrow[or\ Cu,\ \Delta]{h\nu} (C_6H_5)_2\overset{+}{S}-\overset{-}{C}(CO_2CH_3)_2 \qquad (2.48)$$

$$-\ddot{C}- + :SR_2 \rightarrow >\overset{-}{C}-\overset{+}{S}R_2$$

Sulphur ylids do not behave in just the same way as the phosphorus ylids discussed above. In particular, in their reactions with aldehydes and ketones they form epoxides and not alkenes. Two of the most widely used reagents are dimethylsulphonium methylide (27) and dimethyloxosulphonium methylide (28), and the reaction of the former with benzophenone to form 2,2-diphenylethylene oxide is shown in (2.49). Reaction begins in the same way as with the phosphorus ylids by attack of the nucleophilic carbon of the ylid on the electrophilic carbon of the carbonyl group, but since sulphur does not have such a high affinity for oxygen as phosphorus does, reaction subsequently takes a different course and nucleophilic attack on carbon by the oxyanion leads to formation

$$(CH_3)_2\overset{+}{S}-\overset{-}{C}H_2 \quad (27) \qquad (CH_3)_2\overset{+}{S}(=O)-\overset{-}{C}H_2 \quad (28)$$

$$(CH_3)_2\overset{+}{S}-\overset{-}{C}H_2 + O{=}C(C_6H_5)_2 \longrightarrow \left[(CH_3)_2\overset{+}{S}-CH_2-C(C_6H_5)_2-O^-\right] (29) \longrightarrow \underset{O}{CH_2-C}(C_6H_5)_2 \qquad (2.49)$$

of the epoxide with displacement of dimethyl sulphide. Dimethyloxosulphonium methylide reacts in the same way with non-conjugated aldehydes and ketones to form epoxides, and since it is more stable than the sulphonium methylide its use might be preferred. The two reagents differ, however, in their reactions with cyclohexanones: in most cases the sulphonium ylid forms an epoxide with a new axial carbon–carbon bond, whereas the oxosulphonium methylide gives an epoxide with an equatorial carbon–carbon bond. This has been ascribed to the fact that addition of the sulphonium methylide to the carbonyl group to form the intermediate zwitterion (as 29) is irreversible, whereas addition of the oxosulphonium methylide is reversible, allowing accumulation of the thermodynamically more stable zwitterion (Johnson, 1973).

Dimethylsulphonium methylide and dimethyloxosulphonium methylide also differ in their reactions with $\alpha\beta$-unsaturated carbonyl compounds. The sulphonium methylide reacts at the carbonyl group to form, again, an epoxide, but with the oxosulphonium methylide a cyclopropane derivative is obtained by Michael addition to the carbon–carbon double bond. The difference is again due to the fact that the kinetically favoured reaction of the sulphonium ylid with carbonyl groups to give a betaine is irreversible, whereas the corresponding reaction with the oxosulphonium ylid is reversible, allowing preferential formation of the thermodynamically more stable product from the Michael addition. Reaction of $\alpha\beta$-unsaturated carbonyl compounds with dimethyloxysulphonium methylide is now a widely used method for making three-membered rings, along with carbene addition and the Simmons–Smith reaction. Thus, the highly hindered cyclopropane derivative (30) was obtained in 80 per cent yield as in (2.50).

$$\xrightarrow{(CH_3)_2\overset{O}{\overset{\|}{S}}{}^{+}-\bar{C}H_2} \quad (2.50)$$

(30) (80%)

The formation of cyclopropane derivatives from $\alpha\beta$-unsaturated carbonyl compounds and oxosulphonium ylids is limited in practice to reactions involving the parent dimethyloxosulphonium methylid, because of the difficulty of making higher alkylids in this series. This limitation can be circumvented by using the dimethylamino-oxosulphonium ylids (31) formed by deprotonation of *N,N*-dialkyl salts of sulphoximines (Johnson, 1973). These ylids behave very much like dimethyloxosulphonium methylid, forming epoxides by reaction with carbonyl compounds and cyclopropane derivatives with $\alpha\beta$-unsaturated carbonyl compounds, but they have the advangage that higher members of the series can be made. The isopropylid (31, $R^1 = R^2 = CH_3$), for example, may be obtained ultimately from phenyl isopropyl sulphide or by alkylation of the ethylidene ylid (31, $R^1 = H$, $R^2 = CH_3$) with methyl iodide and regeneration of the ylid with base.

$$C_6H_5\overset{O}{\overset{\|}{S}}{}^{+}(N(CH_3)_2)-CHR^1R^2\ BF_4^- \xrightarrow[\text{dimethyl sulphoxide}]{\text{NaH}} C_6H_5\overset{O}{\overset{\|}{S}}{}^{+}(N(CH_3)_2)-\bar{C}R^1R^2 \quad (2.51)$$

(31)

Another valuable sulphonium ylid is diphenylsulphoniumcyclopropylid, which is usually generated *in situ* by action of potassium hydroxide on cyclopropyldiphenylsulphonium fluoroborate (Bogdanowicz and Trost, 1974) (2.52). In contrast to dimethylsulphonium methylid it attacks

$$(C_6H_5)_2\overset{+}{S}-C_3H_5\ BF_4^- \xrightleftharpoons{\text{KOH, dimethyl sulphoxide}} (C_6H_5)_2\overset{+}{S}-\bar{C}_3H_4 \quad (2.52)$$

$\alpha\beta$-unsaturated ketones at the carbon–carbon double bond to form spiropentanes in excellent yield. With aldehydes and ketones, oxaspiropentanes are formed. This reaction is of value in synthesis because of the wide variety of useful transformations which the oxaspiropentanes undergo (Trost, 1974). For example, on treatment with acid or, better, lithium perchlorate or fluoroborate in refluxing benzene, they rearrange to cyclobutanones which are themselves valuable synthetic intermediates (2.53).

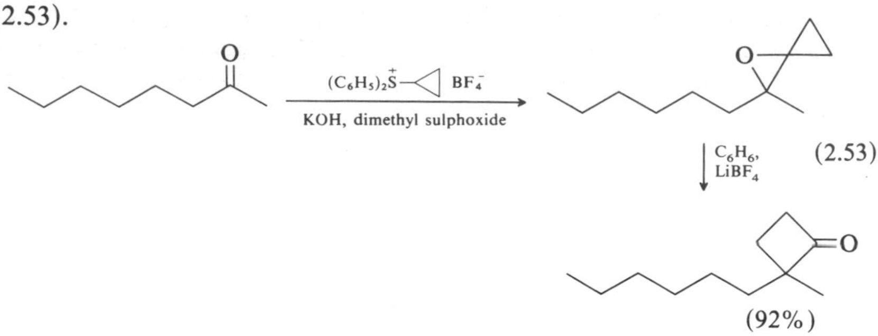

$(C_6H_5)_2\overset{+}{S}$— BF_4^-

KOH, dimethyl sulphoxide

C_6H_6, $LiBF_4$

(2.53)

(92%)

Sulphur ylids undergo a number of rearrangements some of which have been exploited for the formation of carbon–carbon bonds and employed in complex syntheses (Vedejs, 1984) (cf. also p. 143). Thus, reaction of the sulphonium salt (32) with potassium t-butoxide led to

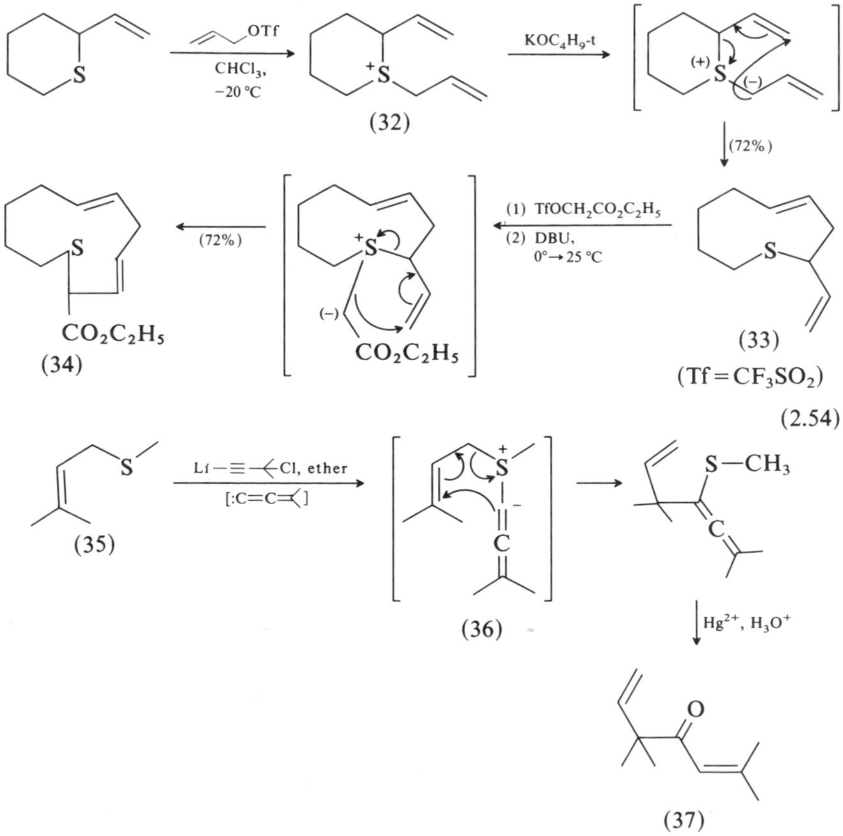

OTf

$CHCl_3$, −20 °C

(32)

KOC_4H_9-t

(72%)

(1) $TfOCH_2CO_2C_2H_5$
(2) DBU, 0°→25 °C

(72%)

$CO_2C_2H_5$

(34)

$CO_2C_2H_5$

(33)

($Tf = CF_3SO_2$)

(2.54)

Li—≡—⟨Cl, ether

[:C=C=⟨]

(35)

(36)

S—CH_3

Hg^{2+}, H_3O^+

(37)

the rearranged product (33) in 72 per cent yield; repetition of the process gave the twelve-membered cyclic sulphide (34) (Vedejs, 1984). Again, a beautiful synthesis of Artemisia ketone (37) proceeded by [2,3]-sigmatropic rearrangement of the intermediate ylid (36) formed by carbene addition to the sulphide (35).

Carbon–carbon double bonds are obtained by Stevens rearrangement of sulphur ylids followed by Hofmann elimination of a derived sul-

$$CH_2\!-\!\overset{+}{S}(CH_3)\!-\!\bar{C}H \longrightarrow -CH_2-CH(SCH_3)- \longrightarrow -CH_2-CH(\overset{+}{S}(CH_3)_2)- \xrightarrow{\text{base}} -CH{=}CH- \quad (2.55)$$

phonium salt (2.55). The method has been used very successfully to prepare highly strained unsaturated cyclophanes, as illustrated in (2.56) for the synthesis of *anti*-[2,2]-metacyclophan-1,9-diene (Mitchell and Boekelheide, 1974). This general strategy of bringing two carbon atoms into close proximity through sulphide linkages, followed by direct linking of the carbon atoms with extrusion of sulphur, has been used with

S S $\xrightarrow{(CH_3O)_2\overset{+}{C}HBF_4^-}$ H_3C-S^+ $^+S-CH_3$ $\xrightarrow[\text{THF}]{t\text{-}C_4H_9OK}$ CH_3S SCH_3

$\swarrow (CH_3O)_2\overset{+}{C}HBF_4^-$

H H $\xleftarrow[\text{THF}]{t\text{-}C_4H_9OK}$ $(CH_3)_2\overset{+}{S}$ $\overset{+}{S}(CH_3)_2$ (2.56)

advantage in other syntheses, notably in the synthesis of vitamin B12 (Eschenmoser, 1970). In one situation the problem was to combine the lactam (38) and the eneamide (39) to give the bicyclic compound (40) and this was beautifully achieved by the 'sulphide contraction' route shown in (2.57). The intermediate (41), on heating with the thiophile triphenylphosphine formed the required product directly, presumably by way of the episulphide.

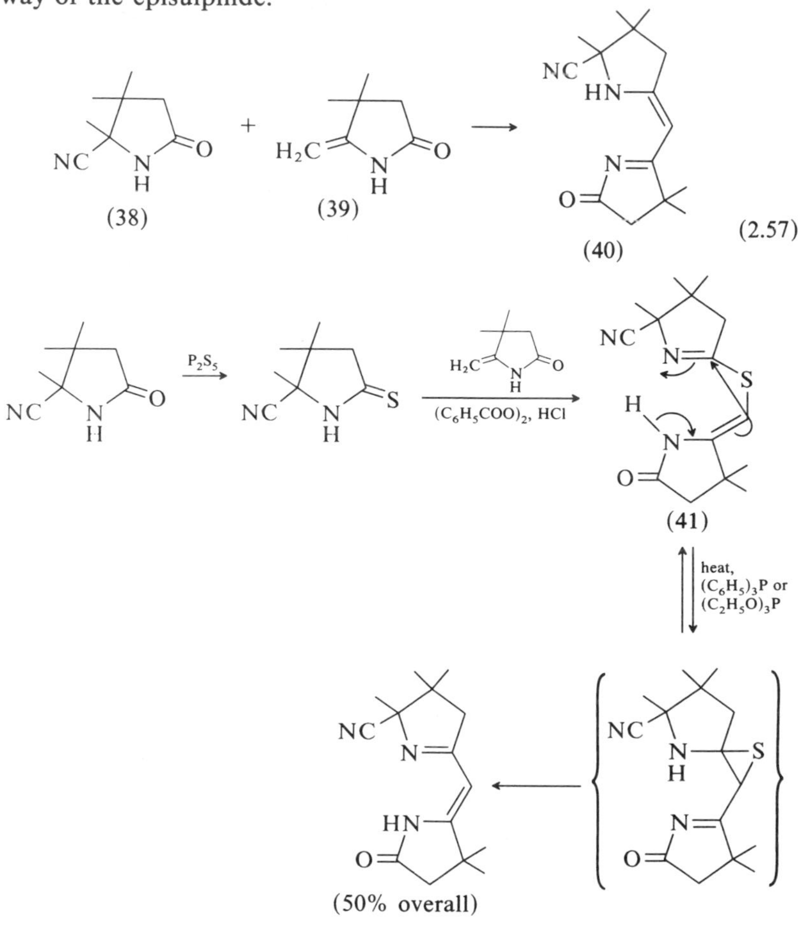

(2.57)

Sommelet–Hauser type rearrangement of arylazasulphonium ylids has been used by Gassman and his co-workers (1974) in a valuable synthesis of *ortho*-alkylanilines and of indole derivatives. The azasulphonium ylids are formed *in situ* by action of an appropriate base on the corresponding sulphonium salt, itself obtained from the *N*-chloroaniline and a sulphide or by reaction of the aniline itself with a halosulphonium halide. The reaction takes the course shown in (2.58).

$$\text{ethyl 4-aminobenzoate} \xrightarrow[CH_2Cl_2,\ -65\,^\circ C]{t\text{-}C_4H_9OCl} \text{ArNHCl} \xrightarrow[-65\,^\circ C]{CH_3SCH_3} \text{ArNH}\overset{+}{S}(CH_3)_2\ Cl^-$$

$$\xrightarrow[-65 \rightarrow 25\,^\circ C]{CH_3ONa,\ CH_3OH} \left[\text{ArNH}\overset{+}{S}(CH_3)\bar{C}H_2 \rightarrow \text{6-}(CH_2SCH_3)\text{-6-H-cyclohexadienimine}\right] \rightarrow \text{2-}(CH_2SCH_3)\text{-4-}(CO_2C_2H_5)\text{aniline} \xrightarrow{\text{Raney Ni}} \text{2-}CH_3\text{-4-}(CO_2C_2H_5)\text{aniline}\ (47\%)$$

(2.58)

A two-fold extrusion process forms the key step in a recent method for the synthesis of highly hindered alkenes. The principle here is that if X and Y are easily extrudable groups then even highly hindered alkenes

$$R^1R^2C\overset{X}{\underset{Y}{\diamond}}CR^3R^4 \longrightarrow R^1R^2C{=}CR^3R^4 + X + Y$$

might be accessible because both the new carbon–carbon bonds are formed intramolecularly. Good results have been obtained with 1,3,4-thiadiazolines (X = —N=N—, Y = S), prepared, for example, by addition of diazo compounds to thiones (Barton, Guziec and Shahak, 1974). Pyrolysis of the diazolines under mild conditions first gives episulphides which subsequently collapse to alkenes in the presence of a suitable thiophile. Thus, thiocamphor and di-t-butyldiazomethane gave an adduct which was converted into the highly hindered alkene (42) on heating with tributylphosphine (2.59) and a similar method was used to prepare (43) the closest compound to tetra-t-butylethylene yet made (Krebs and Ruger, 1979). Even better results have been obtained using selenium analogues (Back *et al.*, 1975). Tetra-t-butylethylene, however, still defies isolation. The tetraisopropyl compound has been obtained by reductive coupling of diisopropyl ketone with an active titanium powder prepared by reduction of titanium trichloride with potassium (McMurry and Fleming, 1975).

S + N_2C —THF, reflux→ N=N, S —→ S —$(C_4H_9)_3P$, 120 °C→

(43) (42) (2.59)

2.5. Alkenes from sulphones

The well-known conversion of α-halosulphones into alkenes on treatment with base (the Ramberg–Bäcklund reaction) is believed to proceed as shown in (2.60) with final extrusion of sulphur dioxide from an episulphone (Paquette, 1977).

$$RR^1CHSO_2CClR^2R^3 \xrightarrow{\text{base}} RR^1\bar{C}\text{–}SO_2\text{–}C(R^2)(R^3)\text{–}Cl \longrightarrow \text{episulphone} \xrightarrow{-SO_2} RR^1C{=}CR^2R^3 \quad (2.60)$$

Sulphones can be used in another way to make alkenes through the corresponding α-lithio derivatives. These react readily with aldehydes and ketones to give α-hydroxysulphones which, in the form of their acetates or *p*-toluenesulphonates, undergo smooth reductive cleavage with sodium amalgam in methanol to form alkenes. The reaction is regioselective and can be used to prepare di-, tri- and tetra-substituted alkenes and conjugated dienes (2.61) (Julia and Paris, 1973; Kocienski,

$$CH_3CH_2SO_2C_6H_5 \xrightarrow[(2)\ (CH_3)_2C{=}CH(CH_2)_2CHO]{(1)\ C_4H_9Li} CH_3{-}CH(SO_2C_6H_5)CHOH(CH_2)_2CH{=}C(CH_3)_2$$

$$\xrightarrow[(2)\ Na{-}Hg,\ C_2H_5OH]{(1)\ CH_3SO_2Cl} H_3C{-}CH{=}CH{-}(CH_2)_2CH{=}C(CH_3)_2\ (80\%)$$

$$\text{1-cyclohexenyl-}CH_2SO_2C_6H_5 \xrightarrow[THF,\ -78\,°C]{C_4H_9Li} \text{1-cyclohexenyl-}CH(Li)SO_2C_6H_5 \xrightarrow[(2)\ (CH_3CO)_2O]{(1)\ C_6H_5CHO} \text{1-cyclohexenyl-}CH(SO_2C_6H_5)CH(OCOCH_3)C_6H_5$$

$$\xrightarrow{Na{-}Hg,\ CH_3OH{-}CH_3CO_2C_2H_5} \text{1-cyclohexenyl-}CH{=}CH{-}C_6H_5\ (93\%)$$

CH_3COO, R, H, H, R^1

(44)

(2.61)

Lythgoe and Ruston, 1978). Reactions leading to 1,2-disubstituted alkenes give very largely the *E*-isomers, irrespective of the configuration of the hydroxysulphone. Possibly the initial reductive cleavage of the phenylsulphonyl group generates an anion which, whatever may have been its original configuration, is long-lived enough to permit it to assume the low energy conformation (44) from which the *E*-alkene is formed by loss of acetate ion. This sequence provides a useful alternative to the Wittig reaction for the preparation of *E*-1,2-disubstituted alkenes. The sulphones required are easily available, even from secondary halides, by reaction with the strongly nucleophilic benzene-thiolate anion and oxidation of the sulphide.

Z-1,2-Disubstituted alkenes can be obtained in another way. Dehydration of the hydroxy sulphones, instead of reductive elimination, by treatment of the tosylate with base, affords the corresponding vinyl sulphones. The *Z*- and *E*-sulphones can be obtained selectively from the appropriate *threo* and *erythro* hydroxy sulphones, and on reductive cleavage of the sulphone by reaction with certain alkyl Grignard reagents in the presence of nickel or palladium catalysts they give the corresponding 1,2-disubstituted alkene with high retention of stereochemistry. The *E*-sulphone (45), for example, gave (*Z*)-2-duodecene in 85 per cent yield. Alternatively, with a different catalyst, substitution of the sulphone may be effected instead, again with high stereoselectivity, to give trisubstituted alkenes (2.62) (Fabre and Julia, 1983; Fabre, Julia and Verpeaux, 1982).

CH_3, $C_6H_5SO_2$, C_9H_{19} (45) $\xrightarrow[\text{Pd(acac)}_2,\ P(C_4H_9)_3,\ \text{THF, 25 °C}]{C_4H_9MgBr}$ CH_3, C_9H_{19} (85%; 98.5% *Z*)

H_3C, CH_3, $t\text{-}C_4H_9SO_2$, H $\xrightarrow[\text{Ni(acac)}_2,\ \text{THF, 25 °C}]{C_6H_5MgBr}$ H_3C, CH_3, C_6H_5, H (73%) (2.62)

H, CH_3, $t\text{-}C_4H_9SO_2$ $\xrightarrow[\text{Fe(acac)}_2,\ \text{THF}]{CH_3MgCl}$ H, CH_3, H_3C (71%; 98% *Z*)

2.6. Decarboxylation of β-lactones

The decomposition of β-lactones to alkenes and carbon dioxide at moderate temperature has been known for many years but has not been much used in synthesis because of the inaccessibility of β-lactones. It has recently been found, however, that β-lactones can be made from β-hydroxy acids by action of benzenesulphonyl chloride in pyridine. β-Lactones decompose readily at 140–160 °C to give the corresponding alkene in virtually quantitative yield. The β-hydroxy acids themselves can be made by the Reformatsky reaction or, better, by condensation of α-metalated carboxylic salts with aldehydes and ketones. The whole sequence makes a useful alternative to the Wittig reaction particularly for the synthesis of tri- and tetra-substituted alkenes (Adam, Baeza and Liu, 1972). The reaction is illustrated in (2.63); it is noteworthy in this example that the double bond does not migrate to give a stilbene, as it does under the basic conditions of the Wittig reaction.

The β-lactones are formed stereospecifically with retention of the geometry of the β-hydroxy acid precursor. The decarboxylation step also is stereospecific, and with stereochemically pure β-hydroxy acids

$$CH_3CH_2CO_2H \xrightarrow[THF]{2LiN(iso\text{-}C_3H_7)_2} CH_3CH{=}C(OLi)O^-\,Li^+ \xrightarrow{C_6H_5CH_2COC_6H_5} H_3C{-}CH(CO_2H){-}C(OH)(CH_2C_6H_5)(C_6H_5)$$

$$\xrightarrow[\text{pyridine, } 0\,^\circ C]{C_6H_5SO_2Cl} \beta\text{-lactone} \xrightarrow{140\,^\circ C} (H_3C)HC{=}C(CH_2C_6H_5)(C_6H_5)\ (>99\%\ E) \qquad (2.63)$$

stereochemically pure alkenes are obtained. But in the ordinary course of events, when the β-hydroxy acid is obtained as a mixture of *threo* and *erythro* isomers reaction affords a mixture of the *Z*- and *E*-alkenes where this is structurally possible.

A way round this difficulty has been found recently (Mulzer *et al.*, 1979). Good yields of *threo* β-hydroxy acids are obtained by addition

$$[R^1CH{=}C(OLi)O^-]\,Li^+ + R^2CHO \longrightarrow HO{-}CH(R^2){-}CH(R^1){-}CO_2H$$

$$\xrightarrow[\text{pyridine}]{C_6H_5SO_2Cl} \beta\text{-lactone} \xrightarrow{140\,^\circ C} HR^1C{=}CR^2H \quad (63\%;\ >99\%\ E \text{ for } R^1{=}C_6H_5,\ R^2{=}CH_3)$$

$$\xrightarrow[(C_6H_5)_3P,\ THF,\ 22\,^\circ C]{C_2H_5O_2CN{=}NCO_2C_2H_5} {}^-O_2C{-}CH(R^1){-}CH(R^2){-}O{-}\overset{+}{P}(C_6H_5)_3\ (46) \xrightarrow[-(C_6H_5)_3P{=}O]{-CO_2} HR^1C{=}CHR^2 \quad (75\%;\ 97\%\ Z \text{ for } R^1{=}C_6H_5,\ R^2{=}CH_3) \qquad (2.64)$$

of the *bis*-lithio anions of carboxylic acids to aldehydes whence *anti* elimination by treating the β-hydroxy acid with triphenylphosphine and diethyl azodicarboxylate gives the *Z*-alkene. *syn*-Elimination to the *E*-alkene is accomplished by the β-lactone procedure. The *Z*-alkene is presumably formed by way of the zwitterion (46) which then fragments with loss of carbon dioxide and triphenylphosphine oxide with the usual *anti* stereochemistry.

2.7. Stereoselective synthesis of tri- and tetra-substituted alkenes

Z- and *E*-1,2-Disubstituted alkenes are readily available by reduction of the corresponding alkynes or by other means (see p. 154), but isomerically pure tri- and tetra-substituted alkenes have been more difficult to obtain. In an early method (Cornforth, Cornforth and Mathew, 1959) the critical step is the reaction of a Grignard reagent with an α-chloro-aldehyde or -ketone. The most reactive conformation of an α-chloroaldehyde or α-chloroketone is that in which the C=O and C—Cl dipoles are antiparallel. Addition of a Grignard reagent to this conformation is highly stereoselective and takes place mainly from the side of the carbonyl group which is less sterically hindered by the groups R^1 and R^2 on on the α-carbon atom, leading predominantly to the chlorohydrin in which the incoming group R^4 and the larger of the groups R^1 and R^2 are *anti* to each other (2.65). The resulting chlorohydrin is

$$R^3\text{—C(=O)—C}(R^1)(R^2)\text{Cl} \xrightarrow{R^4MgBr} \text{HO—C}(R^4)(R^3)\text{—C}(R^1)(R^2)\text{Cl} \xrightarrow{\text{several steps}} (R^4)(R^3)\text{C=C}(R^1)(R^2) \qquad (2.65)$$

converted, by a series of stereoselective reactions into the alkene in which three of the groups on the double bond are derived from the chlorocarbonyl compound and the other from the Grignard reagent. The method is exemplified in (2.66) for the synthesis of (*E*)-3-methyl-2-pentene. A similar series of reactions beginning with the addition of methylmagnesium iodide to 2-chloropentan-3-one gave (*Z*)-3-methyl-2-pentene.

An entirely different procedure which makes possible the stereospecific conversion of a propargylic alcohol into a 2- or 3-alkylated allylic alcohol is based on a remarkable, specific, conversion of propargylic alcohols into β- or γ-iodoallylic alcohols by reduction with a modified lithium aluminium hydride reagent and subsequent reaction of the crude reduction product with iodine (Corey, Katzenellenbogen and Posner, 1967). If the reduction is effected in the presence of sodium methoxide the final product is exclusively the γ-iodoallylic alcohol, whereas reduction with lithium aluminium hydride and aluminium chloride and iodination gives

$$\text{H}_3\text{C}-\text{CO}-\text{CH(Cl)CH}_3 \xrightarrow{C_2H_5MgBr} \text{HO}(\text{C}_2\text{H}_5)(\text{H}_3\text{C})\text{C}-\text{CH(Cl)CH}_3 \xrightarrow{NaOH} \text{epoxide} \xrightarrow{NaI, CH_3CO_2H} \text{HO}(\text{C}_2\text{H}_5)(\text{H}_3\text{C})\text{C}-\text{CH(I)CH}_3 \xrightarrow[\text{pyridine}]{SnCl_2, POCl_3} (\text{C}_2\text{H}_5)(\text{H}_3\text{C})\text{C}=\text{C(H)CH}_3$$

(90%; 85% stereoselective) (91%) (81%)

(2.66)

finally the β-iodoallylic alcohol. Reaction of the resulting iodo compounds with lithium organocuprates (p. 79) then affords the corresponding substituted allylic alcohol in which the substituents originally present in the propargyl alcohol are *trans* to each other (2.67). This method is

$$\text{R}-\text{C}\equiv\text{C}-\text{CH}_2\text{OH} \xrightarrow[(2)\ I_2,\ -78\ ^\circ C]{(1)\ LiAlH_4,\ CH_3ONa,\ THF} (\text{R})(\text{I})\text{C}=\text{C}(\text{H})\text{CH}_2\text{OH} \xrightarrow{Li(CH_3)_2Cu} (\text{R})(\text{CH}_3)\text{C}=\text{C}(\text{H})\text{CH}_2\text{OH}$$

$$\text{R}-\text{C}\equiv\text{C}-\text{CH}_2\text{OH} \xrightarrow[(2)\ I_2,\ -78\ ^\circ C]{(1)\ LiAlH_4,\ AlCl_3,\ THF} (\text{R})(\text{H})\text{C}=\text{C}(\text{I})\text{CH}_2\text{OH} \xrightarrow{Li(CH_3)_2Cu} (\text{R})(\text{H})\text{C}=\text{C}(\text{CH}_3)\text{CH}_2\text{OH}$$

(2.67)

applicable to a variety of synthetic problems in which the stereospecific introduction of trisubstituted carbon–carbon double bonds is involved. For example, it formed a key step in a synthesis of juvenile hormone (2.68).

$$\ldots(\text{CH}_2)_2\text{C}\equiv\text{CCH}_2\text{OH} \xrightarrow[(2)\ I_2,\ (3)\ (CH_3)_2CuLi]{(1)\ LiAlH_4,\ CH_3ONa,\ THF} \ldots\text{CH}_2\text{OH}$$

(54%; 97% *E*)

(2.68)

Tri- and tetra-substituted alkenes can also be obtained stereoselectively from alkynes by addition of organocopper and organoborane reagents. One synthesis leads to $\beta\beta$-dialkylacrylic esters from $\alpha\beta$-acetylenic esters. Lithium dialkylcuprates (p. 78) add rapidly to $\alpha\beta$-acetylenic esters to give the stereoisomeric acrylic esters (47) and (48) in high yield. The

$$R^1{-}C{\equiv}C\ CO_2CH_3 \xrightarrow[(2)\ H_3O^+]{(1)\ LiR^2_2Cu} (R^1)(R^2)C{=}C(H)(CO_2CH_3)\ (47) + (R^1)(R^2)C{=}C(H)(CO_2CH_3)\ (48) \quad (2.69)$$

stereochemistry of the products is dependent on the reaction temperature and the nature of the solvent, and high yields of the *Z*-addition compound (47) are obtained by conducting the reaction at −78 °C in tetrahydrofuran solution. In contrast to the procedure with propargyl alcohols described above, this reaction yields an alkene in which the substituents in the acetylenic precursor are *cis* to each other in the alkene.

Another procedure makes use of organocopper(I) reagents, RCu. Alkyl copper(I) reagents, prepared from Grignard reagents and an equimolecular amount of copper(I) bromide or, better, the copper(I) bromide–dimethyl sulphide complex, add readily to terminal alkynes to give 1-alkenylcopper(I) compounds. The copper adds to the terminal carbon of the alkyne with *syn* addition of the alkyl group of the alkylcopper(I) reagent (2.70). The alkenylcopper(I) compounds thus obtained

$$RMgBr + CuBr.(CH_3)_2S \xrightarrow[-45\,^\circ C]{ether} RCu.(CH_3)_2S.MgBr_2 \xrightarrow{R^1C\equiv CH,\ -25\,^\circ C} (R)(R_1)C{=}C(H)Cu(CH_3)_2S.MgBr_2 \quad (2.70)$$

react with a variety of electrophiles, including alkyl halides, $\alpha\beta$-unsaturated ketones and epoxides, giving trisubstituted alkenes with almost complete retention of configuration (Marfat, McGuirk and Helquist, 1979; Carruthers, 1982). Some examples are shown in (2.71). In the presence of catalytic $Pd(PPh_3)_4$, alkenyl iodides also react, to give conjugated dienes (Jabri, Alexakis and Normant, 1982).

Other useful methods for the stereospecific synthesis of alkenes from alkynes proceed from the derived alkenylalanes and alkenylboranes. Alkenylalanes are readily prepared by hydroalumination of alkynes with, for example, diisobutylaluminium hydride. The reaction takes place by *cis* addition, giving *Z*-alkenylalanes. Reaction of these alkenylalanes with halogens proceeds with retention of configuration to give the corre-

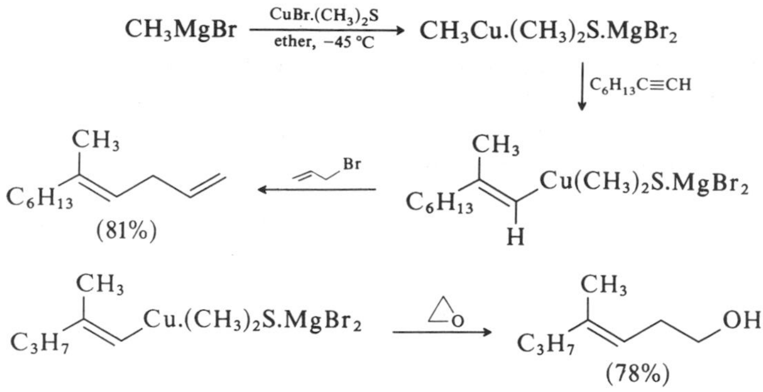

$$CH_3MgBr \xrightarrow[\text{ether, } -45\,°C]{CuBr.(CH_3)_2S} CH_3Cu.(CH_3)_2S.MgBr_2$$

$\downarrow C_6H_{13}C{\equiv}CH$

CH_3, C_6H_{13}, $Cu(CH_3)_2S.MgBr_2$, H $\xleftarrow{\text{allyl Br}}$; CH_3, C_6H_{13} diene (81%)

(2.71)

CH_3, C_3H_7, $Cu.(CH_3)_2S.MgBr_2$ $\xrightarrow{\text{ethylene oxide}}$ CH_3, C_3H_7, OH (78%)

sponding alkenyl halides, the halogen atoms of which are replaceable by alkyl groups by reaction with an organocopper reagent. Thus iodination of the alane from the reaction of 1-hexyne with diisobutylaluminium

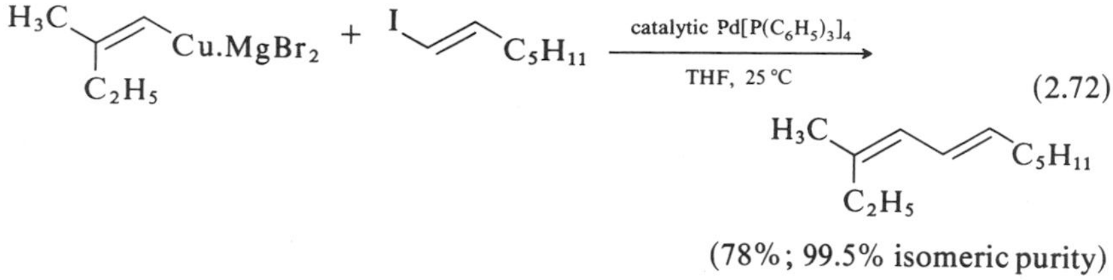

H_3C, C_2H_5, $Cu.MgBr_2$ + I, C_5H_{11} $\xrightarrow[\text{THF, 25 °C}]{\text{catalytic } Pd[P(C_6H_5)_3]_4}$ H_3C, C_2H_5, C_5H_{11}

(78%; 99.5% isomeric purity)

(2.72)

hydride produces the isomerically pure (E)-1-iodo-1-hexene, while 3-hexyne gives (E)-3-iodo-3-hexene (2.73).

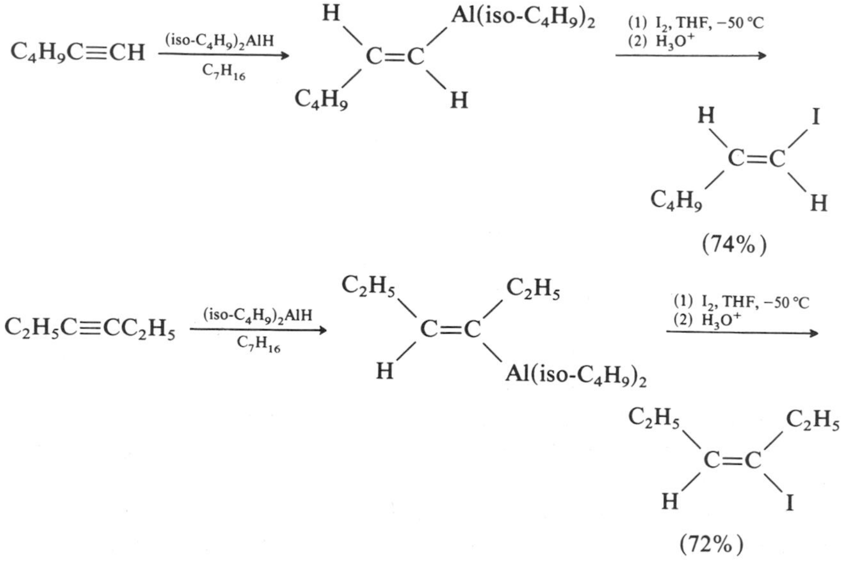

$$C_4H_9C{\equiv}CH \xrightarrow[C_7H_{16}]{(iso\text{-}C_4H_9)_2AlH} (H)(C_4H_9)C{=}C(Al(iso\text{-}C_4H_9)_2)(H) \xrightarrow{(1)\ I_2,\ THF,\ -50\,°C;\ (2)\ H_3O^+} (H)(C_4H_9)C{=}C(I)(H)$$

(74%)

$$C_2H_5C{\equiv}CC_2H_5 \xrightarrow[C_7H_{16}]{(iso\text{-}C_4H_9)_2AlH} (C_2H_5)(H)C{=}C(C_2H_5)(Al(iso\text{-}C_4H_9)_2) \xrightarrow{(1)\ I_2,\ THF,\ -50\,°C;\ (2)\ H_3O^+} (C_2H_5)(H)C{=}C(C_2H_5)(I)$$

(72%)

(2.73)

Another excellent route to *Z*- and *E*-1-halogenoalkenes from terminal acetylenes uses the *E*-alkenylboronic acids which are easily obtained by reaction of terminal acetylenes with catecholborane (p. 292) followed by hydrolysis. Reaction of the boronic acids with iodine and sodium hydroxide results in replacement of the boronic acid group by iodine with retention of configuration to give the *E*-1-iodoalkene. But with bromine, followed by base, substitution is accompanied by inversion and the *Z*-1-bromoalkene is formed. In each case the reaction is highly stereoselective. Thus, 1-octyne was converted into (*Z*)-1-bromo-1-octene in 85 per cent yield with 99 per cent stereoselectivity, and 1-hexyne gave 89 per cent (*E*)-1-iodo-1-hexene (2.74). The inversion of configuration

$C_4H_9C{\equiv}CH$ —(1) catecholborane (BH); (2) H_3O^+→ $C_4H_9(H)C{=}C(H)B(OH)_2$ —(1) I_2; (2) NaOH→ $C_4H_9(H)C{=}C(H)I$
(89%; >99% *E*)

(2.74)

$C_6H_{13}C{\equiv}CH$ —(1) catecholborane (BH); (2) H_3O^+→ $C_6H_{13}(H)C{=}C(H)B\langle$ —(1) Br_2, CH_2Cl_2; (2) $NaOCH_3$, CH_3OH→ $C_6H_{13}(H)C{=}C(Br)H$
(85%; 99% *Z*)

in the bromination reaction can be accounted for by the usual *trans* addition of bromine to the double bond followed by base-induced *trans* elimination of boron and bromine (2.75), but the course of the iodination reaction is not clear.

Reaction of alkenylalanes with methyl-lithium affords the corresponding aluminates which are converted into $\alpha\beta$-unsaturated carboxylic acids in excellent yields on reaction with carbon dioxide. Similarly, paraformaldehyde and other aldehydes afford allylic alcohols (Zweifel and Steele, 1967*a*) (2.76).

In contrast to the reaction with diisobutylaluminium hydride discussed above, hydroalumination of disubstituted acetylenes with lithium hydridodiisobutylmethylaluminate, obtained from diisobutylaluminium

(2.75)

$$CH_3C\equiv CCH_3 \xrightarrow{(iso\text{-}C_4H_9)_2AlH} (CH_3)HC{=}C(CH_3)Al(iso\text{-}C_4H_9)_2 \xrightarrow{CH_3Li,\ ether,\ -30\,°C} [(CH_3)HC{=}C(CH_3)\bar{A}l(iso\text{-}C_4H_9)_2CH_3]\,Li^+ \xrightarrow{CO_2,\ H_3O^+} (CH_3)HC{=}C(CH_3)CO_2H\ (76\%) \quad (2.76)$$

hydride and methyl-lithium, results in *trans* addition to the triple bond, thus opening the way to a convenient synthesis of isomeric series of alkenes from a disubstituted acetylene. Carbonation gives $\alpha\beta$-unsaturated acids, reaction with paraformaldehyde or acetaldehyde affords allylic alcohols and iodine gives alkenyl iodides, all isomeric with the products obtained in the reaction sequences using diisobutylaluminium hydride discussed above (Zweifel and Steele, 1967*b*). Thus the isomeric α-methylcrotonic acids are conveniently obtained from 2-butyne as illustrated in (2.76) and (2.77). In all these reactions the isobutyl groups

$$CH_3C\equiv CCH_3 \xrightarrow{Li[(iso\text{-}C_4H_9)_2AlHCH_3]} [H(CH_3)C{=}C(CH_3)\bar{A}l(iso\text{-}C_4H_9)_2CH_3]\,Li^+$$

(1) paraformaldehyde (2) H_3O^+ → $H(H_3C)C{=}C(CH_3)CH_2OH$ (68%)

CO_2, H_3O → $H(H_3C)C{=}C(CH_3)CO_2H$ (72%) (2.77)

of the hydroaluminating agent are converted into isobutane in the hydrolysis step, and do not interfere with the isolation of the products.

Alkenylboranes, obtained by hydroboration of alkynes (p. 292) do not behave in the same way as alkenylalanes. Iodination does not afford the corresponding iodide but results in stereospecific transfer of one alkyl group from boron to the adjacent carbon, providing another stereospecific synthesis of substituted alkenes. Thus, addition of iodine and sodium hydroxide to the alkenylborane obtained by hydroboration of 1-hexyne with dicyclohexylborane, affords (*Z*)-1-cyclohexyl-1-hexene in 75 per cent yield. The reaction is thought to take the course indicated in (2.78).

$$C_4H_9C{\equiv}CH \xrightarrow{(C_6H_{11})_2BH} (H)(C_4H_9)C{=}C(H)B(C_6H_{11})_2 \xrightarrow{I_2,\ NaOH} \left[\text{iodonium ion; } C_6H_{11}\text{ migrates from B to C, } I^- \text{ to B}\right] \rightarrow C_4H_9CH(I){-}CH(C_6H_{11})B(I)C_6H_{11} \rightarrow (H)(C_4H_9)C{=}C(H)(C_6H_{11}) \quad (2.78)$$

(99% *Z*)

3-Hexyne is similarly converted into (*Z*)-3-cyclohexyl-3-hexyne in 85 per cent yield. The necessity for *anti* elimination of the iodine and boron group ensures the stereoselectivity of the alkene-forming step. A more recent route proceeds by addition of an alkylbromoborane-dimethyl sulphide complex to a 1-alkyne and affords high yields of *Z*-1,2-disubstituted alkenes (Brown and Basavaiah, 1982).

This reaction has been extended to the synthesis of trisubstituted alkenes of defined stereochemistry starting from a disubstituted acetylene (Brown, Basavaiah and Kulkarni, 1982). Dialkylvinylboranes, themselves readily obtained by reaction of dialkylhaloboranes with an internal alkyne in the presence of lithium aluminium hydride, react with iodine in the presence of sodium methoxide to form a trisubstituted alkene stereoselectively in which the two alkyl substituents of the original alkyne are *trans* to each other. (*Z*)-3-Methyl-2-pentene, for example, was obtained exclusively from 2-butyne as shown in (2.79). The mechanism of the reaction is analogous to that invoked in the Zweifel synthesis of *Z*-disubstituted alkenes (2.78).

$$CH_3C{\equiv}CCH_3 \xrightarrow[\frac{1}{4}LiAlH_4,\ THF]{(C_2H_5)_2BBr} (C_2H_5)_2B(H_3C)C{=}C(H)(CH_3) \xrightarrow[CH_3OH,\ -78\ °C]{I_2,\ NaOH} (H_3C)(C_2H_5)C{=}C(H)(CH_3)\ (74\%) \quad (2.79)$$

In another variation trisubstituted alkenes are obtained from terminal alkynes by way of the corresponding 1-alkenyl iodides (2.80). Sequential treatment of the iodide with butyl-lithium, a trialkylborane and iodine leads to good yields of the corresponding trisubstituted alkene in which the iodine of the alkenyl iodide is replaced by an alkyl group of the borane with retention of configuration. The stereochemistry of the products requires *syn* elimination of R_2BI (Lahima and Levy, 1978; Levy, Angelastro and Marinelli, 1980).

$$C_6H_{13}C{\equiv}CH \xrightarrow[CuBr.(CH_3)_2S,\ ether,\ -45\ °C]{C_2H_5MgBr} (C_6H_{13})(C_2H_5)C{=}C(H)(Cu(CH_3)_2S.MgBr_2) \xrightarrow{I_2} (C_6H_{13})(C_2H_5)C{=}C(H)(I)$$

$$\xrightarrow{(1)\ C_4H_9Li;\ (2)\ (C_2H_5)_3B,\ -78\ °C} \left[(C_6H_{13})(C_2H_5)C{=}C(H)(\bar{B}(C_2H_5)_3)\right]Li^+ \xrightarrow[-(C_2H_5)_2BI]{I_2} (C_6H_{13})(C_2H_5)C{=}C(H)(C_2H_5)$$

(75%; >95% stereochemically pure) (2.80)

Another promising stereoselective synthesis of trisubstituted alkenes starting from allylic alcohols depends critically on the stereoselective epoxidation of allylic alcohols described on p. 373. Thus, for the synthesis of (*Z*)-6-methyl-5-undecene, the allylic alcohol (49) was oxidised with t-butyl hydroperoxide and a vanadium catalyst, giving selectively the *erythro*-epoxide (50) which was converted into the vicinal diol (51) by reaction with lithium dibutylcuprate. Stereospecific deoxygenation of the diol was accomplished by reaction with *N,N*-dimethylformamide dimethyl acetal and decomposition of the resulting dioxolane derivative by heating with acetic anhydride and gave the *Z*-alkene with very high stereochemical purity (2.81) (cf. p. 165).

(49) $\xrightarrow[VO(acac)_2]{t\text{-}C_4H_9OOH}$ (50) (96% *erythro*) $\xrightarrow[ether,\ -26\,°C]{Li(C_4H_9)_2Cu}$ (51) $\xrightarrow{(CH_3)_2NCH(OCH_3)_2,\ 25\,°C}$ dioxolane $\xrightarrow[130\,°C]{(CH_3CO)_2O}$ Z-alkene

(2.81)

2.8. Fragmentation reactions

A number of other methods for forming carbon–carbon double bonds have less general application than those discussed above, but are of value in particular circumstances. One of these makes use of the fragmentation of the monotoluene-*p*-sulphonates or methanesulphonates of suitable cyclic 1,3-diols on treatment with base (Grob and Schiess, 1967) (2.82). A feature of these reactions is that when the C—X bond and the $C_{(a)}—C_{(b)}$ bond have the *trans* anti-parallel arrangement the

$\xrightarrow{base}$ (2.82)

X = leaving group, e.g.: $—OSO_2C_6H_4CH_3\text{-}p$, $—OSO_2CH_3$

reaction proceeds very readily by a concerted pathway to give an alkene the stereochemistry of which is governed solely by the relative orientation of groups in the cyclic precursor. For example, the decalin derivative (52) in which the tosyloxy group and the adjacent angular hydrogen atom are *cis* affords (*E*)-5-cyclodecenone in high yield, whereas the isomer (53) in which the tosyloxy substituent and the hydrogen atom are *trans* affords the *Z*-cyclodecenone (i.e. in each case the relative orientation of the hydrogen atoms in the precursor is retained in the alkene). In these derivatives a *trans* anti-parallel arrangement of the breaking

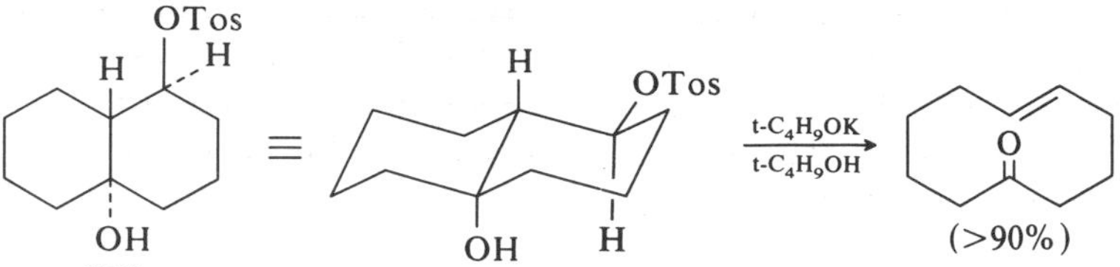

(52)

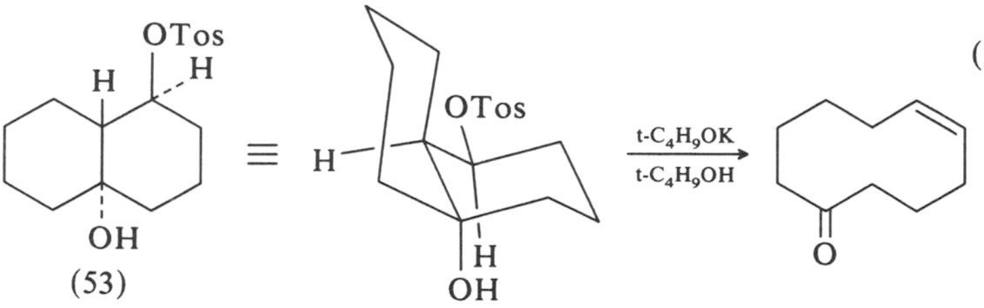

$\xrightarrow[t\text{-}C_4H_9OH]{t\text{-}C_4H_9OK}$ (>90%)

(53)

(2.83)

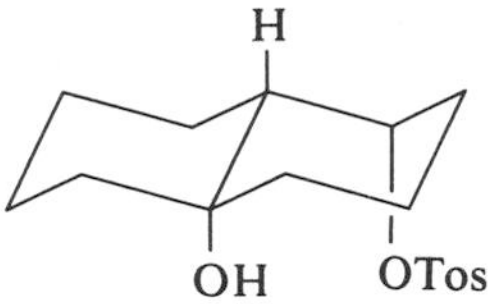

(54)

Tos = H_3C–C_6H_4–SO_2

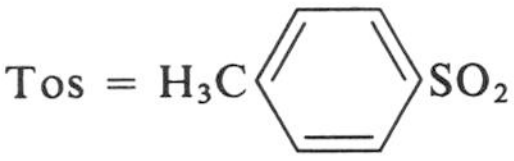

bonds is easily attained, but this is not so in the isomer (54) and this compound on treatment with base gives a mixture of products containing only a very small amount of the *E*-cyclodecenone (2.83). This reaction has been used to prepare a variety of *Z*- and *E*-cyclodecenone and cyclononenone derivatives, notably in the course of the synthesis of caryophyllene (Corey, Mitra and Uda, 1964). Marshall (1969, 1971) has extended the reaction to the preparation of cyclodecadienes by fragmentation of appropriately substituted decalylboranes, themselves easily available by hydroboration of the appropriate alkene (2.84). In a different

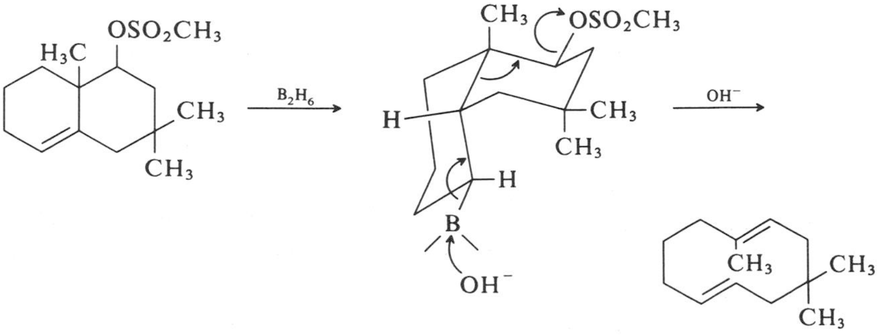

(2.84)

kind of application fragmentation of the bridged, bicyclic compound (55) was used to obtain the product (56) with the correct relative orientation of the substituents in the two rings (Kodama *et al.*, 1980; see also Still and Tsai, 1980) and in an elegant application the twelve-membered macrolide (58) was obtained by a double fragmentation of the acetal-tosylate (57), triggered by loss of carbon dioxide from the carboxylate anion at the melting point (Sternbach *et al.*, 1979). The principle in these examples is the same as before – the breaking bonds are *trans* antiparallel. Thus in the carboxylate anion (57) the central carbon–carbon bridge is antiperiplanar to the equatorial tosylate, as well as to the two 'equatorial'

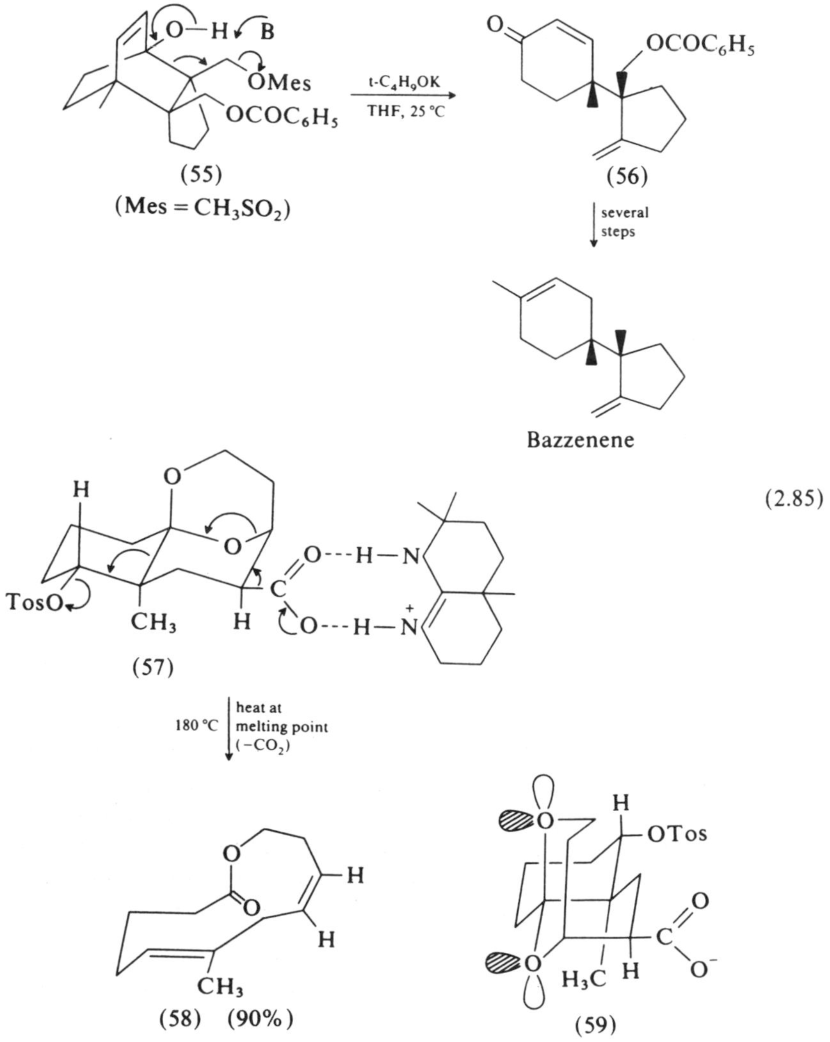

OMes, $OCOC_6H_5$ — $t\text{-}C_4H_9OK$, THF, 25 °C → $OCOC_6H_5$

(55)
(Mes = CH_3SO_2)

(56)

several steps

Bazzenene

(2.85)

(57)

180 °C heat at melting point ($-CO_2$)

(58) (90%)

(59)

electron pair sp^3 axes of the acetal function, and the equatorial carboxylate ion is antiperiplanar to one of the two acetal linkages.

Fragmentation reactions may also be used to prepare acyclic alkenes from cyclic precursors. Control of alkene geometry is thereby transposed to control of relative stereochemistry in cyclic systems. The ketone (62), for example, an intermediate in a synthesis of juvenile hormone, was obtained stereospecifically from the bicyclic compound (60) using two successive fragmentation steps (Zurflüh *et al.*, 1968) (2.86). The geometry of the intermediates (60) and (61) is such as to allow easy fragmentation at each stage.

OH TosO H OH (60) —NaH, THF→ OH O (quantitative) —several steps→ OTos H HO (61) —NaH, THF→ O (80%; 100% *cis, trans*) (62) (2.86)

2.9. Oxidative decarboxylation of carboxylic acids

Another useful method for generating carbon–carbon double bonds is by oxidative decarboxylation of vicinal dicarboxylic acids (2.87)

$$CO_2H,\ CO_2H \longrightarrow C{=}C + 2CO_2 + 2H^+ + 2e \qquad (2.87)$$

(Sheldon and Kochi, 1972). This transformation can be brought about in a number of ways. The familiar procedure with lead tetra-acetate in boiling benzene gives poor and variable yields and is not applicable to bicyclic diacids with nearby double bonds. It has been found, however, that much improved yields are obtained when reaction is effected in presence of oxygen (2.88). An even better procedure is by electrolysis of

$$CO_2H,\ CO_2H \xrightarrow[\text{pyridine, } 67\,^\circ C]{Pb(OCOCH_3)_4,\ O_2} \quad (76\%) \qquad (2.88)$$

the acid in pyridine solution in presence of trimethylamine. This method gives good yields and since reaction proceeds under mild conditions this is an attractive procedure for preparing highly strained unsaturated small and bridged ring compounds. Dewar benzene, for example, is best prepared by this route from bicyclo[2,2,0]hex-2-en-5,6-dicarboxylic acid (2.89). Another useful method which proceeds smoothly under mild

$$\text{bicyclo[2,2,0]hexene-}CO_2II,\ CO_2H \xrightarrow[\text{pyridine, }(CH_3)_3N]{\text{electrolysis}} \text{Dewar benzene}\ (35\%)$$

$$\text{norbornane-}CO_2OC_4H_9\text{-t},\ CO_2OC_4H_9\text{-t} \xrightarrow[C_6H_6,\ 25\ ^\circ C]{h\nu} \text{norbornene}\ (34\%) \qquad (2.89)$$

conditions is the thermal or photolytic decomposition of di-t-butyl peresters, which are readily obtained from the diacid chlorides and t-butyl hydroperoxide. This photolytic process can be used for the synthesis of thermally labile alkenes. The vicinal dicarboxylic acids required as starting materials in these reactions are readily available by Diels–Alder or photosensitised addition of maleic anhydride to dienes or alkenes.

Related to these reactions is the oxidative decarboxylation of monocarboxylic acids with lead tetra-acetate. Under ordinary conditions the course of the reaction depends on the structure of the acid and poor yields, and mixtures of products are often obtained. In the presence of catalytic amounts of copperII acetate, however, preparatively useful yields of alkenes are formed from primary and secondary carboxylic acids, either photolytically or thermally, in boiling benzene. Thus, nonanoic acid is converted into 1-octene in quantitative yield and cyclobutanecarboxylic acid gives cyclobutene in 77 per cent yield (2.90).

$$CH_3(CH_2)_7CO_2H + Pb(OAc)_4 \xrightarrow[C_6H_6,\ h\nu,\ 30\ ^\circ C]{Cu(OCOCH_3)_2} \underset{\text{quantitative}}{CH_3(CH_2)_5CH{=}CH_2}$$

$$\text{cyclobutane-}CO_2H \xrightarrow[C_6H_6,\ \text{reflux}]{Pb(OCOCH_3)_4,\ Cu(OCOCH_3)_2,} \text{cyclobutene}\ (77\%) \qquad (2.90)$$

The following radical chain mechanism is proposed.

$$\begin{aligned} RCO_2Pb^{III} &\longrightarrow R^{\cdot} + CO_2 + Pb^{II} \\ R^{\cdot} + Cu^{II} &\longrightarrow \text{alkene} + H^+ + Cu^{I} \\ Cu^{I} + RCO_2Pb^{IV} &\longrightarrow Cu^{II} + RCO_2Pb^{III},\ \text{etc.} \end{aligned} \qquad (2.91)$$

This reaction features in a useful sequence leading to alkylation-decarboxylation of aromatic acids. The carbanion obtained from the acid with lithium and ammonia (p. 000) is alkylated directly by reaction with an alkyl halide, and the resulting dihydroaromatic carboxylic acid is decarboxylated with lead tetra-acetate, regenerating the aromatic nucleus in which, however, the carboxyl group is now replaced by an alkyl group. Rosefuran, for example, was obtained from 3-methyl-2-furoic acid as shown in (2.92).

(1) Li, NH_3; (2) prenyl bromide; then $Pb(OCOCH_3)_4$, $Cu(OCOCH_3)_2$, C_6H_6, 35 °C (70%) (2.92)

A different mode of oxidative decarboxylation, resulting in the conversion of a carboxylic acid, not into an alkene, but into a ketone, with the loss of one carbon atom, can be effected as illustrated in (2.93). Sulphenylation of the dianion of the acid with dimethyl disulphide and reaction

(1) $2LiN(iso\text{-}C_3H_7)_2$; (2) CH_3SSCH_3; N-chlorosuccinimide, C_2H_5OH, $NaHCO_3$; C_2H_5OH; H_3O^+ (2.93)

(63) → (64)

of the sulphenylated acid with *N*-chlorosuccinimide in ethanol in presence of sodium hydrogen carbonate results in rapid loss of carbon dioxide to give a diacetal, which is readily hydrolysed to the ketone with dilute acid (Trost and Tamaru, 1975). The bicyclo-octenecarboxylic acid (63), for example, is readily converted into the ketone (64) at 0 °C. The acid is obtained from cyclohexadiene and acrylic acid, and using this sequence acrylic acid thus becomes equivalent to ketene in the Diels–Alder reaction (cf. p. 189).

2.10. Alkenes from arylsulphonylhydrazones

Alkenes are readily obtained from aliphatic and alicyclic ketones with at least one α-hydrogen atom by reaction of the toluene-*p*-sulphonylhydrazones with two equivalents of an alkyl-lithium or of lithium diisopropylamide (Kolonko and Shapiro, 1978; Adlington and Barrett, 1983). The reactions proceed under mild conditions without rearrangement of the carbon skeleton and, under the appropriate conditions (see below), lead to the less substituted alkene where there is a choice. Thus, pinacolone gives exclusively 3,3-dimenthyl-1-butene and phenylacetone provides 3-phenylpropene and not the corresponding styrene (2.94).

$$CH_3-C(CH_3)_2-C(CH_3)=NNHTos \xrightarrow[(2)\ H_2O]{(1)\ 2\text{ equivs } C_4H_9Li,\ C_6H_6} CH_3-C(CH_3)_2-CH=CH_2 \quad \text{'quantitative'} \qquad (2.94)$$

$$C_6H_5CH_2C(=NNHTos)CH_3 \xrightarrow[(2)\ D_2O]{(1)\ 2\text{ equivs } C_4H_9Li,\ C_6H_{14}\text{-TMEDA}} C_6H_5CH_2C(D)=CH_2 \quad (74\%)$$

(Tos = $pCH_3C_6H_4SO_2$; TMEDA = tetramethylethylenediamine)

Reaction proceeds by way of the dianion (66) which rapidly fragments to the lithio derivative (67). The best yields of alkene are obtained in tetramethylethylenediamine as solvent, or in the presence of tetramethylethylenediamine, and under these conditions the less substituted alkene is the predominant product from an unsymmetrical ketone. In hydrocarbon or ethereal solvents, in contrast, the position of the double bond in the alkene produced is determined by the stereochemistry of the hydrazone. Abstraction of the second proton from (65) takes place *syn* to the ($ArSO_2\overset{(-)}{N}-$) arylsulphonamido anion (Lipton and Shapiro, 1978).

The vinyl-lithium intermediates (67) may be protonated by the solvent or an added source of protons, or substituted by reaction with another electrophile. In tetramethylethylenediamine, quenching the reaction

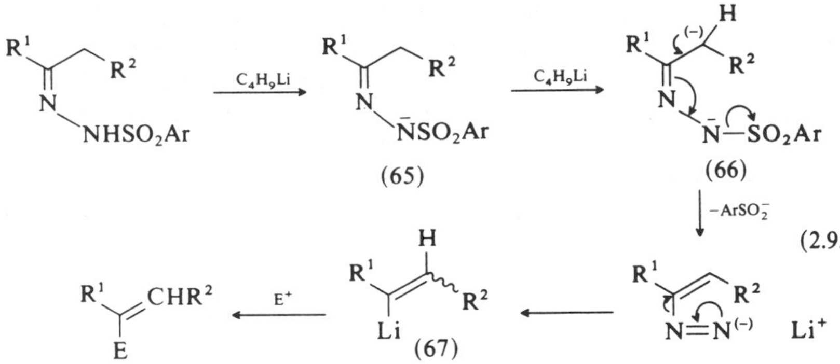

(2.95)

mixture with deuterium oxide provides an excellent method for the preparation of specifically deuterated alkenes (2.94). Thus, 2-methylcyclohexanone was converted into 2-deuterio-3-methylcyclohexene in 98 per cent yield and octan-2-one gave entirely 2-deuterio-1-octene. Protonation of the lithio compound (67) to give a 1,2-disubstituted alkene generally leads to a preponderance of the *Z*-isomer.

The vinyl-lithium intermediates (67) can be trapped with a variety of other electrophiles besides protons giving a range of substituted alkenes including vinyl bromides, acrylic acids and allyl alcohols (2.96). In these transformations it is advantageous to use 2,4,6-triisopropylbenzenesulphonylhydrazones as the source of the vinyl-lithiums rather than the

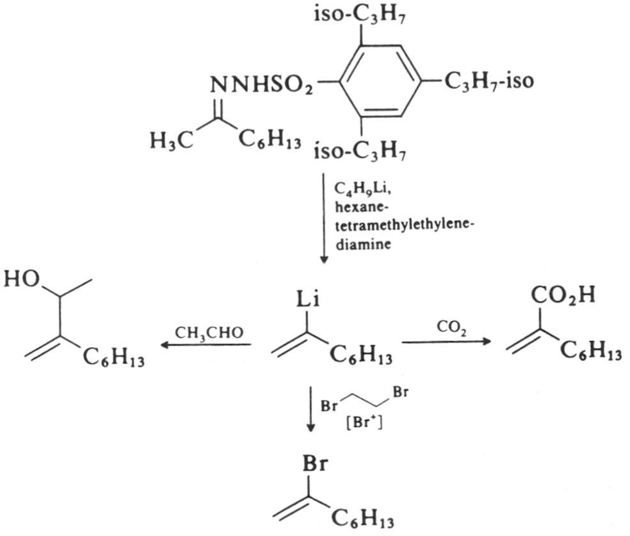

(2.96)

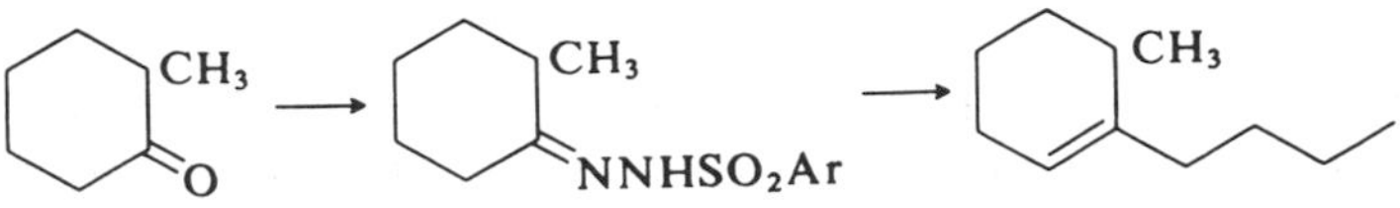

p-toluenesulphonyl compounds. With the latter reagents yields of substituted alkene are reduced because of *ortho* lithiation of the benzenesulphonyl group; this is prevented in the triisopropyl derivative (Chamberlin, Stemke and Bond, 1978; Chamberlin, Liotta and Bond, 1983).

The reduction of carbonyl tosylhydrazones to hydrocarbons with sodium cyanoborohydride in acidic media, or with catecholborane is a mild and selective alternative to Wolf–Kishner deoxygenation (p. 476). With $\alpha\beta$-unsaturated derivatives, alkenes are usually formed in which the double bond has migrated to the position formerly occupied by the carbonyl group. The reaction is most conveniently effected with sodium borohydride in acetic acid, in which the reducing agent is probably $NaBH(OCOCH_3)_3$ (Hutchins and Natale, 1978). Reaction is believed to proceed by way of a diazene intermediate, followed by 1,5-migration of hydride from nitrogen to carbon.

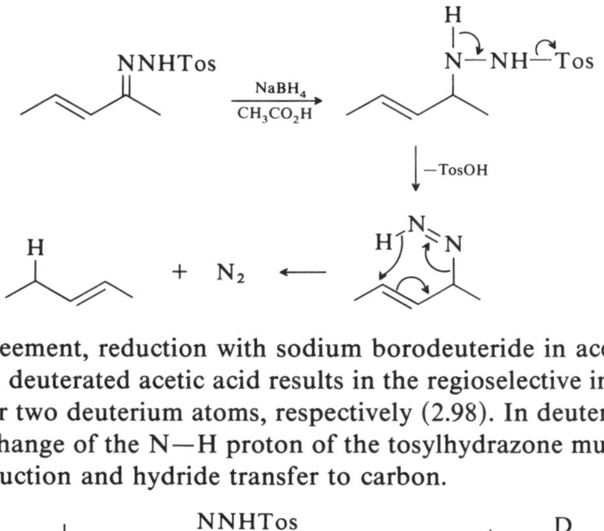

NNHTos → ($NaBH_4$, CH_3CO_2H) → H–N–NH–Tos → ($-$TosOH) → diazene → H + N_2

(2.97)

In agreement, reduction with sodium borodeuteride in acetic acid or carboxyl deuterated acetic acid results in the regioselective introduction of one or two deuterium atoms, respectively (2.98). In deuterated acetic acid exchange of the N—H proton of the tosylhydrazone must be faster than reduction and hydride transfer to carbon.

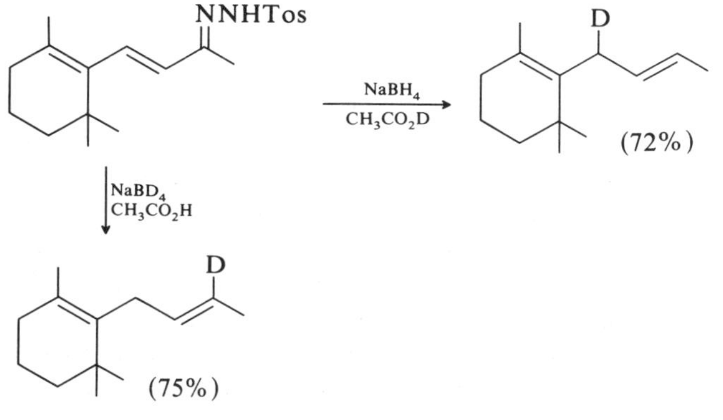

NNHTos → ($NaBH_4$, CH_3CO_2D) → D (72%)

NNHTos → ($NaBD_4$, CH_3CO_2H) → D (75%)

(2.98)

Alkenes can also be made from ketones by way of derived enol ethers or esters, by reductive cleavage or by coupling with organocuprates. The reduction of *N,N,N′,N′*-tetramethylphosphordiamidates with lithium and amines is a general method for cleaving the C—O bond; with enol derivatives it affords good yields of the corresponding alkene (2.99) (Ireland, O'Neil and Tolman, 1983).

$OC(CH_3)_3$

(1) $LiN(iso\text{-}C_3H_7)_2$
(2) $[(CH_3)_2N]_2POCl$
(3) Li, NH_3, t-C_4H_9OH

$OC(CH_3)_3$

(2.99)

TfO C_4H_9 / C_4H_9 H — $(CH_3)_2CuLi$ → H_3C C_4H_9 / C_4H_9 H

($Tf = CF_3SO_2$)

The oxygen atom of the enol can be replaced by an alkyl group rather than a hydrogen atom by reaction of the enol diphenylphosphate or, better, the enol trifluoromethanesulphonate (triflate) with a lithium dialkylcuprate (McMurry and Scott, 1980). The reaction is highly stereoselective. (*Z*)-5-Trifluoromethanesulphonyloxy-5-decene, for example, gave only (*E*)-5-methyl-5-decene on reaction with lithium dimethylcuprate.

2.11. Stereospecific synthesis from 1,2-diols

Several approaches to the regiospecific and stereospecific generation of double bonds from 1,2-diols have been devised, all proceeding by decomposition of an intermediate of the type (68). One of the best methods uses the cyclic thionocarbonates (70) which are readily obtained from the diol with thiophosgene. As originally conceived decomposition of the thionocarbonates to the alkenes required heating with triethylphosphite but in a recent modification the conversion is effected at 25–40 °C by reaction of the thionocarbonate with 1,3-dimethyl-2-phenyl-1,3,2-diazophospholidine (Corey and Hopkins, 1982), possibly by a concerted process or by way of an intermediate carbene (69). Under these mild conditions the reaction is applicable to complex and sensitive molecules containing a variety of functional groups. Thus, the 3,5-acetonide of the complex macrolide erythronolide A was converted into the 11,12-deoxy compound in 70 per cent yield, and the diol (71) gave (72) in 75 per cent yield.

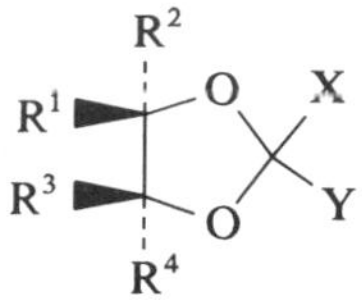

(68) 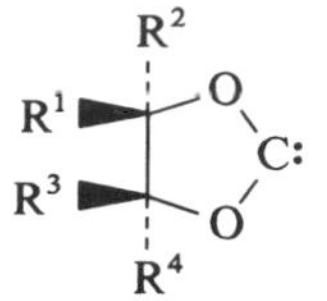 (69)

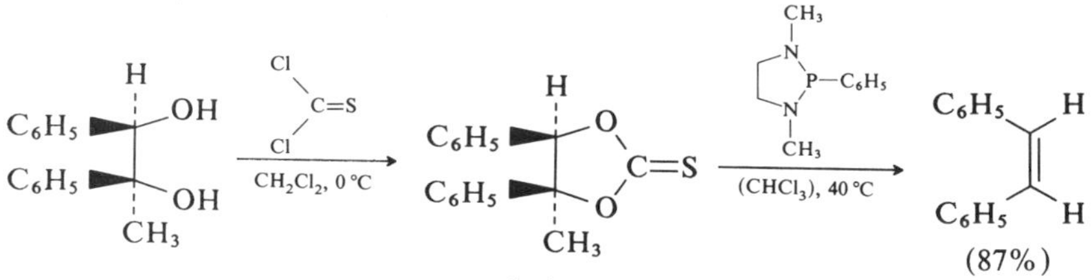

$Cl_2C{=}S$, CH_2Cl_2, 0 °C → (70) → (1,3-dimethyl-2-phenyl-1,3,2-diazaphospholidine, $(CHCl_3)$, 40 °C) (87%)

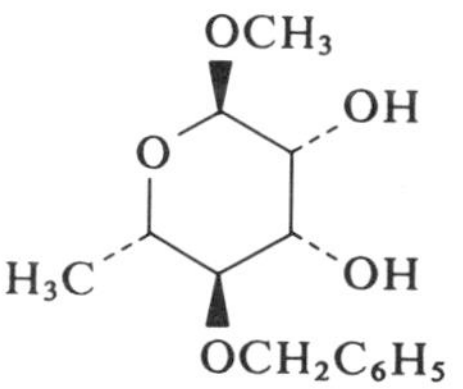

(71) 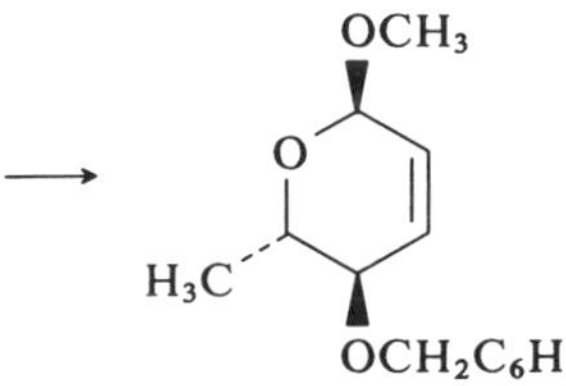 → (72) (2.100)

The reactions proceed with complete stereospecificity by a *syn* elimination pathway, allowing the stereospecific synthesis of strained cycloalkenes and, together with the *trans* perhydroxylation reaction (p. 364) providing a general and unambiguous method for the interconversion of *Z*- and *E*-alkenes. Thus, *meso*-1,4-diphenylbutan-2,3-diol was converted into (*Z*)-1,4-diphenyl-2-butene in 96 per cent yield while the *dl*-compound gave the *E*-alkene, and (*Z*)-cyclooctene was converted into the *E*-isomer as shown in (2.101). In an alternative procedure the benzylidene derivative of the diol is treated with butyl-lithium, but this method, although valuable, is of limited application because of the susceptibility of many functional groups to butyl-lithium, and it is said to be suitable for only lightly substituted alkenes (Hines *et al.*, 1973). Other good stereospecific routes proceed through acetals formed from the diol with ethyl orthoformate (Hiyama and Nozaki, 1973) or *N,N*-dimethyl-

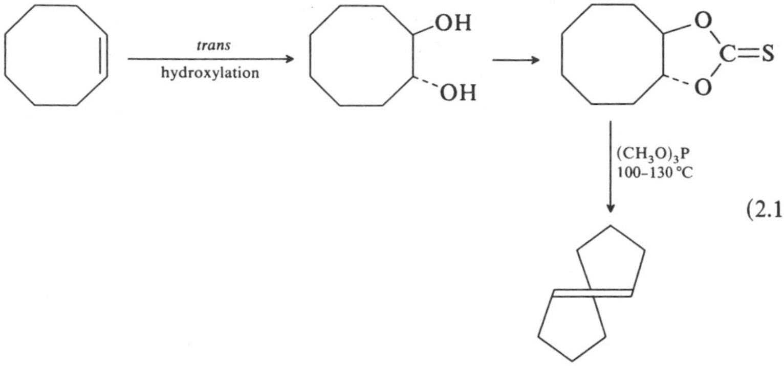

trans hydroxylation

$(CH_3O)_3P$ 100–130 °C

(2.101)

(75%; 99% *E*)

formamide dimethyl acetal (Hanessian, Bargiotti and LaRue, 1978). 1,2-Diols have also been converted into alkenes by reaction with active titanium metal (see p. 181) or with tungsten salts, K_2WX_6, but neither method is stereospecific.

2.12 Claisen rearrangement of allyl vinyl ethers

The Claisen rearrangement of allyl vinyl ethers provides an excellent stereoselective route to $\gamma\delta$-unsaturated carbonyl compounds (aldehydes, ketones, esters and amides) from allyl alcohols, and has formed an important step in the synthesis of a number of natural products (Rhoads and Raulins, 1975; Ziegler, 1977; Bartlett, 1980). The reaction involves a [3,3]-sigmatropic rearrangement, and takes place by a concerted mechanism through a cyclic six-membered transition state (2.102). In the course of the reaction both a new carbon–carbon double bond and a new carbon–carbon single bond are formed. Much of its value in

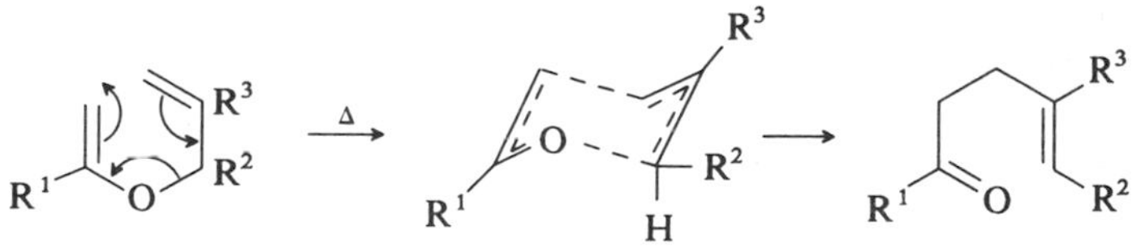

(73)

$R^1 = H$, alkyl, OR, NR_2, $OSi(CH_3)_3$ [2.102)

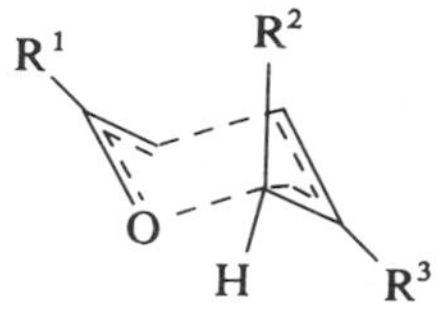

(74)

synthesis stems from the fact that it is highly stereoselective, particularly when $R^1 \neq H$, leading predominantly to the *E*-configuration of the newly formed double bond, and to the controlled stereochemical disposition of substituents on the single bond (see below). A chair conformation (73) is preferred for the cyclic transition state with the substituent R^2 in the equatorial conformation, and the high stereoselectivity favouring the *E*-double bond is a consequence of non-bonded interaction between the substituents R^1 and R^2 in the alternative transition state (74) which would give the *Z*-alkene.

The allyl vinyl ethers used in the reaction are prepared directly from allyl alcohols by acid-catalysed ether exchange. Thus, reaction of an allyl alcohol with ethyl vinyl ether and pyrolysis of the resulting vinyl ether affords a $\gamma\delta$-unsaturated aldehyde (2.103); with appropriately substituted vinyl ethers, $\gamma\delta$-unsaturated ketones are obtained. Thus, reaction of the allylic alcohol (75) with 2-methoxy-3-methyl-1,3-butadiene gave the *E*-$\gamma\delta$-unsaturated ketone (76) directly without isolation of the intermediate allyl vinyl ether. None of the *Z*-isomer was detected (cf. Daub, *et al.* 1982).

CH_2OH $\xrightarrow[\text{reflux}\ Hg(OCOCH_3)_2]{C_2H_5OCH{=}CH_2}$ $\xrightarrow{320\ °C}$ CHO (85%)

HO (75) + OCH_3 $\xrightarrow[RCO_2H]{110\ °C}$ [O] $\longrightarrow$ O (76) (67%; 100% *E*)

(2.103)

$\gamma\delta$-Unsaturated esters are obtained by heating the allylic alcohol with an orthoester in the presence of a weak acid (propionic acid is often used), as in the following synthesis of the insect pheromone (78) from the allyl alcohol (77) and methyl orthoacetate (2.104). A mixed orthoester is first formed and loses methanol to form a ketene acetal which rearranges to the $\gamma\delta$-unsaturated ester. $\gamma\delta$-Unsaturated *N,N*-dimethyl amides are similarly obtained by heating allyl alcohols with 1-dimethylamino-1-methoxyethene or its precursor *N,N*-dimethylacetamide dimethyl acetal (Felix *et al.*, 1969).

A disadvantage of the above procedures is the relatively high temperatures needed and the fact that, requiring acid, they are unsuitable

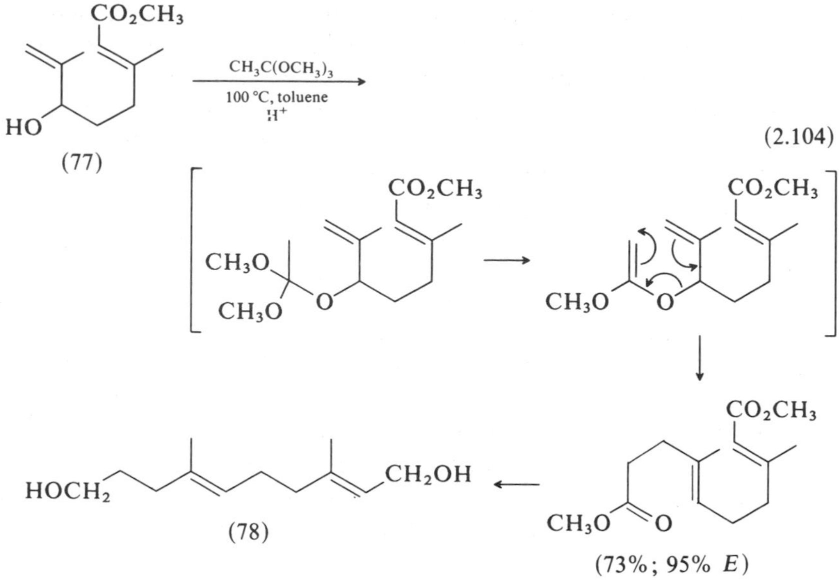

CO_2CH_3

HO

(77)

$\xrightarrow[\text{100 °C, toluene, } H^+]{CH_3C(OCH_3)_3}$

(2.104)

CO_2CH_3 CH_3O CH_3O O $\longrightarrow$ CO_2CH_3 CH_3O O

$HOCH_2$ CH_2OH $\longleftarrow$ CO_2CH_3 CH_3O O

(78)

(73%; 95% *E*)

for compounds containing acid-sensitive functional groups. A valuable alternative route to $\gamma\delta$-unsaturated acids which overcomes these difficulties proceeds from carboxylic esters of allyl alcohols by Claisen rearrangement of the lithium enolates or, better, the derived trimethylsilyl or t-butyldimethylsilyl enol ethers. High yields of the unsaturated acids are obtained (after hydrolysis of the trimethylsilyl esters) under very mild neutral or weakly alkaline conditions, again with predominant formation of the *E*-double bond (Ireland, Mueller and Willard, 1976). Thus the allylic acetate (79) was converted into the $\gamma\delta$-unsaturated acid (80) in 80 per cent yield with almost complete *E*-selectivity.

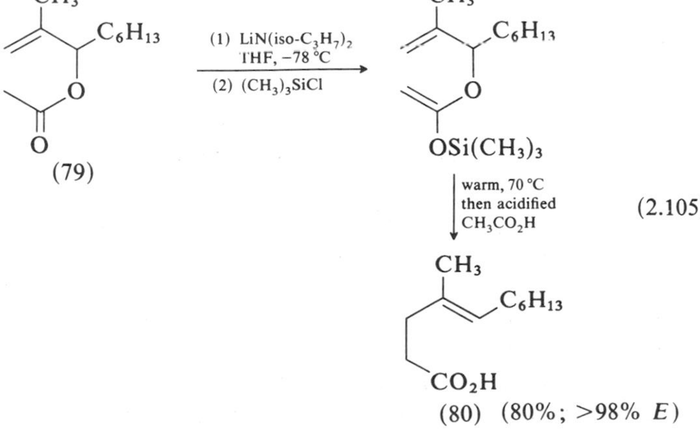

CH_3 C_6H_{13} O O

(79)

(1) $LiN(iso\text{-}C_3H_7)_2$ THF, −78 °C
(2) $(CH_3)_3SiCl$

CH_3 C_6H_{13} O $OSi(CH_3)_3$

warm, 70 °C
then acidified
CH_3CO_2H

(2.105)

CH_3 C_6H_{13} CO_2H

(80) (80%; >98% *E*)

Another valuable feature of the Claisen rearrangement is the stereochemical control it allows of substituents on the newly formed *single* bond. There may be up to three chiral centres involved in the reaction, that in the starting material (81) and those in the product (82). Since the reaction proceeds suprafacially with regard to the allylic fragment, the configuration at the allylic centre C-3 of the product is directly related to that of the starting material, as shown in (2.106). Either

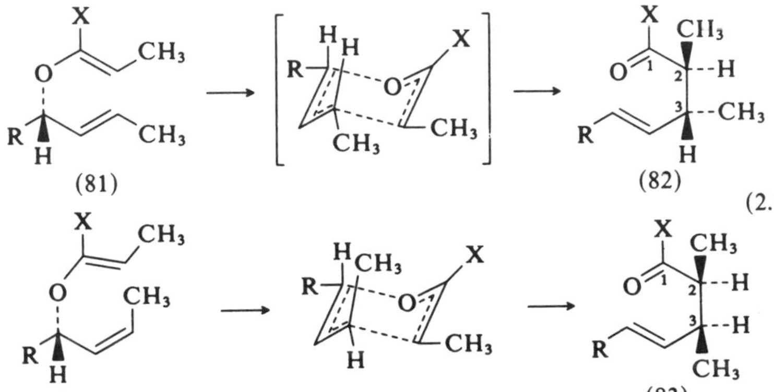

(81) (82) (83) (2.106)

configuration of the new chiral centre at C-3 may be obtained by changing the configuration of the allylic double bond in the starting material. A chiral centre is destroyed in the reaction and a new one is formed in a controlled way at the allylic position in the product. The reaction can thus be used to transmit chirality along a carbon chain. For example, in a key step in a synthesis of (+)-15(*S*)-prostaglandin A2 the optically active allylic alcohol (84) was converted into the optically active γδ-unsaturated ester (85) by the orthoester procedure (Stork and Raucher, 1976).

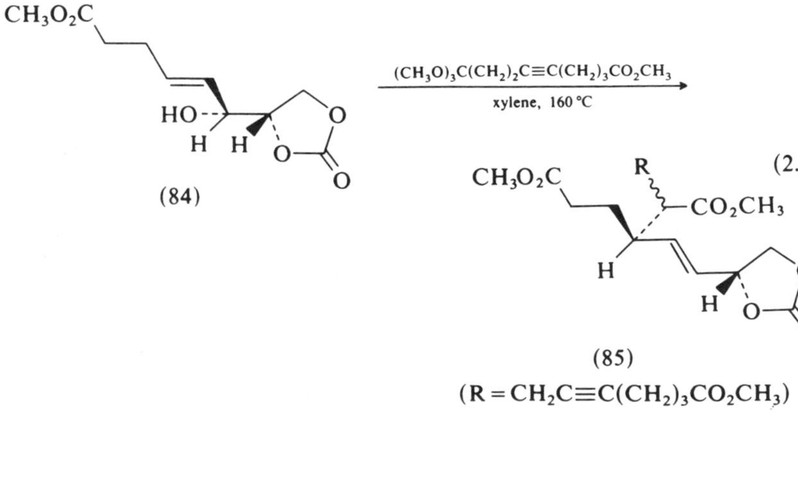

$(CH_3O)_3C(CH_2)_2C{\equiv}C(CH_2)_3CO_2CH_3$, xylene, 160 °C

(84) (85) (2.107)

$(R = CH_2C{\equiv}C(CH_2)_3CO_2CH_3)$

The relationship of the chiral centre at C-2 to that at C-3 in the products (82) and (83) is established by the chair-like transition state and depends on the geometry of the double bonds in the starting material, as shown in (2.108) for the isomeric 2-butenyl-1-propenyl ethers. This feature of the reaction, combined with the suprafacial rearrangement, allows the configuration of both chiral centres at the newly formed single bond in the product to be related to that of the allylic hydroxyl group in the starting material in a controlled manner, by selecting the appropriate geometry of the double bonds.

143–160 °C, 96% stereoselective; 160–190 °C, 95% stereoselective

145–165 °C, 96% stereoselective; 145–175 °C, 95% stereoselective

(2.108)

One of the best methods for controlling the geometry of the enol ether double bond in the Claisen rearrangement is by Ireland's ester–enolate procedure described above. It is found that conversion of the allylic ester (86) to the corresponding silyl ketene acetal with lithium diisopropylamide in tetrahydrofuran followed by silylation with chlorotrimethylsilane leads preferentially to the *E*-ketene acetal (87) and thence to the acid (88). Reaction with lithium diisopropylamide in tetrahydrofuran

(86) → (1) $LiN(iso\text{-}C_3H_7)_2$, THF, −78 °C; (2) $t\text{-}C_4H_9(CH_3)_2SiCl$ → (87) → (88)

(86) → (1) $LiN(iso\text{-}C_3H_7)_2$, THF-hexamethyl phosphoric triamide, −78 °C; (2) $t\text{-}C_4H_9(CH_3)_2SiCl$ → (89) → (90)

$[X = t\text{-}C_4H_9(CH_3)_2Si]$ (2.109)

containing 23 per cent of the polar co-ordinating solvent hexamethylphosphoric triamide followed by silylation, leads on the other hand, to the *Z*-ketene acetal (89) and the acid (90). Thus, *E*-crotyl propanoate (91) leads predominantly to the *erythro* acid (92) when enolisation is carried out in tetrahydrofuran, but to the *threo* acid (93) when the solvent tetrahydrofuran contains hexamethylphosphoric triamide. With *Z*-crotyl acetate (94) the results are reversed. In the less co-ordinating tetrahydrofuran alone at −78 °C association of lithium with the ester carbonyl

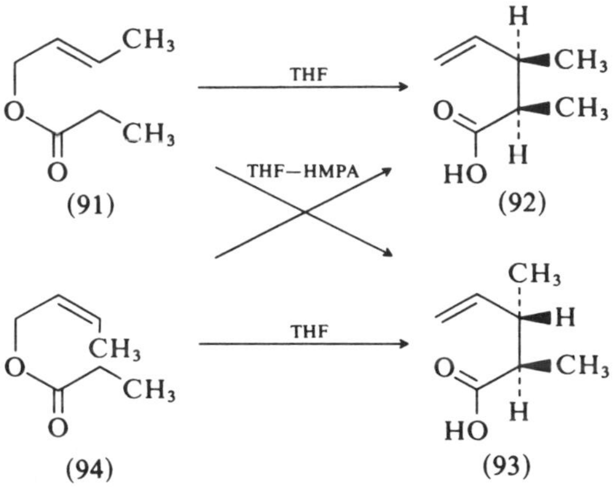

(91) THF → (92); (91) THF–HMPA → (93); (94) THF → (93); (94) THF–HMPA → (92) (2.110)

group is important and the carbonyl group has a larger steric effect than the ether oxygen. Non-bonded interactions favour transition state (95*a*) and formation of the *Z*-enolate. In the presence of hexamethylphosphoric triamide, co-ordination is less important, the ether oxygen becomes the more sterically demanding one, and the *E*-enolate is favoured.

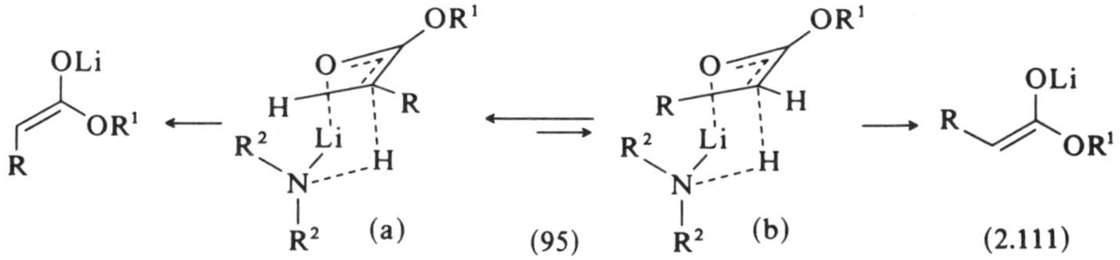

(95) (a) ⇌ (b) (2.111)

The ester-enolate (ketene acetal) variation of the Claisen rearrangement has been employed in the stereocontrolled synthesis of a number of natural products. Thus, in a synthesis of the irregular monoterpene methyl santolinate (98) the propionate ester (96) gave mainly the *Z*-ketene acetal (97) by reaction with lithium isopropylcyclohexylamide in tetrahydrofuran followed by t-butyldimethylsilyl chloride. Rearrangement at 65 °C, hydrolysis and esterification with diazomethane then gave methyl santolinate with high stereoselectivity (Boyd, Epstein and Frater, 1976).

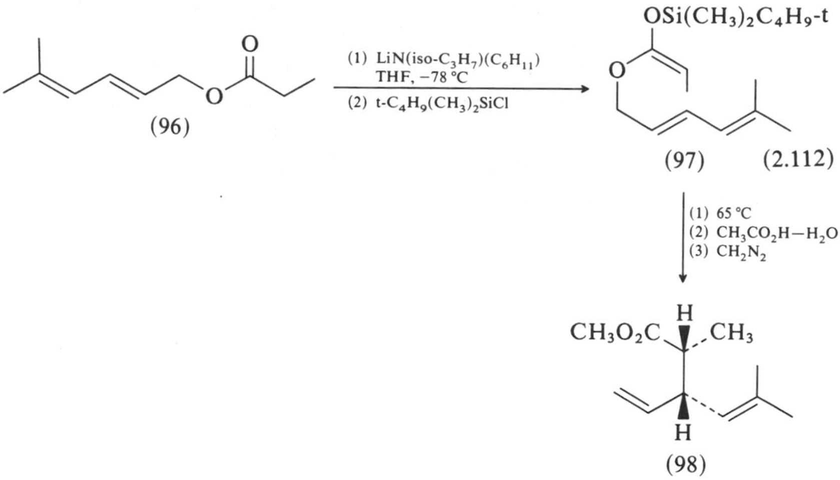

(96) (97) (2.112)

(1) $LiN(iso\text{-}C_3H_7)(C_6H_{11})$, THF, −78 °C; (2) $t\text{-}C_4H_9(CH_3)_2SiCl$

(1) 65 °C; (2) $CH_3CO_2H{-}H_2O$; (3) CH_2N_2

(98)

In another interesting synthesis the *Z*-crotyl glycine ester (99) was converted into the $\gamma\delta$-unsaturated amino-acid derivative (101a) and thence into the lactone hydrochloride (102), a product isolated after hydrolysis of α- and β-amanitin, toxic cyclic peptides from the Green Death Cap Toadstool *Amanita phalloides* (Bartlett, Tanzella and Barstow, 1982). Assuming a chair-like transition state (103) the stereochemical course of

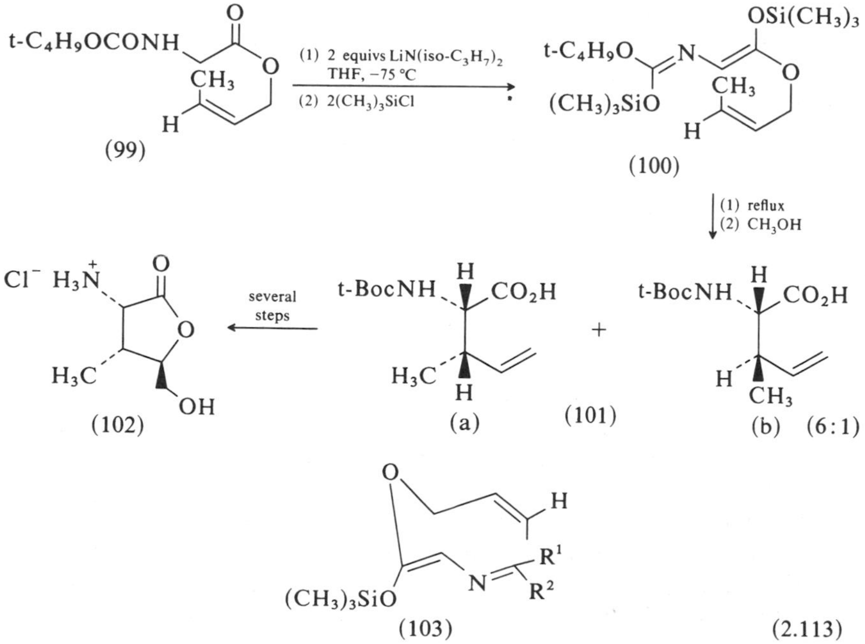

(99) (100)

(1) 2 equivs $LiN(iso\text{-}C_3H_7)_2$, THF, −75 °C; (2) $2(CH_3)_3SiCl$

(1) reflux; (2) CH_3OH

(102) ← several steps ← (a) (101) + (b) (6:1)

(103) (2.113)

the rearrangement indicates that the intermediate ketene acetal has mainly the *E*-configuration (100) and not the expected *Z*-configuration. This may be general for esters of acids carrying a heterosubstituent on the α-carbon, such as α-amino and α-hydroxy acids (Bartlett and Barstow, 1982).

While it is true that rearrangement of vinyl ethers of acyclic allylic alcohols normally takes place through a chair-like transition state, a number of examples are known, involving substrates in which the double bond of the vinyl ether or the allyl alcohol forms part of a ring, which clearly take place through a boat-like transition state. In these reactions the alternative chair form would involve considerable strain. A case in point is the useful reaction between allylic alcohols and cyclic orthoesters which takes place with Claisen rearrangement to give lactones containing the inverted allyl group as an α-substituent. Reaction of the cyclic orthoester (104) with the allyl alcohol (105), for example, leads to the lactone (106) in 80 per cent yield with complete stereoselectivity. In this

C_2H_5O OC_2H_5 (104) + $C_6H_5CO_2$ OH (105) $\xrightarrow[\text{reflux, 1,2-dimethylbenzene,}]{\text{catalytic } CH_3CH_2CO_2H}$ $C_6H_5CO_2$ H O O (106)

(2.114)

reaction the configuration at the new chiral centre of the cyclohexene ring is controlled directly by the configuration of the original allylic hydroxyl group and that at the new centre of the lactone ring by the conformation (chair or boat) of the transition state. The stereochemistry of the product shows that the reaction must have proceeded entirely through a boat-like transition state (Chapleo *et al.*, 1977). In a related series of reactions vinyl lactones are converted into cycloalkenes by Claisen rearrangement of the derived cyclic enol ethers; in these reactions also the stereochemistry of the products is indicative of a boat-like transition state (Danishefsky, Funk and Kerwin, 1980) (2.115).

The Claisen rearrangement of allyl vinyl ethers may be regarded as a particular case of the well-known Cope [3,3]-sigmatropic rearrangement of 1,5-hexadienes (2.116) (Rhoads and Raulins, 1975; Wehrli *et al.*, 1976). A disadvantage of the Cope rearrangement from a preparative standpoint is the high temperature often required. It has now been found that the reaction can be performed at room temperature in many cases, in the

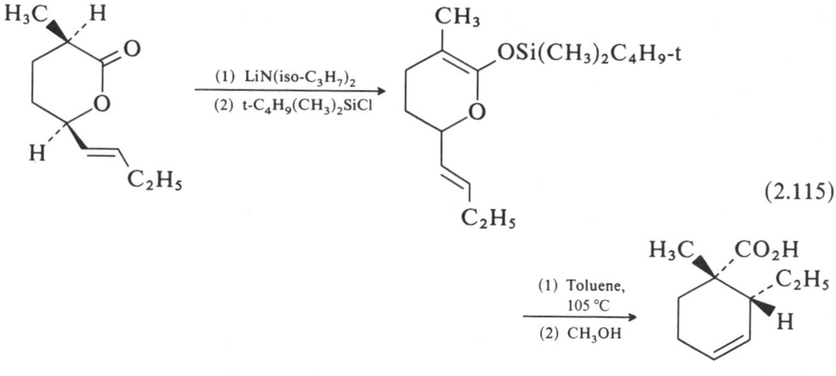

(1) $LiN(iso\text{-}C_3H_7)_2$
(2) $t\text{-}C_4H_9(CH_3)_2SiCl$

CH_3 $OSi(CH_3)_2C_4H_9\text{-}t$ C_2H_5

(2.115)

(1) Toluene, 105 °C
(2) CH_3OH

H_3C CO_2H C_2H_5 H

(71%)

(2.116)

presence of a catalytic amount of palladium chloride *bis*(benzonitrile) complex (Overman and Renaldo, 1983). For example, treatment of 2-methyl-3-phenyl-1,5-hexadiene (107) with catalytic $PdCl_2(PhCN)_2$ in tetrahydrofuran at room temperature for 24 hours gave the dienes (108) and (109) in 87 per cent yield in a ratio of 97:3; in benzene, reaction was complete in one hour. In contrast, thermal Cope rearrangement of (107) requires elevated temperatures and is less stereoselective.

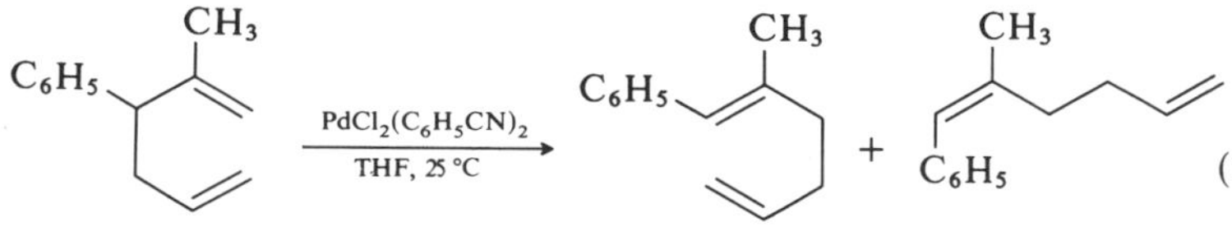

$PdCl_2(C_6H_5CN)_2$
THF, 25 °C

(107) (108) (109) (2.117)

The so-called oxy-Cope rearrangement of 1,5-hexadienes with a hydroxyl substituent at C-3 provides a useful route to $\delta\varepsilon$-unsaturated aldehydes and ketones; with a hydroxyl substituent at C-3 and C-4 of the diene 1,6-dicarbonyl compounds are produced [2.118].

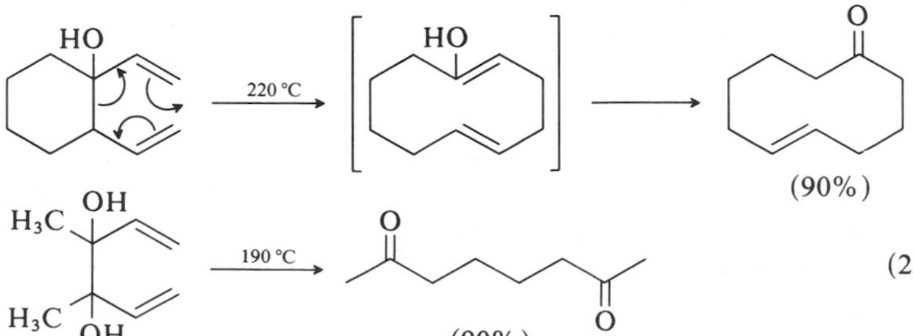

220 °C

(90%)

190 °C

(90%)

(2.118)

Very large increases in rate in these rearrangements are obtained by using the potassium alkoxides, in which the metal ion is free of the oxyanion, rather than the hydroxy compounds themselves (Evans and Golob, 1975; Evans and Nelson, 1980). The rearrangements proceed in a concerted fashion via chair-like transition states and are highly stereoselective. Thus, while the diene (110) rearranged smoothly to the unsaturated ketone (111), the isomer (112) was unaffected, and in a key step of a synthesis of the sesquiterpene (±)-juvabione the diene (113) gave the cyclohexanone stereoselectively in 77 per cent yield (2.119).

CH_3O — OH (110) —[KH, hexamethyl phosphoric triamide, 10 °C]→ CH_3O — OK (111); CH_3O — OH (112)

HO — OCH_3 — CH_3 (113) —[KH, diglyme, 110 °C]→ O — OCH_3 (77%) (2.119)

The reaction has been frequently employed in the synthesis of natural products, for example of germacrane sesquiterpenes, using transformations similar to that in (2.120) (Still, 1977, 1979).

OH —[KH, THF, 25 °C]→ O (75%) (2.120)

The alkaline conditions in the reactions described above are unsuitable for use with alkali-sensitive compounds, but it has recently been found that the rearrangement can frequently be effected under neutral conditions, at room temperature, in the presence of a catalytic amount of palladium chloride *bis*(benzonitrile) complex (Bluthe, Malacria and Gore, 1983), or with mercury *bis*(trifluoroacetate) in aqueous tetrahydrofuran with subsequent demercuration with sodium borohydride (Bluthe, Malacria and Gore, 1982) or lithium trifluoroacetate (Bluthe, Malacria and Gore, 1984).

The thio Claisen rearrangement of allyl vinyl sulphides takes place easily, giving products which can be hydrolysed to $\gamma\delta$-unsaturated carbonyl compounds. The thio reaction has an added degree of flexibility in that alkylation can be effected at the carbon atom adjacent to the sulphur before rearrangement, as in the preparation of 4-tridecenal from allyl vinyl sulphide shown in (2.121).

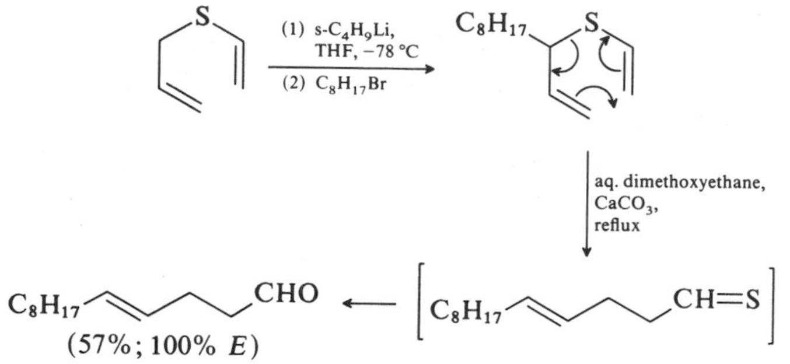

(1) s-C_4H_9Li, THF, −78 °C
(2) $C_8H_{17}Br$

C_8H_{17} S

aq. dimethoxyethane, $CaCO_3$, reflux

C_8H_{17} CHO ← [C_8H_{17} CH=S]

(57%; 100% *E*)

(2.121)

Related to the rearrangement of allyl vinyl sulphides is the [2,3]sigmatropic rearrangement of allylsulphonium ylids, illustrated in the synthesis of γ-cyclocitral (2.122). α-Allylthio carbanions and α-allylthio

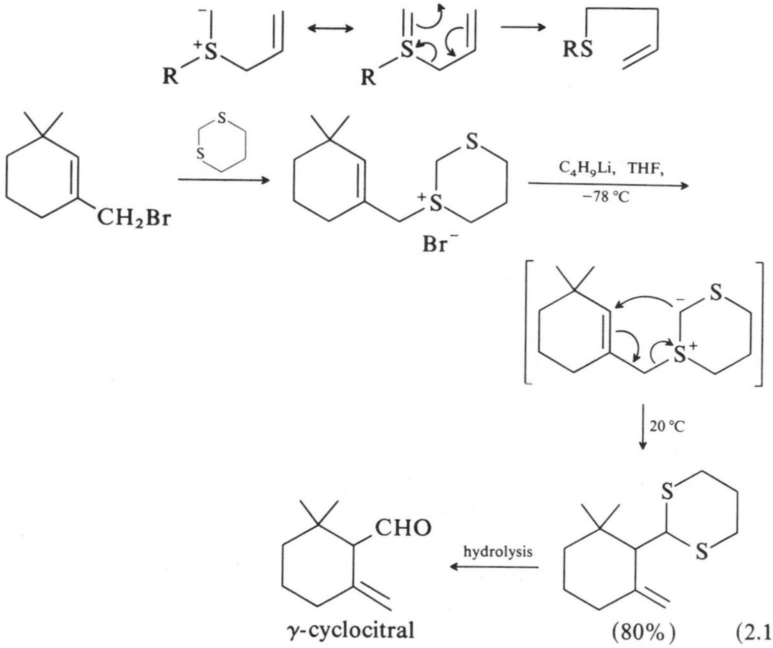

R $^+$S ↔ R S → RS

CH_2Br → Br$^-$ C_4H_9Li, THF, −78 °C

20 °C

CHO ← hydrolysis

γ-cyclocitral (80%) (2.122)

carbenes will also rearrange. The essential requirement is the availability of six electrons for the electrocyclic reaction (Kreiser and Wurziger, 1975; Baldwin and Walker, 1972).

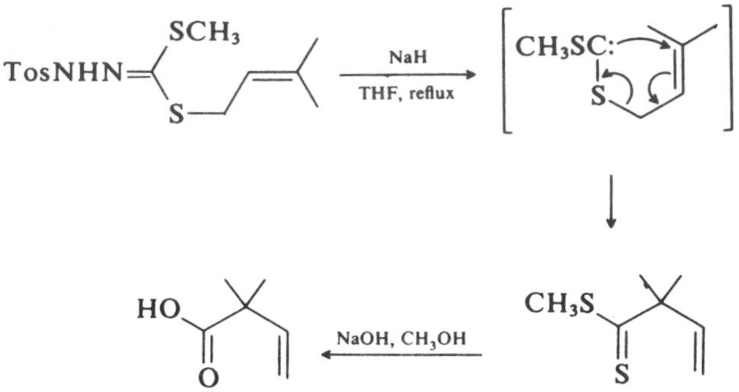

NaH, THF, reflux; NaOH, CH_3OH

(quantitative)

(2.123)

There is also an aza-Cope rearrangement proceeding from 4-butenyliminium ions (2.124). This useful carbon–carbon bond-forming reaction occurs under mild conditions but hitherto it has not been much employed in synthesis because of its reversibility. It has recently been found,

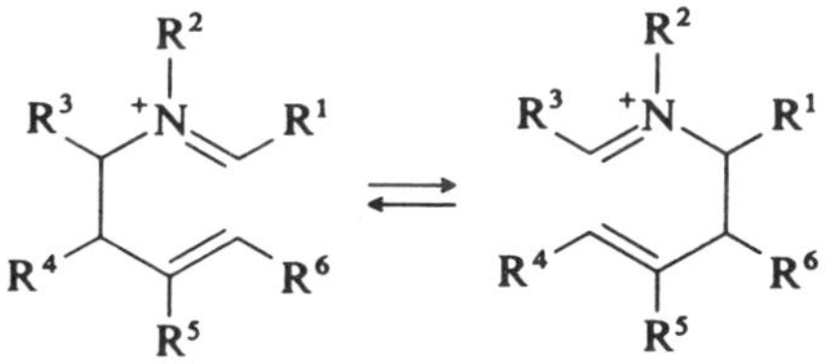

(114)

(2.124)

however, that with a properly placed hydroxyl or alkyl substituent at C-3 of the butenyl group (114, $R^4 = OH$) the reaction can be driven in the forward direction by capture of the rearranged iminium salt in an intramolecular Mannich reaction (Overman, Kakimoto, Okazaki and Meier, 1983), providing an excellent synthesis of 2-substituted pyrrolidine derivatives. Thus, pyridine 3-aldehyde heated in benzene solution with the *N*-methyl-2-hydroxybutenamine (115) and an acid catalyst (camphor sulphonic acid) gave the acetylnicotine derivative (118) directly in 84 per cent yield. The initial iminium salt (116) rearranges to (117) which is irreversibly trapped in an intramolecular Mannich reaction to give (118).

When the hydroxyl and amino groups are neighbouring substituents on a ring an interesting conversion takes place to give a bicyclic pyrrolidine derivative in which the original ring is expanded by one carbon atom (2.126). If the hydroxyl and amino groups are *cis* to each other in the starting material the reaction takes place with high stereoselectivity to give the *cis*-fused bicyclic compound. The reaction with the *trans* hydroxy amine is not so selective.

Both reactions have been exploited in the synthesis of natural products. Thus, in a key step in a synthesis of perhydrogephyrotoxin, the butanylamine (119) was converted into the tricyclic (120) in 79 per cent yield (Overman and Fukaya, 1980) and in model studies for a synthesis of *Aspidosperma* alkaloids the tricyclic ketone (121) was obtained in 83 per cent yield (Overman *et al.*, 1981*a*).

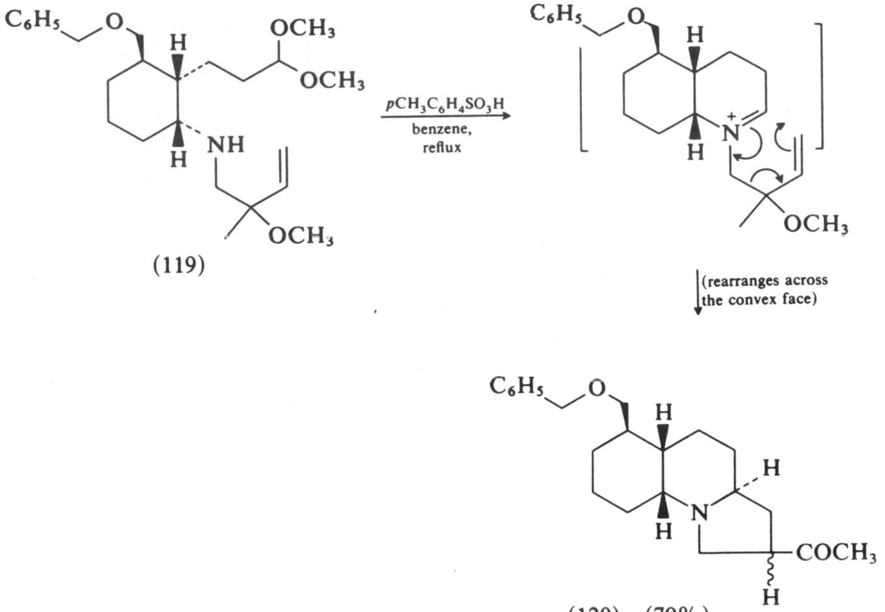

C_6H_5 O H OCH_3 OCH_3 NH H OCH_3 (119)

$pCH_3C_6H_4SO_3H$, benzene, reflux

C_6H_5 O H $\overset{+}{N}$ H OCH_3

(rearranges across the convex face)

C_6H_5 O H H N H $COCH_3$ H

(120) (79%)

(2.127)

R N H O O C_6H_5 → HN C_6H_5 HO O C_6H_5

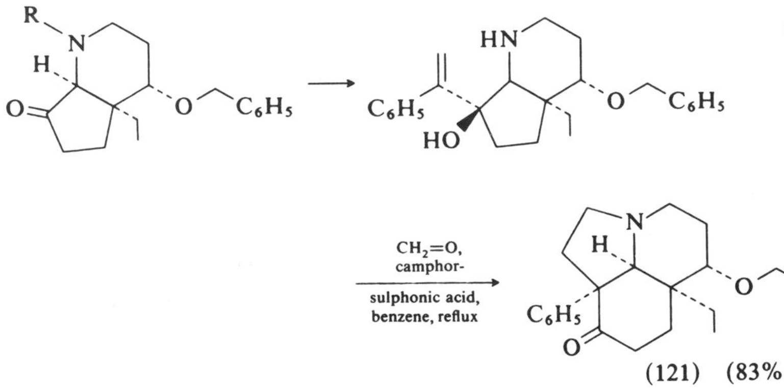

$CH_2{=}O$, camphor-sulphonic acid, benzene, reflux

N H O C_6H_5 C_6H_5 O

(121) (83%)

2.13. Reductive dimerisation of carbonyl compounds

Alkenes can be obtained from aldehydes and ketones on reductive dimerisation by treatment with a reagent prepared from titanium(III) chloride and lithium aluminium hydride or zinc copper couple, or with a species of active titanium metal formed by reduction of titanium(III) chloride with potassium or lithium (McMurry, 1983). The reaction is of wide application but in intermolecular reactions generally affords a mixture of the *E*- and *Z*-alkenes (2.128).

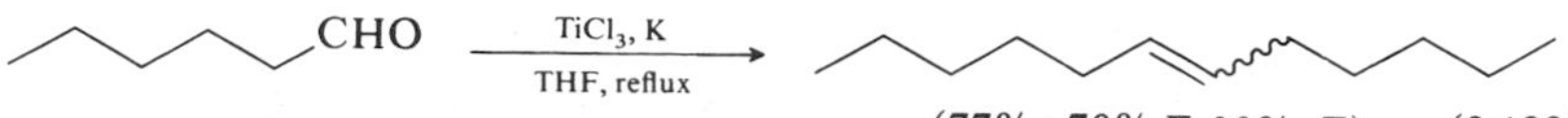

CHO $\xrightarrow[\text{THF, reflux}]{TiCl_3, K}$

(77%; 70% *E*, 30% *Z*) (2.128)

The reaction takes place in two steps on the surface of the active titanium particles. The first stage, leading to the formation of a new carbon–carbon bond is simply a pinacol reduction. The titanium reagent donates an electron to the carbonyl compound generating a ketyl which dimerises to give the pinacol. The intermediacy of pinacols in the reaction is supported by the fact that pinacols are smoothly converted into alkenes on treatment with the titanium reagent. In the second stage de-oxygenation is effected by way of a species formed by co-ordination of the pinacol to the surface of a titanium particle. Cleavage of the two carbon–oxygen bonds then occurs, yielding the alkene and an oxidised titanium surface (2.129).

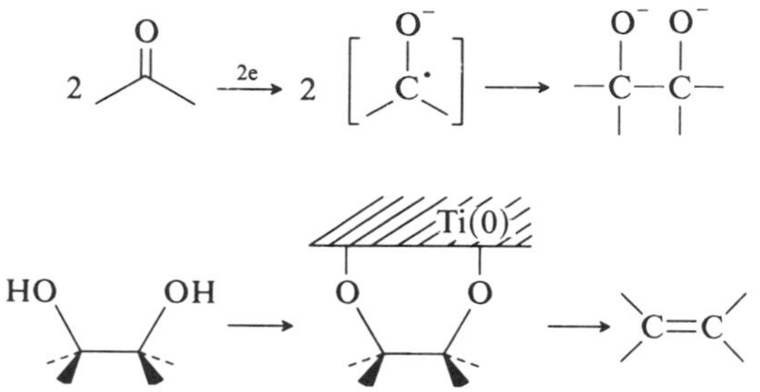

2 → 2e → 2 [ ] → —C—C—

(2.129)

Ti(0)

HO OH → O O → C=C

Mixed coupling reactions, using two different carbonyl compounds, can be effected but they generally lead to mixtures of products and are of limited use in synthesis. Intramolecular reactions with dicarbonyl compounds, on the other hand, provide a good route to cyclic alkenes; cycloalkenes containing four to sixteen carbon atoms have been prepared in good yield. The keto-aldehyde (122), for example, gave the fifteen-membered cyclic diterpene flexibilene, (123), in 52 per cent yield.

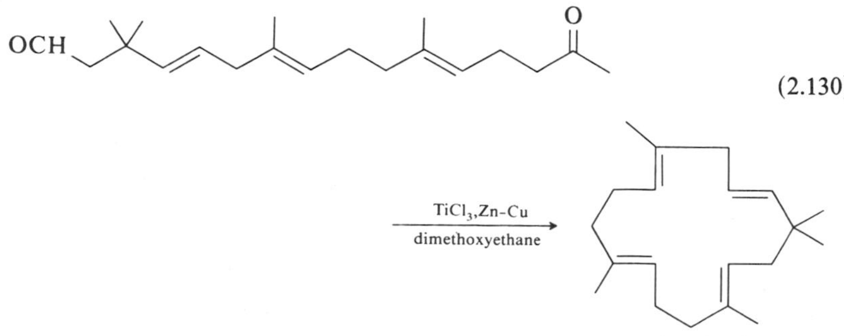

OCH O

(2.130)

$\xrightarrow[\text{dimethoxyethane}]{TiCl_3, Zn\text{-}Cu}$

(123) (52%)

Keto-esters also can be cyclised to give cyclic ketones; even medium ring ketones can be made this way. Isocaryophyllene was readily synthesised by cyclisation of the keto-ester (124) to (125) followed by Wittig methylenation. The unusual $E \rightarrow Z$-isomerisation of the double bond in this conversion is believed to be induced by strain in the cyclic intermediate in which the two oxygen atoms are bound to titanium.

$$\text{(124)} \xrightarrow[\text{dimethoxyethane, reflux}]{TiCl_3-LiAlH_4} \text{(125)} \qquad (2.131)$$

$C_2H_5O_2C$ (124) (125)

3 The Diels–Alder and related reactions

3.1. General

The Diels–Alder reaction, one of the most useful synthetic reactions in organic chemistry, is one of a general class of cyclo-addition reactions (Huisgen, Grashey and Sauer, 1964; Huisgen, 1968*b*; Schmidt, 1973) which includes also 1,3-dipolar cyclo-addition reactions (p. 256) and $[2+2]\pi$ cyclo-additions (p. 107). In the Diels–Alder reaction a 1,3-diene reacts with a dienophile to form an adduct with a six-membered hydroaromatic ring (3.1). In the reaction two new σ-bonds and a new π-bond are formed at the expense of two π-bonds in the starting materials (Onischenko, 1964; Sauer, 1966, 1967).

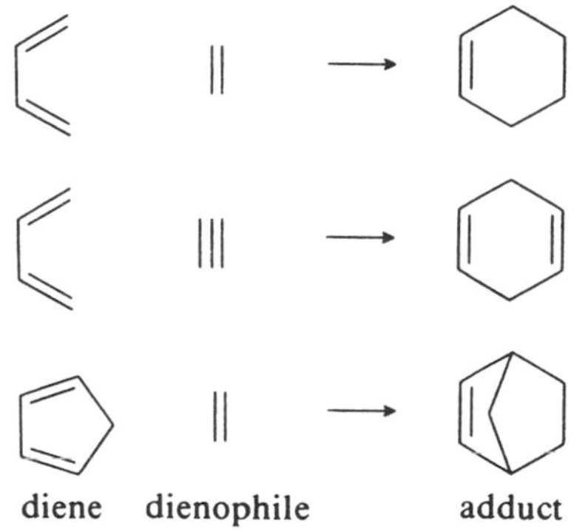

(3.1) (*a*)

(*b*)

(*c*)

diene dienophile adduct

In general the reaction takes place easily, simply by mixing the components at room temperature or by gentle warming in a suitable solvent, although in some cases with unreactive dienes or dienophiles more vigorous conditions may be necessary. The Diels–Alder reaction is reversible, and many adducts dissociate into their components at quite low temperatures. In these cases heating is disadvantageous and the forward reaction is facilitated and better yields are obtained by using an excess of one of the components, or a solvent from which the adduct separates readily. Many Diels–Alder reactions are accelerated by Lewis acid catalysts (see p. 221). In a few cases high pressures have been used to facilitate reactions which otherwise take place only slowly or not at all at room temperature (Dauben and Krabbenhoft, 1977; Dauben, Kessel and Takemura, 1980).

The usefulness of the Diels–Alder reaction in synthesis arises from its versatility and from its remarkable stereoselectivity. By varying the nature of the diene and the dienophile many different types of ring structures can be built up. In the majority of cases all six atoms involved in forming the new ring are carbon atoms but this is not necessary and ring closure may also take place at atoms other than carbon, giving rise to heterocyclic compounds. It is very frequently found, moreover, that although reaction could conceivably give rise to a number of structurally or stereo-isomeric products, one isomer is formed exclusively or at least in preponderant amount.

Many dienes can exist in a *cisoid* and a *transoid* conformation, and it is only the *cisoid* form which can undergo addition. If the diene does not have, or cannot adopt a *cisoid* conformation no reaction occurs.

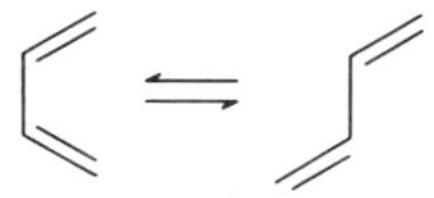

(3.2)

cisoid *transoid*

3.2. The dienophile

Many different kinds of dienophile can take part in the Diels–Alder reaction. They may be derivatives of ethylene or acetylene (the majority of cases) or reagents in which one or both of the reacting atoms is a heteroatom. All dienophiles do not react with equal ease; the reactivity depends on the structure. In general, the greater the number of electron-attracting substituents on the double or triple bond the more reactive is the dienophile, due to the lowering of the energy of the lowest unoccupied molecular orbital of the dienophile by the substituents (see p. 222). Thus, whereas maleic anhydride and 1,3-butadiene afford a quantitative yield of adduct in boiling benzene or, more slowly, at room temperature, tetracyanoethylene, with four electron-attracting substituents reacts extremely rapidly even at 0 °C. Similarly, acetylene reacts with electron-rich dienes only under severe conditions, but propiolic acid, phenylpropiolic acid, and acetylenedicarboxylic acid react readily and have frequently been used as dienophiles in the Diels–Alder reaction (Holmes, 1948; Fuks and Viehe, 1969). Table 3.1 gives some values for the rates of addition of a number of dienophiles to cyclopentadiene and 9,10-dimethylanthracene in dioxan at 20 °C.

Table 3.1. *Reaction of dienophiles with cyclopentadiene and 9,10-dimethylanthracene*

Dienophile	Cyclopentadiene $10^5\ k_1/\mathrm{l\ mol^{-1}\ s^{-1}}$	9,10-Dimethylanthracene $10^5\ k_1/\mathrm{mol^{-1}\ s^{-1}}$
Tetracyanoethylene	*c.* 43 000 000	*c.* 1 300 000 000
Tricyanoethylene	c. 480 000	590 000
1,1-Dicyanoethylene	45 500	12 700
Acrylonitrile	1.04	0.089
Dimethyl fumarate	74	215
Dimethyl acetylene dicarboxylate	31	140

It should be noted, however, that there are a number of Diels–Alder reactions for which the above generalisation does not hold, in which reaction takes place between an electron-rich dienophile and an electron-deficient diene (see Sauer, 1967). The essential feature is that the two components should have complementary electronic character. These Diels–Alder reactions with inverse electron demand, as they are called, also have their uses in synthesis, but the vast majority of reactions involve an electron-rich diene and an electron-deficient dienophile.

The most commonly encountered activating substituents for the 'normal' Diels–Alder reaction are CO, CO_2R, CN and NO_2 and dienophiles which contain one or more of these groups in conjugation with a double or triple bond react readily with dienes.

$\alpha\beta$-Unsaturated carbonyl compounds are reactive dienophiles and are probably the most widely used dienophiles in synthesis. Typical examples are acrolein (propenal), acrylic acid and its esters, maleic acid and its anhydride, 2-butyn-1,4-dioic acid (acetylenedicarboxylic acid) and numerous derivatives of 2-cyclohexenone. Thus, acrolein reacts rapidly with butadiene in benzene solution at 0 °C to give tetrahydrobenzaldehyde in quantitative yield, and 2-butyn-1,4-dioic acid and butadiene give 3,6-dihydrophthalic acid (3.3). Substituents exert a pronounced steric effect on the reactivity of dienophiles. Comparative experiments show

CHO + → CHO

COOH ||| COOH + → CO_2H CO_2H

(3.3)

that the yields of adducts obtained in the condensation of butadiene and 2,3-dimethylbutadiene with derivatives of acrylic acid decrease with the introduction of substituents into the α-position of the dienophile, and $\alpha\beta$-unsaturated ketones with two alkyl substituents in the β-position react very slowly. Similarly, the reactions of butadiene and 2,3-dimethylbutadiene with methylmaleic anhydride require more vigorous conditions than those with maleic anhydride itself.

Another important group of dienophiles of the $\alpha\beta$-unsaturated carbonyl class are quinones (Butz and Rytina, 1949). *p*-Benzoquinone itself reacts readily with butadiene at room temperature to give a high yield of the mono-adduct, tetrahydronaphthaquinone; under more vigorous conditions a bis-adduct is obtained which can be converted into anthraquinone by oxidation of an alkaline solution with atmospheric oxygen (3.4). As with other dienophiles alkyl substitution on the double bonds leads to a decrease in activity and monoalkyl *p*-benzoquinones preferentially add dienes at the unsubstituted double bond.

(3.4)

In contrast to these reactive dienophiles in which the double or triple bond is activated by conjugation with electron-withdrawing groups, ethylenic compounds such as allyl alcohol and its esters, allyl halides and vinyl esters are relatively unreactive, although they can frequently be induced to react with dienes under forcing conditions. Vinyl ethers and esters have been used in the synthesis of dihydropyrans and chromans by reaction with $\alpha\beta$-unsaturated carbonyl compounds in a Diels–Alder reaction with inverse electron demand. 2-Alkoxydihydropyrans are obtained in good yields at temperatures between 150 and 200 °C (Colonge and Descotes, 1967) (3.5). They are useful intermediates for the preparation of glutaraldehydes. Enamines also react readily as dienophiles with $\alpha\beta$-unsaturated carbonyl compounds, although it is not certain that these reactions are concerted cycloadditions; they may be stepwise ionic reactions (Fleming and Kargar, 1967). The products are easily hydrolysed to glutaraldehydes. Diels–Alder reactions with cyclic enamines have been used in the synthesis of alkaloids (cf. Ziegler and Spitzer, 1970). With

200 °C / 400 °C; H_3O^+; O; OR; CHO CHO

(3.5)

ynamines 1-amino-1,4-cyclohexadienes are obtained and can be hydrolysed to cyclohexenones (Genet and Ficini, 1979).

Isolated carbon–carbon double or triple bonds do not usually take part in intermolecular Diels–Alder reactions under normal conditions, but a number of cyclic alkenes and alkynes with pronounced angular strain are reactive dienophiles. The driving force for these reactions is thought to be the reduction in angular strain associated with the transition state for the addition. Thus, cyclopropene reacts rapidly and stereospecifically with cyclopentadiene at 0 °C to form the *endo* adduct (1) in 96 per cent yield, and butadiene gives norcarene (2) in 37 per cent yield. Many cyclopropene derivatives behave similarly (Deem, 1972).

CH_2Cl_2, 0 °C; 96% H H CH_2

(1) (2) (3.6)

Some cyclic alkynes are also powerful dienophiles. Because of its linear structure a carbon–carbon triple bond can only be incorporated without strain into a ring with nine or more members. The increasing strain with decreasing ring size in the sequence cyclo-octyne to cyclopentyne is shown in an increasing tendency to take part in 1,4-cycloaddition reactions. Cyclo-octyne has been prepared as a stable liquid with pronounced dienophilic properties. It reacts readily with diphenylisobenzofuran to give an adduct in 91 per cent yield (3.7). The lower cycloalkynes have not been isolated but their existence has been shown by trapping them with diphenylisobenzofuran (Wittig, 1962).

$(CH_2)_6$ C≡C + O (C_6H_5, C_6H_5) ⟶ $(CH_2)_6$ O (C_6H_5, C_6H_5) (3.7)

Arynes, such as dehydrobenzene, also readily undergo Diels–Alder addition reactions. Cyclopentadiene, cyclohexadiene and even benzene and naphthalene add to the highly reactive species C_6H_4 (Wittig, 1962; Hoffman, 1967) (3.8). Analogous addition reactions are shown by dehydroaromatics in the pyridine and thiophen series.

(3.8)

+ other products

Indirect methods have been developed for engaging unactivated alkenes as dienophiles in Diels–Alder reactions by temporary introduction of activating groups. Thus, the readily available phenyl vinyl sulphone serves very conveniently as an ethylene equivalent as shown in (3.9). Here reductive cleavage of the sulphone group from the initial

$C_6H_5SO_2$; 135 °C ; $SO_2C_6H_5$; Na—Hg ; (76%)

(1) $LiN(iso\text{-}C_3H_7)_2$
(2) $C_6H_5CH_2Br$

(3.9)

$SO_2C_6H_5$; $CH_2C_6H_5$; Na—Hg ; C_6H_5

(83% overall)

adduct with sodium amalgam leads to 1,2-dimethylcyclohexene, produced, formally, by addition of ethylene to 2,3-dimethylbutadiene. Alkylation of the sulphone before reductive cleavage provides access to other derivatives. In the example shown, 4-benzyl-1,2-dimethylcyclohexene is obtained by formal addition of styrene to the butadiene (Carr and Paquette, 1980; Paquette and Crouse, 1983).

The low reactivity of acetylene in Diels–Alder reactions coupled with the synthetic usefulness of the cyclohexa-1,4-dienes formed as adducts, has led to the development of a number of dienophilic acetylene equivalents. Ethynyl sulphones serve very well in this way. Thus, *p*-tolyl ethynyl sulphone reacts with 2,3-dimethylbutadiene to give the adduct (3) in nearly quantitative yield. Reductive cleavage of the sulphone with sodium amalgam then afforded the corresponding 1,4-cyclohexadiene (3.10) (Davis and Whitham, 1980). Phenyl vinyl sulphoxide and 1-benzenesulphonyl-2-(trimethylsilyl)ethylene have also been employed as acetylene equivalents (Paquette and Williams, 1981; Paquette *et al.*, 1978).

+ $SO_2C_6H_4CH_3$ → SO_2Ar (3) ; Na—Hg →

(3.10)

It is evident that, under the appropriate conditions, most olefinic and acetylenic compounds can function as dienophiles. A notable exception is ketene. The C=C linkage in the system C=C=O does indeed react with dienes but the additions are not diene syntheses; the products are *four-membered* ring compounds formed by 1,2-addition to the diene. For example, cyclopentadiene and dimethylketene form the cyclobutanone derivative (4). These reactions are concerted $\pi 2s + \pi 2a$ cyclo-additions,

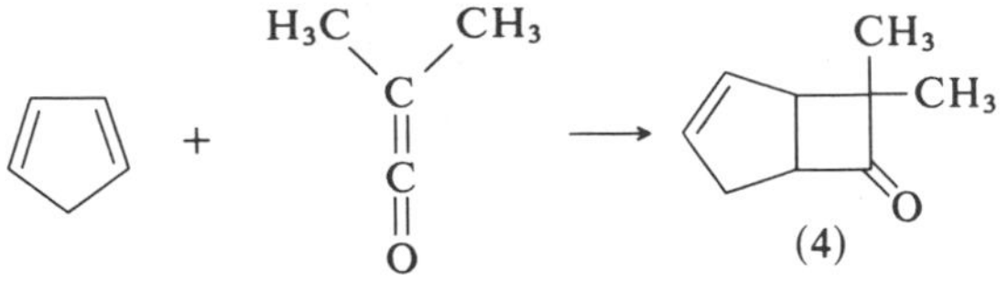

H_3C CH_3 C=C=O → CH_3 CH_3 O (4) (3.11)

with the ketene acting as the antarafacial component (Woodward and Hoffmann, 1970; Fleming, 1976). However, indirect methods have been developed to achieve the conversion corresponding to Diels–Alder addition of ketene to 1,3-dienes. Most of these methods involve addition of

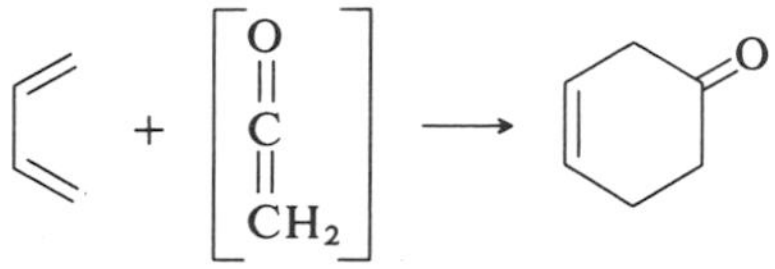

+ [O=C=CH_2] → O (3.12)

a suitably chosen acrylic acid derivative to the diene followed by conversion of the initial adduct into the required ketone. A number of reagents have been used; the best so far appears to be 2-chloroacrylonitrile (Evans, Scott and Truesdale, 1972). Conversion of the initial adducts into the desired ketones is easily effected by hydrolysis (Madge and Holmes, 1980) (3.13). Nitroethylene and vinyl sulphoxides have also been

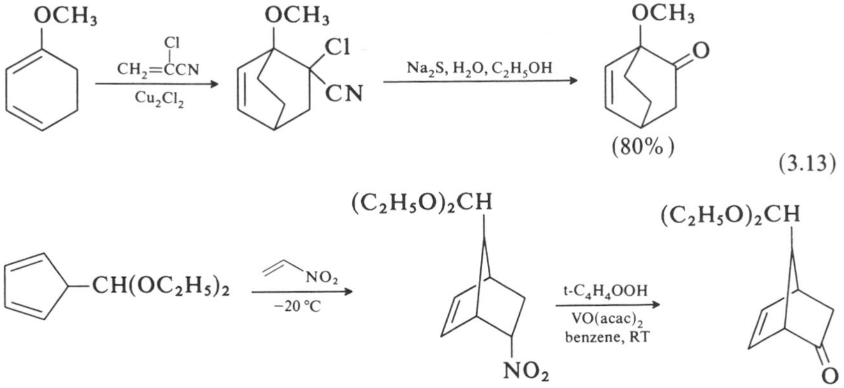

OCH_3 $\xrightarrow[Cu_2Cl_2]{CH_2=C(Cl)CN}$ OCH_3 Cl CN $\xrightarrow{Na_2S, H_2O, C_2H_5OH}$ OCH_3 O (80%)

(3.13)

$CH(OC_2H_5)_2$ $\xrightarrow[-20\,°C]{NO_2}$ $(C_2H_5O)_2CH$ NO_2 $\xrightarrow[VO(acac)_2 \text{ benzene, RT}]{t\text{-}C_4H_4OOH}$ $(C_2H_5O)_2CH$ O

employed as ketene equivalents. Nitroethylene is an excellent dienophile and oxidation of the initial nitro-adducts readily affords the corresponding ketone (3.13) (Bartlett, Green and Webb,1977; Ranganathan *et al.*, 1980; Kornblum *et al.*, 1982). An attractive feature of vinyl sulphoxides as ketene equivalents is that they can be obtained in optically active form because of the chirality of the sulphoxide group, thus allowing the possibility of enantioselective Diels–Alder additions (cf. Maignan and Raphael, 1983). Methylthiomaleic anhydride (Trost and Lunn, 1977) and methyl(phenylthio)propiolate (Gupta and Yates, 1982) have been employed as synthetic equivalents of carbomethoxyketene in Diels–Alder reactions.

A number of diene additions involving allene derivatives have been recorded. Allene itself only reacts with electron-deficient dienes but allene carboxylic acid, in which a double bond is activated by conjugation with the carboxyl group, reacts readily with cyclopentadiene to give a 1:1 adduct in 84 per cent yield.

$CHCO_2H{=}C{=}CH_2$ ... CO_2H, H, CH_2 (3.14)

An 'allene equivalent' is vinyltriphenylphosphonium bromide, which is reported to react with a number of dienes to form cyclic phosphonium salts which can be converted into methylene compounds by the usual Wittig procedure (Ruden and Bonjouklian, 1974).

$\overset{+}{P}(C_6H_5)_3Br^-$ + H_3C, H_3C —CH_3CN, 150 °C→ H_3C, H_3C, $\overset{+}{P}(C_6H_5)_3Br^-$ ↓ $LiN(iso\text{-}C_3H_7)_2$ (3.15)

H_3C, H_3C, CH_2 ←CH_2O— H_3C, H_3C, $P(C_6H_5)_3$

Heterodienophiles

One or both of the carbon atoms of the dienophile multiple bond may frequently be replaced by a heteroatom without significant loss of reactivity (Weinreb and Staib, 1982; Weinreb and Levin, 1979).

Carbonyl groups in aldehydes and ketones add to dienes and the reaction has been used to prepare derivatives of 5,6-dihydropyran. Formaldehyde reacts only slowly but reactivity increases with reactive carbonyl compounds such as chloral or esters of glyoxalic acid (3.16) (Jurczak and Zamojski, 1972; Chmielewski and Jurczak, 1981; David and Eustache, 1979).

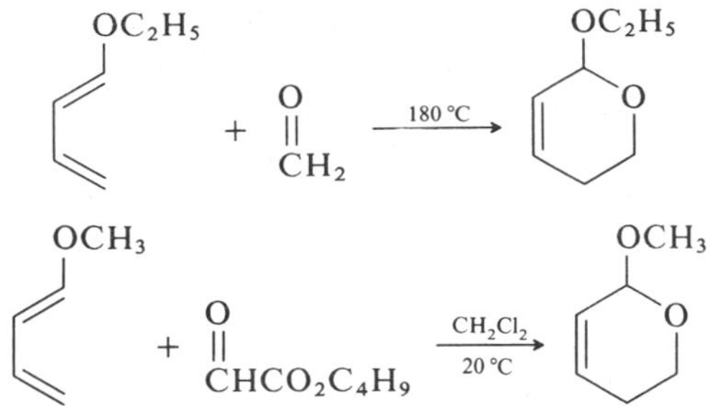

(3.16)

In the presence of zinc chloride or boron trifluoride etherate catalysts, however, the scope of this reaction is greatly extended. Under these conditions oxygenated butadiene derivatives react readily with a wide variety of aldehydes to give dihydro-γ-pyrones in good yield (Danishefsky, Kerwin and Kobayashi, 1982; Danishefsky *et al.*, 1982) (3.17). The reactions catalysed by zinc chloride are true cyclo-additions

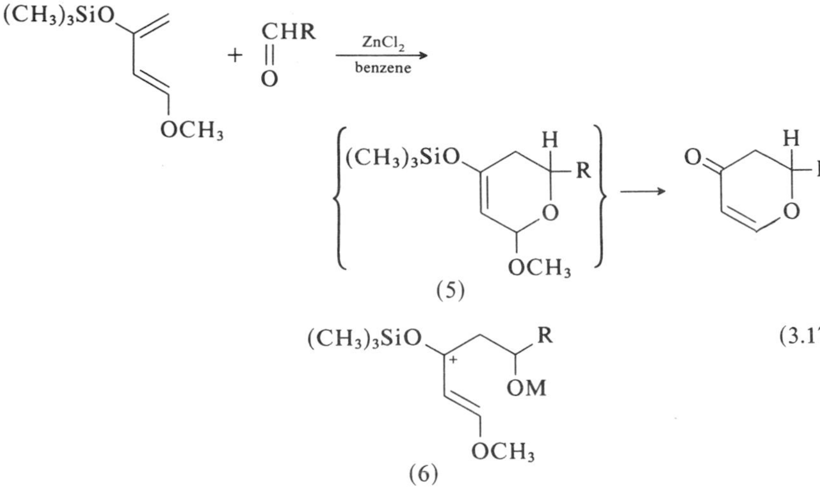

(5)

(6)

(3.17)

and proceed through 1:1 cyclo-adducts such as (5), but the boron trifluoride reactions appear to proceed in stepwise fashion by way of an open-chain aldol-like-product (6) (Larson and Danishefsky, 1982). Imines react in an analogous way to give 5,6-dihydro-4-pyridones (Kerwin and Danishefsky, 1982) (3.18) and, remarkably, $\alpha\beta$-unsaturated

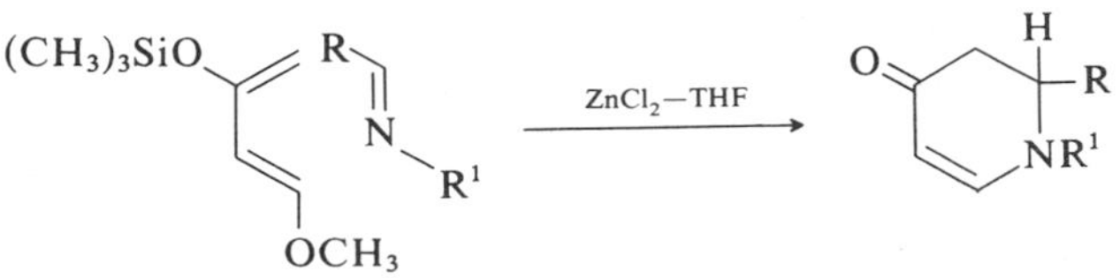

(3.18)

aldehydes and $\alpha\beta$-unsaturated imines react exclusively at the carbonyl or imino group (Danishefsky and Kerwin, 1982).

The dihydro-γ-pyrones obtained in these reactions are useful synthetic intermediates. They have been used in the synthesis of 4-deoxyhexoses, which are otherwise difficultly accessible, and of C-1 branched glycosides (Danishefsky and Kerwin, 1982). Addition to the carbonyl group of chiral aldehydes is stereoselective, following Cram's rule, and this was exploited in highly stereoselective syntheses of the Prelog-Djerassi lactone (9) from (S)-2-phenylpropanal (Danishefsky *et al.*, 1982) and of the novel β-hydroxy-γ-amino acid, statine, in the form of its *N*-t-butyloxycarbonyl derivative (8), from *N*-t-butyloxycarbonyl-leucine (Danishefsky, Kobayashi and Kerwin, 1982). In the latter synthesis reaction of the aldehyde (6) with 1-methoxy-3-trimethylsilyloxybutadiene in the presence of zinc chloride gave the dihydropyrone almost entirely as the '*threo*' isomer (7) through attack on the carbon of the aldehyde from the less hindered side. Oxidative degradation of the pyrone ring then led to *N*-t-butyloxycarbonylstatine with preservation of stereochemistry.

OCH$_3$ (CH$_3$)$_3$SiO + H–C(=O)–CH(NHBoc)CH$_2$CH(CH$_3$)$_2$ (6) —ZnCl$_2$, benzene→ (7) —(1) O$_3$ (2) H$_2$O$_2$→ (8)

(9)

(3.19)

Thio-aldehydes have also been reported to be excellent dienophiles (Vedejs, Eberlein and Varie, 1982; Baldwin and Lopez, 1982; Vedejs, 1984).

Another useful group of heterodienophiles are the imines, containing the group —C=N—. These react with 1,3-dienes to form 1,2,3,6-tetrahydropyridines (Weinreb and Levin, 1979; Weinreb and Staib, 1982;

CH_3 + Tos–N=$CHCCl_3$ (10) —benzene, heat→ CH_3, NTos, CCl_3 (11) (72%) (3.20)

Oppolzer, 1972). Thus, the *N*-tosylimine (10) and 1,3-pentadiene readily gave the tetrahydropyridine (11). Dihydropyridines are obtained from iminochlorides (available from amides and phosphorus oxychloride) and aliphatic dienes. For example 2,3-dimethylbutadiene and acetamide gave 3,4,6-trimethyl-1,2-dihydropyridine by elimination of hydrogen chloride from the initial adduct. Hitherto it has been believed

H_3C, H_3C + NH=CCH_3–Cl → H_3C, H_3C, NH, CH_3, Cl → H_3C, H_3C, NH, CH_3 (3.21)

that only 'activated' imines in which either the nitrogen or carbon atom of the imine linkage, or both, carried an electron-withdrawing group, would undergo the intermolecular imino-Diels–Alder reaction readily. It has recently been found, however, that 'ordinary' Schiff bases react readily with the reactive butadiene derivative (12) in the presence of zinc chloride to form 2,3-dihydro-4-pyridones in good yield (Kerwin and Danishefsky, 1982) (3.22). Intramolecular versions of the imino Diels–Alder reaction are being increasingly applied in the synthesis of alkaloids

OCH_3, $(CH_3)_3SiO$ (12) + C_6H_5–N=CHC_6H_5 —(1) $ZnCl_2$, THF (2) H_2O→ [OCH_3, NC_6H_5, $(CH_3)_3SiO$, C_6H_5] → NC_6H_5, O, C_6H_5 (62%) (3.22)

and other polycyclic nitrogenous compounds (e.g. Oppolzer, Francotti and Bättig, 1981; Weinreb, 1985). The intramolecular reactions in many cases are highly stereoselective (cf. Nader, Franck and Weinreb, 1980; Nader *et al.*, 1981) and frequently take place in the absence of activating groups on the imino double bond. Thus, the acetate (13), on pyrolysis, affords the bicyclic lactam (14), subsequently converted into the alkaloid δ-coniceine (Weinreb, Khatri and Shringarpure, 1979; Schmitthenner and Weinreb, 1980; Khatri, Schmitthenner, Shringarpure and Weinreb, 1981) and the diene (15) gave exclusively the stereoisomer (16).

HN, CH_3COO, O (13); 370–390 °C toluene; N, H_2C, O; N, O (14); (3.23)

CO_2CH_3, $OCOCH_3$, NH, O, O (15); 215 °C toluene; H, CO_2CH_3, N, O, H, O (16); N

Nitriles also react with dienes affording pyridine derivatives, but in general the reactions are of limited preparative value because of the high temperatures required (cf. Weinreb and Staib, 1982). Sulphonyl cyanides react with a variety of cyclic and acyclic dienes under much milder conditions, giving, initially, dihydropyridines which are oxidised to the corresponding pyridines under the reaction conditions. Intramolecular reactions have again proved useful. The nitrile (17), for example, gave the dihydroquinoline derivative (19) in 76 per cent yield, by way of the *ortho*-xylylene (18) (cf. page 207) (Oppolzer, 1972).

O, NCH_3, C, N (17); 180 °C (76%); O, NCH_3, C, N (18); O, NCH_3, NH (19) (3.24)

Nitroso compounds also react with dienes to form oxazine derivatives (Hamer and Ahmad, 1967; Weinreb and Staib, 1982). Aromatic nitroso compounds are the most reactive. Thus, butadiene and nitrosobenzene react readily at 0 °C to give *N*-phenyl-3,6-dihydro-oxazine in 95 per cent yield (3.25).

$$C_6H_5NO + \text{(butadiene)} \xrightarrow[0\,°C]{C_2H_5OH} C_6H_5N\text{–O (dihydro-oxazine)} \quad (95\%) \qquad (3.25)$$

Aliphatic nitroso compounds with an electron-withdrawing group on the α-carbon atom, such as α-chloro- or α-cyano nitroso compounds react readily with butadienes, and electron deficient *N*-acyl nitroso compounds, generated *in situ*, are also excellent dienophiles (Kirby, 1977; Kirby and Sweeny, 1981; Baldwin *et al.*, 1983). α-Chloronitroso compounds react with dienes in alcoholic solvents in an interesting way; in the presence of an alcohol the initial adduct reacts further and the product actually isolated is the dihydro-1,2-oxazine formed, in effect, by addition of HN=O to the diene. This was exploited in a neat synthesis of the natural compound conduramine-F1 (20) (Kresze and Dittel, 1981).

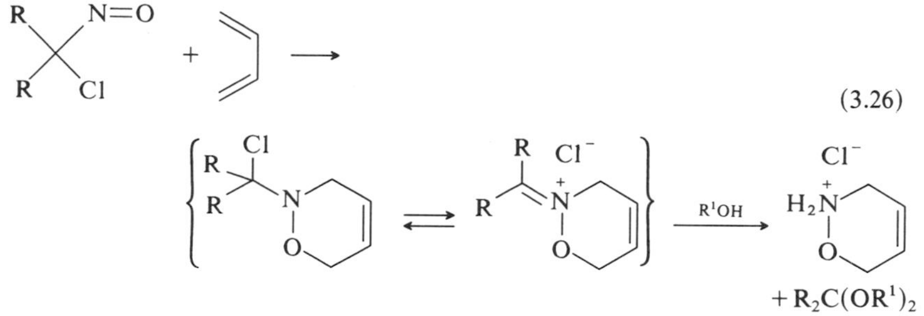

R^1OH ; $+ R_2C(OR^1)_2$ (3.26)

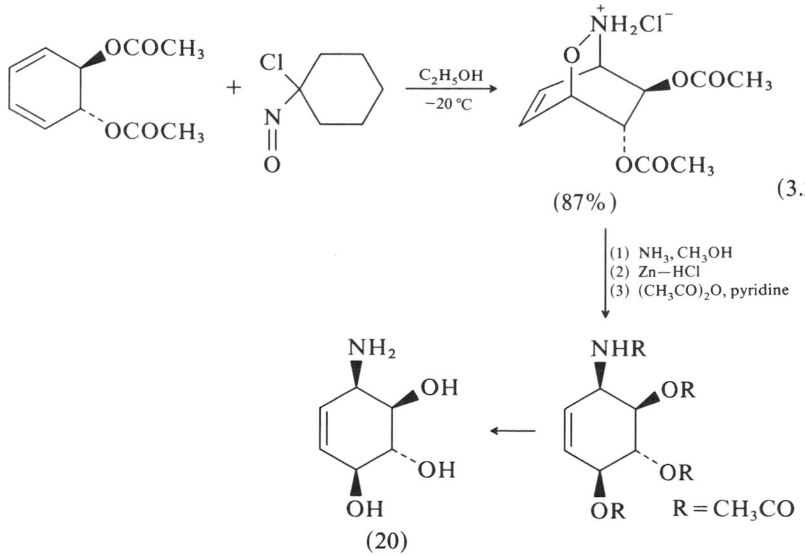

C_2H_5OH, $-20\,°C$; (87%) (3.27)

(1) NH_3, CH_3OH
(2) Zn—HCl
(3) $(CH_3CO)_2O$, pyridine

$R = CH_3CO$

(20)

Some azo compounds with electron-attracting groups attached to the nitrogen atoms are reactive dienophiles (Gillis, 1967). The most commonly used reagent of this type is ethyl azodicarboxylate which reacts with a variety of dienes to give tetrahydropyridazines. Thus, 2,3-dimethylbutadiene and ethyl azodicarboxylate react readily at room temperature to give an almost quantitative yield of the tetrahydropyridazine, which is readily converted into the cyclic hydrazine by the steps shown (3.28).

H_3C, H_3C + $NCO_2C_2H_5$ ‖ $NCO_2C_2H_5$ ⟶ H_3C, H_3C, $NCO_2C_2H_5$ | $NCO_2C_2H_5$

H_2, Pt

H_3C, H_3C, NH | NH ⟵ KOH, H_2O ⟵ H_3C, H_3C, $NCO_2C_2H_5$ | $NCO_2C_2H_5$ (3.28)

Oxygen as a dienophile

The reaction of oxygen with 1,3-dienes to form endoperoxides is of outstanding interest (Gollnick and Schenck, 1967; Wasserman and Ives, 1982). Generally, addition of oxygen to dienes is effected under the influence of light, either directly or in the presence of a photosensitiser, but the reaction can also be brought about with sodium hypochlorite and hydrogen peroxide and by other means, and it is generally believed that singlet oxygen is the reactive species in each case (Foote, 1968; Kearns, 1971). The reaction bears many similarities to the conventional thermal Diels–Alder reaction. It appears to be concerted, and it has been suggested that it proceeds through a six-membered cyclic transition state as does the Diels–Alder reaction, although a pathway via a polar per-epoxide has also been proposed (cf. Gorman and Rodgers, 1981).

+ O_2 ⟶ (3.29)

The light-induced 1,4-addition of oxygen to dienes was discovered independently by Clar and by Windaus. Clar found in 1930 that when a solution of the linear pentacyclic aromatic hydrocarbon pentacene in benzene was irradiated with ultra-violet light in the presence of oxygen the transannular peroxide (21) was obtained (3.30). More than a hundred photoperoxides of this type have now been prepared in the anthracene and tetracene series.

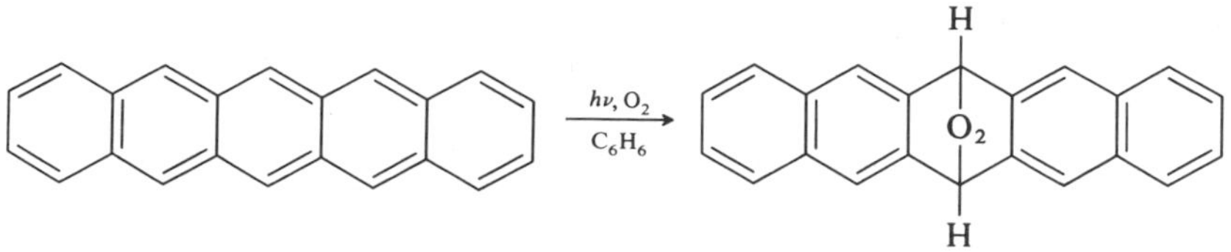

(3.30)

The photosensitised addition of oxygen to conjugated alicyclic dienes was discovered by Windaus in 1928 in the course of his classical studies of the conversion of ergosterol into vitamin D. Irradiation of a solution of ergosterol in alcohol in the presence of oxygen and a sensitiser led to the formation of a peroxide which was subsequently shown to have been formed by 1,4-addition of a molecule of oxygen to the conjugated diene system. Numerous other endoperoxides have since been obtained by sensitised photo-oxygenation of other steroids with cisoid 1,3-diene systems, and of a variety of other cyclic and open-chain conjugated dienes. Thus, irradiation of cyclohexadiene in the presence of oxygen with chlorophyll as a sensitiser leads to the endoperoxide norascaridole and thence, by catalytic hydrogenation to *cis*-1,4-dihydroxycyclohexane (3.31).

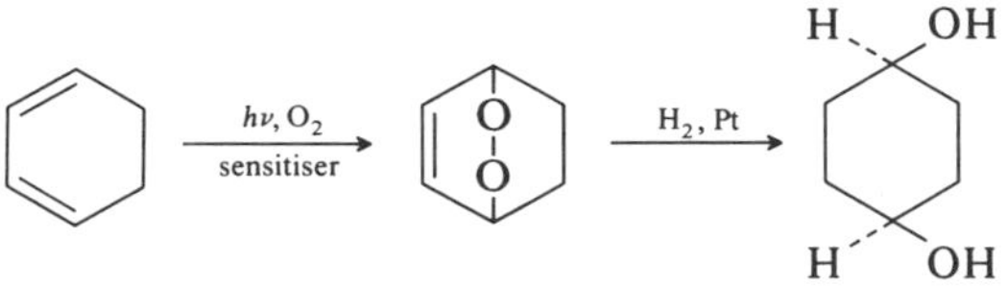

(3.31)

Photosensitised oxygenation of 1,3-dienes has been used with conspicuous success in the synthesis of a number of other natural products (Wasserman and Ives, 1982). Acyclic 1,3-dienes give endoperoxides which can be rearranged and dehydrated to furans and this has been exploited in the synthesis of a number of naturally-occurring furan derivatives (cf. Demole, Demole and Berthet, 1973). Another useful reaction of 1,4-endoperoxides is their thermal rearrangement to *cis*-diepoxides (3.32) (Carless *et al.*, 1985) and a reaction of this kind formed the key step in the synthesis of the naturally occurring tumour inhibitory compound crotepoxide (22). Some endoperoxides can be rearranged in

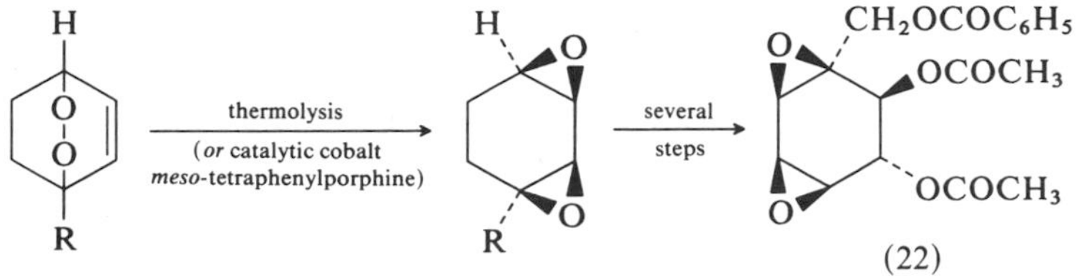

(3.32)

a different way to give 4-hydroxyenones. In the synthesis of the growth-inhibitory substance abscissic acid (23) the unsaturated ketol system was introduced in this way (3.33). Endo-peroxides can also be prepared from some 1,3-dienes by reaction with *triplet* oxygen in the presence of triphenylmethyl cation or a Lewis acid. Thus, α-terpinene is converted into ascaridole in high yield with oxygen and trityl cation in methylene chloride at −78 °C. The precise mechanism of these reactions is uncertain. Light is not always required, and it appears that two pathways are available, but it is believed that in each case the catalyst provides a mechanism for overcoming the spin barrier (Barton *et al.*, 1975).

$h\nu$, O_2, eosin / benzene–ethanol

dil, NaOH / 100 °C

(23) (3.33)

3.3. The diene

It has already been pointed out (p. 184) that the diene component must have or must be able to adopt the *cisoid* conformation before it can take part in Diels–Alder reactions with dienophiles. The majority of dienes which satisfy this condition undergo the reaction more or less easily depending on their structure.

Acyclic dienes

Acyclic conjugated dienes react readily often forming the adducts in almost quantitative yield. For example, butadiene itself reacts quantitatively with maleic anhydride in benzene at 100 °C in 5 h, or more slowly at room temperature, to form *cis*-1,2,3,6-tetrahydrophthalic anhydride.

Examples have already been given (p. 188) of 'dienophile equivalents' in Diels–Alder reactions. 'Diene equivalents' are also known. A useful 'equivalent' for butadiene is the crystalline 2,5-dihydrothiophen sulphone (sulpholene), the cyclic adduct of butadiene and sulphur dioxide, which regenerates butadiene *in situ* on heating; in the presence of dienophiles cyclo-adducts are obtained. By this means the formation of polymeric byproducts from butadiene can often be avoided. Substituted sulpholenes have also been used in a number of neat syntheses (McIntosh and Sieler, 1978; Schmitthenner and Weinreb, 1980; Martin *et al.*, 1982; Bloch and Abecassis, 1982; Bloch and Abecassis, 1983). Thus the substituted sulpholene (24) on heating in xylene solution gave the cyclo-addition product (26) by way of the butadiene (25).

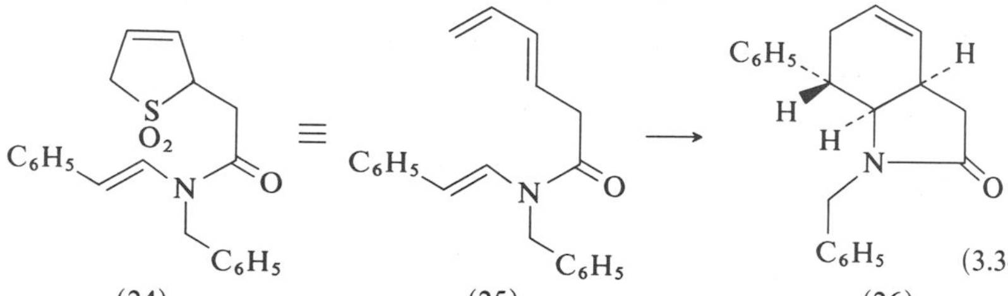

(24) (25) (26) (3.34)

Substituents in the butadiene molecule influence the rate of cycloaddition both through their electronic nature and by a steric effect on the conformational equilibrium. The rate of the reaction is often increased by electron-donating substituents (e.g. $-NMe_2$, $-OMe$, $-Me$) in the diene as well as by electron-attracting substituents in the dienophile. Bulky substituents which discourage the diene from adopting the *cisoid* conformation hinder the reaction. Thus, whereas 2-methyl-, 2,3-dimethyl- and 2-t-butylbutadiene react normally with maleic anhydride the 2,3-diphenyl compound is less reactive and 2,3-di-t-butylbutadiene is completely unreactive. Apparently the molecule of the 2,3-di-t-butyl compound is prevented from attaining the necessary planar *cisoid* conformation by the steric effects of the two bulky t-butyl substituents. In contrast, 1,3-di-t-butylbutadiene, in which the substituents do not interfere with each other even in the *cisoid* form, reacts readily with maleic anhydride.

Z-Alkyl or aryl substituents in the 1-position of the diene reduce its reactivity by sterically hindering formation of the *cisoid* conformation through non-bonded interaction with a hydrogen atom at C-4. Accordingly, an *E*-substituted 1,3-butadiene reacts with dienophiles much more readily than the *Z*-isomer. Thus, (*Z*)-1,3-pentadiene gave only a 4 per cent yield of adduct when heated with maleic anhydride at 100 °C, whereas the *E*-isomer formed an adduct in almost quantitative yield in benzene at 0 °C (3.35).

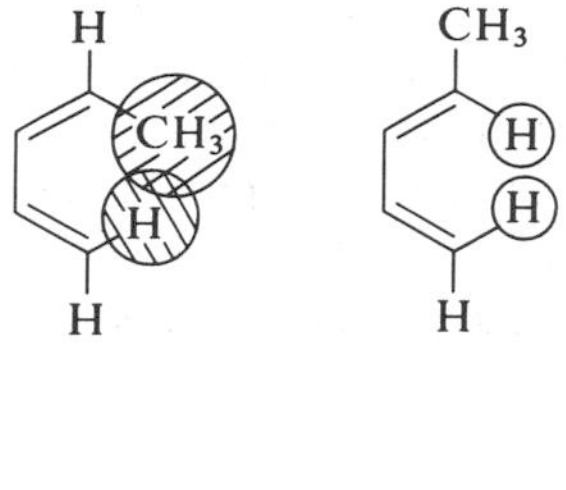

(3.35)

Similarly, (*E*),(*E*)-1,4-dimethylbutadiene reacts readily with many dienophiles, but the *Z,E*-isomer yields an adduct only when the components are heated in benzene at 150 °C. *Z* substituents in both the 1- and 4-positions prevent reaction. 1,1-Disubstituted butadienes also react with difficulty, and with such compounds addition may be preceded by isomerisation of the diene to a more reactive species. Thus, in the reaction of 1,1-dimethylbutadiene with acrylonitrile the diene first isomerises to 1,3-dimethylbutadiene which then reacts in the normal way.

Very many Diels–Alder reactions with alkyl- and aryl-substituted butadienes have been effected (Onischenko, 1964). Derivatives of hydroxybutadienes have come into prominence recently. Thus, 2-alkoxy- and 2-trimethylsilyloxy-butadienes react easily with dienophiles to form adducts which, as enol ethers, are readily hydrolysed to cyclohexanone derivatives (3.36) (Jung *et al.*, 1981).

$ClSi(CH_3)_3$, $(C_2H_5)_3N$; $(CH_3)_3SiO$; CO_2CH_3 ; $(CH_3)_3SiO$, CO_2CH_3 ; H_3O^+ ; CO_2CH_3 (3.36)

One of the most widely used dienes of this class is 1-methoxy-3-trimethylsilyloxybutadiene (27), the so-called Danishefsky diene, which has been employed in several elegant syntheses of natural products (cf. Danishefsky, 1981). Cyclohexenones, cyclohexadienones or benzene derivatives may be prepared using this reagent depending on the precise mode of operation. Thus, reaction with 2-methylpropenal gave 4-formyl-4-methylcyclohexenone after acid hydrolysis of the initial adduct (28), and a useful route to *para* acylphenols is provided by reaction of the diene with β-phenylsulphinyl-$\alpha\beta$-unsaturated carbonyl compounds (3.37) (Danishefsky, Harayama and Singh, 1979). If the $\alpha\beta$-unsaturated carbonyl compound in the latter sequence carries an α-substituent then aromatisation is not possible and the product obtained after elimination of phenylsulphinic acid is a 4,4-disubstituted cyclohexadienone. A sequence of this kind was employed in an elegant synthesis of sodium prephenate and of optically active pretyrosine from L-glutamic acid (3.38) (Danishefsky, Morris and Clizbe, 1981).

A related group of oxy-substituted butadienes are the vinylketene acetals such as 1,1-dimethoxy-3-trimethylsilyloxybutadiene (29). It reacts readily with acetylenic dienophiles, for example, providing a convenient

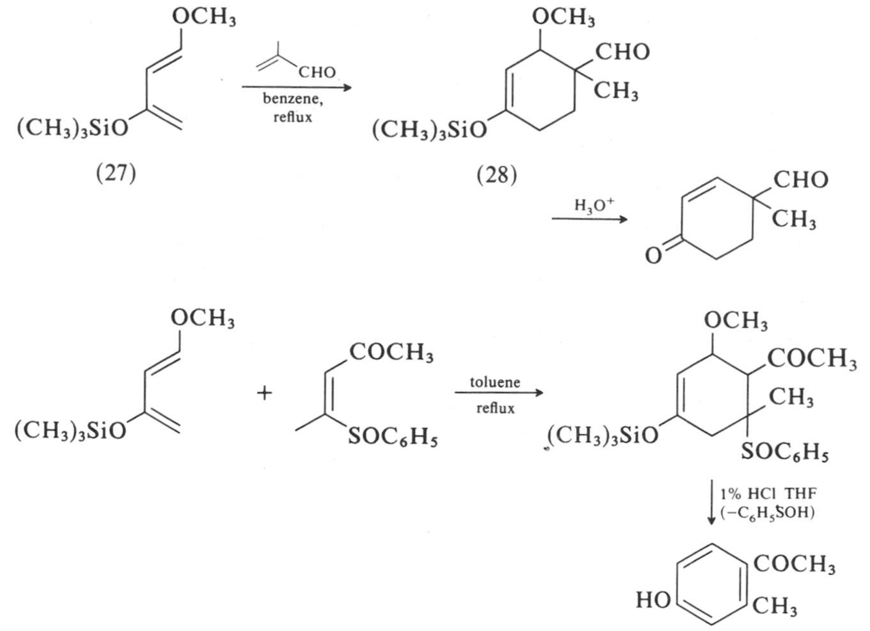

(27) (28)

(3.37)

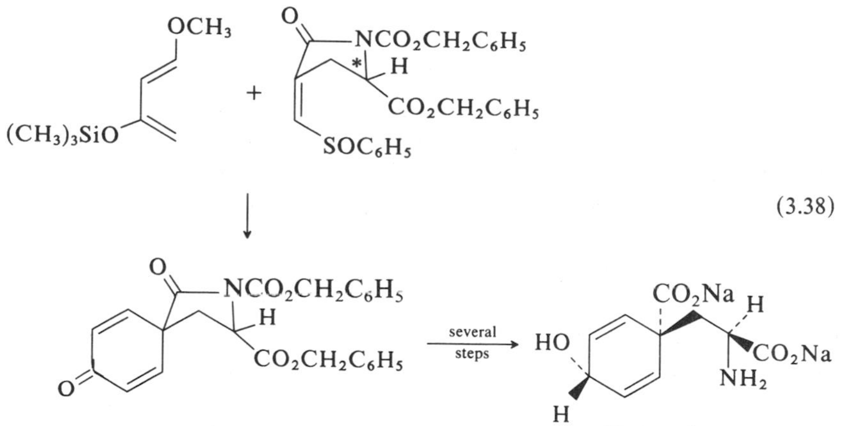

(3.38)

Pretyrosine

synthesis of otherwise difficultly accessible derivatives of resorcinol monomethyl ether (3.39) (Danishefsky, Singh and Gammill, 1978). Benzene derivatives are also obtained by reaction of 1,4-diacetoxybutadiene with olefinic or acetylenic dienophiles; the initial adducts readily eliminate acetic acid to give the aromatic compound. In many cases it is unnecessary to isolate the intermediate adduct and the benzene derivative

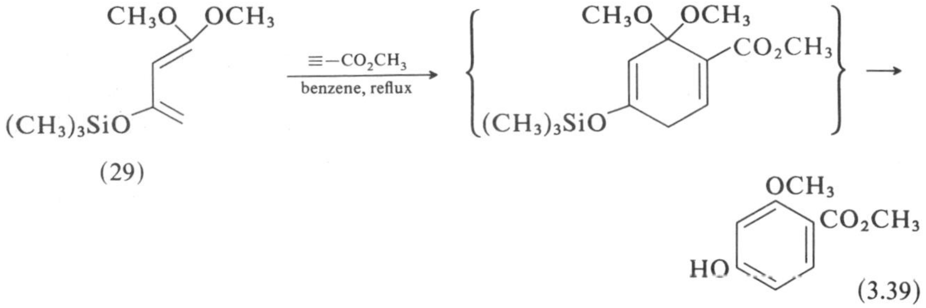

(3.39)

is obtained directly by heating the components together at 100–120 °C (Hill and Carlson, 1965). 1,4-Dimethoxybutadiene has been used in a similar way to prepare derivatives of anisole and of 1,4-dimethoxybenzene (Hiranuma and Miller, 1982).

Another valuable group of heterosubstituted dienes are the 1- and 2-acylaminobutadienes (Oppolzer, Bieber and Francotti, 1979; Overman *et al.*, 1981*a*). They react readily with dienophiles to give substituted amino-cyclohexanes. The acylamino group is a powerful directing group and most reactions proceed with high regio- and stereo-selectivity. Thus, styrene reacted readily with the diene (30) to give the 1:2 ('ortho') addition product (31) almost exclusively in over 90 per cent yield, and with *trans*-crotonaldehyde the adduct (32) was obtained; the latter reaction formed the key step in an elegant synthesis of the poison arrow alkaloid pumiliotoxin (33) (Overman and Jessup, 1978). The synthetic

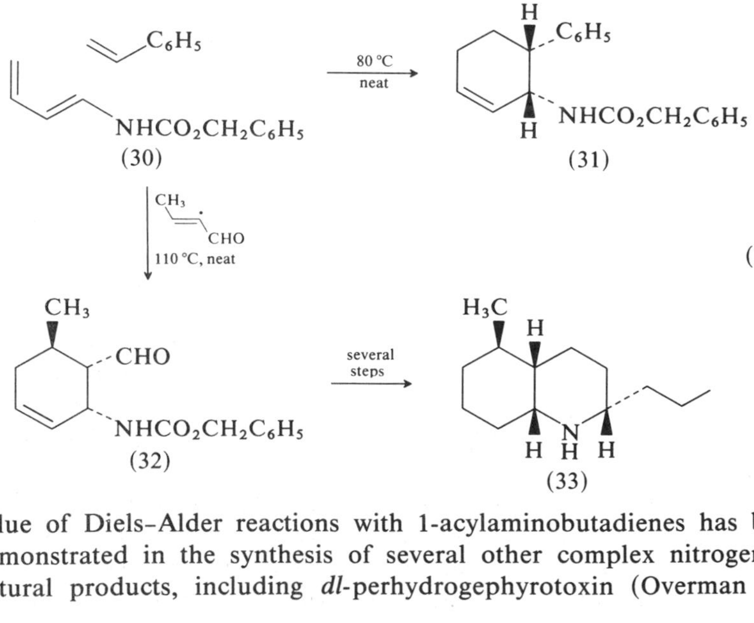

(3.40)

value of Diels–Alder reactions with 1-acylaminobutadienes has been demonstrated in the synthesis of several other complex nitrogenous natural products, including *dl*-perhydrogephyrotoxin (Overman and

Fukaya, 1980) and *dl*-isogabaculine (Danishefsky and Hershenson, 1979). Intramolecular reactions have been employed with advantage. Thus, in a synthesis of optically active (−)-pumiliotoxin C the dieneamide (34), itself prepared in optically active form from (*S*)-norvaline, cyclised to the octahydroquinoline (35) with high selectivity. Here the chiral centre in (34) controls the formation of the three new developing centres in (35) (Oppolzer and Flashkamp, 1977).

215 °C
toluene,
sealed tube

(34) (35)

(3.41)

2-Acylamino-1,3-dienes have also been used in synthesis, although so far less extensively than the 1-substituted isomers. They lead to cyclohexylenamides which on hydrolysis give cyclohexanones (cf. Overman, 1980).

1- and 2-Phenylthiobutadienes are also useful. They form adducts which are versatile synthetic intermediates since the sulphide group can be reductively removed or, after oxidation to the sulphoxide eliminated to form a new double bond or converted into an allyl alcohol by a [2,3]-sigmatropic rearrangement (cf. p. 124) (Petrzilka and Grayson, 1981). In reactions with unsymmetrical dienophiles the regiochemistry of the additions is controlled by the sulphur substituent and advantage can be taken of this in reaching adducts of unusual orientation. Thus, 2-methoxy-3-(phenylthio)-butadiene reacts with methyl vinylketone to give predominantly the adduct (36). Hydrolysis of the enol ether and reductive removal of the phenylthio substituent then affords an acetylcyclohexanone of different orientation from that obtained with 2-methoxybutadiene itself (Trost, Vladuchick and Bridges, 1980; cf. also Blatcher, Grayson and Warren, 1978):

Trimethylsilyl-1,3-dienes have also been finding some use in synthesis. 1-Trimethylsilylbutadiene itself reacts with a variety of olefinic and acetylenic dienophiles (Carter, Fleming and Percival, 1981; Jung and Gaede, 1979). The cyclic allylsilanes formed with dienophiles undergo

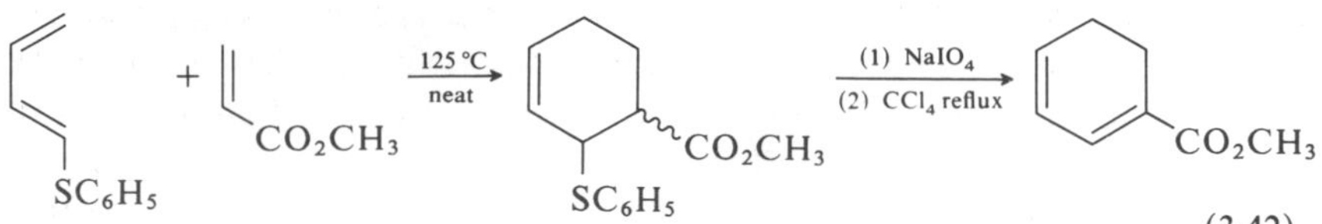

(3.42)

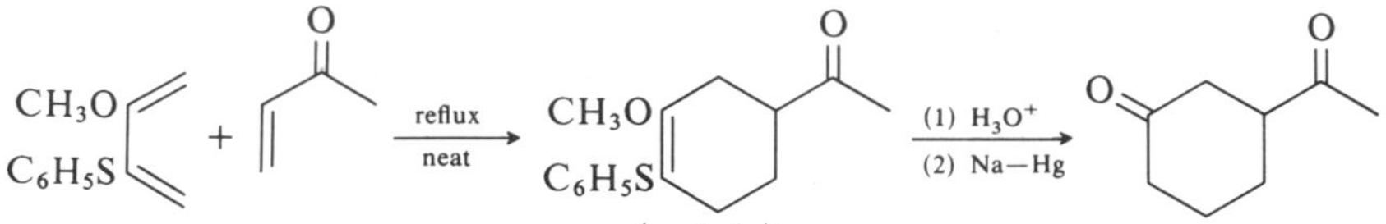

(mainly)
(36) (75%)

a range of useful conversions; protodesilylation with acid leads to cyclohexenes, and with peroxy acids cyclic allylic alcohols are obtained (cf. p. 330). The trimethylsilyl substituent is only weakly directing so that in reactions with unsymmetrical dienophiles mixtures of regioisomers are often obtained and if another substituent is present in the diene as well, it controls the regiochemistry of the cyclo-addition. Thus, 4-acetoxy-1-(trimethylsilyl)butadiene and ethyl acrylate give largely the adduct in which the direction of addition is controlled by the acetoxy group (Koreeda and Ciufolini, 1982). With 1-trimethylsilylbutadiene itself a mixture of 'ortho' and 'meta' adducts was obtained (Jung and Gaede, 1979).

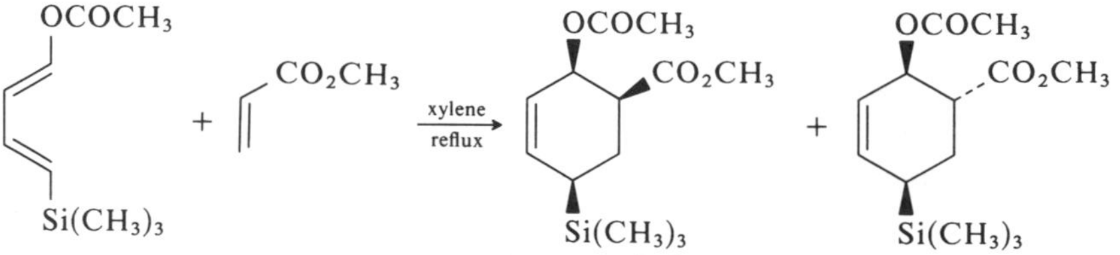

(37) (72%) (8%)

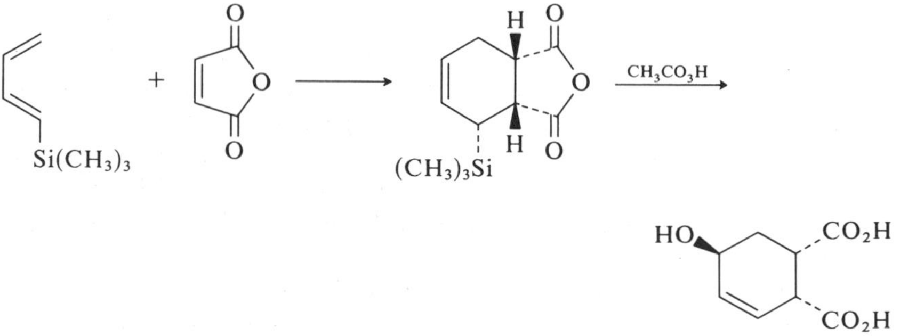

(3.43)

Two other derivatives of butadiene with some promise in synthesis are the pyrone derivatives (38) and (39). The pyrone (38) reacts with dienophiles with loss of carbon dioxide to form cyclic dienes and with some electron-rich alkynes benzene derivatives are formed directly (Bryson and Donelson, 1977; Ireland *et al.*, 1983*a*). The crystalline hydroxypyrone (39) is a useful synthetic equivalent of vinyl-ketene. It reacts with a variety of dienophiles to give adducts which are converted easily into dihydrophenols or cyclohexenones by loss of carbon dioxide. Thus, reaction with methyl acrylate affords the cyclohexenone (41) in 85 per cent yield via the adduct (40), and with benzoquinone the naphthalene (42) is obtained after acetylation. Acetylenic dienophiles give aromatic compounds.

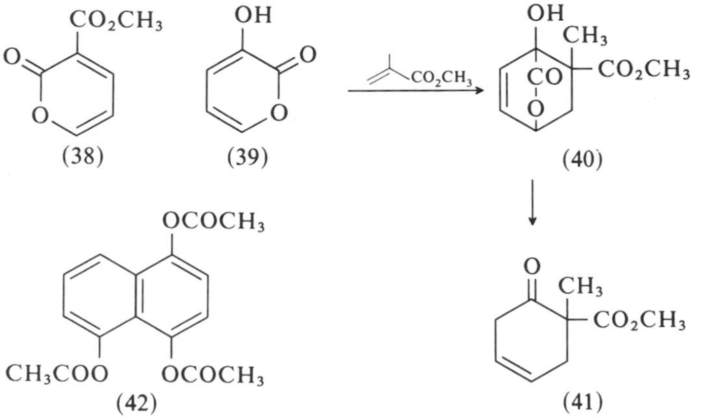

(3.44)

Heterodienes

Heterodienes, in which one or more of the atoms of the conjugated diene is a heteroatom, have not been so extensively employed in synthesis as heterodienophiles (p. 190) and only the 1,4-addition reactions of $\alpha\beta$-unsaturated carbonyl compounds have been used to any extent so far (Colonge and Descotes, 1967). They react most readily as dienes with electron rich dienophiles such as enol ethers (Ireland and Daub, 1983) and enamines. With less reactive dienophiles dimerisation of the $\alpha\beta$-unsaturated carbonyl compound is a competing reaction. The reaction with enol ethers and with *N*-vinyl compounds such as enamines and *N*-vinyl carbamates, proceeds readily to yield derivatives of dihydropyran which can be converted into synthetically useful 1,5-dicarbonyl compounds (3.45).

Hitherto diene condensations of heterodienes containing nitrogen atoms in the diene system have been much less studied but recently it has been found that Diels–Alder reactions of properly chosen derivatives

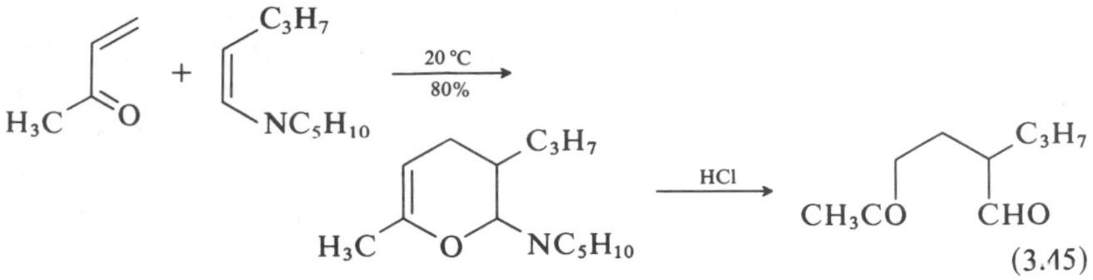

(3.45)

of 1- and 2-aza-butadienes provide very convenient access to some six-membered nitrogen heterocycles (Boger, 1983). In the 2-aza series, substituted 1-dimethylamino-2-azabutadienes react with dienophiles with loss of dimethylamine to give pyridines or dihydropyridines (Demoulin *et al.*, 1975) (3.46) and the *bis*-silyloxy derivative (43) has been used to make derivatives of 2-pyridone and 2-piperidone (Sainte *et al.*, 1982).

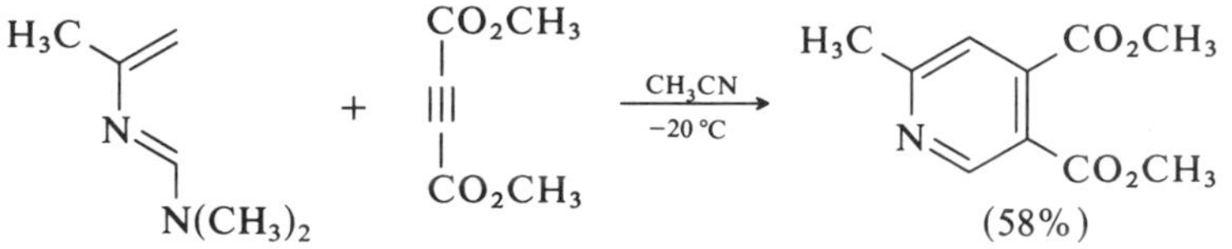

(58%)

(3.46)

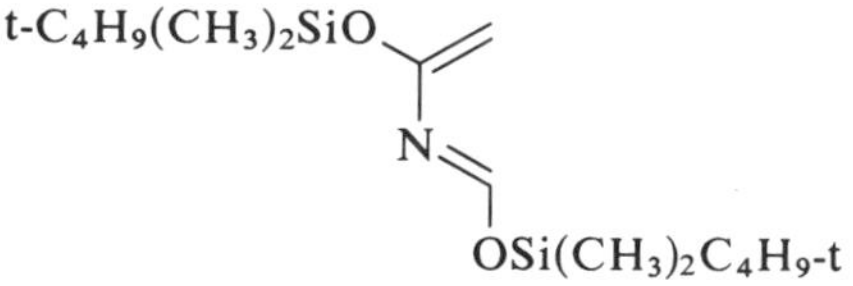

(43)

1-Azabutadienes have traditionally been regarded as less reactive dienes than their 2-aza analogues, but the $\alpha\beta$-unsaturated hydrazone (44) at any rate reacts regioselectively with a range of dienophiles to give adducts which are smoothly converted into the corresponding piperidine derivatives by reductive cleavage with zinc and acetic acid (Serckx-Poncin *et al.*, 1982). The corresponding oxime did not react and the reactivity of (44) is ascribed to interaction between the lone pair of the NMe_2 group and the π system.

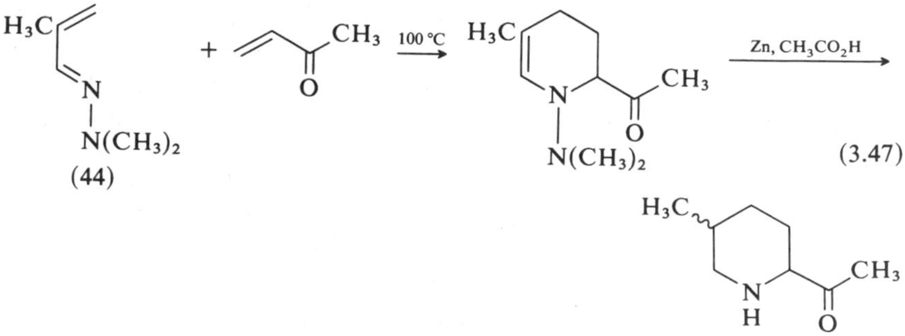

(44)

(3.47)

1,2-Dimethylenecycloalkanes

In 1,2-dimethylenecycloalkanes the *cisoid* conformation of the double bonds necessary for 1,4-addition is fixed and dienes of this type react readily with dienophiles, often forming the adducts in almost quantitative yield. This and the ready availability of the compounds make 1,4- additions to dimethylenecycloalkanes a convenient route to polycyclic compounds. Thus, 1,2-dimethylenecyclohexane itself reacts exothermally with maleic anhydride, and with benzoquinone the *bis*-adduct (45) is obtained and is readily converted into the corresponding aromatic hydrocarbon pentacene. A variety of alkyl derivatives of this and other linearly condensed polycyclic aromatic compounds has been obtained by this general route.

CH_2 CH_2 + O O $\xrightarrow[77\%]{100\,°C,\ 2\ h}$ O O (3.48)

(45)

Related to the 1,2-dimethylenecycloalkanes are the *ortho*-xylylenes or *ortho*-quinodimethanes (as 46). These are very reactive dienes and readily form adducts with a variety of dienophiles. The *ortho*-xylylenes used in

CH_2 CH_2 + R^1 R^2 ⟶ R^1 R^2 (3.49)

(46)

these reactions are generated *in situ* by thermal ring opening of benzocyclobutenes, by photoenolisation of *ortho* alkyl aromatic aldehydes or ketones (Sammes, 1976) or by elimination from appropriate derivatives of 1,2-dialkylbenzenes (cf. Funk and Vollhardt, 1980; Oppolzer, 1978). The most commonly used method hitherto has been by thermolysis of derivatives of benzocyclobutene (3.50). With derivatives having a substituent on the cyclobutene ring the thermally allowed conrotatory ring opening takes place preferentially outward to give the sterically less hindered *E*-diene (as 47). The ease of ring opening varies widely with

OH $\underset{}{\overset{110\,°C}{\rightleftarrows}}$ OH CH_2 $\overset{h\nu}{\rightleftarrows}$ CHO CH_3 (3.50)

(47)

the nature of the substituent on the cyclobutene ring. 1-Hydroxybenzocyclobutene and its acyl derivative undergo ring opening particularly easily to form the *ortho*-xylylene (47) in which the hydroxyl substituent adopts the *E*-configuration. The same product is formed by irradiation of *ortho*-methylbenzaldehyde, and both procedures have been used in the synthesis of polycyclic compounds by trapping the *ortho*-xylylenes with dienophiles. Thus, heating 1-hydroxybenzocyclobutene with naphthaquinone in benzene leads directly and stereoselectively to the adduct (48) (Arnold, Sammes and Wallace, 1974). Similarly, irradiation of the *ortho*-benzylbenzaldehyde (49) in the presence of dimethyl acetylenedicarboxylate leads to the adduct (50) (Arnold, Mellows and Sammes, 1973).

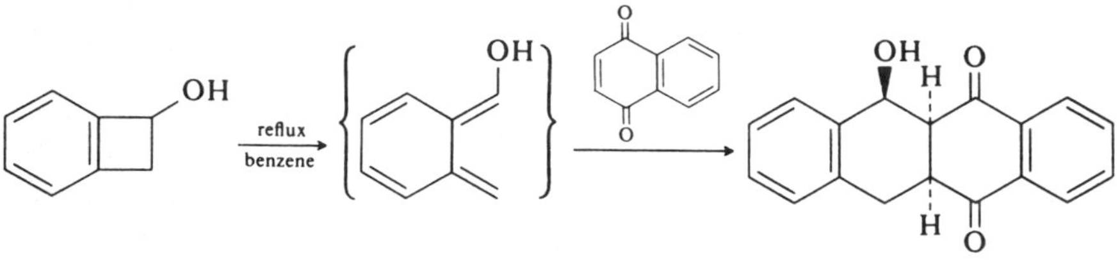

OH reflux benzene OH O O OH H O H O

(48) (3.51)

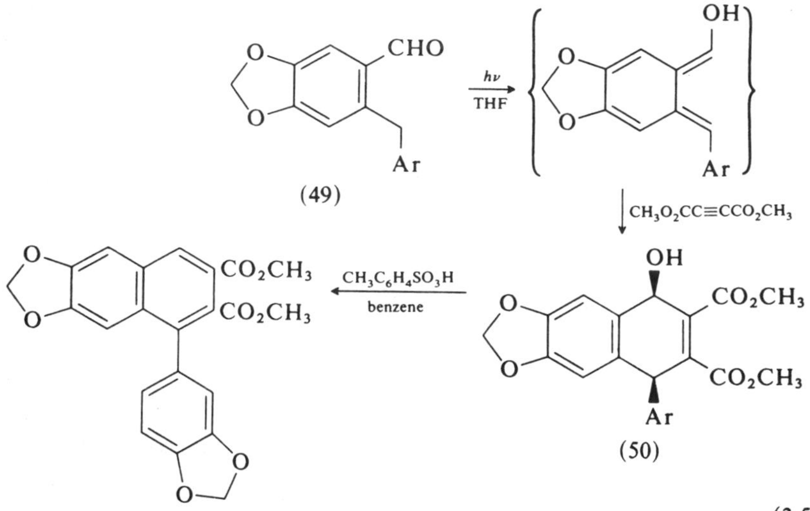

O O CHO Ar (49) $h\nu$ THF O O OH Ar

$CH_3O_2CC{\equiv}CCO_2CH_3$

O O CO_2CH_3 CO_2CH_3 O O $CH_3C_6H_4SO_3H$ benzene OH O O CO_2CH_3 CO_2CH_3 Ar (50)

(3.52)

ortho-Xylylenes have been used in the synthesis of natural products in the alkaloid, steroid and terpene series (Funk and Vollhardt, 1980; Oppolzer, 1978; Kematani, 1979; Magnus *et al.*, 1984). Intramolecular reactions take place easily. Thus, pyrolysis of the amine (51) readily gave the tricyclic product (52).

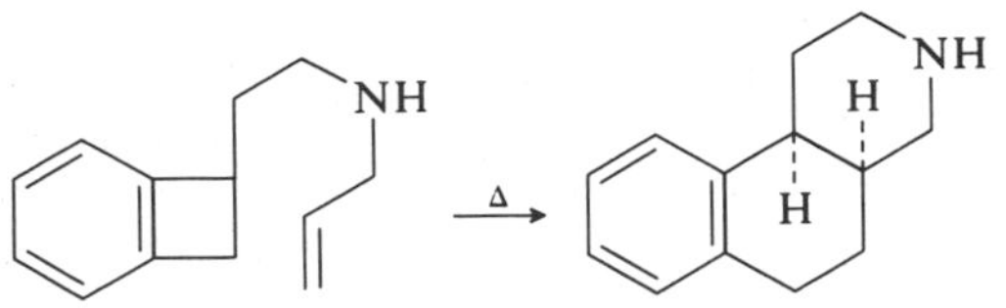

(51) (52) (3.53)

A disadvantage of the approach to *ortho*-xylylenes by thermolysis of benzocyclobutenes is the comparative inaccessability of appropriately substituted benzocyclobutenes. A novel route to these compounds is provided by the discovery that hexa-1,5-diynes may be co-cyclised with substituted acetylenes, particularly with bis(trimethylsilyl)acetylene, in the presence of the cobalt(I) catalyst (η^5-cyclopentadienyldicarbonyl-cobalt) to give substituted benzocyclobutenes (3.54) (Vollhardt, 1984). With a suitably substituted hexa-1,5-diyne which incorporates a dienophile, intramolecular reaction can take place via the benzocyclobutene and derived *ortho*-xylylene, to give polycyclic compounds. Thus, *bis*(trimethylsilyl)-acetylene and the 1,5-diyne (53) gave the tricyclic compound (56) and thence the octahydrophenathrene (57) by way of the benzocyclobutene (54) and the *ortho*-xylylene (55) (Funk and Vollhardt, 1979).

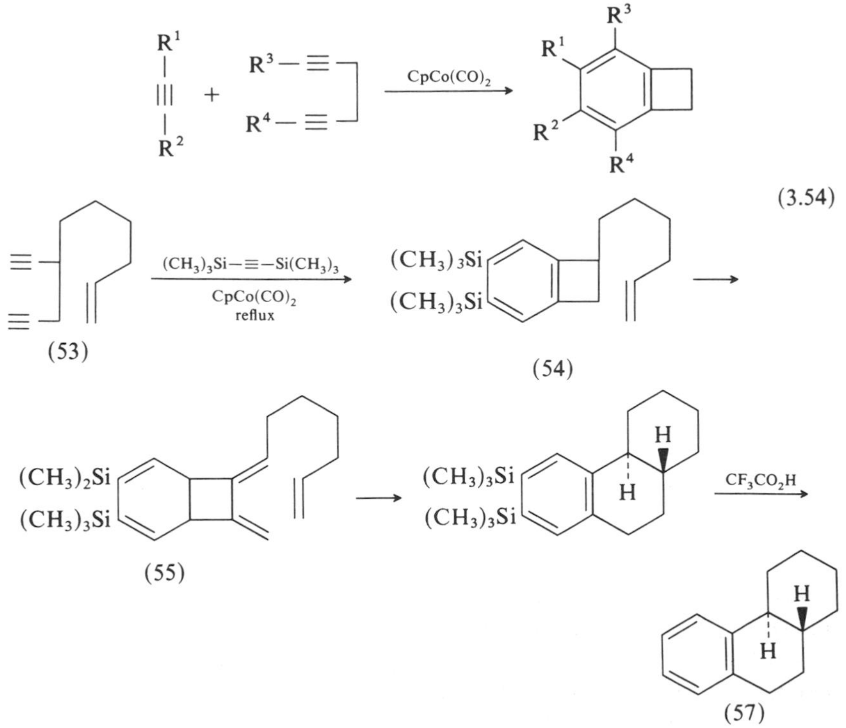

R^1, R^2, R^3, R^4; $CpCo(CO)_2$

(3.54)

$(CH_3)_3Si-\equiv-Si(CH_3)_3$, $CpCo(CO)_2$, reflux

(53) (54)

$(CH_3)_2Si$ $(CH_3)_3Si$ (55)

$(CH_3)_3Si$ $(CH_3)_3Si$ H H CF_3CO_2H

(57)

ortho-Xylylenes can also be obtained by pyrolytic elimination of sulphur dioxide from 1,3-dihydrobenzo[*c*]thiophen sulphones (Oppolzer and Roberts, 1980), and by pyrolysis of *ortho*-alkylbenzyl chlorides (Gray *et al.*, 1978; Schiess *et al.*, 1978) but these methods require high temperatures. A much milder approach is by 1,4-elimination from appropriate α,α'-substituted 1,2-dialkylbenzenes, for example by Hofmann elimination from (*ortho*-methylbenzyl)trimethylammonium hydroxide or by iodide-induced debromination of *ortho* xylylene dibromides (e.g. Wiseman *et al.*, 1980) (3.55). Thus, in a synthesis of (±)-4-demethoxy-

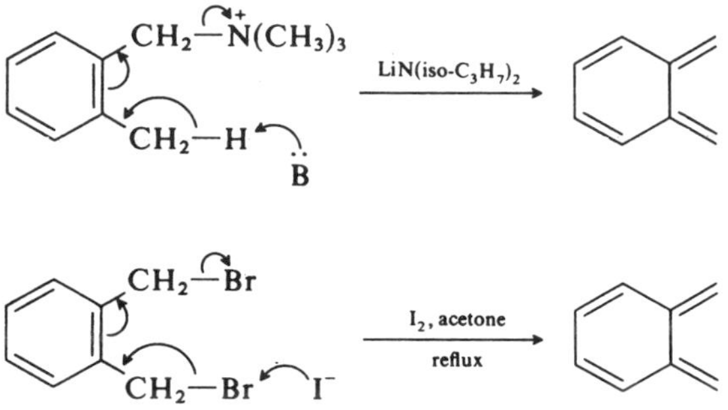

(3.55)

daunomycinone the tetracyclic compound (60) was obtained by cycloaddition of methyl vinylketone to the *ortho*-xylylene (59), itself obtained by iodide induced elimination from the corresponding *ortho*-bisbromomethyl compound (58) (Kerdesky *et al.*, 1981).

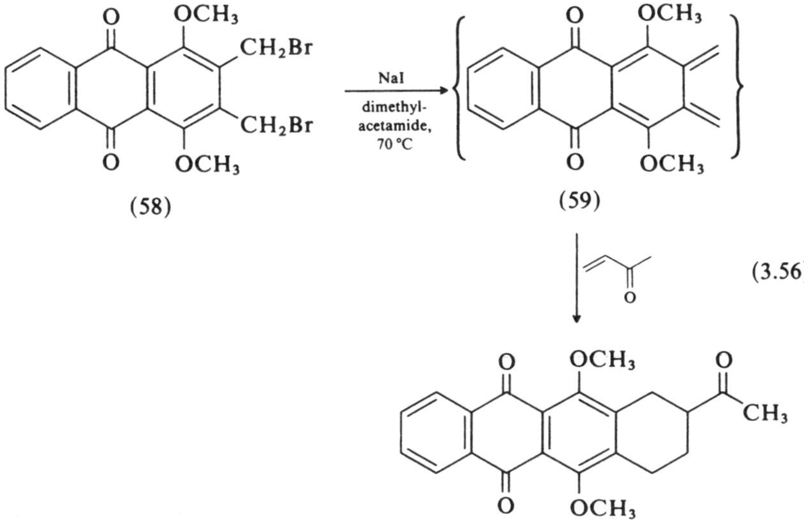

(58) (59) (3.56)

(60) (75%)

Another method proceeds by elimination from *ortho*-silylmethylammonium salts, triggered by fluoride ion. Thus, generated in the presence of dimethyl fumarate the *ortho*-xylylene (61), gave the tetralin derivative (62) in quantitative yield (Ito, Nakatsuka and Saegusa, 1981, 1982).

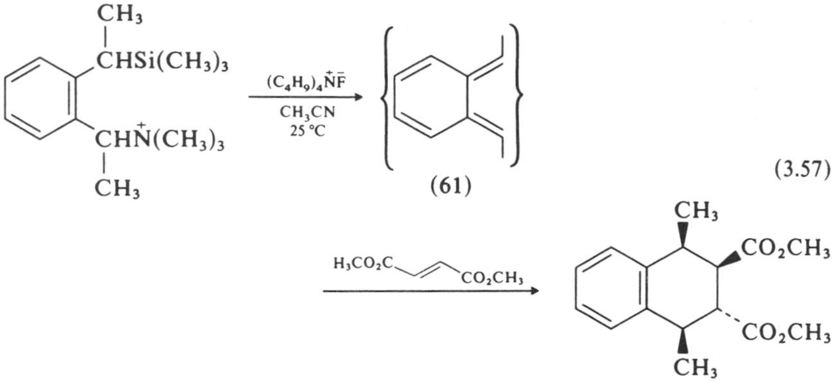

CH_3 | $CHSi(CH_3)_3$ ... $CH\overset{+}{N}(CH_3)_3$ | CH_3 $\xrightarrow[CH_3CN,\ 25\,°C]{(C_4H_9)_4\overset{+}{N}\overset{-}{F}}$ (61)

$\xrightarrow{H_3CO_2C\text{–CH=CH–}CO_2CH_3}$ CH_3, CO_2CH_3, CO_2CH_3, CH_3 (62)

(3.57)

Vinylcycloalkenes and vinylarenes

Many vinylcycloalkenes react readily with dienophiles provided they have or can adopt the necessary *cisoid* conformation of the double bonds. 1-Vinylcyclohexene itself reacts exothermally with maleic anhydride to form a high yield of the adduct (63). Reactions of this kind with cyclopentenones and cyclohexenones as dienophiles have been extensively used to build up tetracyclic structures related to steroid systems (cf. Onischenko, 1964). In contrast, dienes of the types (3.59) in which the double bonds are constrained in the *transoid* conformation do not react.

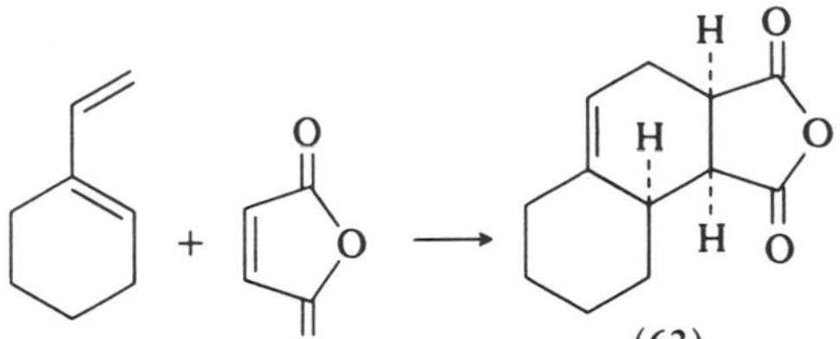

(63) (3.58)

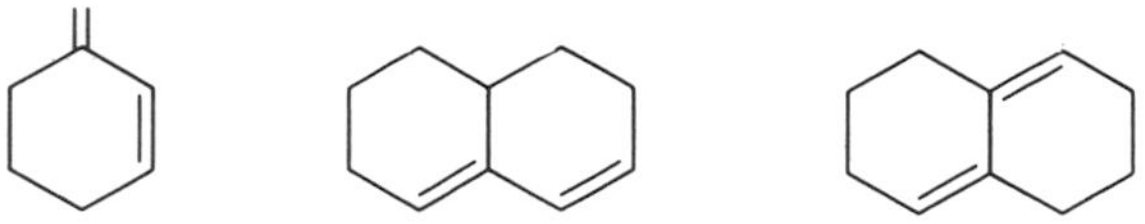

(3.59)

Aromatic hydrocarbons

Some polycyclic aromatic hydrocarbons react with dienophiles by 1,4-addition, but the reaction is particularly characteristic of anthracene and the higher linear acenes. Anthracene reacts with maleic anhydride in boiling xylene to form the 9,10-addition product (64) in quantitative yield. Many anthracene derivatives and anthracene benzologues behave similarly, although the ease of reaction varies with the structure of the hydrocarbon. Tetracene and the higher acenes react with

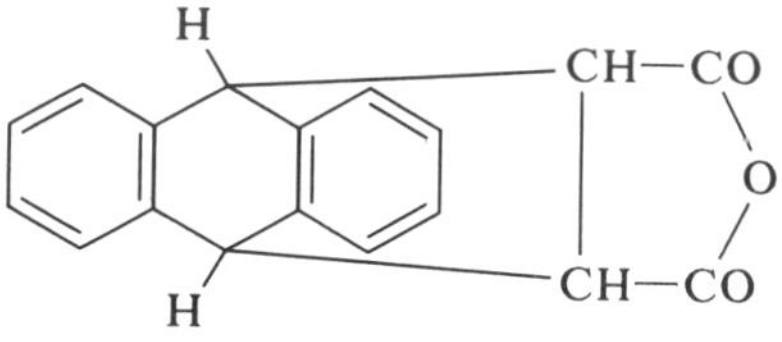

(64)

dienophiles even more readily than anthracene. Since the reaction results in loss of aromatic stabilisation it is not surprising that benzene does not normally undergo thermal Diels–Alder reactions. A *bis*-adduct has been obtained with maleic anhydride under the influence of ultra-violet light (see p. 244). Adducts have also been obtained with some acetylene derivatives in the presence of aluminium chloride catalyst (Ciganek, 1967).

Cyclic dienes

Cyclopentadiene, in which the double bonds are constrained in a planar *cisoid* conformation, reacts easily with a variety of dienophiles. 1,3-Cyclohexadiene is also reactive but with increase in the size of the ring the reactivity of cyclic dienes rapidly decreases. In the larger rings the double bonds can no longer easily adopt the necessary coplanar configuration because of non-bonded interaction of methylene groups in the planar molecule. *Cis,cis-* and *cis,trans-*1,3-cyclo-octadienes form only copolymers when treated with maleic anhydride and *cis,cis-* and *cis,trans-*1,3-cyclodecadienes similarly do not form adducts with maleic anhydride. Dienes with fourteen- and fifteen-membered rings again react with dienophiles but only under relatively severe conditions.

Cyclopentadiene is a very reactive diene and reacts easily with dienophiles to form bridged compounds of the bicyclo[2,2,1]heptane series (3.60). The reaction of cyclopentadiene with mono- and *cis*-disubstituted ethenes could apparently give rise to two stereochemically distinct

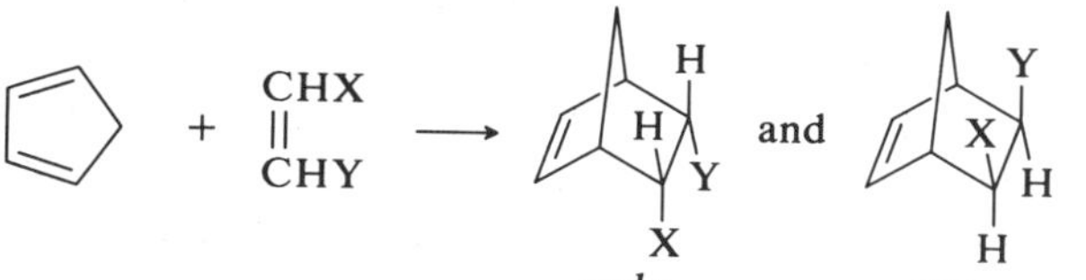

endo *exo* (3.60)

products, the *endo*- and *exo*-bicyclo[2,2,1]heptene derivatives (3.60). It is found in practice, however, that the *endo* isomer always predominates, except under conditions where isomerisation of the first-formed adduct occurs (see p. 231).

Compounds of the bicyclo[2,2,1]heptane type are widely distributed in nature among the bicyclic terpenes, and the Diels–Alder reaction provides a convenient method for their synthesis. Thus, cyclopentadiene and vinyl acetate react smoothly when warmed together to form the acetate (65) which is easily transformed into norcamphor (66) and related compounds.

$OCOCH_3$ → H, $OCOCH_3$ (65) → (1) hydrolysis (2) oxidation (3) hydrogenation → O (66)

(3.61)

1,3-Cyclohexadienes react with ethylenic dienophiles to form derivatives of bicyclo[2,2,2]octene. In general, the additions proceed more slowly than the corresponding reactions with cyclopentadiene. With acetylenic dienophiles, derivatives of bicyclo[2,2,2]octadiene are formed initially, but these often undergo a retro Diels–Alder reaction (see p. 219) with elimination of an alkene and formation of a benzene derivative, as illustrated in (3.62).

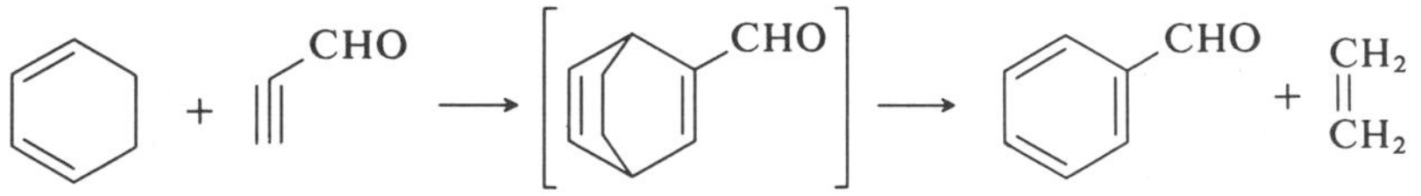

CHO → [CHO] → CHO + $CH_2{=}CH_2$

(3.62)

Cyclopentadienones and ortho-quinones

Derivatives of cyclopentadienone and of *ortho*-quinones also form adducts with ethylenic and acetylenic dienophiles. Many cyclopentadienones can act as dienes and dienophiles, and this is particularly true of cyclopentadienone itself which dimerises spontaneously. The adducts obtained from cyclopentadienone lose carbon monoxide easily on heating, with formation of dihydrobenzene or benzene derivatives.

Ortho-quinones can also react both as dienes and dienophiles. *o*-Benzoquinone spontaneously forms a stable crystalline dimer in acetone at room temperature. But it appears that, in general, their capacity for acting as dienes predominates (Ansell *et al.*, 1971).

Furans

Many furan derivatives react with ethylenic and acetylenic dienophiles to form bicyclo compounds with an oxygen bridge (3.63). Most of the adducts obtained from furans are thermally labile and dissociate readily into their components on warming. The adduct from furan and maleic anhydride has been shown to have the *exo* structure (67), apparently violating the rule (p. 231) that the *endo* isomer predominates. The reason for this is found in the related observation that the

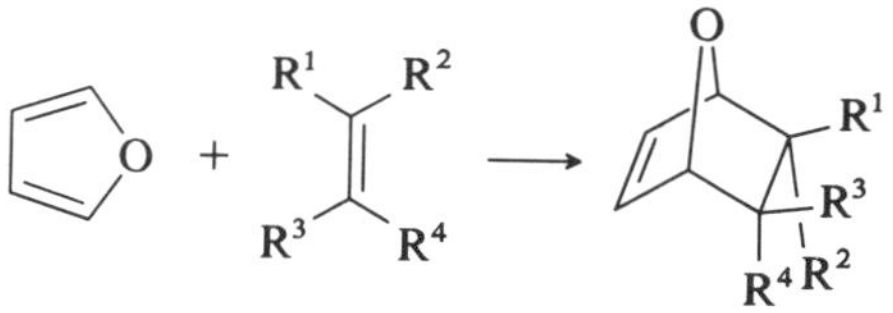

(3.63)

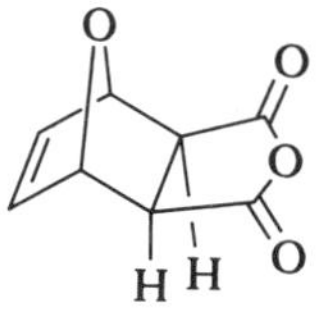

(67)

normal *endo* adduct formed from maleimide and furan at 20 °C dissociates at temperatures only slightly above room temperature and more rapidly on warming, allowing conversion of the *endo* adduct formed in the kinetically controlled reaction into the thermodynamically more stable *exo* isomer. With the maleic anhydride adduct, equilibration takes place below room temperature so that the *endo* adduct formed under kinetic control is not observed.

Pyrrole and its derivatives are unsuitable as dienes in the Diels–Alder reaction because the susceptibility of the nucleus to electrophilic substitution leads to side reactions. Thiophen and some of its simple derivatives react with acetylenic dienophiles under vigorous conditions to form benzene derivatives by extrusion of sulphur from the initially formed Diels–Alder adducts (Kuhn and Gollnick, 1972) (3.64). Thiophen-1,1-dioxide behaves both as a diene and dienophile and spontaneously dimerises on preparation.

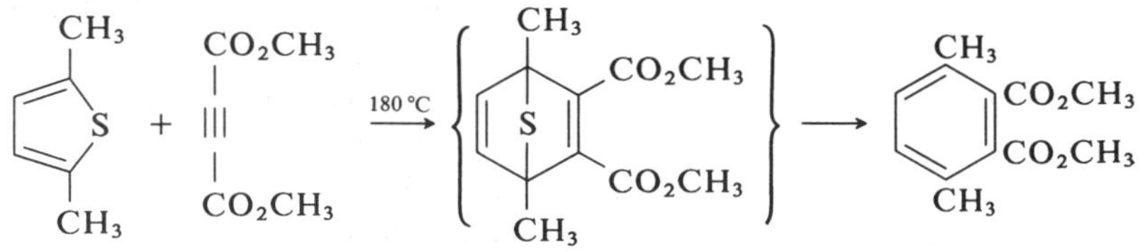

(3.64)

3.4. Intramolecular Diels–Alder reactions

The usefulness of intermolecular Diels–Alder reactions in synthesis is evident from the examples which have already been given. The *intra*molecular reaction, in which the diene and dienophile form part of the same molecule (3.65), has also been widely used in the synthesis of natural products in the alkaloid, steroid and terpenoid series (Bridger and Bennett, 1980; Oppolzer, 1981; Kametani and Nemoto, 1981; Fallis,

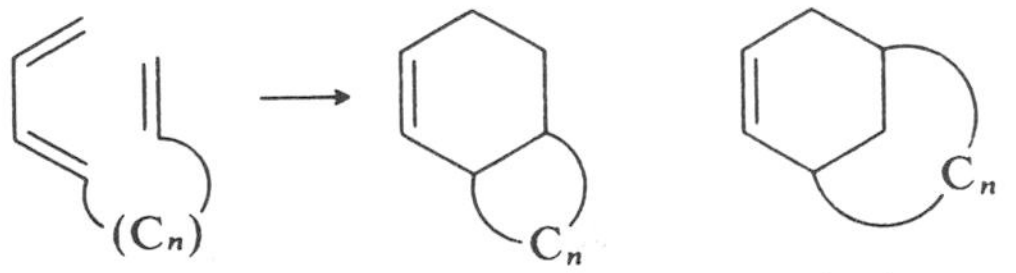

(3.65)

(68) (69) (70)

1984). Because of the favourable entropy factor the intramolecular reactions often proceed more easily than comparable intermolecular reactions leading to the formation of two new rings with several chiral centres of predetermined configurations in one operation. Thus, the *trans*-dienylacrylic ester (71) gave the *trans*-hydrindane (72) while the *cis* ester gave the *cis*-hydrindane (74). In contrast to the comparatively easy cyclisation of (73) *inter*molecular Diels–Alder reactions with *Z*-1-substituted butadienes generally take place only with difficulty.

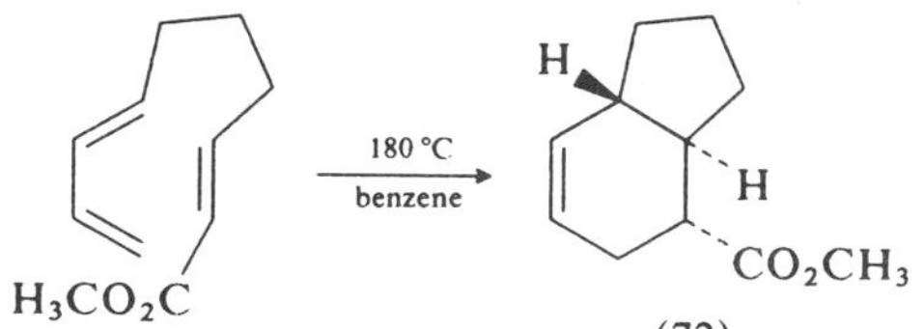

(3.66)

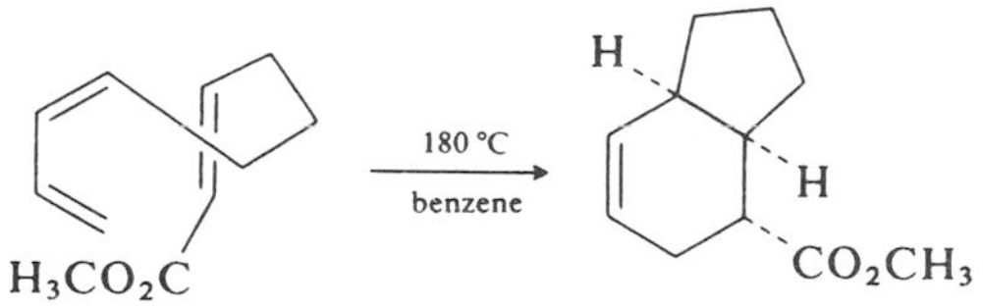

(71) → 180 °C, benzene → (72)

(73) → 180 °C, benzene → (74)

In intramolecular Diels–Alder reactions involving 1-substituted butadienes (68) reaction could apparently take place in two ways to give the fused ring product (69) or the bridged ring compound (70). Generally, cyclisation to give the fused ring product is highly favoured and very few examples of the formation of bridged ring compounds from 1-substituted butadienes (68) are known. With 2-substituted butadienes, however, bridged ring compounds are necessarily formed, and reactions of this kind have been used to make a series of bridgehead alkenes (Shea *et al.*, 1982) and bridgehead enol lactones. Further manipulation of the latter provides a useful route to stereochemically defined substituted cyclohexanones (Shea and Wada, 1982).

385 °C

O

O

(70%)

(3.67)

O

O

$C_2H_5O_2C$

185 °C

toluene

O

O

several

steps

$C_2H_5O_2C$

O

$C_2H_5O_2C$

$CO_2C_2H_5$

Many intramolecular Diels–Alder reactions involving 2-substituted butadienes are catalysed by Lewis acids and in this way can be effected in high yield at ordinary temperatures (Shea and Gilman, 1983).

A valuable feature of the intramolecular reactions, as of intermolecular Diels–Alder reactions, is their stereoselectivity. Thus, the triene (75) forms the indane derivative (77) in a reaction in which four new chiral centres are set up selectively in one step, by way of the sterically preferred transition state (76) (Edwards *et al.*, 1984; Nicolaou and Magolda, 1981; Roush and Myers, 1981). Frequently, products formed by way of an *endo* transition state predominate, as in the above example, but this is not universally so. In some cases non-bonding interactions of substituent groups and other steric constraints in the transition state, may lead to the formation of the *exo* adduct or mixtures of stereoisomers (compare Roush, Gillis and Ko, 1982; White and Sheldon, 1981). Much greater stereocontrol with preferential formation of the *endo* adduct, can be achieved by carrying out the reactions at ordinary temperatures in the presence of Lewis acid catalysts (cf. Roush and Myers, 1981; Roush, Gillis and Ko, 1982). It appears also that greater control of stereochemistry may be possible in some intramolecular Diels–Alder reactions by using *Z*-1-substituted 1,3-dienes rather than the *E*-dienes commonly employed (Boeckman and Alessi, 1982; Pyne, Hensel and Fuchs, 1982), a further illustration of how much more readily the

$C_2H_5O_2C$ RO (75) $\xrightarrow[\text{toluene}]{130\,°C}$ $C_2H_5O_2C$ RO (76)

(3.68)

RO H $C_2H_5O_2C$ H H H (77)

intramolecular reactions proceed than their intermolecular counterparts; intermolecular Diels–Alder reactions involving *Z*-1,3-dienes are rare.

This can be understood by consideration of the geometry of the transition states. Models show that in intramolecular additions to *E*-dienes the diene and dienophile are necessarily oriented as in *A* and *B*; the orientation *C* is too strained unless the bridging chain is (considerably) longer than the three or four atoms normally encountered (3.69). In *Z*-dienes the bridge is forced into the *exo* positions *F* or *G* of which *F* appears to be entropically more favourable; transition state *D* is too strained. *cis*-Adducts thus arise with high stereoselectivity from *Z*-dienes, via transition state *F*, as long as the experimental conditions are sufficiently mild to preclude isomerisation of the diene to the *E*-form. In the thermally more stable *E*-dienes the transition state orientations *A* and *B* are both strain-free. As a result the energy difference between them depends on bonding and non-bonding interactions of substituents and on other conformational influences which are not easily predictable, and mixtures of products sometimes result.

The stereochemistry of the products formed in intramolecular Diels–Alder reactions may be affected also by the conformation of the bridge between the diene and dienophile. Thus, whereas the amine (78) gave the *trans*-fused product (79), the closely related amide (80), in which we now have an sp^2 carbon atom in the bridge, gave the *cis*-fused product (81) (Oppolzer, 1974). This is rationalised by supposing that in the case of the amide the reaction passes predominantly through the *endo*

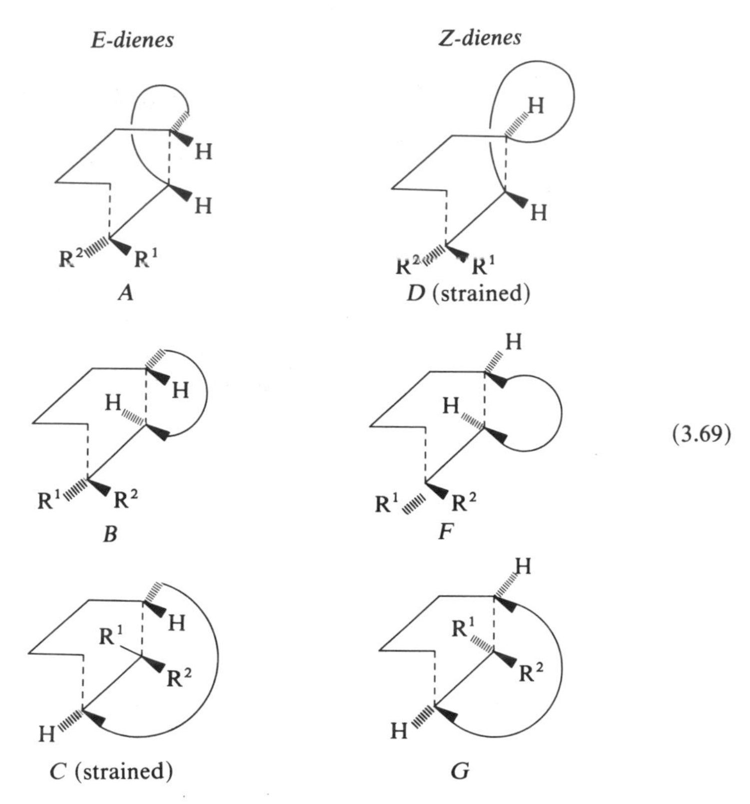

(3.69)

transition state (82), leading to the *cis* product, whereas the *exo* orientation (83) is favoured for the cyclisation of the amine. These factors apply, of course, only to the kinetically controlled reactions.

Δ

(78) (79) (3.70)

Δ

(80) (81)

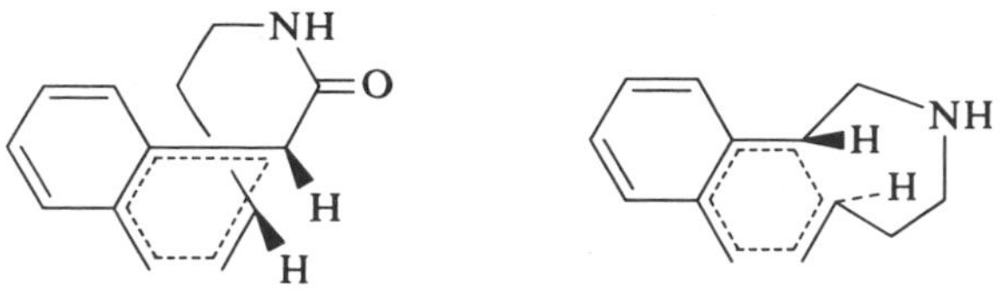

(82) (83) (3.71)

The nature of the bridge may also affect the ease of reaction. Thus, the ester (84) gave only small yields of cyclised products even at 220 °C whereas the closely related ether (85) cyclised readily at 170 °C. It is suggested that the geometry of the transition state required for cyclisation of (84) prevents complete overlap of the lone pair electrons of the ethereal oxygen with the π-system of the carbonyl group and that reaction is thereby kinetically disfavoured in (84) compared with (85) where no such effect operates (Boeckman and Demko, 1982).

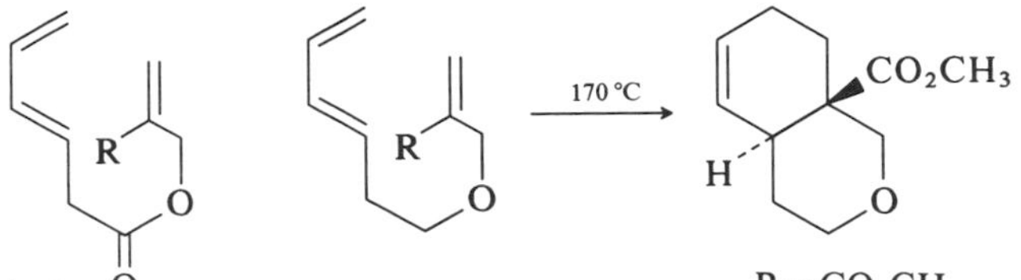

(84) (85) 170 °C $R = CO_2CH_3$ (3.72)

3.5. The retro Diels–Alder reaction

Diels–Alder reactions are reversible, and on heating many adducts dissociate into their components, sometimes under quite mild conditions. This can be made use of as in, for example, the separation of anthracene derivatives from mixtures with other hydrocarbons through their adducts with maleic anhydride, and in the preparation of pure D vitamins from the mixtures obtained by irradiation of the provitamins.

More interesting are reactions in which the original adduct is modified chemically and subsequently dissociated to yield a new diene or dienophile (Kwart and King, 1968; Ripoll, Rouessac and Rouessac, 1978). Thus, catalytic hydrogenation of the adduct obtained from 4-vinylcyclohexene and anthracene followed by thermal dissociation affords vinylcyclohexane, whereas direct hydrogenation of vinylcyclohexene itself results in reduction of both double bonds. In this case adduct formation has been used to protect one of the double bonds of the vinylcyclohexene.

It is not always the bonds formed in the original diene addition which are broken in the retro reaction. Examples have already been noted for the adducts formed from cyclohexadiene and acetylenic dienophiles (p. 213). A reaction of this kind was used in an ingenious synthesis of benzocyclopropene (88) (3.73). Addition of methyl acetylenedicarboxylate to 1,6-methanocyclodecapentaene (86) afforded the adduct (87) from

(86) + CO_2CH_3–C≡C–CO_2CH_3 → (87) (3.73)

$\xrightarrow[400\,°C]{vacuum}$ (88) + dimethyl phthalate

which benzocyclopropene was obtained in 45 per cent yield on pyrolysis at 400 °C, with elimination of dimethyl phthalate. In another example furan-3,4-dicarboxylic acid was readily prepared by decomposition of the hydrogenated adduct from furan and acetylenedicarboxylic ester (3.74), and substituted fumaric esters were obtained in good yield by retro Diels–Alder reaction from compounds (90), themselves prepared by alkylation of the readily available adduct (89) (Girard and Bloch, 1982).

$\xrightarrow{H_2,\ Pd}$ $\xrightarrow{heat}$ $CH_2{=}CH_2$ + furan-3,4-(CO_2CH_3)$_2$

(89) $\xrightarrow[(2)\ RHal]{(1)\ 2LiN(iso\text{-}C_3H_7)_2}$ (90) (3.74)

$\downarrow$ 650 °C

CH_3O_2C–C(R)=CH–CO_2CH_3 + cyclopentadiene

Most retro Diels–Alder reactions are brought about by heat, but some photo-induced reactions have been observed (Nozake, Kato and Nyori, 1969; Ripoll, Rouessac and Rouessac, 1978).

3.6. Catalysis by Lewis acids

Catalysis of the Diels–Alder reaction by acids has been known for some time, but the influence of catalysts on the rate has generally been small. It has been found, however, that some Diels–Alder condensations are accelerated remarkably by aluminium chloride and other Lewis acids such as boron trifluoride and tinIV chloride (Yates and Eaton, 1960; Sauer, 1967). Thus equimolecular amounts of anthracene, maleic anhydride and aluminium chloride in methylene chloride solution gave a quantitative yield of adduct in 90 s at room temperature. It is estimated that reaction in absence of the catalyst would require 4800 h for 95 per cent completion. Similarly, butadiene and methyl vinyl ketone react in 1 h at room temperature in presence of tinIV chloride to give a 75 per cent yield of acetylcyclohexene. In absence of a catalyst no reaction takes place.

Pure *cis* addition is observed in reactions carried out in presence of Lewis acids just as in the uncatalysed reaction (see p. 230), that is the relative orientation of substituents in the diene and dienophile is preserved in the adduct. But the ratio of structurally isomeric or stereoisomeric 1:1 adducts may be different in the catalysed reactions and generally greatly increased selectivity is observed. Thus on addition of methyl vinyl ketone to isoprene two structural isomers are formed, of which the predominant one is that with the greatest separation between the two groups (3.75). The proportion of this product is even greater in the catalysed reaction. Many other examples of this effect have been recorded (Kreiser, Haumesser and Thomas, 1974).

$COCH_3$ + ⟶ $COCH_3$ + $COCH_3$ (3.75)

no catalyst, toluene, 120 °C; ratio 71:29

$SnCl_4.5H_2O$, benzene, <25 °C; ratio 93:7

Similarly, in the addition of acrylic acid to cyclopentadiene the proportion of *endo* adduct was found to increase noticeably in presence of aluminium chloride etherate (Sauer and Kredel, 1966) (3.76). Other Lewis acid catalysts had the same effect. These effects are ascribed to complex formation between the Lewis acid and the polar groups of the dienophile which brings about changes in the energies and orbital coefficients of the frontier orbitals of the dienophile.

The influence of Lewis acids on the course of Diels–Alder reactions has been rationalised by application of frontier orbital theory. In a normal Diels–Alder reaction, that is one involving an electron-deficient dienophile and an electron-rich diene, the main interaction is that

$$\text{cyclopentadiene} + CH_2{=}CHCO_2H \longrightarrow \textit{endo} + \textit{exo} \quad (3.76)$$

		endo	*exo*
0 °C, no catalyst;	ratio	84	16
0 °C, +47% $AlCl_3.O(C_2H_5)_2$;	ratio	93	7
−70 °C, +47% $AlCl_3.O(C_2H_5)_2$;	ratio	97	3

between the highest occupied molecular orbital (HOMO) of the diene and the lowest unoccupied orbital (LUMO) of the dienophile, and the smaller the energy difference between these orbitals and the better the overlap the more readily the reaction occurs (Houk, 1975; Sustmann, 1974). Coordination of a Lewis acid with, say, the non-bonding electrons of a carbonyl or cyano group in the dienophile lowers the energies of the frontier orbitals of the dienophile and alters the distribution of the atomic orbital coefficients. The energy difference between the HOMO of the diene and the LUMO of the dienophile is thus reduced and, while the energy difference between the HOMO of the dienophile and the LUMO of the diene is increased at the same time, the former effect predominates resulting in stabilisation of the reaction complex and faster reaction. This is illustrated diagrammatically below for the reaction between butadiene and methyl acrylate.

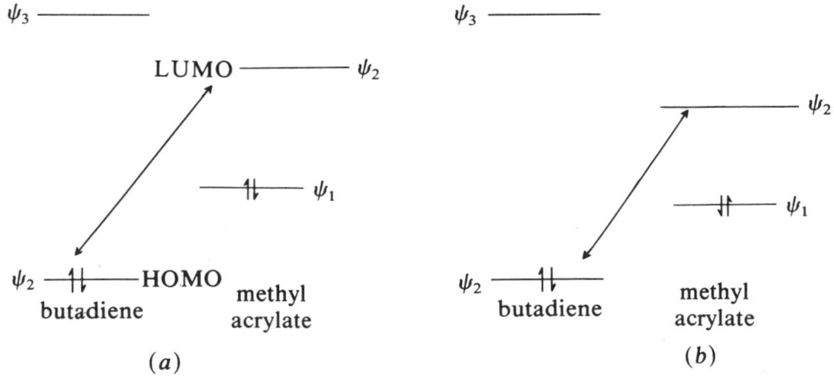

Effect of Lewis acid catalysts on the interaction of the frontier orbitals of butadiene and methyl acrylate (*a*) non-catalysed reaction, (*b*) catalysed reaction.

In agreement with these general ideas, lowering of the HOMO of the diene by conjugation with an electron-attracting substituent results in a slower reaction. In Diels–Alder reactions with inverse electron demand, interaction of the LUMO of the *diene* and the HOMO of the *dienophile* is the controlling factor, and reaction is facilitated by electron release in the dienophile which raises the energy of the frontier orbitals.

The increased regio-selectivity shown in catalysed reactions has also been accounted for by consideration of frontier orbital interactions. The orientation of the compounds in the transition state of a Diels–Alder reaction is governed by the orbital coefficients of the reacting atoms, the atom with the larger coefficient in the dienophile interacts preferentially with that with the larger coefficient in the diene, since this leads to more efficient overlap of orbitals. For the reaction of 2-phenylbutadiene with methyl acrylate, for example, the interaction (*a*) is favoured over (*b*),

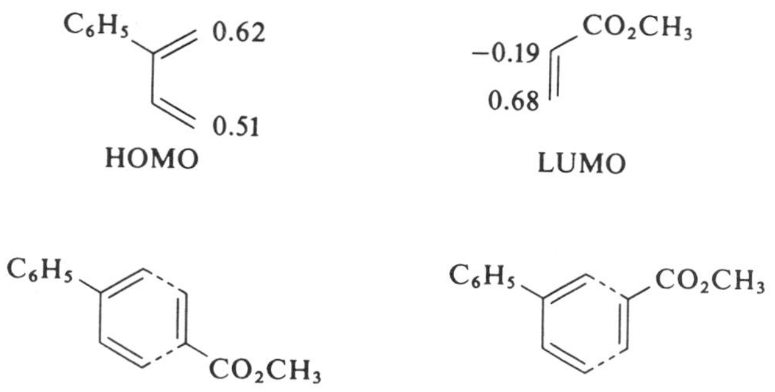

C_6H_5 0.62 0.51 HOMO

CO_2CH_3 −0.19 0.68 LUMO

(3.77)

C_6H_5 CO_2CH_3 (*a*)

C_6H_5 CO_2CH_3 (*b*)

and the ratio of *para* to *meta* products found is 80:20. In the catalysed reaction, interaction of the catalyst with the ester group of the methyl acrylate increases the difference between the coefficients at C-2 and C-3 with the result that the reaction becomes more selective; transition state (*a*) is favoured even more and the *para*: *meta* ratio rises to 97:3. Secondary orbital interactions may also play a part in determining the orientation of the adducts obtained in some catalysed reactions (Alston and Ottenbrite, 1975).

The increase in *endo*: *exo* ratio in catalysed reactions is also ascribed to increased secondary orbital interactions. In a typical non-catalysed reaction, e.g. that between cyclopentadiene and acrolein, the preferred formation of the *endo* product is due to secondary interactions involving the carbonyl group of the acrolein and C-2 of the diene (see figure overleaf) (Hoffmann and Woodward, 1965). This interaction is greatly increased in the catalysed reaction because of a large increase in the coefficient of the carbonyl carbon atom.

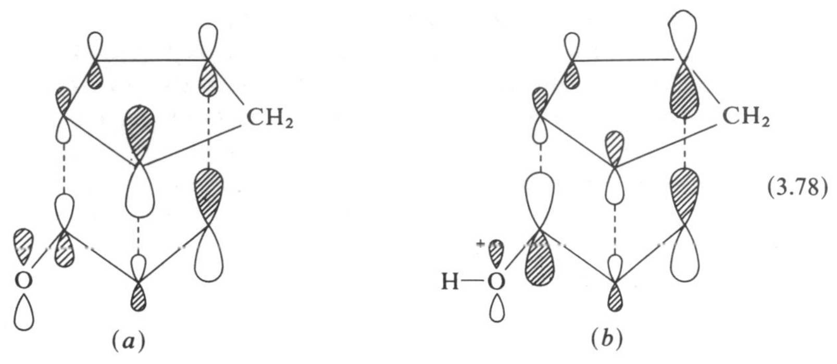

(3.78)

Diene HOMO–dienophile LUMO interactions in *endo* transition states with (*a*) acrolein and (*b*) protonated acrolein.

Hitherto it has been assumed that all Lewis acids which promote regioselectivity in a Diels–Alder reaction would enhance the formation of the same regio-isomer. It now appears that this is not so and that the orientation of the product may depend on the catalyst used. Thus, juglone (91) gave either (92) or (93) depending on whether the catalyst was boron trifluoride or boron triacetate (Gupta, Jackson and Stoodley, 1982). No adequate general explanation for these effects is yet available.

(93) CH_3, HO ← (1) [isoprene], $B(OCOCH_3)_3$, benzene, 20 °C; (2) $(C_2H_5)_3N$, CH_2Cl_2 (aromatisation) — (91)

(91) → (1) [isoprene], BF_3, $(C_2H_5)_2O$, toluene, 0 °C; (2) $(C_2H_5)_3N$, CH_2Cl_2 (aromatisation) → (92) HO, CH_3

(3.79)

Lewis acids, in general, enhance the reactivity of dienophiles containing oxygen-bearing substituents. Usefully, it has now been found that reactions involving hydrocarbon dienophiles are catalysed by the aminium radical cation, tris(*p*-bromophenyl)aminium hexachlorostibnate, thus allowing the preparation of some Diels–Alder adducts previously available in only poor yield. Thus, the Diels–Alder dimerisation of 1,3-cyclohexadiene has been effected in 30 per cent yield after 20 h at 200 °C. In the presence of the cation-radical the dimerisation occurs in 70 per cent yield within 15 minutes (Belleville and Bauld, 1982).

3.7. Regiochemistry of the Diels–Alder reaction

Foreknowledge of the regiochemistry to be expected in a Diels–Alder cyclo-addition is important in synthesis. Addition of an unsymmetrical diene to an unsymmetrical dienophile could apparently take place in two ways to give two structurally isomeric products. It is found in practice, however, that formation of one of the isomers is strongly favoured (Sauer, 1967). Thus, in the additon of acrylic acid derivatives to 1-substituted butadienes the '*ortho*' (1,2)-adduct is favoured, irrespective of the electronic nature of the substituent (see Table 3.2). The orientating forces are relatively weak, however, and with greater steric demand of the substituents the proportion of the '*meta*' (1,3)-isomer approaches the statistical value. The very strong '*ortho*' orientation observed with acrylic acid and butadiene-1-carboxylic acid is lost when the anions are used, presumably because of the Coulomb repulsion of the two charged groups.

TABLE 3.2 *Proportions of structural isomers formed in addition of acrylic acid derivatives to 1-substituted butadienes*

R^1 + R^2 CO_2R → R^1 H CO_2R R^2 (1,2-) R^1 H R^2 CO_2R (1,3-)

R^1	R^2	R	T °C	Ratio of products 1,2-	1,3
$N(C_2H_5)_2$	H	C_2H_5	20	1,2-only	—
CH_3	H	CH_3	20	18	1
CO_2H	H	H	70	1,2-only	—
CO_2Na	H	Na	220	1	1
$C(CH_3)_3$	$C(CH_3)_3$	CH_3	200	0·9	1
CH_2OH	H	CH_3	190	1	1

Correspondingly, in the addition of methyl acrylate to 2-substituted butadienes the '*para*' (1,4)-adduct is formed predominantly irrespective of the electronic nature of the substituent (see Table 3.3). Addition of acetylenic dienophiles to 1- and 2-substituted butadienes also results respectively in preferential formation of '*ortho*' and '*para*' adducts.

As would be expected, in 1,3-disubstituted butadienes, the directive influence of the substituents is additive. Thus, in the reaction of 1,3-dimethylbutadiene and acrylic acid, the adduct (94) is formed almost exclusively. However, the magnitude of the orienting effect differs from one substituent to another and also with their position on the diene; a

TABLE 3.3 *Proportions of structural isomers formed in addition of methyl acrylate to 2-substituted butadienes*

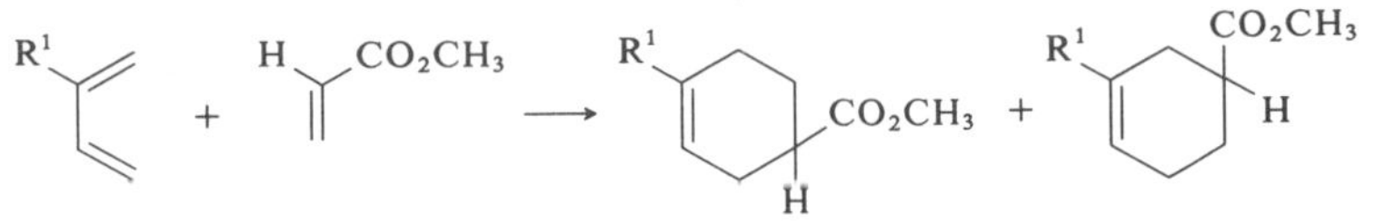

(Reaction scheme: 2-R^1-butadiene + $H_2C{=}CH{-}CO_2CH_3$ ⟶ 1,4-adduct + 1,3-adduct)

R_1	T °C	Ratio 1,4- to 1,3-
OC_2H_5	160	1,4-only
C_6H_5	150	4·5 to 1
CN	95	1,4-only

substituent at C-1 generally has a more pronounced directing effect than it has at C-2. With 1,2-and 2,3-disubstituted butadienes, therefore, the structure of the adducts obtained will depend on the nature and disposition of the substituents. Thus, reaction of 2-methyl-3-phenylbutadiene with acrylic acid gives mainly the adduct (95) and only a minor amount of the isomer (96); here clearly the phenyl substituent has a stronger

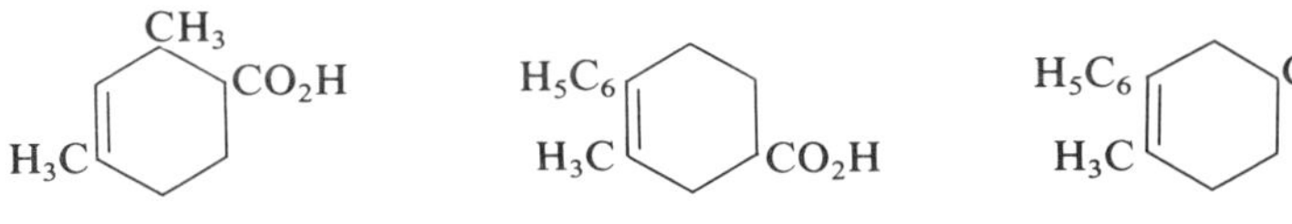

(94) (95) (96) (3.80)

directing effect than the methyl group. This hierarchical order of substituent effects can be exploited to obtain adducts of unusual orientation through the use of dienes and dienophiles containing 'temporary' substituent groups which control the orientation of addition and are then removed after the adduct has been formed. The phenylsulphide and nitro substituent groups are particulary useful in this way.

Thus, with both 1-acetoxy-4-(phenylthio)butadiene (Trost, Ippen and Vladuchick, 1977) and 2-methoxy-1-(phenylthio)butadiene (Cohen and Kosarych, 1982) the direction of addition to dienophiles is controlled by the sulphide substituent. 2-Methoxy-1-phenylthiobutadiene, for example, reacts with methyl vinyl ketone in the presence of magnesium bromide catalyst to give the adduct (97) in which the methoxy group is '*meta*' to the carbonyl group of the dienophile; with 2-methoxybutadiene itself the product is the '*para*' isomer. Mild hydrolysis of the enol ether in (97) provides the corresponding β-keto sulphide. Regio-control by the sulphur substituent is also observed in the thermal Diels–Alder reactions of 2-acetoxy- and 2-methoxy-3-phenylthio-1,3-butadiene

CH_3O ... SC_6H_5 + ... $COCH_3$ $\xrightarrow[25\,°C]{MgBr_2}$ (3.81)

CH_3O ... H ... $COCH_3$ ... H SC_6H_5 (97) $\longrightarrow$ O ... H ... $COCH_3$ ... H SC_6H_5

(Trost, Vladuchick and Bridges, 1980). The sulphur substituent in the products of these reactions can be removed, if desired, by reductive cleavage or by conversion into the sulphoxide and elimination. Thus, the adduct (98) from 2-methoxy-3-phenylthio-1,3-butadiene and methyl vinyl ketone was used in a synthesis of the terpene carvone as illustrated in (3.82).

CH_3O, C_6H_5S (98) — (1) $CH_2{=}P(C_6H_5)_3$ (2) H_3O^+ → C_6H_5S — NaH, CH_3I → H_3C, C_6H_5S — Na—Hg, methanol → H_3C (3.82)

(1) m-$ClC_6H_4CO_3H$, CH_2Cl_2 ($C_6H_5S \rightarrow C_6H_5SO$) (2) reflux, CH_2Cl_2 → H_3C carvone

A useful feature of dienes carrying sulphide substituents is that oxidation to the corresponding sulphoxide reverses the electronic affinity of the diene so that it can then react with electron-rich dienophiles such as enamines. A reaction of this kind formed a key step in a synthetic approach to hasubanan alkaloids (Evans, Bryan and Sims, 1972).

Regio-control in Diels–Alder reactions can be effected by proper choice of activating groups in the dienophile as well as in the diene. Thus, in β-nitro-$\alpha\beta$-unsaturated ketones and β-nitro-$\alpha\beta$-unsaturated esters the nitro group controls the orientation of addition; reaction of 3-nitrocyclohexenone and 1,3-pentadiene, for example, readily affords the adduct

(99). Reductive removal of the nitro group with tri-n-butyltin hydride gives the bicyclic ketone (100) with orientation opposite to that in the product obtained from reaction of pentadiene with cyclohexenone itself (Ono, Miyaki and Kaji, 1982). Alternatively, elimination of nitrous acid with 1,5-diazabicyclo[4,3,0]non-5-ene (DBN) leads to the dienone (101).

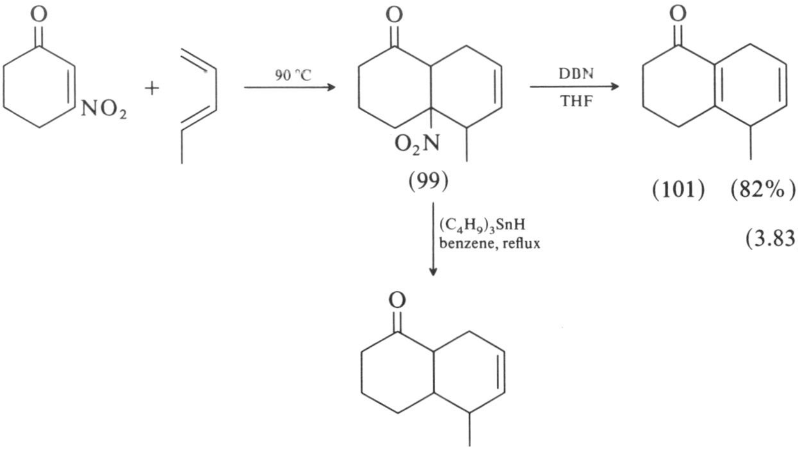

O NO_2 + → 90 °C O O_2N (99) → DBN THF O (101) (82%)

$(C_4H_9)_3SnH$ benzene, reflux

(3.83)

O

(100) (80%)

The reason for these orientation effects has long been puzzling, but they have now been rationalised using frontier orbital theory (Houk, 1975; Sustman, 1974; Fleming, 1976).

In a 'normal' Diels–Alder reaction (i.e. one involving a conjugated or electron-rich diene and an electron-deficient dienophile), the main interaction in the transition state is between the HOMO of the diene and the LUMO of the dienophile and the orientation of the product obtained from an unsymmetrical diene and an unsymmetrical dienophile is governed largely by the atomic orbital coefficients at the termini of the conjugated systems concerned. The atoms with the larger terminal coefficients on each addend bond preferentially in the transition state, since this leads to better orbital overlap. It turns out that in most cases this leads mainly to the 1,2('*ortho*') adduct with 1-substituted butadienes and to the 1,4('*para*') adduct with 2-substituted butadienes. Thus, for butadiene-1-carboxylic acid and acrylic acid (Table 3.2) the frontier orbitals are polarised as shown below, where the size of the circles is roughly proportional to the size of the coefficients. Shaded and empty circles represent lobes of opposite sign, an 'allowed' reaction involving overlap of lobes of the same sign. Similarly for reaction of 2-phenylbutadiene and methyl acrylate, preferential formation of the '*para*' adduct is predicted and found (Table 3.3). With 2-methyl-1,3-butadiene, however, where

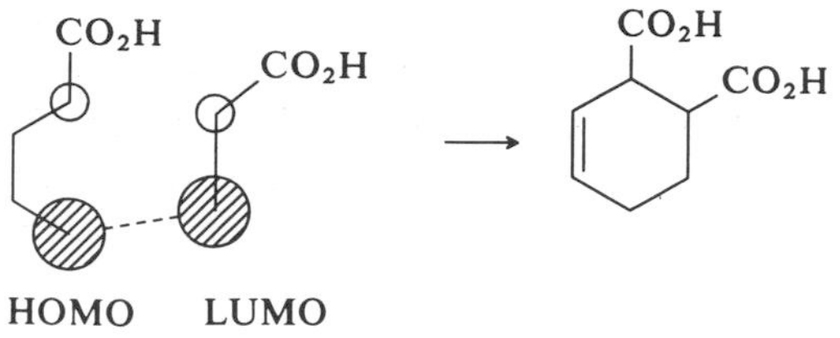

CO₂H CO₂H → (3.84)

HOMO LUMO

the coefficients of the terminal carbon atoms in the HOMO do not differ from each other so much as they do in the 2-phenyl compound, reaction with methyl acrylate or acrolein gives larger amounts of the '*meta*'(1,3)-adduct.

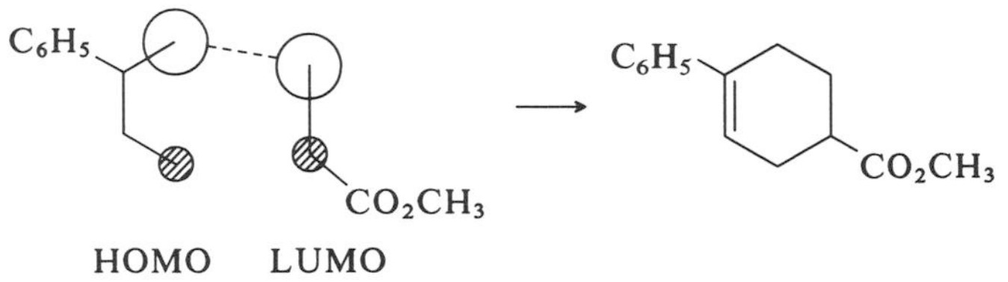

(3.85)

HOMO LUMO

Secondary orbital interactions may also play a part in determining the course of the reaction particularly when primary interactions provide little preference for one regio-isomer (Houk *et al.*, 1978; Alston, Ottenbrite and Cohen, 1978).

The relative amounts of the structurally isomeric products formed in these reactions are strongly influenced by the presence of Lewis acid catalysts, and when these conditions are applicable very high yields of a single isomer can be obtained. Thus, for the addition of methyl vinyl ketone or acrolein to isoprene the proportion of the '*para*' adduct was increased in presence of tinIV chloride, so that it became almost the exclusive product of the reaction (3.86). Similar effects were noted for the addition of isoprene to methyl acrylate.

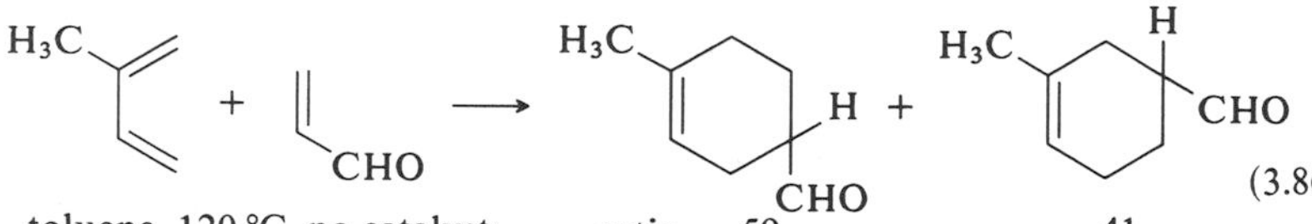

(3.86)

toluene, 120 °C, no catalyst;	ratio	59	41
benzene, 25 °C, $SnCl_4.5H_2O$;	ratio	96	4

3.8. Stereochemistry of the Diels–Alder reaction

The great synthetic usefulness of the Diels–Alder reaction depends not only on the fact that it provides easy access to a variety of six-membered ring compounds, but also on its remarkable stereoselectivity. This factor more than any other has contributed to its successful application in the synthesis of a number of complex natural products. It should be noted, however, that the high stereoselectivity applies only to the kinetically controlled reaction and may be lost by epimerisation of the product or starting materials, or by easy dissociation of the adduct allowing thermodynamic control of the reaction. These factors are fully discussed by Martin and Hill (1961).

The cis *principle*

The stereochemistry of the adduct obtained in many Diels–Alder reactions can be selected on the basis of two empirical rules formulated by Alder and Stein in 1937. According to the '*cis* principle', which is very widely followed, the relative stereochemistry of substituents in both the dienophile and the diene is retained in the adduct. That is, a dienophile with *trans* substituents will give an adduct in which the *trans* configuration of the substituents is retained, while a *cis* disubstituted dienophile will form an adduct in which the substituents are *cis* to each other. For example, in the reaction of cyclopentadiene with dimethyl maleate the *cis* adducts (102) and (103) are formed while in the reaction with dimethyl fumarate the *trans* configuration of the ester groups is retained in the adduct (104) (3.87).

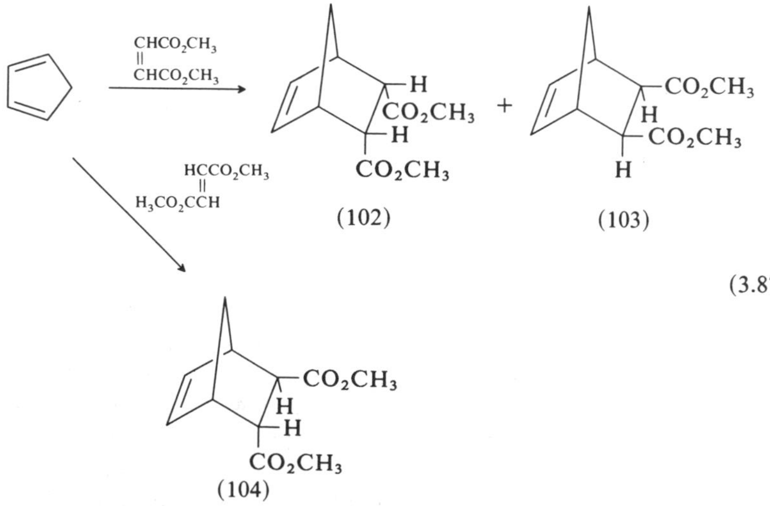
CHCO2CH3
||
CHCO2CH3
HCCO2CH3
||
H3CO2CCH
H
CO2CH3
H
CO2CH3
+
CO2CH3
H
CO2CH3
H
(102)
(103)
CO2CH3
H
H
CO2CH3
(104)

(3.87)

Similarly with the diene component, the relative configuration of the substituents in the 1- and 4-positions is retained in the adduct; a *trans,trans*-1,4-disubstituted diene gives rise to adducts in which the

1-and 4-substituents are *cis* to each other, and a *cis,trans*-disubstituted diene gives adducts with *trans* substituents (3.88).

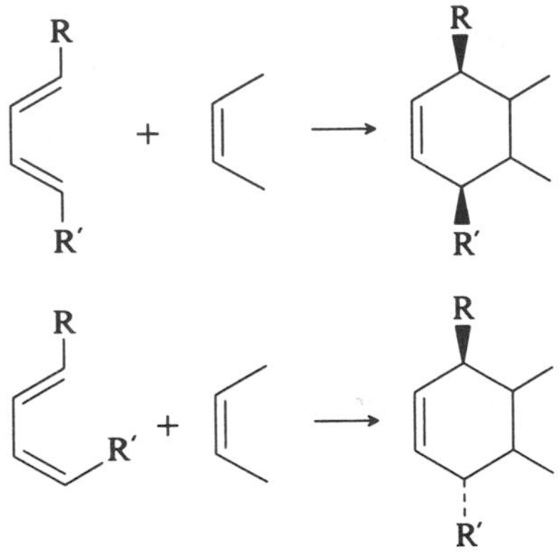

R R R' R' (3.88)

R R R' R'

The almost universal application of the *cis* principle provides strong evidence for a mechanism for the Diels–Alder reaction in which both the new bonds between the diene and the dienophile are formed at the same time. But a two-step mechanism is not completely excluded, for the same stereochemical result would obtain if the rate of formation of the second bond in the (diradical or zwitterionic) intermediate (105), (3.89), were faster than the rate of rotation about a carbon–carbon bond (compare Bartlett, 1970, 1971).

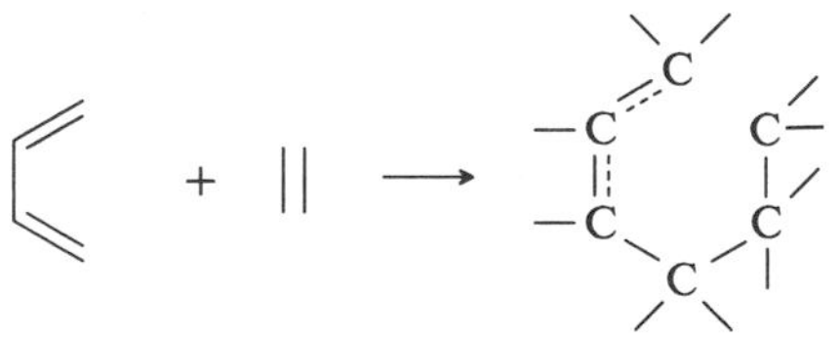

(3.89)

(105)

The endo *addition rule*

In the addition of maleic anhydride to cyclopentadiene, two different products, the *endo* and the *exo*, might conceivably be formed depending on the manner in which the diene and the dienophile are disposed in the transition state. According to Alder's *endo* addition rule, in a diene addition reaction the two components arrange themselves in parallel planes, and the most stable transition state arises from the orientation in which there is 'maximum accumulation of double bonds'. Not only the double bonds which actually take part in the addition are taken into account, but also the π-bonds of the activating groups in the dienophile. The rule appears to be strictly applicable only to the addition of cyclic dienophiles to cyclic dienes, but it is a useful guide in many other additions as well.

Thus, in the addition of maleic anhydride to cyclopentadiene the *endo* product, formed from the orientation wtih maximum accumulation of double bonds, is produced almost exclusively (3.90). The thermodynamically more stable *exo* compound is formed in yields of less than 1.5 per

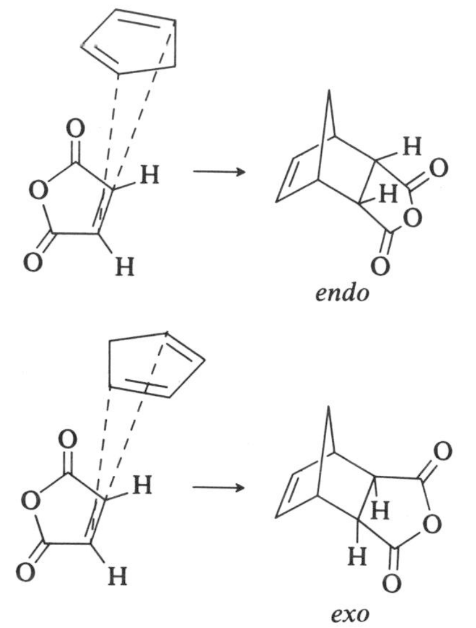

endo (3.90)

exo

cent. From benzoquinone and cyclopentadiene again only the *endo* adduct (106) was isolated, the configuration of the product being shown by its conversion into the 'caged' compound (107) with ultra-violet light (3.91).

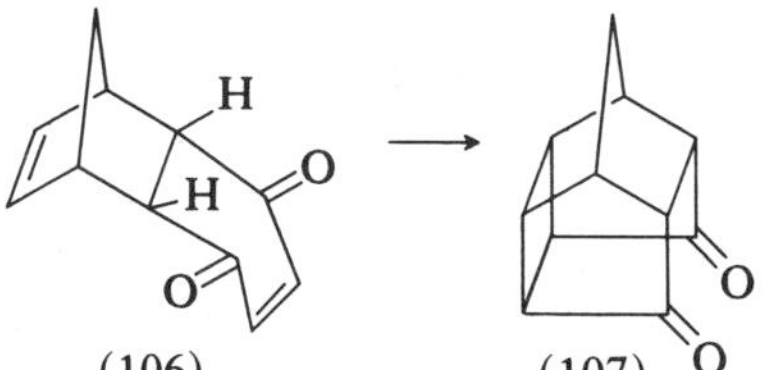

(106) (107) (3.91)

The products obtained from the cyclic diene furan and maleic anhydride and from diene addition reactions of fulvene do not obey the *endo* rule. The reason is that the initial *endo* adducts easily dissociate at moderate temperatures, allowing conversion of the kinetic *endo* adduct into the thermodynamically more stable *exo* isomer (cf. p. 219). In other cases prolonged reaction times may lead to the formation of some *exo* isomer at the expense of the *endo*.

TABLE 3.4 *Proportion of* endo *and* exo *acids formed in addition of α-substituted acrylic acids to cyclopentadiene*

$H_2C{=}C(X)CO_2H$ + cyclopentadiene → (endo CO_2H adduct) + (exo CO_2H adduct)
(108)

X	*endo* CO_2H	*exo* CO_2H
	endo CO_2H	*exo* CO_2H
H	75	25
CH_3	35	65
C_2H_5	—	100
C_6H_5	60	40
Br	30	70

In the addition of open-chain dienophiles to cyclic dienes, the *endo* rule is not always obeyed and the composition of the mixture obtained may depend on the precise structure of the dienophile and on the reaction conditions. Thus, in the addition of acrylic acid to cyclopentadiene the *endo* and *exo* products were obtained in the ratio 75:25 but in the α-substituted acrylic acids, (108) the product ratio varied depending on the nature of the group X (Martin and Hill, 1961, p. 550) (see Table 3.4). Equally variable ratios are observed in reactions with β-substituted acrylic acids. With acrylic acid itself, the proportion of *endo* adduct formed was noticeably increased by the presence of Lewis acid catalysts (p. 222).

Solvent and temperature may also affect the product ratio. Thus, in the kinetically controlled addition of cyclopentadiene to methyl acrylate, methyl methacrylate and methyl *trans*-crotonate in different solvents, the proportion of *endo* product increased with the polarity of the solvent, and the product ratio was also slightly affected by the temperature of the reaction. In all cases mixtures of the *endo* and *exo* products were obtained and with methyl methacrylate the *exo* isomer was the predominant product under all experimental conditions. With methyl crotonate the *exo* adduct was predominant in some solvents (e.g. trimethylamine at 30 °C) and the *endo* in others (ethanol, acetic acid) (see Fig. 3.1).

The adducts obtained from acyclic dienes and cyclic dienophiles are frequently formed in accordance with the *endo* rule. Thus, in the addition of maleic anhydride to *trans,trans*-1,4-diphenylbutadiene the *cis* adduct (109) is formed almost exclusively (3.92) through the orientation of diene and dienophile with 'maximum accumulation of double bonds'. Again in accordance with the *endo* rule *cis*-1-ethyl-1,3-butadiene and maleic

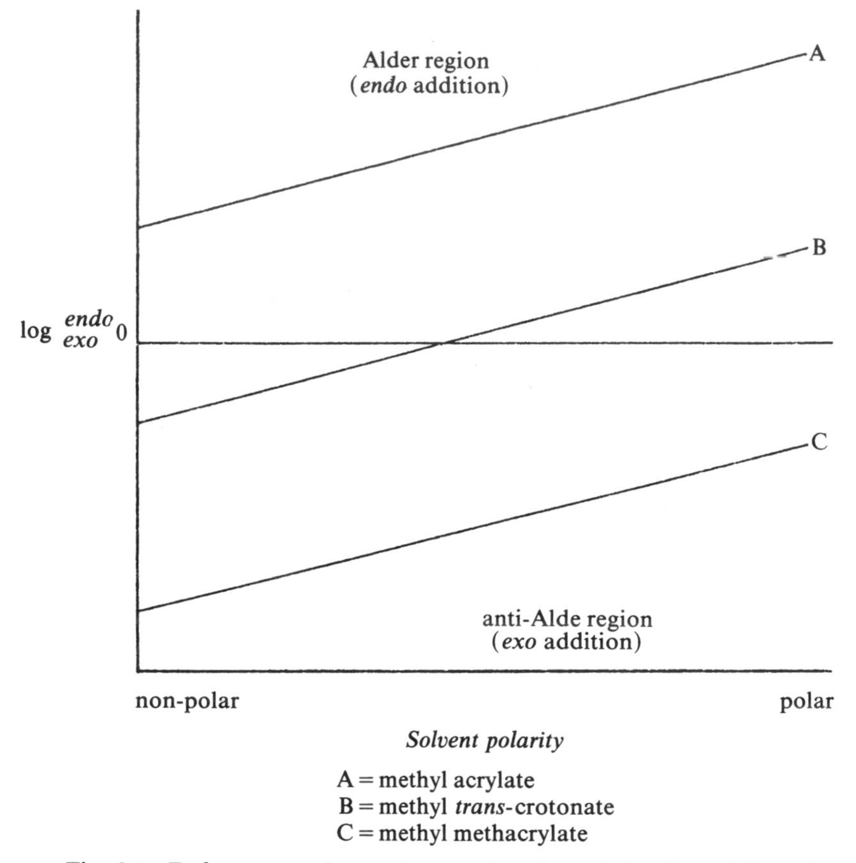

Fig. 3.1. *Endo*: *exo* product ratio as a function of the dienophile and solvent polarity in addition of methyl acrylate derivatives to cyclopentadiene (schematic). (From Berson, Hamlet and Mueller, 1962.)

anhydride afford the adduct (110) in which in this case the ethyl substituent is *trans* to the anhydride group.

In the addition reactions of open-chain dienes and open-chain dienophiles the *endo* adduct is the main product at moderate temperatures, but in a number of cases it has been found that the proportion of the *exo* isomer increases with rise in temperature. Thus, in the addition of acrylonitrile to *trans*-1-phenylbutadiene at 100 °C the *cis*(*endo*) isomer (111) was obtained, whereas at reflux temperature the main product was the *trans* (*exo*) isomer (112) (3.93). Table 3.5 shows the effect of temperature on the ratio of *cis* and *trans* adducts obtained by reaction of *trans*-butadiene-1-carboxylic acid and acrylic acid.

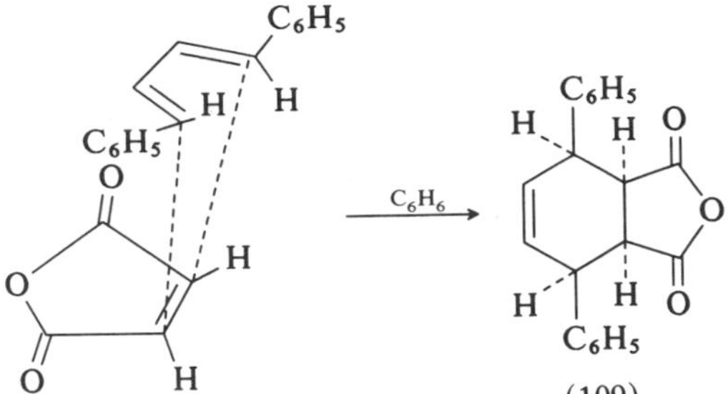

(109) (3.92)

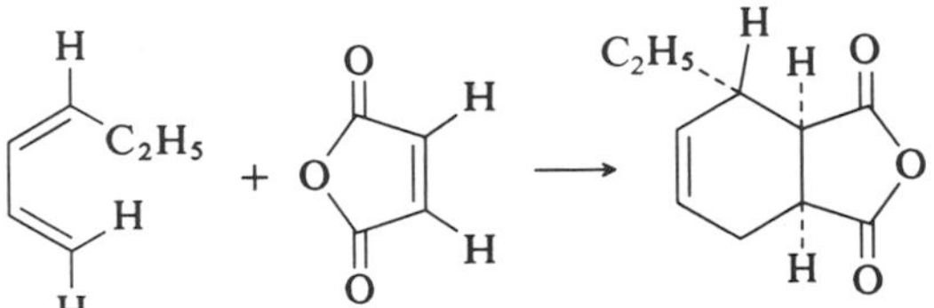

(110)

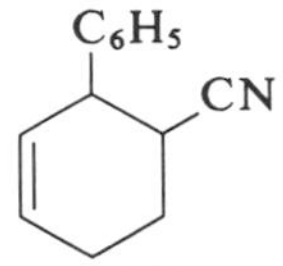

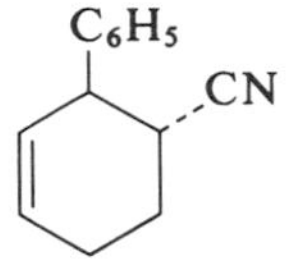

(111) (112) (3.93)

TABLE 3.5 Effect of temperature on ratio of cis- *and* trans-*adducts formed in reaction of* trans-*butadiene*-1-*carboxylic acid with acrylic acid* (from Sauer, 1967)

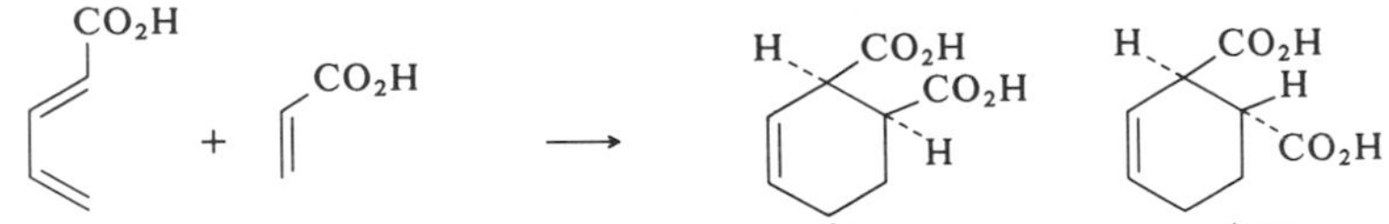

Temperature °C	75	90	100	110	130
Ratio *cis* : *trans*	*cis* only	7:1	4.5:1	2:1	1:1

The factors which determine the steric course of diene additions are still not completely clear. It appears that a number of different forces operate in the transition state and the precise steric composition of the product depends on the balance among these (see Fleming, 1976). The predominance of *endo* addition has been ascribed to the tendency of dienophile substituents to be so oriented in the favoured transition state of the reaction that they lie directly above the residual unsaturation of the diene, either for reasons of spatial orbital overlap or for reasons of steric accommodation. Non-bonded charge-transfer type interactions may also play a part in the stabilisation of the transition state in some cases. Alder and Stein originally summarised the steric nature of the transition state by their principle of 'maximum accumulation of double bonds'.

This has been rationalised by Woodward and Hoffmann (1969) as a stabilisation of the *endo* transition state by secondary orbital interaction. Thus, in the reaction of cyclopentadiene with maleic anhydride, secondary orbital interactions, represented by the dashed lines in (3.94), lower the energy of the *endo* transition state (shown) relative to that of the *exo* transition state, where these secondary interactions are absent; hence the *endo* adduct is one obtained under kinetically controlled conditions.

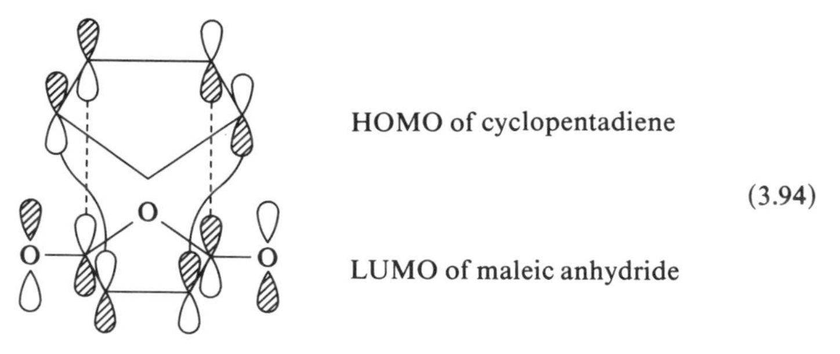

(3.94)

However, this is clearly not the whole story and it appears that these attractive forces are easily overweighed by steric factors and, in some cases, by changes in the experimental conditions (Kakushima, 1979; Franck, John, Objiniazak and Blount, 1982).

It appears that the transition state which is best stabilised by spatial orbital overlap and by non-bonded interactions and simultaneously least destabilised by unfavourable steric repulsions has the lowest energy and consequently predominates in the kinetically controlled product. For reactions with cyclic dienophiles this is often the *endo* adduct, but with open-chain dienophiles the interplay of the different factors makes it more difficult to predict precisely the steric course of the additions.

In *intra*molecular reactions the *endo* rule is sometimes a good guide to the stereochemistry of the product and sometimes not. Thus the *trans*-hexahydroindene (114) was obtained specifically in 70 per cent yield from the triene (113) via the sterically favoured *endo* transition state. Nevertheless a growing body of evidence supports the view that the Alder *endo* rule is not universally valid for intramolecular Diels–Alder reactions. A number of cases are known in which cyclisation has clearly proceeded to a considerable extent, or even predominantly, by way of

H_3CO_2C RO OR toluene, 130 °C H_3CO_2C H RO H OR (3.96)

[113, R = $Si(CH_3)_2C_4H_9$-t] (114)

an *exo* transition state. Thus the trienoic ester (115) gave the cyclised products (116) and (117) formed via the corresponding *exo* and *endo* transition states, in the ratio 2:1. It is suggested that non-bonded interactions in the transition states play an important part in determining the stereochemical course of intramolecular reactions, and in some cases may outweigh the effect of secondary orbital interactions which would favour an *endo* transition state (Boeckman and Ko, 1982).

H_3CO_2C 180 °C CO_2CH_3 H H + CO_2CH_3 H H (3.97)

(115) (116) (117)

The presence of an sp^2 carbon atom in the chain connecting the diene and dienophile can also affect the stereochemical outcome of an intramolecular Diels–Alder reaction (cf. p. 217).

3.9 Asymmetric Diels–Alder reactions

In reactions of disymmetric dienes and dienophiles in which the plane of the double bonds is not a plane of symmetry, the diene and dienophile will normally approach each other from the less hindered side of each. If one of the components is chiral and reaction gives rise to a new asymmetric centre the interesting possibility arises that preferential approach from one direction will result in formation of the two enantiomeric forms of the new chiral centre in a non-statistical ratio, resulting in an asymmetric synthesis. This possibility has been realised in a number of instances. In general, the normal thermal reaction gives only low optical yields and best results have been obtained in reactions catalysed by Lewis acids at low temperatures. The absolute stereochemistry of the predominant isomer formed can often be predicted from that of the chiral starting component and a knowledge of the orientation of the transition state (Sauer and Kredel, 1966; Oppolzer, 1984).

Most work has been done with optically active dienophiles, particularly with esters of acrylic acid and optically active alcohols. Having performed their 'directing' function the optically active 'auxiliary' alcohol groups are removed from the product and reused. Several optically active alcohols have been employed in this sequence including menthol and, better, (−)-8-phenylmenthol, but the best results so far have been obtained with the neopentyl esters (118) and (119), themselves derived ultimately

OH O (118) HO O (119) (3.98)

from (R)-(+)- and (*S*)-(−)-camphor (Oppolzer, Chapius, Dao, Reichlin and Godel, 1982; Helmchen and Schmierer, 1981). In catalysed addition of the derived acrylate esters to cyclopentadiene the (2*R*)- or (2*S*)-adducts (120) were obtained at will with almost complete asymmetry induction. Reduction with lithium aluminium hydride regenerated the auxiliary alcohol and gave the optically pure *endo* alcohol (121). It is suggested that asymmetric reaction takes place by addition of the diene to the ester in the conformation (122) in which access to the *re*-face of the double bond (i.e. the front face as drawn) is hindered by the t-butyl group. This forces addition to take place preferentially from the back, in the present case clearly almost exclusively. It will be interesting to see if equally high optical yields are obtained in reactions of the esters with open-chain dienes. Reactions of acrylic esters with open chain dienes are generally not very selective (cf. p. 234).

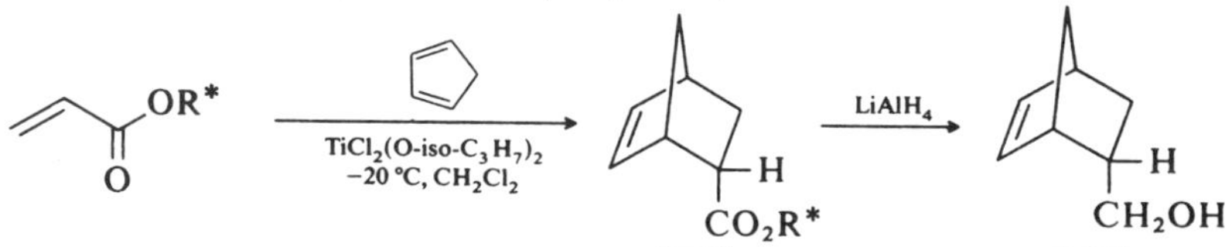

OR* O $TiCl_2(O\text{-iso-}C_3H_7)_2$ −20 °C, CH_2Cl_2 H CO_2R^* (120) (96% *endo* 99.4% enantiomeric excess) $LiAlH_4$ H CH_2OH (121) (3.99)

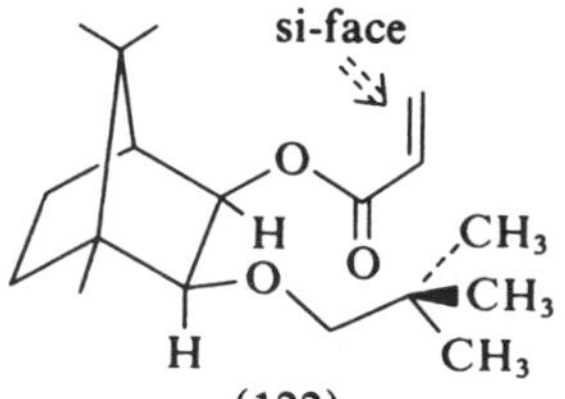

si-face O H O O CH_3 CH_3 H CH_3 (122)

Promising results have been obtained recently with chiral ketols (123) in which the chiral auxiliary group is attached one atom closer to the three-carbon enone unit than in the acrylate esters described above. With these reagents excellent diastereofacial selectivity was achieved in additions to cyclopentadiene *even in the absence of a Lewis acid catalyst.* Thus, with R = t-butyl at −20 °C the *endo* product (124*a*) was obtained with a diastereotopic selectivity of more than 100:1 (Choy, Reed and Masamune, 1983).

(123) —(−20 °C, toluene)→ (*a*) (100) + (*b*) (1) (124) (3.100)

Reactions of dienes containing optically active auxiliary groups have not been so widely studied as those of chiral dienophiles. In one example reaction of the (*S*)-*O*-methylmandelyl ester of 1-hydroxy-1,3-butadiene with juglone in the presence of boron triacetate gave the adduct (125) in 98 per cent yield with virtually complete asymmetric induction. The absolute configuration of the product corresponds to reaction of the diene in the conformation (126) in which one face of the diene is shielded by the phenyl substituent (Trost, O'Krongly and Bellatire, 1980; see also Gupta, Harland and Stoodley, 1983).

OCOR* + juglone —($B(OCOCH_3)_3$, $CHCl_3$, 0 °C)→ (125)

($R^* = -C(OCH_3)(H)C_6H_5$) (3.101)

(126)

Attempts are being made to achieve asymmetric Diels–Alder reactions with non-chiral substrates in the presence of optically active catalysts. Lewis acids and lanthanide complexes bearing chiral ligands have been used as catalysts but so far results have been variable (Roush, Gillis and Ko, 1982; Bednarski, Maring and Danishefsky, 1983).

Not many examples of intramolecular asymmetric Diels–Alder reactions have been reported so far. In one the 8-phenylmenthyl ester (127) gave predominantly the *endo* adduct (128) with diastereomeric excess of 50 per cent in the presence of (menthyloxy)aluminium dichloride; somewhat better results were obtained using (1-bornyloxy)aluminium chloride. The poorer optical yields obtained here in comparison with those achieved in intermolecular reactions are ascribed to the higher temperatures required.

R*
O
O
(menthyloxy)$AlCl_2$
CH_2Cl_2-toluene
23 °C
CO_2R^*
CO_2R^*
+
3
1
(127)
(128)
CH_3
R =
H_3C
CH_3
C_6H_5

(3.102)

3.10 Mechanism of the Diels–Alder reaction

The precise mechanism of the Diels–Alder reaction has been the subject of much debate. There is general agreement that the rate-determining step in adduct formation is bimolecular and that the two components approach each other in parallel planes roughly orthogonal to the direction of the new bonds about to be formed. Formation of the two new σ-bonds takes place by overlap of molecular π-orbitals in a direction corresponding to endwise overlap of atomic p-orbitals. But there has been uncertainty about the nature of the transition state and in particular, about the timing of the changes in covalency that result in the formation of the new bonds.

Two main views have been considered: (a) The reaction is a concerted addition in which both of the new single bonds are formed at the same time (3.103). (b) The reaction takes place in two steps, the first of which, the formation of a single bond between atoms of the reactants, is rate controlling; the addition is then completed by formation of the second bond in a fast reaction. The intermediate in the second alternative may have either zwitterionic or diradical character.

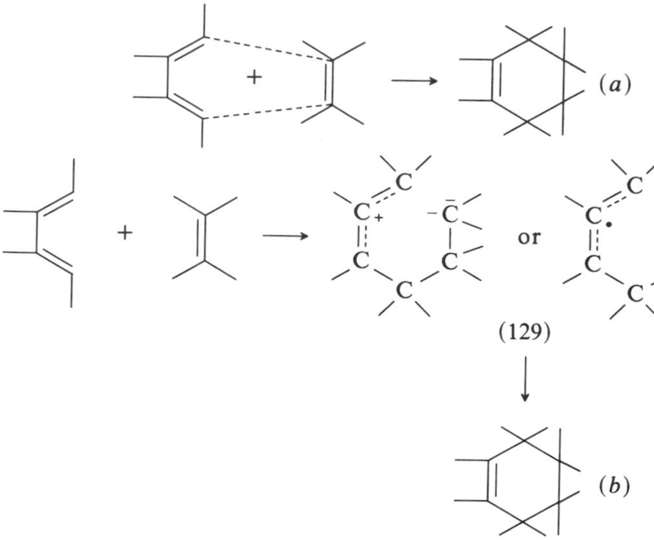

(a)

or (3.103)

(129)

(b)

It is now generally believed that most thermal Diels–Alder additions are concerted (e.g. Woodward and Hoffmann, 1970; Sauer and Sustmann, 1980). A major factor bringing about acceptance of this view has been the high stereoselectivity of the reactions, although a two-step mechanism is not entirely ruled out by this evidence if rotation about carbon–carbon single bonds in the intermediate (129) is slow compared with the rate of ring closure. In this connection it is noteworthy that cyclo-addition of *cis*- and *trans*-1,2-dichloroethylene to cyclopentadiene is completely stereospecific (3.104). A two-step mechanism would have given the diradicals (130) and (131) which might have been expected to be long-lived

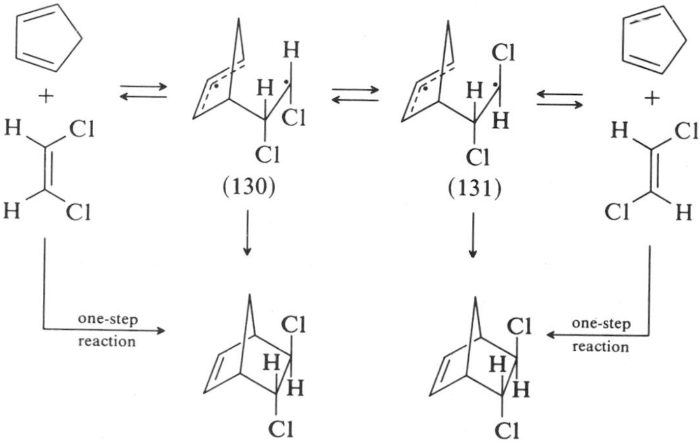

(130) (131) (3.104)

one-step reaction

one-step reaction

enough to undergo some interconversion, resulting in a mixture of products. Addition of dichlorodifluoroethylene to *cis,cis-* and *trans,trans-*2,4-hexadiene, which certainly proceeds by a two-step mechanism with a biradical intermediate, is not stereospecific because the rate of rotation about a carbon–carbon single bond in the intermediate (132) is comparable with the rate of ring closure (Bartlett, 1970).

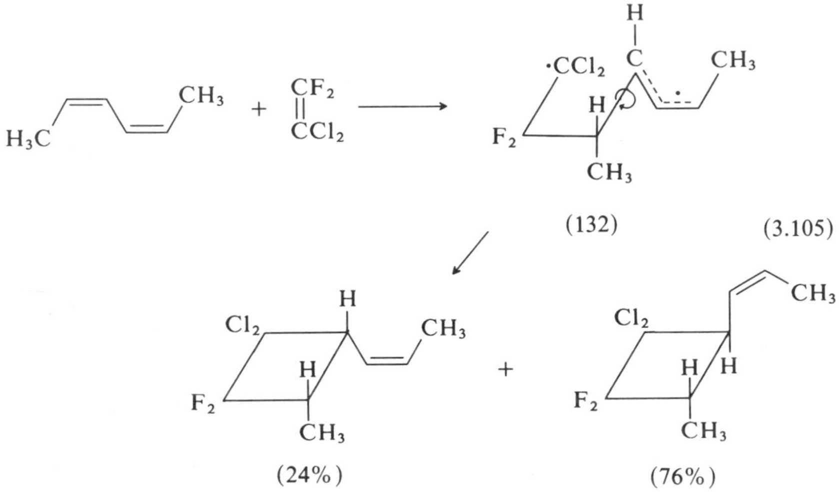

(132) (3.105)

(24%) (76%)

Attempts to show the involvement of free radicals in the Diels–Alder reaction have been negative. No biradical intermediates have been detected, and compounds that catalyse singlet–triplet transitions have no effect on the reaction. Similarly, the kinetic effects of *para* substituents in 1-phenylbutadiene, although large in absolute terms, are considered much too small for a rate-determining transition state corresponding to a zwitterion intermediate.

Whether or not both of the new bonds in the concerted mechanism are formed to the same extent in the transition state is an open question. A current view is that although they both begin to be formed at the same time the process may take place at different rates in the two bonds so that the transition state is 'lop-sided', with one bond formed to a greater extent than the other. There may be a gradation of mechanisms for different Diels–Alder reactions, extending from a completely concerted four-centre mechanism with a symmetrical transition state at one extreme to something approaching a two-step process at the other. Thus in reactions involving a heterodiene or heterodienophile, the heteroatoms will probably be able to stabilise polar intermediates to a greater extent than carbon so that hetero Diels–Alder reactions are more likely to be non-concerted or at least to proceed through an unsymmetrical transition state. It has been generally assumed, however, that in the reaction between a symmetrical diene and a symmetrical dienophile the transition state is symmetrical or nearly so, but this view is challenged by calculations which suggest that even the prototypical reaction between butadiene and ethylene proceeds via a highly unsymmetrical transition state in which one new δ bond is almost completely formed while the other is hardly formed at all, and it is held that there is no evidence that any Diels–Alder reaction takes place in a synchronous manner (Dewar and Pierini, 1984; Dewar, 1984).

It has always been difficult to understand why addition of alkenes to conjugated systems to form six-membered rings takes place so easily, while thermal addition to form cyclobutanes is so uncommon, although it is easily effected photochemically. From a consideration of molecular orbital symmetries, Woodward and Hoffmann (1969, 1970) have been able to explain why this is so. Using extended Hückel calculations, they have shown that the hypothetical concerted reaction of two ethylene molecules to form cyclobutane is a highly unfavourable ground state process, because of symmetry restrictions, and requires an activation energy comparable to that of electronic excitation of ethylene, an energy unavailable in ordinary thermal reactions. On the other hand, concerted formation of a six-membered ring by addition of an alkene to a conjugated system involves ground-state molecules only and is a thermally allowed reaction. Woodward and Hoffmann have embodied their results in a set of selection rules, according to which cyclo-addition reactions involving ethylenic bonds and conjugated systems take place easily in a concerted fashion when six-membered rings are being formed, but not when the products are four- or eight-membered rings.

Another way of looking at this is based on the idea that the transition state for the Diels–Alder reaction, which involves six π-electrons in a cyclic array, may be considered as 'aromatic' and thus energetically favoured, while for a $(2+2)\pi$ cyclo-addition the transition state, with only four π-electrons, is anti-aromtic and energetically disfavoured, at least for the thermal reaction (Dewar, 1971; Gilchrist and Storr, 1979).

Not all Diels–Alder reactions proceed with equal ease. Frontier molecular orbital theory has been used to show how the ease of reaction with substituted dienes and dienophiles can be correlated with the effect of the substituents on the energies of the frontier orbitals concerned in the reaction (cf. p. 222). Fleming, 1976; Sauer and Sustmann, 1980).

3.11. Photosensitised Diels–Alder reactions

A number of Diels–Alder reactions are induced by light. It is probable that the mechanism of these photoreactions differs from that of thermal Diels–Alder additions and none is known to be concerted. Concerted suprafacial $(4+2)\pi$ photo cyclo-additions are forbidden by the Woodward–Hoffmann selection rules and a different mechanism is indicated also by the fact that the products of the photo and thermal reactions are not the same. Thus, irradiation of butadiene in the presence of ketonic sensitisers affords a mixture of *cis*- and *trans*-divinyl-cyclobutane along with the normal Diels–Alder adduct vinylcyclohexene (133)

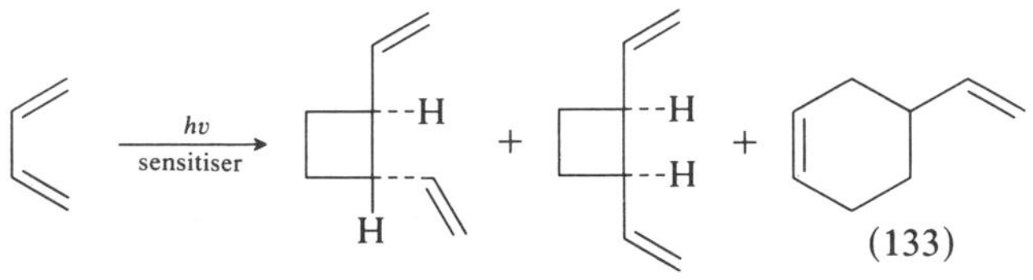

(3.106)

(3.106). Cyclohexadiene also affords a mixture of three products on irradiation; none of the thermal *endo* adduct is produced in this case. It is suggested (Hammond, Turro and Liu, 1963) that these photo products are formed by way of diallylic biradicals which themselves arise by addition of triplet diene to another molecule of diene.

The formation of a stable 2:1 adduct by ultraviolet irradiation of a solution of maleic anhydride in benzene, formerly thought to involve a photochemical Diels–Alder reaction is now known to proceed by way of a photochemical [2+2] addition followed by thermal Diels–Alder addition of maleic anhydride to the cyclohexadiene thus produced (Bryce-Smith, Deshpande and Gilbert, 1975).

3.12. The ene reaction

Formally, related to the Diels–Alder reaction is the 'ene' reaction in which an alkene bearing an allylic hydrogen atom reacts thermally with a dienophile (now called the enophile) with formation of a new σ-bond to the terminal carbon of the allyl group, 1,5-migration of the allylic hydrogen and change in the position of the allylic double bond (3.107); Hoffmann, 1969; Keung and Alper, 1972). Like the Diels–Alder reaction this also is a 6π-electron electrocyclic reaction, but here the two electrons of the allylic C-H σ bond take the place of two π-electrons of the diene in the Diels–Alder reaction. The activation energy is thus greater and higher temperatures are required than in most Diels–Alder reactions.

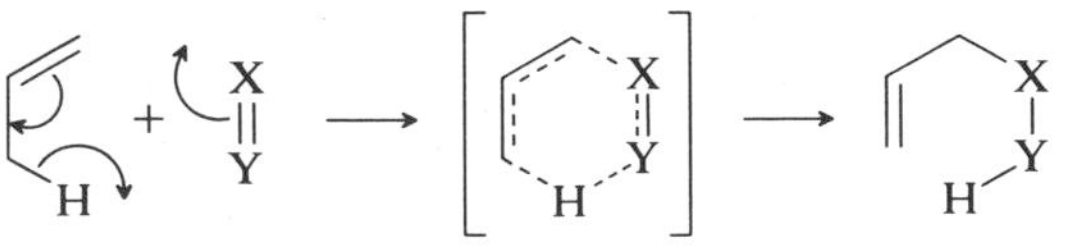

(3.107)

X=Y=C=C, C=O, C=S, N=O, N=N etc.

Thus, maleic anhydride and propene react only at 200 °C to give the product (134), and cyclohexene forms (135). Fortunately, many ene reactions can be catalysed by Lewis acids and then proceed under milder conditions, often with much improved stereoselectivity. A number of

200 °C

(134) (135)

(3.108)

catalysts have been used ($AlCl_3$, $SnCl_4$, $TiCl_4$) but the best results have been obtaned with alkylaluminium halides. These act as proton scavengers as well as Lewis acids and this helps to prevent the occurrence of unwanted side reactions (Snider, 1980). Thus, the uncatalysed reaction of methyl acrylate with 2-methyl-2-propene requires a temperature of 230 °C, but with ethylaluminium dichloride as catalyst good yields of adducts are obtained from reactive alkenes at 25 °C.

Like the Diels–Alder reaction itself the ene reaction is reversible. This is shown, for example, in the decomposition of 1-pentene at 400 °C to ethene and propene and by the well-known thermal decomposition of β-hydroxyalkenes to alkenes and aldehydes.

400 °C → $H_2C{=}CH_2$ + propene

$\xrightarrow{\Delta}$ methylenecyclohexane + $C_6H_{13}CH{=}O$

(3.109)

The ene reaction resembles the Diels–Alder reaction also in its stereoselectivity, showing *cis* addition and a preference for the formation of *endo* products. *Cis* addition is exemplified in the reaction of 1-heptene with dimethyl acetylenedicarboxylate to give the adduct (136); none of the corresponding fumaric acid derivative was obtained. That is to say, the hydrogen atom and the alkyl residue add to the same side of the triple bond of the enophile.

H_9C_4 + CO_2CH_3–C≡C–CO_2CH_3 $\xrightarrow[(80\%)]{170\text{–}190\ °C}$ (136) (3.110)

Many ene reactions take place preferentially by way of an *endo* transition state, that is one in which the two components overlap as much as possible. This is important in synthesis because it allows some control over the stereochemistry in a planned sequence. Thus, in the reaction of *trans*-2-butene with maleic anhydride the major product was the *erythro* adduct (137) whereas with *cis*-2-butene the *threo* product (139) was obtained almost exclusively through the *endo* transition state (138). The poorer selectivity in the reaction with *trans*-2-butene is presumably due to the steric effect of the *trans* methyl group in the *endo* transition state

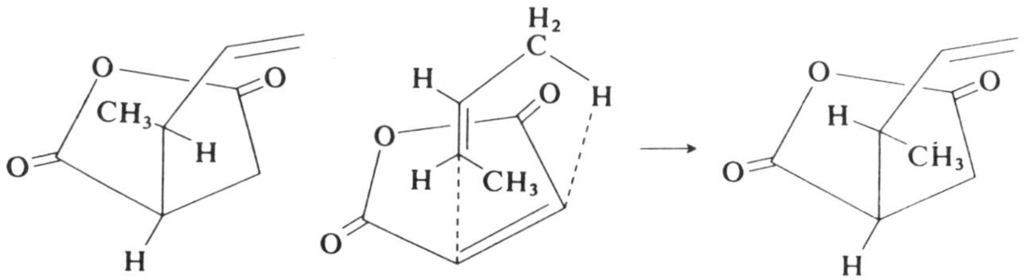

(137) (*erythro*) (138) (*endo*) (139) (*threo*)

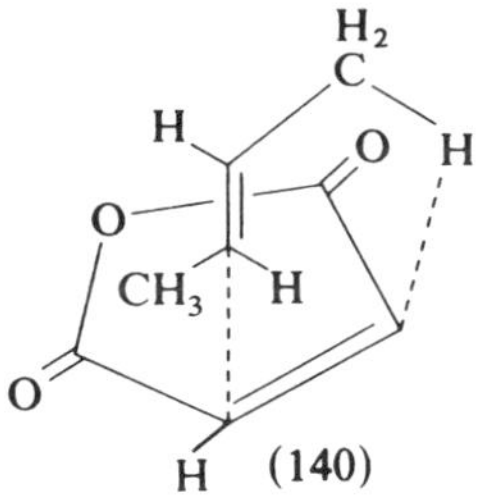

(140)

(3.111)

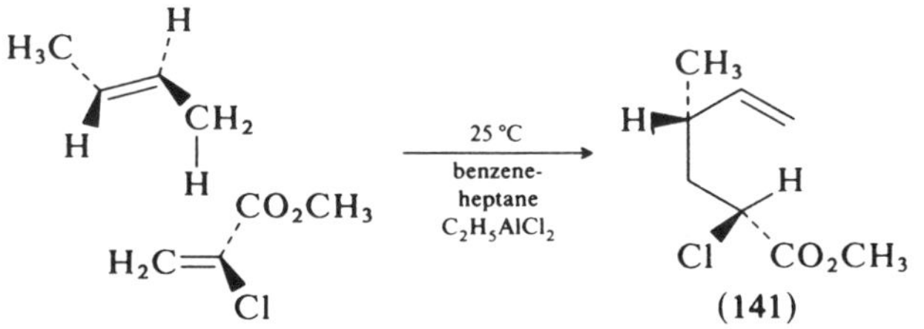

25 °C, benzene-heptane, $C_2H_5AlCl_2$

(141)

(140). Again, in the catalysed reaction of methyl α-chloroacrylate with *trans*-2-butene the product is largely that (141) formed by way of the transition state in which the methoxycarbonyl group is disposed in the *endo* orientation (Duncia, Lansbury, Miller and Snider, 1982). *Endo* addition in ene reactions is favoured by orbital symmetry relationships, but the preference is not so marked as in the Diels–Alder reaction (Berson, Wall and Perlmutter, 1966) and may be easily overcome by other factors. In some intramolecular reactions, for example, steric constraints enforce an *exo* transition state (Oppolzer, 1981).

It is now generally believed that in most cases thermal ene reactions proceed by a concerted mechanism through a cyclic six-membered transition state, although the transition state may be unsymmetrical, with the new C—X bond (3.107) more highly developed than the Y—H bond. This view is supported by theoretical work and by the observation that the new C—X and Y—H bonds are formed *cis* to each other (equation 3.110) and by the fact that optically active products are formed in reactions where the hydrogen atom transferred is initially attached to a chiral centre. Thus, reaction of maleic anhydride with the optically active alkenes (142) gave the optically active adducts (143). The degree of retention of optical activity could not be determined but the transfer of optical asymmetry from one of the reactants to a different site in the adducts is best accounted for by a concerted process in which all the new bonds are formed essentially simultaneously (Hill and Rabinovitz, 1964). A concerted mechanism is not universal, however, and it appears that some catalysed reactions, particularly those in which carbonyl compounds act as the enophile, proceed by a stepwise mechanism involving a zwitterionic intermediate (Snider, 1980).

(3.112)

[142; R = C_6H_5—, $(CH_3)_2CH(CH_2)_3$—] (143)

The enophile need not be a compound with a carbon–carbon multiple bond; hetero-enophiles are also known. In the nitrogen series diethyl azodicarboxylate, $EtO_2CN{=}NCO_2Et$ reacts readily with a variety of alkenes (cf. Hoffmann, 1969) and thermal and catalytic ene reaction of butyl *N*-(*p*-tolylsulphonyl)imino-acetate (144) with alkenes gives adducts which are readily converted into $\gamma\delta$-unsaturated α-amino acids (Achmatowicz and Pietraszkiewicz, 1981).

$SnCl_4$, CH_2Cl_2, 0 °C

(144) (3.113)

Tos = *p*-$CH_3C_6H_4SO_2$

Carbonyl compounds can also act as enophiles under thermal or Lewis acid-catalysed conditions (cf. Snider, 1980). Reaction of aldehydes with alkenes in the presence of alkylaluminium halides provides a convenient route to some homoallylic alcohols (Snider, Rodini, Kirk and Cordova,

1982; Snider and Phillips, 1983). Thus, in the presence of dimethylaluminium chloride limonene reacts selectively with acetaldehyde to give the alcohol (145). These catalysed reactions of aldehydes are believed to proceed through a zwitterionic intermediate (146).

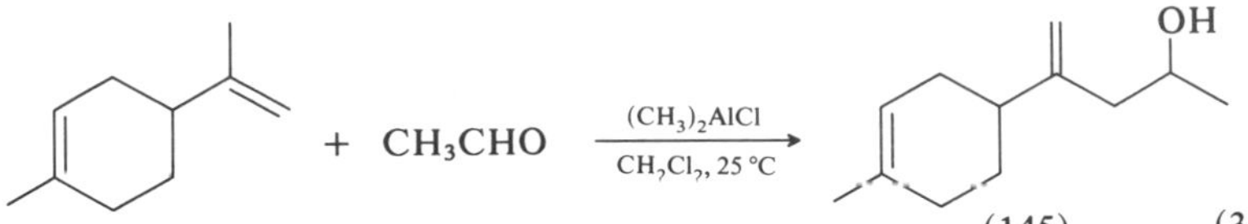

+ CH_3CHO $\xrightarrow[CH_2Cl_2,\ 25\ °C]{(CH_3)_2AlCl}$ (145) (3.114)

$\xrightarrow{(CH_3)_2AlCl}$ [$\bar{O}Al(CH_3)_2Cl$] $\rightarrow$ OH

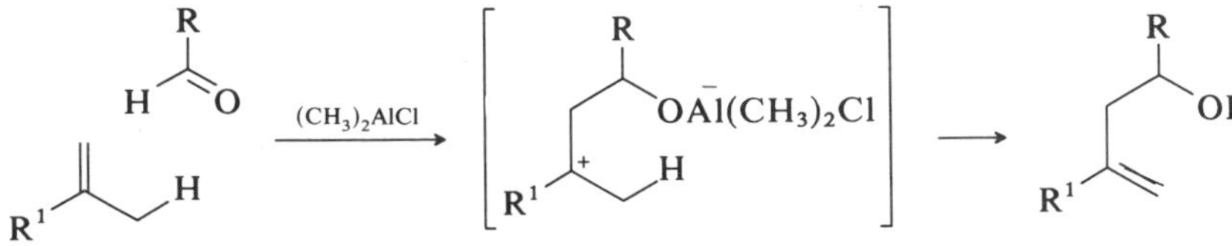

(146)

The ene reaction has not been so extensively used in synthesis as the Diels–Alder reaction, but it is evident from recent work that the intramolecular reaction has great potential for the synthesis of cyclic compounds, and particularly for the synthesis of five-membered rings from 1,6-dienes. These reactions in some cases are highly stereoselective. Thus, kinetically controlled cyclisation of the *cis*-diene (147) gave the *cis*-disubstituted cyclopentane (148) with complete stereoselectivity by way of the *exo* transition state (149); the *endo* transition state is highly

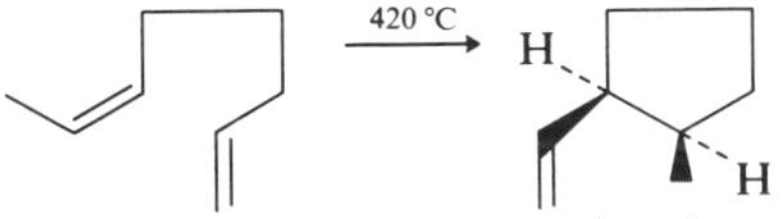

$\xrightarrow{420\ °C}$

(147) (148)

(3.115)

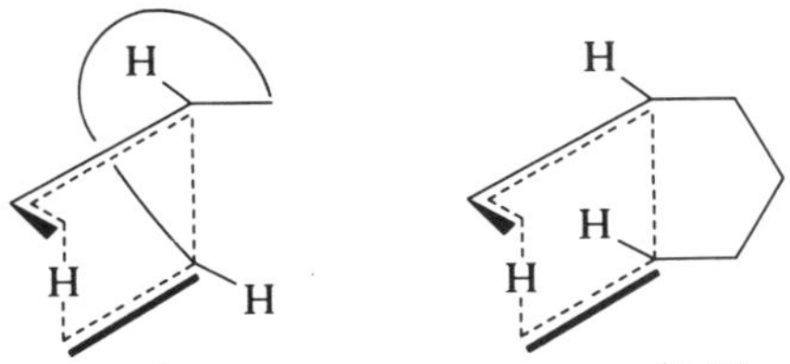

endo *exo* (149)

strained here. Analogous cyclisation of the corresponding *trans*-diene again gave mainly the *cis*-cyclopentane (150), this time clearly by way of the *endo* transition state.

Reactions of this kind are being increasingly exploited in the synthesis of natural products (Oppolzer, 1981). Thus, a recent synthesis of the

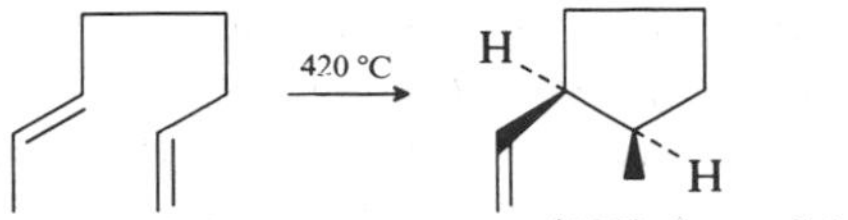

420 °C

(150) (*mainly*)

(3.116)

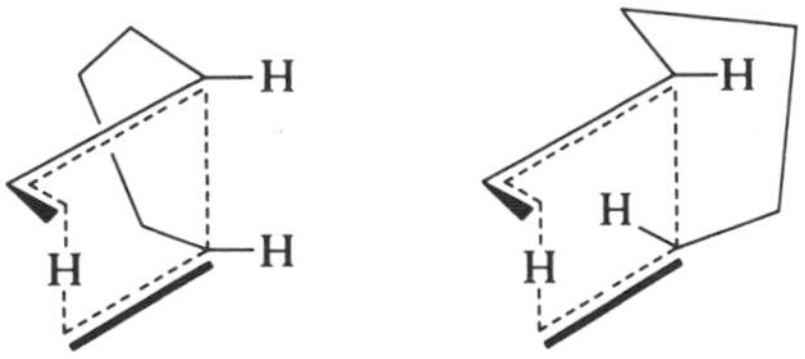

endo *exo*

sesquiterpene modhephene proceeded by intramolecular ene cyclisation of the bicyclic 1,6-diene (151) to the tricyclic propellane (152), and in the synthesis of (−)-(α)-kainic acid (155) the optically active diene (153), itself prepared from (S)-glutamic acid, was cyclised to the pyrrolidine derivative (154) in 75 per cent yield with almost complete stereoselectivity (Oppolzer and Thirring, 1982).

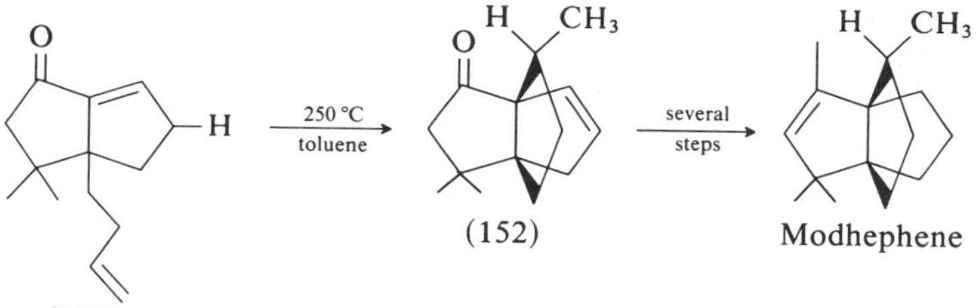

250 °C toluene; several steps

(151) (152) Modhephene

(3.117)

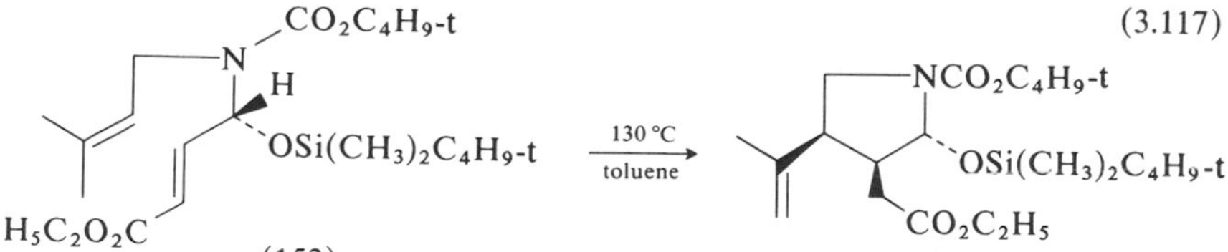

130 °C toluene

(153) (154) (70%)

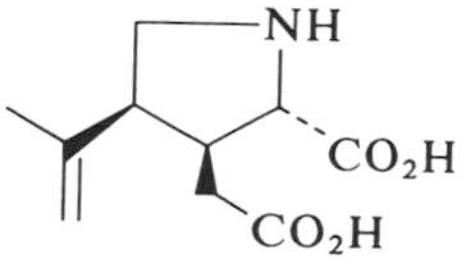

(155)

Six-membered rings can also be formed by ene cyclisation of 1,7-dienes, but in general lower yields are obtained than in the reactions with 1,6-dienes and higher temperatures are required.

Thermal cyclisation of 5,6-ethylenic ketones provides a useful route to cyclopentanones and cyclopentyl ketones; reaction proceeds by intramolecular ene reaction in the corresponding enols (3.118) (Leyendecker, Drouin and Conia, 1974; Conia and Le Perchec, 1975).

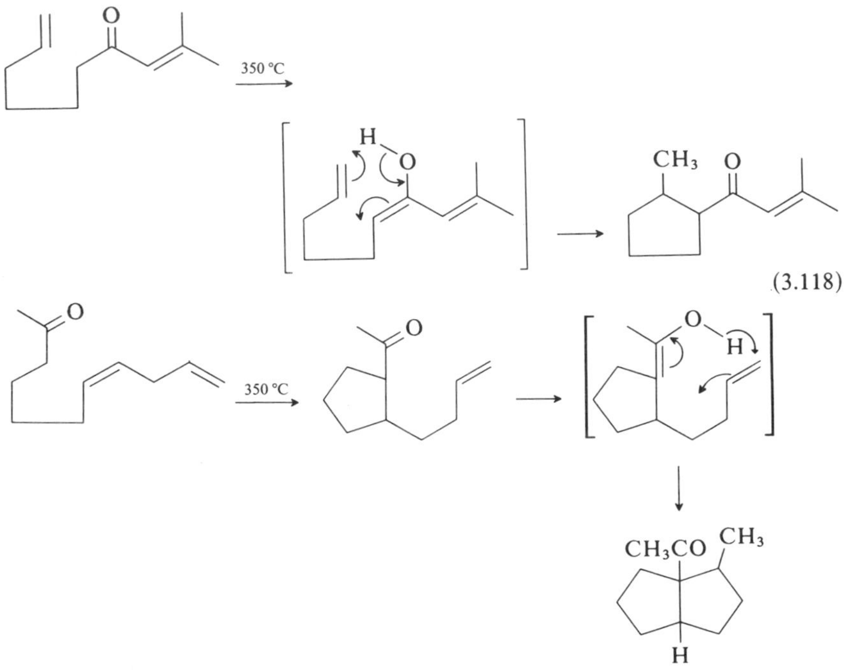

(3.118)

This sequence is of wide application and can be used to make fused ring, spiro and bridged ring compounds, in addition to simple cyclopentane derivatives. The main drawback is the high temperatures required. Dienes in which the ene component is contained in a ring give rise to bicyclic systems stereospecifically and in high yield. An illustration is provided by the elegant synthesis of the sesquiterpene alcohol (±)-acorenol shown in (3.119), where the spiro[5,4]decane system was constructed by an ene reaction with the appropriate 1,6-diene and, at a later stage in the synthesis, a shift in the position of a double bond was effected through a retro ene reaction of an allylic acetal (Oppolzer, 1973).

Some very efficient enantioselective reactions have been effected using optically active esters as the eneophile component. Best results so far have been obtained with esters of 8-phenylmenthol. Acid-catalysed reaction of 8-phenylmenthyl glyoxalate with 1-hexene, for example, gave the (*S*)-alcohol (156) in nearly 98 per cent diastereomeric excess (Whitesell, Bhattacharya, Aguilar and Henke, 1982), and in an intramolecular

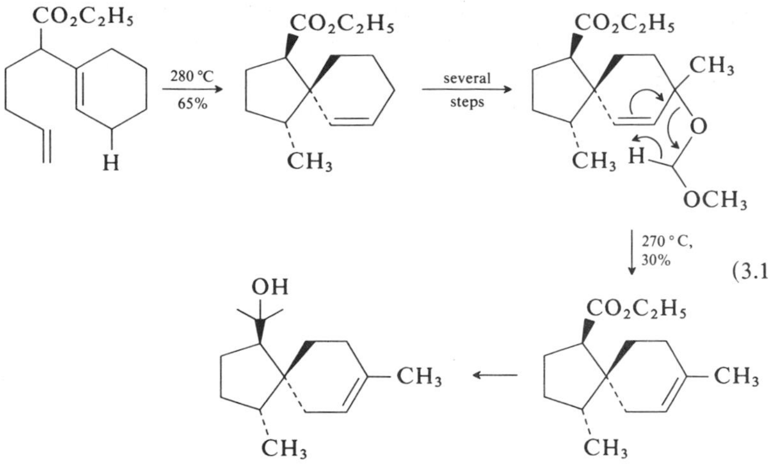

$CO_2C_2H_5$ — 280 °C, 65% → several steps → CH_3, O, OCH_3 — 270 °C, 30% ↓ (3.119)

OH — CH_3 ← $CO_2C_2H_5$ — CH_3

(±)-acorenol

example, catalysed cyclisation of the 8-phenylmenthyl ester (157) gave mainly the product (158) converted subsequently into (+)-α-allokainic acid (Oppolzer, 1981).

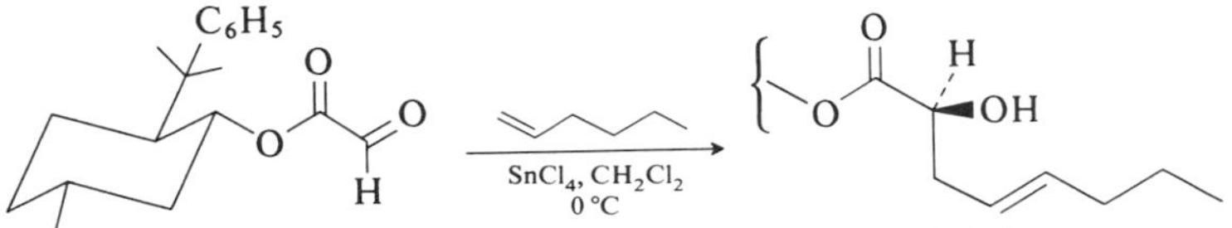

C_6H_5 — $SnCl_4$, CH_2Cl_2, 0 °C → H, OH (156)

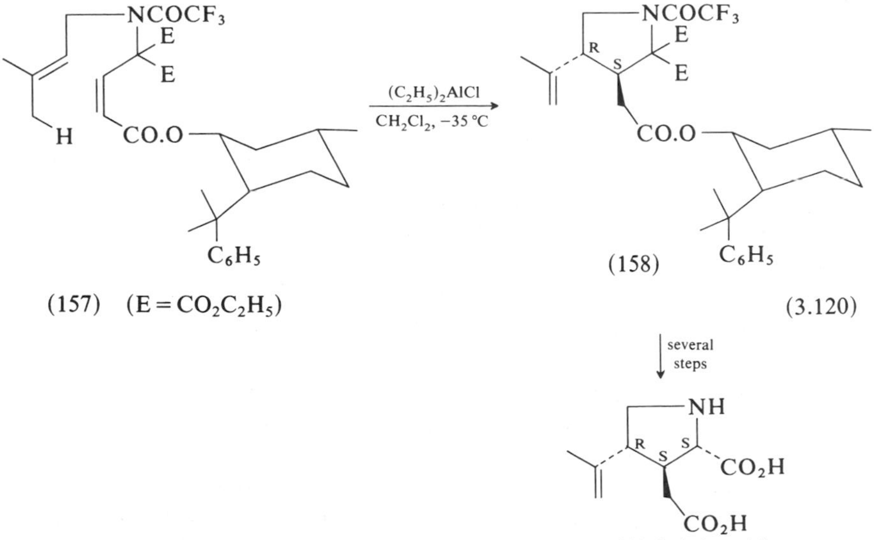

$NCOCF_3$, E, E, CO.O, C_6H_5 — $(C_2H_5)_2AlCl$, CH_2Cl_2, −35 °C → $NCOCF_3$, E, E, R, S, CO.O, C_6H_5

(157) ($E = CO_2C_2H_5$) (158)

(3.120)

↓ several steps

NH, R, S, S, CO_2H, CO_2H

α-Allokainic acid

The scope of the ene reaction has been extended by the discovery that allylic Grignard reagents take part in the reaction with remarkable ease by migration of magnesium instead of hydrogen and formation of a new carbon–magnesium bond.

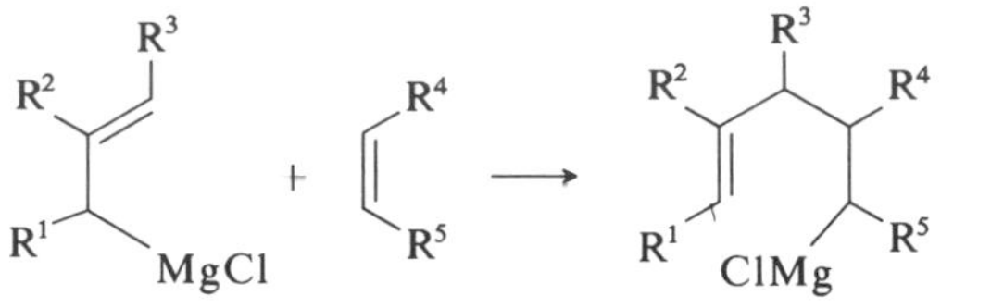

(3.121)

Intramolecular 'magnesio-ene' reactions have been cleverly exploited in the synthesis of natural products. Two modes of reaction have been distinguished in which the enophile is linked at either end of the double bond of the allyl group.

Thus, in a recent synthesis of the marine sesquiterpene $\Delta^{9(12)}$-capnellene the allylic Grignard reagent (159) on warming to 60 °C and reaction of the cyclised product with acrylic aldehyde gave the cyclopentane derivative (160) in 57 per cent yield (Oppolzer and Bättig, 1982) generating a quaternary carbon atom with high stereochemical control. The 'magnesio-ene' reaction was repeated with the allylic Grignard reagent derived from (160) to construct the second five-membered ring of $\Delta^{9(12)}$-capnellene.

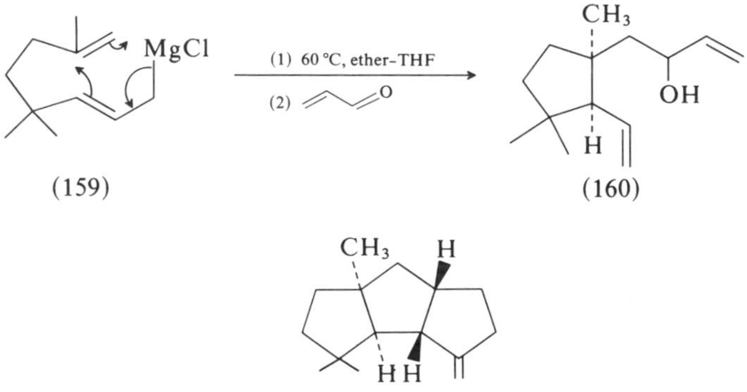

(159) (160)

(3.122)

$\Delta^{9(12)}$-Capnellene

Again, illustrating the other mode of reaction in which the enophile is linked to the central carbon of the allylic Grignard reagent, the key step in a synthesis of the norsesquiterpene (±)-khusimone was the 'magnesio-ene' cyclisation of (161) to (162) which was trapped by reaction with carbon dioxide to give the carboxylic acid (163) (Oppolzer and Pitteloud, 1982).

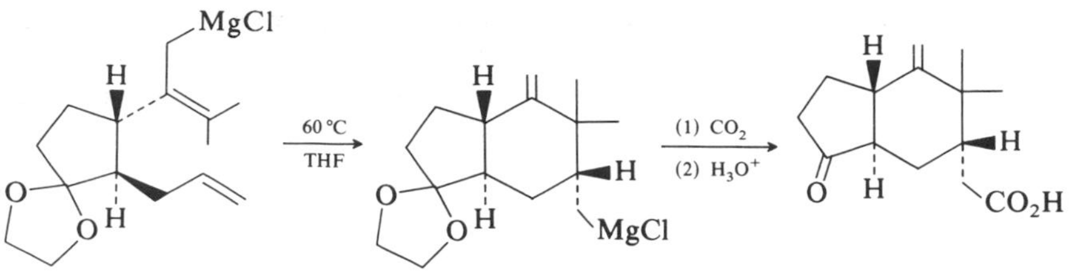

MgCl

H H H

60 °C THF (1) CO_2 (2) H_3O^+

O O H O H H MgCl O H CO_2H

(161) (162) (163)

(3.123)

3.13. Cyclo-addition reactions with allyl cations and allyl anions

The Diels–Alder reactions discussed above are concerted (4+2) cyclo-additions involving six π-electrons and lead readily to the formation of six-membered rings. The possibility of analogous six π-electron cyclo-additions involving allyl anions and allyl cations leading to five- and seven-membered rings respectively is predicted by the Woodward–Hoffmann rules (Woodward and Hoffmann, 1970). Examples of both

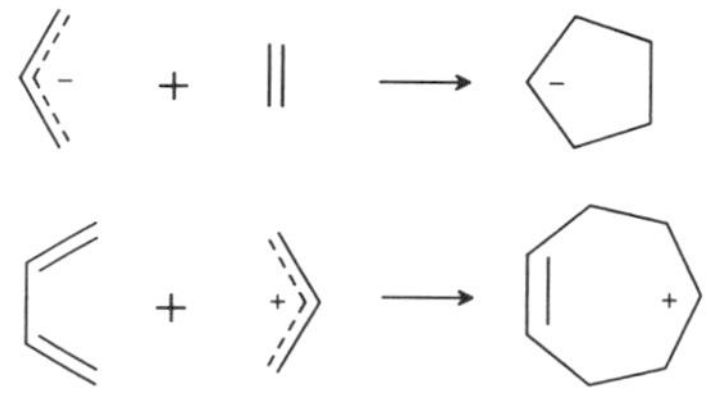

(3.124)

processes have been observed, although the synthetic scope of the reactions is as yet limited. In the anion series best results have been obtained with the 2-aza-allyl anion, which adds to a variety of alkenes to form pyrrolidine derivatives (Kauffmann, 1974). Allyl-lithium compounds themselves seem to undergo cyclo-addition to carbon–carbon double bonds only when an electron-attracting group is attached at C-2 of the allyl system.

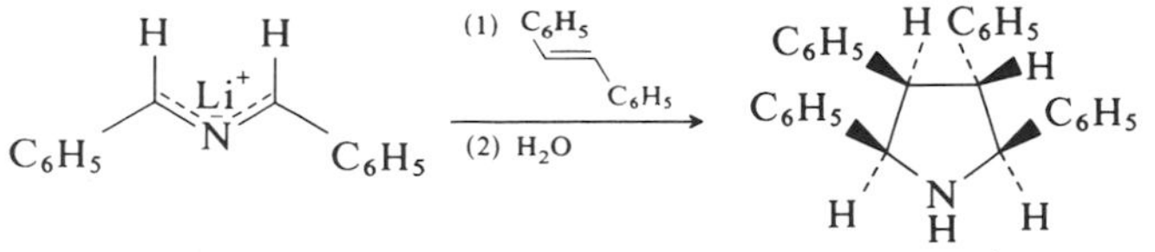

H H Li$^+$ N C_6H_5 C_6H_5 (1) C_6H_5–CH=CH–C_6H_5 (2) H_2O → C_6H_5 H C_6H_5 H C_6H_5 C_6H_5 H N H H (83%) (3.125)

Cyclo-addition of allyl cations to conjugated dienes provides a route to seven-membered carbocycles (Hoffmann, 1984) and is being increasingly used in synthesis. Several methods are used to generate the allyl cations; one convenient route is from an allyl halide and silver trifluoroacetate. Reaction proceeds best with cyclic dienes. Thus, cyclohexadiene and methylallyl cation gave 3-methylbicyclo[3,2,2]-nona-2,6-diene (164) in 30 per cent yield, and in an intramolecular example addition of an allyl cation, generated *in situ*, to a cyclopentadiene, formed the key step in a synthesis of zizaene sesquiterpenes (Hoffmann, Henning and Lalko, 1982). In this sequence the trimethylsilyl group serves both to stabilise the allyl cation and as a trigger for the formation of the exocyclic methylene group (3.126).

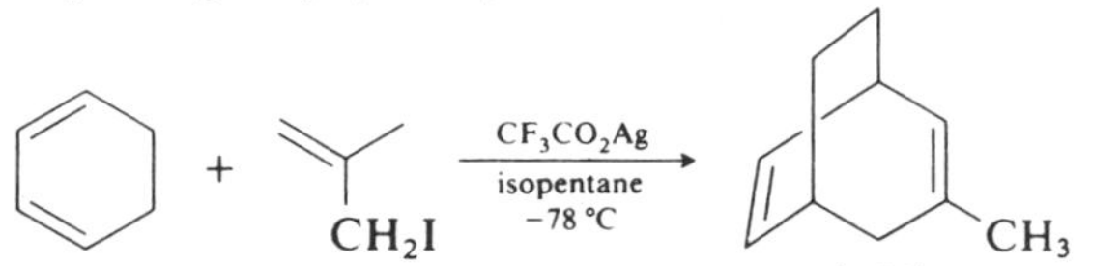

CH_2I ; CF_3CO_2Ag, isopentane, −78 °C ; CH_3 (164) (3.126)

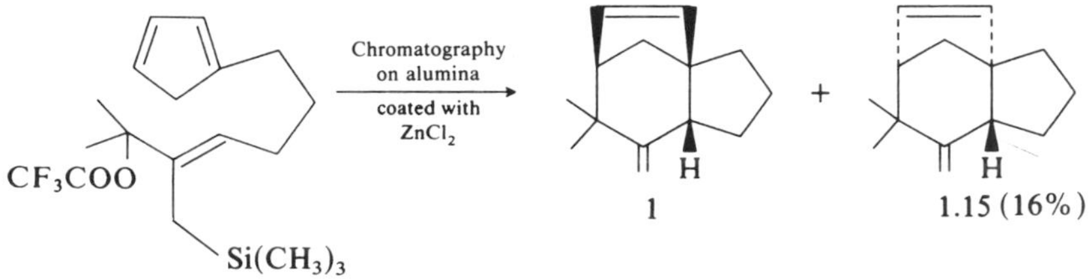

CF_3COO ; $Si(CH_3)_3$; Chromatography on alumina coated with $ZnCl_2$; H 1 + H 1.15 (16%)

There has been increasing interest in the application of 2-oxy-allyls in synthesis. These species can be produced from α,α'-dibromoketones or from α-chloro-trimethylsilyl enol ethers (Takaya *et al.*, 1978). Generated in the presence of open chain or some cyclic dienes they react readily to form seven-membered ring ketones. Thus, 2,3-dimethylbutadiene and

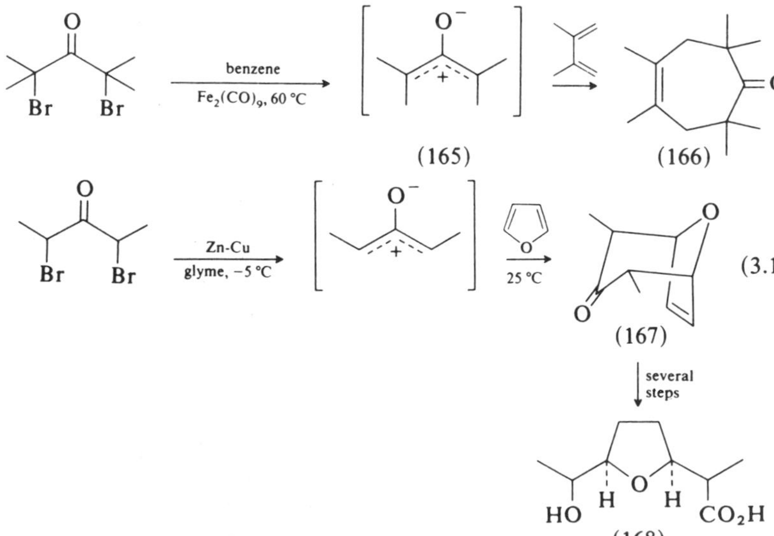

O ; Br Br ; benzene, $Fe_2(CO)_9$, 60 °C ; O^- + (165) ; O (166)

O ; Br Br ; Zn-Cu, glyme, −5 °C ; O^- + ; O, 25 °C ; O O (167) (3.127)

several steps

HO H O H CO_2H (168)

the oxy-allyl (165) gave the cycloheptanone (166) in 71 per cent yield and in a synthesis of (±)-nonactic acid the tetrahydrofuran derivative (168) was prepared stereospecifically from the adduct (167) itself obtained by reaction of 2,4-dibromo-3-pentanone with furan in the presence of zinc–copper couple (Arco, Trammell and White, 1976).

Some alkylcyclopropanones readily undergo cyclo-addition reactions with dienes to form seven-membered ring ketones (Turro, 1969) and it has been suggested that these reactions too may proceed by way of an oxyallyl species (as 169).

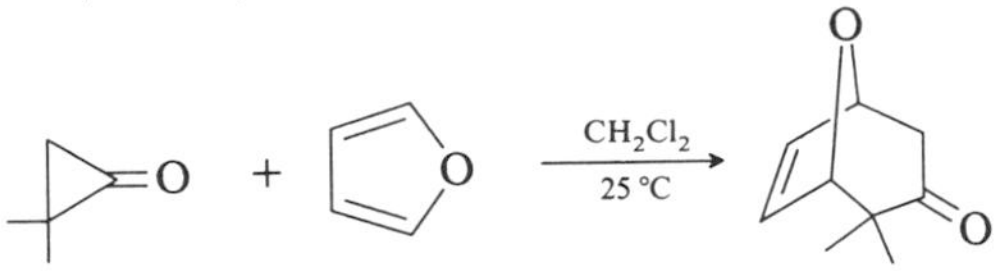

CH_2Cl_2, 25 °C (3.128)

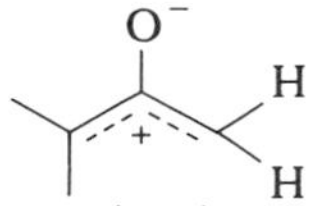

(169)

Cyclo-addition reactions of ionic dienes may also be envisaged (3.129).

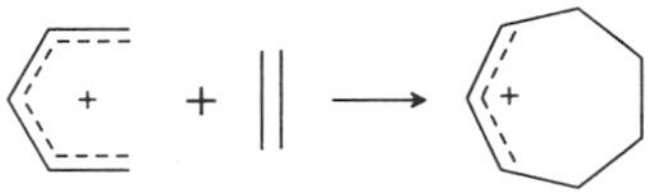

(3.129)

Reactions of this kind have formed the key step in several elegant syntheses. In these the pentadienyl cation is formed *in situ* from a *p*-quinone-monoketal. Thus, the sesquiterpene gymnomitral (172) was neatly synthesised from the adduct (171) formed in the reaction of 1,2-dimethylcyclopentene with the hemi-ketal (170) in the presence of tinIV chloride (Büchi and Chu, 1979).

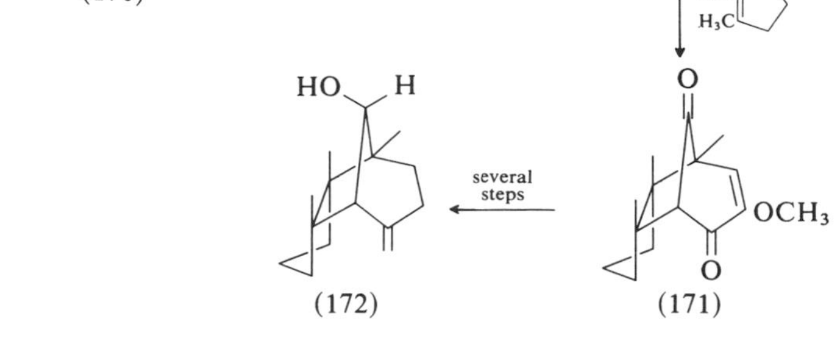

$SnCl_4$, CH_3NO_2–CH_2Cl_2, −20 °C

(170) (171) (172) several steps (3.130)

3.14. 1,3-Dipolar cycloaddition reactions

Closely related to the Diels–Alder and ene reactions are the so-called 1,3-dipolar cyclo-addition reactions. These also are 6π-electron pericyclic reactions, but here the 4π-electron component, called the 1,3-dipole, is a three atom system *abc* containing at least one hetero-atom and represented by a zwitterionic octet structure, and the 2π-electron component, here called the dipolarophile, is a compound containing a double or triple bond. The product of the reaction is a *five*-membered heterocyclic compound.

$$\overset{+}{a}-b-\overset{-}{c} \longleftrightarrow a=\overset{+}{b}-\overset{-}{c} \;(+\; d\equiv e) \longrightarrow \text{a–b–c–e–d (five-membered ring)} \qquad (3.131)$$

A typical example is the well-known reaction between diazomethane, the 1,3-dipole, and an $\alpha\beta$-unsaturated ester to form a pyrazoline (3.132).

$$CH_2{=}\overset{+}{N}{=}\overset{-}{N} + CH_2{=}C(CH_3){-}CO_2CH_3 \longrightarrow \text{pyrazoline: } H_2C{-}N{=}N{-}C(CH_3)(CO_2CH_3){-}CH_2 \qquad (3.132)$$

The name '1,3-dipole' is perhaps unfortunate and should not be taken to imply that the 4π-electron component necessarily has a dipole moment. A considerable number of 1,3-dipoles containing various combinations of carbon and hetero-atoms is theoretically possible (Huisgen, 1963, 1964, 1968*b*) and many of these have been made and their reactions with dipolarophiles studied. Some of the more commonly encountered classes among those containing carbon, oxygen and nitrogen are shown in (3.133). All 1,3-dipoles contain 4 π-electrons in three parallel *p*-orbitals (3.134).

$HC{\equiv}\overset{+}{N}{-}\overset{-}{C}H_2$ nitrile ylids; $HC{\equiv}\overset{+}{N}{-}\overset{-}{N}H$ nitrile imines; $HC{\equiv}\overset{+}{N}{-}\overset{-}{O}$ nitrile oxides

$H_2C{=}\overset{+}{N}{=}\overset{-}{N}$ diazo-alkanes; $HN{=}\overset{+}{N}{=}\overset{-}{N}$ azides; $R^1CH{=}\overset{+}{N}(R){-}O^-$ nitrones (3.133)

1,3-Dipolar cycloadditions provide a versatile route to a wide variety of five-membered heterocyclic compounds. In most reactions the 1,3-dipole is not isolated but generated *in situ* in presence of the dipolarophile. In many cases the dipolarophile is an alkene or an alkyne but this is not essential and dipolarophiles containing hetero-multiple bonds have also been used, for example imines, nitriles and carbonyl compounds. Thus, elimination of hydrogen chloride from *N*-(*p*-nitrobenzyl)benzimidoyl chloride (173) with triethylamine gives a purple

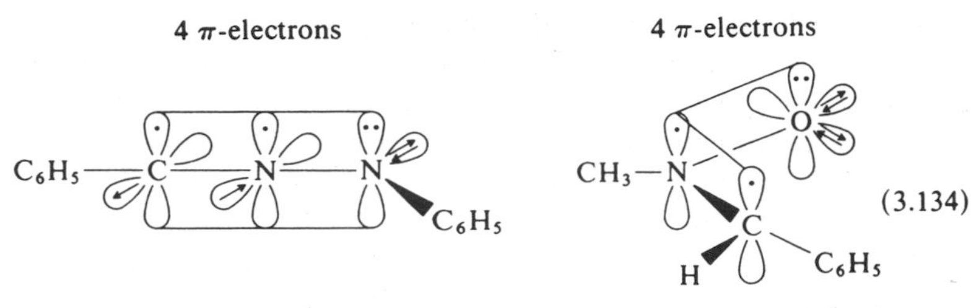

solution which contains the nitrile ylid (174). Generated in the presence of methyl acrylate it readily forms the Δ^1-pyrroline derivative (175). Dehydrogenation of the latter with chloranil yields the pyrrole (176) which can also be produced directly by reaction of the nitrile ylid with methyl propynoate.

C_6H_5—C(=$NCH_2C_6H_4NO_2$)Cl (173) $\xrightarrow[\text{benzene, 25 °C}]{(C_2H_5)_3N}$ $C_6H_5C\equiv\overset{+}{N}-\overset{-}{C}HAr$ (174)

(174) $\xrightarrow{HC\equiv C.CO_2CH_3}$ (176); (174) $\xrightarrow{CH_2=CH.CO_2CH_3}$ (175) (3.135)

(176): H, N, C_6H_5, Ar, CO_2CH_3

(175): C_6H_5, N, Ar, H, CO_2CH_3, H

Again, nitrile oxides, which are conveniently generated *in situ* from α-chloro-oximes or by dehydration of primary nitro compounds with phenylisocyanate, react with alkenes to form Δ^2-isoxazolines; with alkynes aromatic isoxazoles are formed directly. The Δ^2-isoxazolines

$CH_3CH_2NO_2$ $\xrightarrow[\substack{(C_2H_5)_3N \\ \text{benzene}}]{C_6H_5NCO}$ $[CH_3-C\equiv\overset{+}{N}-\overset{-}{O}]$ $\xrightarrow{CH_3CH=CHCH_3}$ (N, O, CH_3, H, CH_3, H_3C, H)

(3.136)

C_6H_5—C(=NOH)Cl $\xrightarrow[(C_2H_5)_2O,\ 20\ °C]{(C_2H_5)_3N}$ $[C_6H_5-C\equiv\overset{+}{N}-\overset{-}{O}]$ $\xrightarrow{C_6H_5C\equiv CH}$ (N, O, C_6H_5, C_6H_5) (97%)

formed in these reactions, as well as being useful in their own right, are valuable synthetic intermediates, for on catalytic hydrogenolysis they afford 1,3-amino-alcohols and on hydrolytic reduction with Raney nickel in aqueous acetic acid they yield β-hydroxy ketones, which are readily dehydrated to $\alpha\beta$-unsaturated ketones (Curran, 1982).

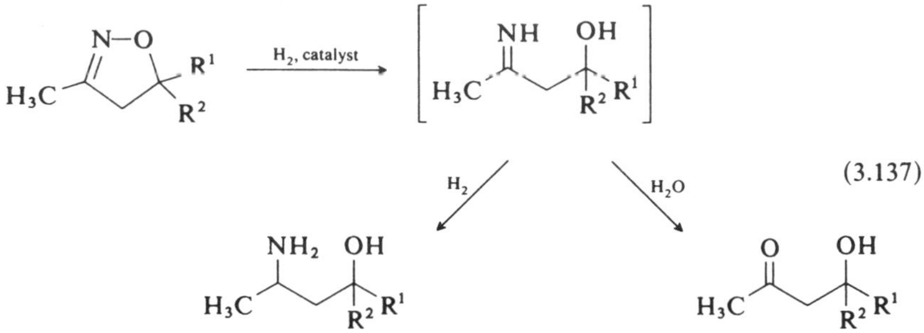

(3.137)

Intramolecular reactions take place readily and some of the most useful applications of 1,3-dipolar cyclo-additions in synthesis have involved reactions of this kind (Padwa, 1976; Oppolzer, 1977; Kozikowski, 1984). This is nicely illustrated in a recent synthesis of the antitumour agent sarcomycin (181, R = H) (3.138) (Kozikowski and Stein, 1982). Dehydration of the nitro-alkene (177) gave the isoxazoline (179) directly in 55 per cent yield by way of the nitrile oxide (178). Hydrogenolysis with Raney nickel in aqueous acetic acid then led to the β-hydroxy ketone (180) which by dehydration gave (181, R = Et) and thence sarcomycin (181, R = H).

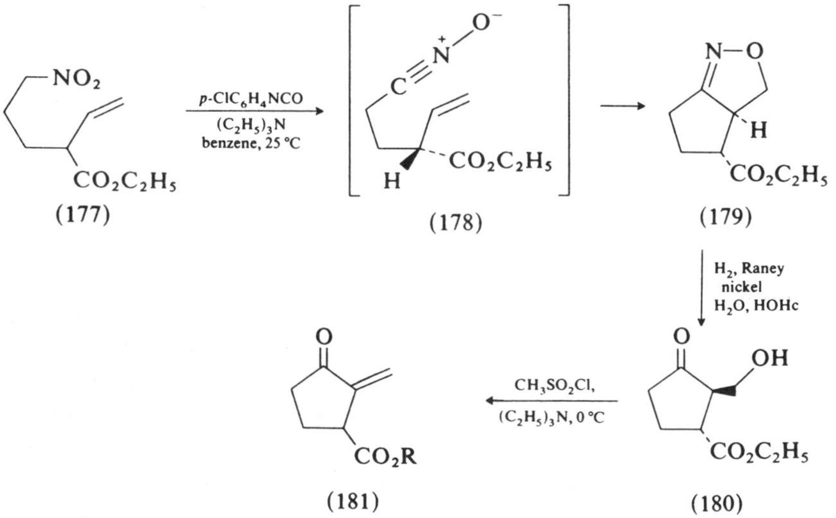

(177) (178) (179) (180) (181)

(3.138)

Intramolecular cyclo-addition reactions of nitrones have also been widely employed (Tufariello, 1979; Baggiolini *et al.*, 1982). The required unsaturated nitrones are readily obtained by oxidation of *N*-alkenylhydroxylamines or by condensation of an aldehyde with a monosubstituted hydroxylamine. Thus, the *cis* bicyclic isoxazolidine (183) was obtained directly from *N*-(5-hexenyl)-*N*-methyl hydroxylamine by oxidation with mercuric oxide, or by reaction of 5-hexenal with *N*-methylhydroxylamine, by way of the intermediate nitrone (182).

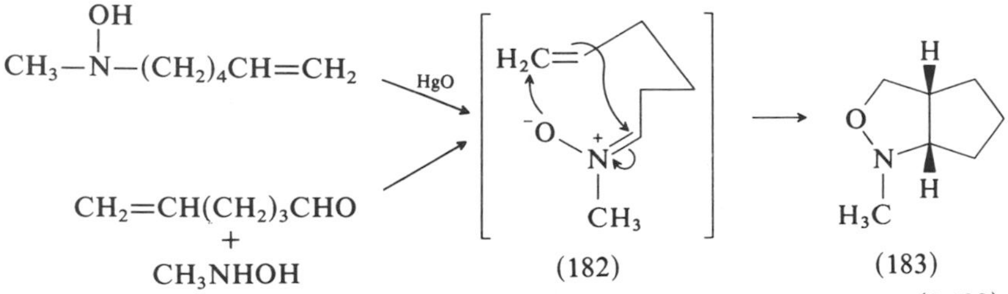

$CH_3-N(OH)-(CH_2)_4CH=CH_2$ (HgO) or $CH_2=CH(CH_2)_3CHO + CH_3NHOH$ → (182) → (183)

(3.139)

Bridged-ring structures are sometimes produced, depending on the structure of the unsaturated nitrone (3.140).

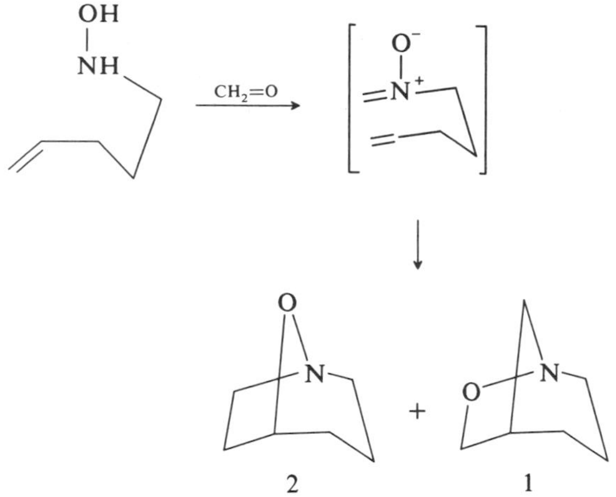

$CH_2=O$

2 + 1

(3.140)

Like the Diels–Alder reaction 1,3-dipolar cyclo-addition reactions are both regio- and stereo-selective, features which enhance their synthetic usefulness. Thus, although reaction of an unsymmetrical 1,3-dipole with an unsymmetrical dipolarophile could apparently give two products, in general, one is formed predominantly or even exclusively, and in reactions with geometrically isomeric alkenes the '*cis* rule' holds, that is the steric disposition of the substituents about the double bond in the alkene is retained in the adduct. But, again as in the Diels–Alder reaction, there is not always a clear distinction between the relative stabilities of the '*exo*' and '*endo*' transition states, to use Diels–Alder terminology. Thus, in the reaction of the nitrone 3,4,5,6-tetrahydropyridine oxide (184) with propene, the adduct (185) was formed with high stereo and regioselectivity through the '*exo*' transition state (187) but in many other reactions, particularly those with open-chain 1,3-dipoles, mixtures of products are formed by way of the corresponding '*exo*' and '*endo*' transition states.

toluene 110 °C; $LiAlH_4$

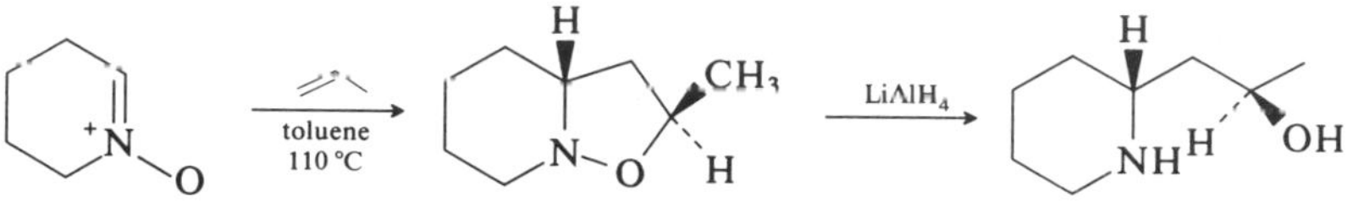

(184) (185) (186)

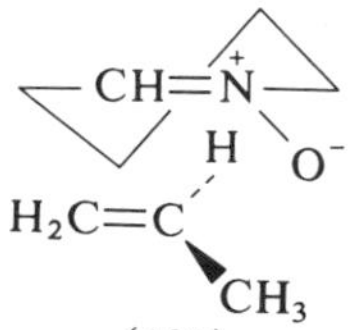

(187) (3.141)

Enantioselective cyclisations of several optically active unsaturated nitrones have been achieved. Thus, in a synthesis of the alkaloid (+)-luciduline (190) reaction of the optically active hydroxylamine (188) with formaldehyde in boiling toluene gave the adduct (189) directly in nearly 90 per cent yield, converted in several steps into optically pure (+)-luciduline (Oppolzer and Petrzilka, 1978). This synthesis illustrates a

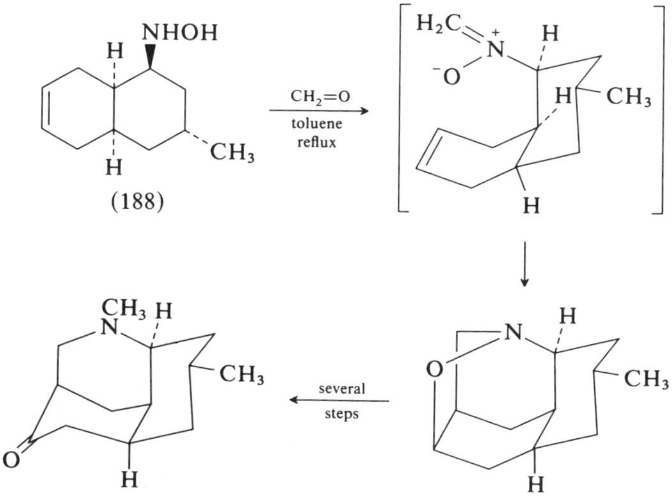

$CH_2{=}O$, toluene reflux; several steps

(188) (189) (190) (3.142)

useful feature of the 1,3-dipolar addition of nitrones; it provides an alternative to the Mannich reaction as a route to β-amino-ketones through reductive cleavage of the N—O bond in the isoxazolidine and oxidation of the 1,3-hydroxy-amine produced.

As in the Diels–Alder reaction, the orientation of addition of an unsymmetrical 1,3-dipole to an unsymmetrical dipolarophile is controlled by the coefficients of the relevant highest occupied and lowest unoccupied molecular orbitals at the reaction sites, although it is often more difficult to make predictions for these reactions than it is for Diels–Alder reactions. In practical terms it seems to be the case that for addition of nitrones and nitrile oxides, two of the most frequently used 1,3-dipoles, to monosubstituted and 1,1-disubstituted alkenes, except those carrying powerfully electron-withdrawing groups, the oxygen of the 1,3-dipole attacks the more highly substituted carbon of the double bond. Thus, 3,4,5,6-tetrahydropyridine 1-oxide reacts with both methyl acrylate and propene to give adducts (191), and the nitrile oxide (193) reacted with the optically active alkene (192) to give the isoxazoline (194). The latter reaction formed a key step in an asymmetric synthesis of 2-deoxy-D-ribose (Kozikowski and Ghosh, 1982) (3.143). As illustrated in the reaction between tetrahydropyridine oxide and propene, 1,3-dipolar compounds are reactive compounds and frequently form adducts with 'non-activated' dipolarophiles. Like Diels–Alder reactions [3+2] cyclo-additions are reversible and the rule given above for the regioselectivity only

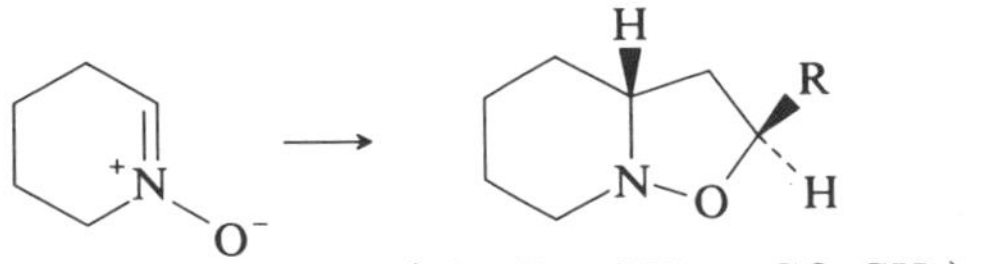

(191; R = CH_3 or CO_2CH_3)

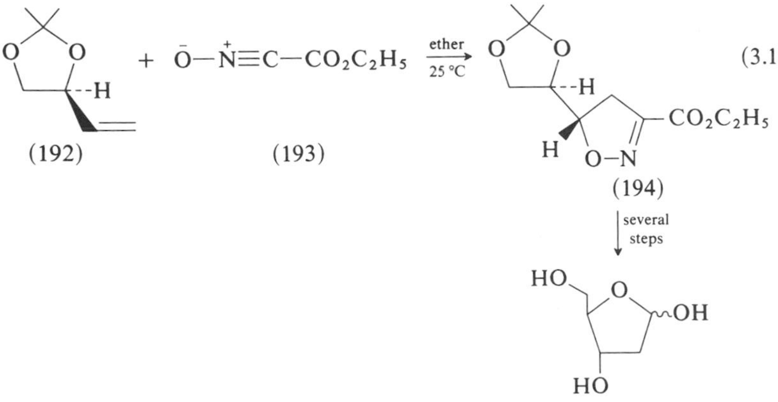

(192) + (193) $\bar{O}-\overset{+}{N}\equiv C-CO_2C_2H_5$ $\xrightarrow[25\,°C]{\text{ether}}$ (194) (3.143)

several steps

2-Deoxy-D-ribose

applies to reactions carried out under kinetically controlled conditions; selectivity may be lost in thermodynamically controlled reactions.

There has been some controversy about the mechanism of 1,3-dipolar cyclo-addition reactions. It has been suggested that they proceed in two steps through a biradical intermediate, but it is now generally agreed that, like the Diels–Alder reaction they are concerted 6π-electron reactions and take place through a five-membered cyclic transition state in which the two components approach each other in parallel planes (Huisgen, 1968*a*; Fleming, 1976; but see also Dewar, 1984). This view is supported by the relative insensitivity of the reactions to solvent polarity, and by their stereoselectivity. Numerous examples with a variety of 1,3-dipolarophiles show that the stereochemistry of substituents in the dipolarophiles is preserved in the adducts. This is much more satisfactorily accounted for by a concerted pathway then by a two-step radical sequence.

4 Reactions at unactivated C—H bonds

Most synthetically useful organic reactions at saturated carbon atoms take place either by displacement of a suitable leaving group, or by replacement of a hydrogen atom at a carbon atom which is 'activated' by the presence of a neighbouring activating group. Ionic attack at unactivated C—H bonds is uncommon. Free radicals, on the other hand, can often be obtained with enough energy to break unactivated C—H bonds, but intermolecular reactions of this type are of limited value synthetically, because the reagents are unselective and mixtures of products generally result. It has been found, however, that in many molecules which meet certain structural and geometrical requirements *intramolecular* free radical attack at unactivated C—H bonds can become quite specific, leading to the introduction of functional groups at the site of specific C—H bonds. A number of reactions of this type have become of synthetic importance.

The key step in these reactions is an intramolecular abstraction of hydrogen from a carbon atom, resulting in the transfer of a hydrogen atom from carbon to the attacking free radical in the same molecule. Because of the geometrical requirements of the transition state for this step, the most frequently observed intramolecular hydrogen transfers of this type are 1,5-shifts, corresponding to specific attack on a hydrogen atom attached to a δ-carbon. The reactions can be generally represented as in (4.1). Homolytic cleavage of the Y—X bond is followed by hydrogen

$$\begin{array}{ccc} & H & \\ C & & Y{-}X \\ | & & | \\ C & & C \\ & C & \end{array} \xrightarrow{-X\cdot} \begin{array}{ccc} & H & \\ C & & Y\cdot \\ | & & | \\ C & & C \\ & C & \end{array} \longrightarrow \begin{array}{ccc} C\cdot & & YH \\ | & & | \\ C & & C \\ & C & \end{array} \xrightarrow{+X'\cdot} \begin{array}{ccc} & X' & \\ C & & YH \\ | & & | \\ C & & C \\ & C & \end{array} \qquad (4.1)$$

transfer from the δ-carbon atom to Y·, and the resulting carbon radical finally reacts with a free radical X′· which may or may not be identical with X·.

A number of important reactions of this type in which Y = O are discussed below. Another useful reaction in which Y = N is the well-known Hofmann-Loeffler-Freytag reaction.

4.1. The Hofmann–Loeffler–Freytag reaction

This reaction provides a convenient and useful method for the synthesis of pyrrolidine derivatives from *N*-halogenated amines (Wolff, 1963) (4.2). The reaction is effected by warming a solution of the halogenated amine in strong acid (concentrated sulphuric acid or trifluoroacetic

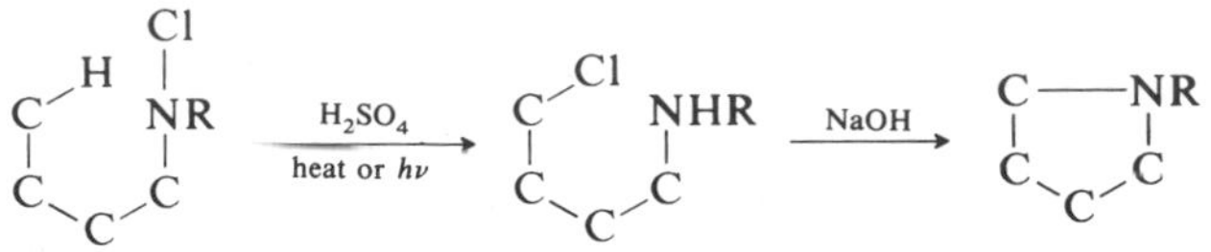

H_2SO_4, heat or $h\nu$; NaOH (4.2)

acid are often used), or by irradiation of the acid solution with ultraviolet light. The immediate product of the reaction is the δ-halogenated amine, but this is not generally isolated, and by basification of the reaction mixture it is converted directly into the pyrrolidine. Both *N*-bromo- and *N*-chloroamines have been used as starting material, but the *N*-chloroamines are said to give better yields. They are easily obtained from the amines by action of sodium hypochlorite or *N*-chlorosuccinimide.

The first example of this reaction was reported by A. W. Hofmann in 1883. In the course of a study of the reactions of *N*-bromoamides and *N*-bromoamines, he treated *N*-bromoconiine (1) with hot sulphuric acid and obtained, after basification, a tertiary base which was later identified as δ-coneceine (2) (4.3). Later, further examples of the reaction were

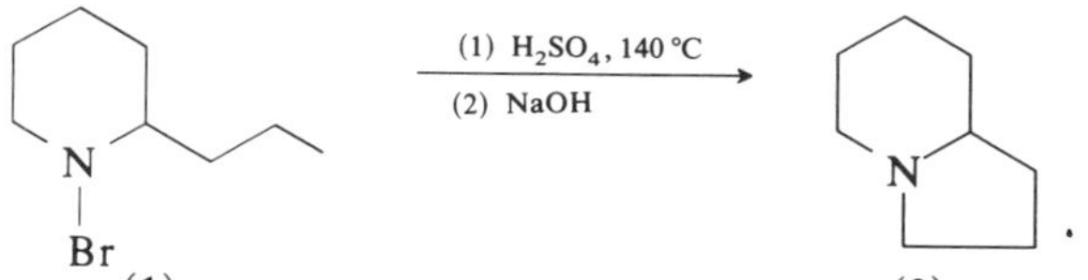

(1) H_2SO_4, 140 °C (2) NaOH (1) (2) (4.3)

reported by Loeffler, including a neat synthesis of the alkaloid nicotine (3) (4.4). Numerous other cyclisations leading to both simple pyrrolidines and to more complex polycyclic structures have since been recorded.

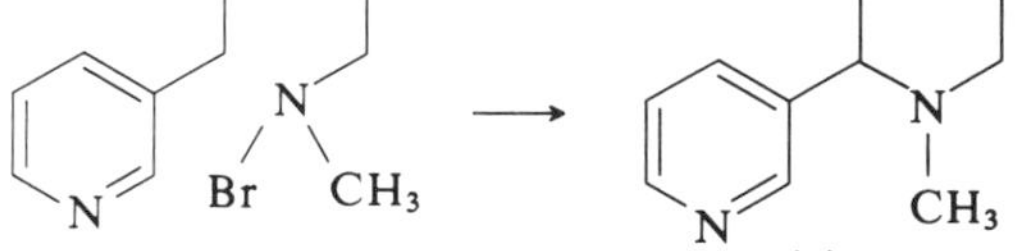

(3) (4.4)

It was long believed that only *N*-halogenated derivatives of secondary amines would undergo the reaction, but it is now known that *N*-chloro primary amines will also undergo the reaction in strongly acid solution by using ironII ions as initiators. Thus, 2-propylpyrrolidine is readily obtained from both *N*-chloro-1-aminoheptane and *N*-chloro-4-aminoheptane (4.5).

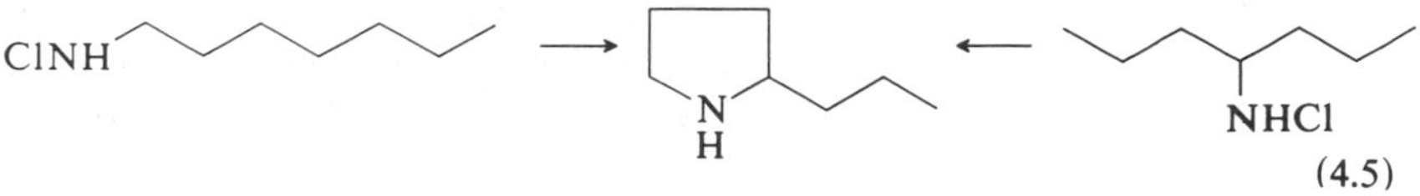

ClNH → N H ← NHCl

(4.5)

Although the majority of applications of the reaction have employed *N*-haloamines as starting materials, *N*-haloamides have been used occasionally and these also give rise to secondary pyrrolidines by loss of the acyl group. *N*-Butyl-*N*-chloroacetamide, for example, when heated with sulphuric acid and subsequently treated with alkali, is converted into pyrrolidine in 50 per cent yield.

With *N*-halocycloalkylamines cyclisation leads to bridged-ring structures, but in these cases the products may not be exclusively pyrrolidine derivatives. Thus, whereas *N*-bromo-*N*-methylcycloheptylamine was converted into tropane (4) in 40 per cent yield when warmed with sulphuric acid, *N*-chloro-4-ethylpiperidine gave a mixture of the azabicycloheptane (5) and quinuclidine (6), and *N*-chloro-*N*-methylcyclooctylamine is reported to yield *N*-methylgranatamine (7) exclusively (4.6). The reasons for the preferential formation of a six-membered ring

NCH_3

(4)

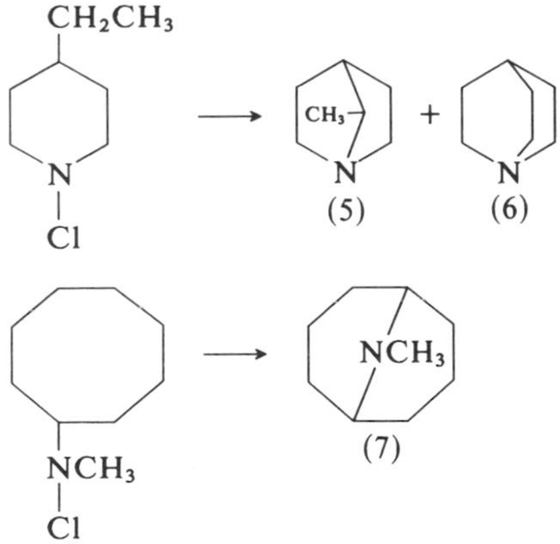

CH_2CH_3

N
|
Cl

→ CH_3 N (5) + N (6)

(4.6)

→ NCH_3 (7)

NCH_3
|
Cl

in this case are not clearly understood. In general, yields obtained in these 'bicyclo' reactions, and the ease of reaction, are noticeably less than with open-chain amines. This is because it may not be so easy in the cycloalkylamines for the molecule to adopt the necessary conformation in the transition state to allow the nitrogen radical to abstract a hydrogen atom from the δ-carbon.

The reaction is believed to proceed by a free radical chain process, and the reaction path shown in (4.7) has been proposed (Corey and Hertler, 1960).

$$\begin{array}{c} CH_2-CH_2 \\ | \quad\quad | \\ R-CH_2 \quad CH_2 \\ \overset{+}{Cl}NHR' \end{array} \longrightarrow \begin{array}{c} CH_2-CH_2 \\ | \quad\quad | \\ R-CH_2 \quad CH_2 \\ \cdot\overset{+}{N}HR' \end{array} + Cl\cdot \longrightarrow \begin{array}{c} CH_2-CH_2 \\ | \quad\quad | \\ R-CH\cdot \quad CH_2 \\ \overset{+}{N}H_2R' \end{array}$$

$$\downarrow R''\overset{+}{N}HClR' \qquad (4.7)$$

$$\begin{array}{c} CH_2-CH_2 \\ | \quad\quad | \\ R-CH \quad CH_2 \\ N \\ | \\ R' \end{array} \xleftarrow{\text{base}} \begin{array}{c} CH_2-CH_2 \\ | \quad\quad | \\ R-CHCl \quad CH_2 \\ \overset{+}{N}H_2R' \end{array}$$

The reacting species may be either the free chloramine or the chlorammonium ion. This dissociates under the reaction conditions to form the ammonium radical, which then abstracts a suitably situated hydrogen atom on the δ-carbon to give the corresponding carbon radical. This in turn abstracts a chlorine atom from another molecule of chloramine, thus propagating the chain and at the same time forming the δ-chloroamine, from which the cyclic amine is subsequently obtained.

The free radical nature of the reaction is suggested by the fact that it does not proceed in the dark at 25 °C, and that it is initiated by heat, light or iron(II) ions and inhibited by oxygen. The hydrogen abstraction step must be intramolecular for only thus can the specificity of reaction at the δ-carbon be understood. Strong evidence that the decomposition of *N*-chloroamines in acid involves an intermediate in which the δ-carbon atom is trigonal is provided by the observation that the optically active chloroamine (8) on decomposition in sulphuric acid at 95 °C gave a 43 per cent yield of pure 1,2-dimethylpyrrolidine (9) which was optically inactive (4.8). The intermediacy of δ-chloroamines in the reaction has been confirmed by their isolation in a number of cases.

$$\underset{(8)}{H_3C(H)(D)C\text{-}CH_2CH_2CH_2\text{-}N(Cl)CH_3} \longrightarrow \underset{(9)}{\text{1,2-dimethylpyrrolidine}} \qquad (4.8)$$

As in other free radical reactions, secondary hydrogen atoms react more readily than primary. In nearly all cases reaction at the δ-carbon atom with formation of a pyrrolidine is favoured. Thus, in the reaction with *N*-chlorobutylpentylamine, attack by the nitrogen radical on the δ-methyl group would lead subsequently to 1-pentylpyrrolidine, whereas attack on the δ-methylene would result in formation of 1-butyl-2-methylpyrrolidine. Only the latter compound was formed (4.9). Tertiary hydrogens react very readily, but the resulting tertiary chlorides are rapidly solvolysed under the conditions of the reaction, and no cyclisation products are formed.

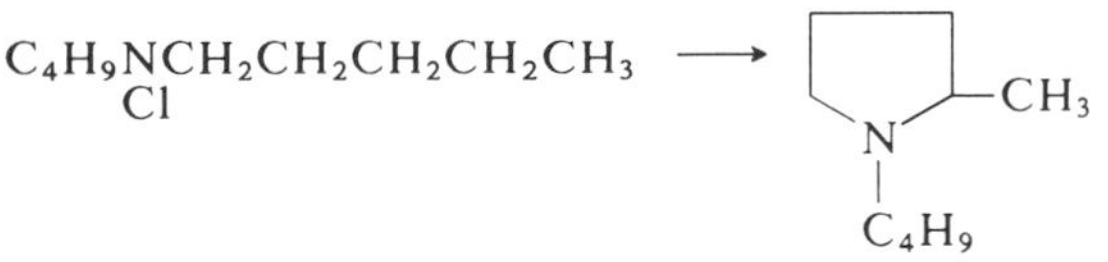

$$C_4H_9N(Cl)CH_2CH_2CH_2CH_2CH_3 \longrightarrow \text{1-butyl-2-methylpyrrolidine} \quad (4.9)$$

An outstanding application of the Hofmann–Loeffler–Freytag reaction is found in the synthesis of the steroidal alkaloid derivative dihydroconessine (10) illustrated in (4.10). In this synthesis the five-membered

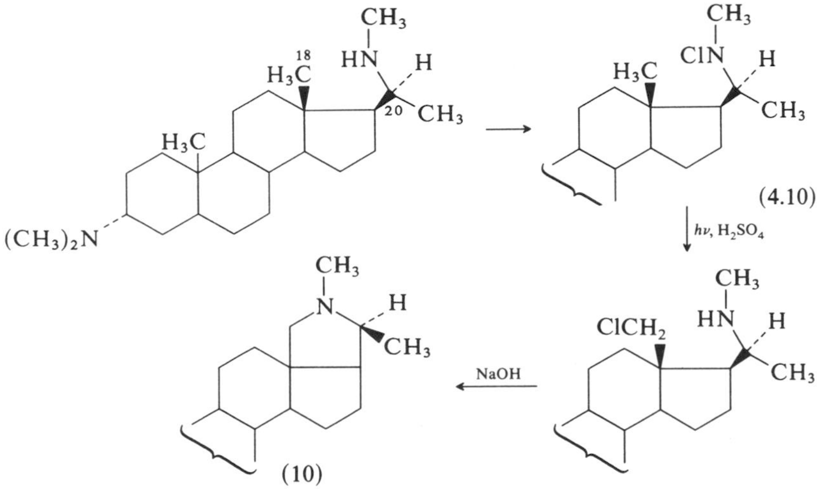

(4.10)

(10)

nitrogen ring is constructed by attack on the unactivated C-18 angular methyl group of the precursor by a suitably placed nitrogen radical at C-20 (Corey and Hertler, 1959). The ease of this reaction is due to the fact that in the rigid steroid framework the β-C-18 angular methyl group and the β-C-20 side chain carrying the nitrogen radical are suitably disposed in space to allow easy formation of the six-membered transition state necessary for 1,5-hydrogen transfer from the methyl group to nitrogen (4.11).

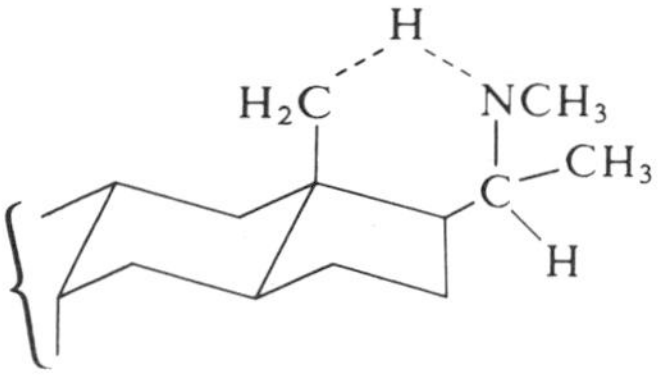

(4.11)

4.2. Cyclisation reactions of nitrenes

Intramolecular reactions leading to the formation of five-membered rings containing nitrogen atoms also take place on thermal decomposition of azides (Abramovitch and Davis, 1964). Thus, the azide (11) on heating in boiling diphenyl ether is converted into the indoline (12) in 45–50 per cent yield. Carbazoles are formed by thermal or photolytic decomposition of 2-azidodiphenyls.

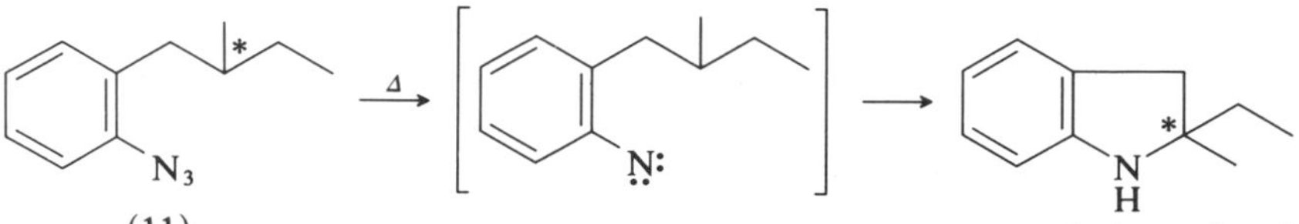

N_3 (11) $\xrightarrow{\Delta}$ [N:] $\longrightarrow$ N H (12) (4.12)

There is much evidence to suggest that these reactions involve the intermediate formation of nitrenes (electron deficient species in which nitrogen has only six electrons in its outer shell; they may have either a singlet R—N̈: or a triplet diradical structure R—N̈·). The cyclisation step is thought to take place by direct insertion of singlet nitrene into the C—H bond, and not through a diradical intermediate, for reaction with the optically active azide (11) gave the optically active indoline (12). A diradical intermediate would have been expected to give an optically inactive product.

Acylnitrenes can also be obtained, most readily by photolysis of acyl azides (Lwowski, 1967). In general they rearrange readily to isocyanates, but in suitable cases where the geometry of the molecule brings a δC—H bond into close proximity to the nitrogen atom, cyclisation may take place. This has been used to prepare compounds related to the diterpene alkaloids. Thus, the azide of podocarpic acid methyl ether (13) on photolysis in cyclohexane afforded the lactam (14) by attack on the unactivated angular methyl group (4.13), a synthesis which completed the structural proof of the atisine family of alkaloids.

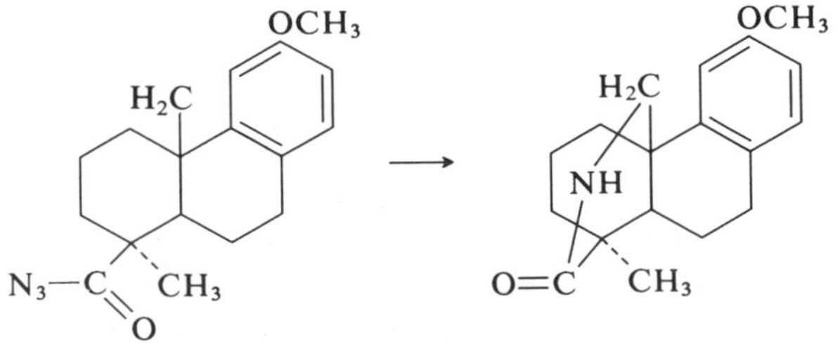

OCH_3 H_2C N_3—C(=O) CH_3 (13) $\longrightarrow$ OCH_3 H_2C NH O=C CH_3 (14) (4.13)

4.3. The Barton reaction and related processes

In the general equation for intramolecular free-radical hydrogen transfer reactions (4.1), Y may be an oxygen atom, and a number of synthetically useful reactions of this type, leading to attack on unactivated C—H bonds, have Y = O and X = NO, Cl, I or OH.

Photolysis of organic nitrites

It has been known for many years that vapour phase pyrolysis or photolysis of organic nitrites (Y = O, X = NO) affords alkoxy radicals and nitrogen monoxide. But in many cases the radicals produced are consumed in synthetically useless reactions such as fragmentation, disproportionation and non-selective intermolecular hydrogen abstraction. It has been found, however, that when the structure of the molecule is such as to bring the $-\overset{|}{\underset{|}{C}}-O-NO$ group and a C—H bond into close proximity, or potentially close proximity, the alkoxyl radicals produced by photolysis of the nitrites in solution have sufficient energy to bring about selective *intramolecular* hydrogen abstraction according to the general scheme (4.1), with subsequent capture of nitrogen monoxide by the carbon radical and formation of a nitroso-alcohol, which may be isolated as the dimer or, where the structure permits, rearranged to an oxime (4.14). The nitroso (oximino) compounds produced may be further transformed into other functional derivatives such as carbonyl compounds, amines or cyano derivatives. The photolytic conversion of organic nitrites into nitroso alcohols has become known as the Barton reaction, after its inventor, and the sequence has been widely used (see Nussbaum and Robinson, 1962), particularly in the synthesis of biologically important steroid derivatives.

$$-\overset{H}{\overset{|}{\underset{|}{C}}}-(C_2)-\overset{ONO}{\overset{|}{\underset{|}{C}}}- \xrightarrow{h\nu} -\overset{NO}{\overset{|}{\underset{|}{C}}}-(C_2)-\overset{OH}{\overset{|}{\underset{|}{C}}}- \longrightarrow \text{dimer} \qquad (4.14)$$

$$\downarrow\uparrow$$

$$-\overset{NOH}{\overset{\|}{C}}-(C_2)-\overset{OH}{\overset{|}{\underset{|}{C}}}-$$

The reaction is effected by irradiation under nitrogen of a solution of the nitrite in a suitable non-hydroxylic solvent with light from a high pressure mercury arc lamp. A pyrex filter is usually employed to limit the radiation to wavelengths greater than about 300 nm, thus avoiding deleterious side-reactions induced by more energetic lower wavelength radiation. This is possible because of the multiplicity of weak absorption bands of organic nitrites in the region 320–380 nm. It is the absorption of the light due to these weak bands which brings about the dissociation of the nitrite.

Detailed examination of a range of examples leads to the conclusion that the reaction proceeds in discrete steps by an intramolecular radical mechanism, and is strongly favoured by the possibility of a cyclic six-membered transition state. Mechanistic studies using ^{15}N have shown that photolysis of a nitrite gives, in a reversible step, an alkoxyl radical and nitrogen monoxide which are completely dissociated from each other. The alkoxyl radical rearranges rapidly, by abstraction of hydrogen, to a carbon radical which can be captured by radical trapping reagents to give 'transfer products'. The normal fate of the carbon radical in the absence of trapping reagents is to react relatively slowly with nitrogen monoxide to give the nitroso alcohol. The sequence is pictured in (4.15).

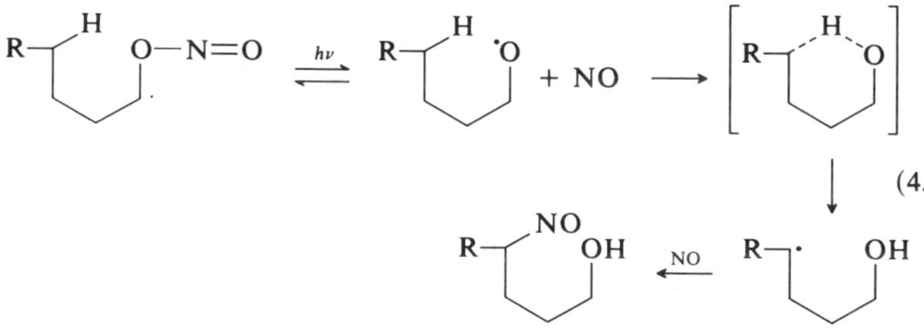

(4.15)

There is ample evidence to support the view that the hydrogen transfer step takes place through a six-membered cyclic transition state. In practice, reaction occurs almost exclusively by abstraction of a hydrogen atom from the δ-carbon atom. Only a very few examples in which this is not the case have been found. Thus, photolysis of 1-octyl nitrite in benzene solution gave a 45 per cent yield of the dimer of 4-nitroso-1-octanol, by way of the transition state (4.16). No evidence for the

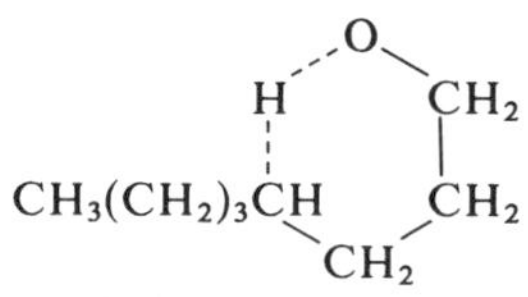

(4.16)

formation of any other nitroso-octanol was found. Similarly, 4-phenyl-1-butyl nitrite and 5-phenyl-1-pentyl nitrite are readily converted into nitroso dimers whose structures can be rationalised by assuming a six-membered transition state in the hydrogen transfer step. In the latter example, none of the product formed by abstraction of a benzylic hydrogen atom, which would have required a seven-membered transition state, was obtained. In the same way, photolysis of 3-phenylpropyl nitrite did not yield any product corresponding to abstraction of a benzylic hydrogen atom through a five-membered transition state, even although abstraction of the benzylic hydrogen should be highly favourable thermodynamically.

The exact spatial arrangement of the six atomic nuclei forming the transition state is not known with certainty and may not be the same in every case. The chair, boat and semi-chair forms have all been invoked, as well as the form in which the C, H and O atoms are approximately linear, but reaction in many cases appears to be favoured by the availability of a chair-form transition state. But even in geometrically favourable cases reaction only takes place provided the distance between the reacting centres is not too great. This may become the deciding factor in rigid molecules. For example, in the radical (15) the distance between the δ-methyl group and the oxygen atom is too large for hydrogen abstraction to take place and the alkoxyl radical undergoes fragmentation instead (4.17). On the other hand, in the pinane derivative (16), in which the

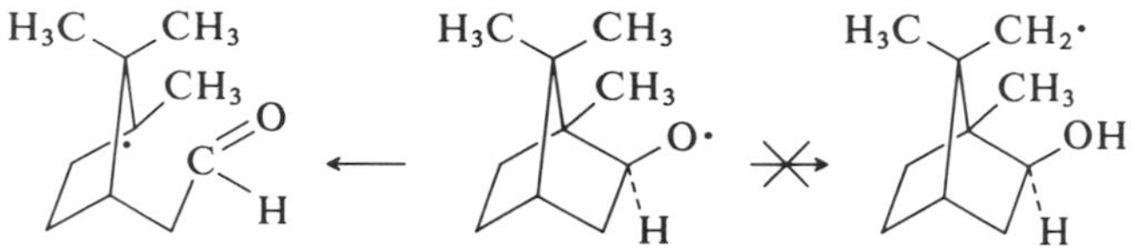

(4.17)

(15)

methyl and hydroxyl groups are nearer each other, attack of the oxy radical on the methyl hydrogen atoms takes place readily (4.18), thus providing a convenient method for functionalising the unactivated bridge methyl groups of the pinane structure (Bosworth and Magnus, 1972).

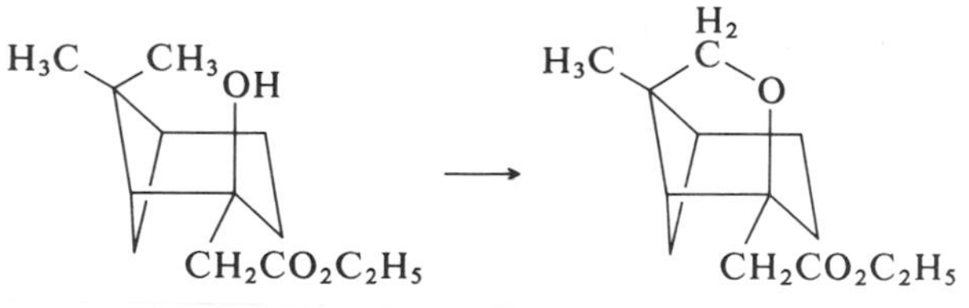

(4.18)

(16)

Other similar cases are found in the steroid series. It is estimated that the activation energy for the intramolecular abstraction of hydrogen reaches a minimum in fixed systems with an O—C distance of 0.25 to 0.27 nm in the starting material. For distances exceeding 0.28 nm the rate of this reaction falls below that of intermolecular hydrogen abstraction of fragmentation reactions.

Interesting results which further support the proposed reaction pathway for the Barton reaction have been obtained in the photolysis of certain cyclohexyl nitrites. Nitroso dimers or monomers were obtained in only small amounts when cyclohexyl nitrite and *cis*- or *trans*-3-methylcyclohexyl nitrite were irradiated. In these compounds, formation of a six-membered cyclic transition state is unfavourable on conformational grounds. In contrast, rearrangement of both *cis*- and *trans*-2-ethylcyclohexyl nitrites proceeded readily and in good yield, clearly facilitated by the fact that the transition states for the hydrogen transfer steps can adopt stable conformations similar to the *cis*- and *trans*-decalins (4.19).

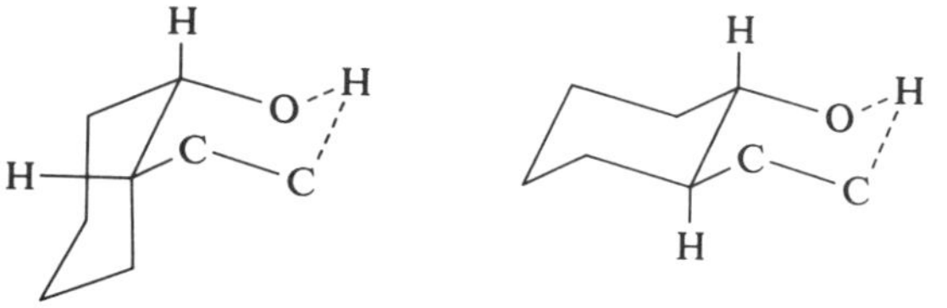

(4.19)

Other factors being equal, it is found that in general, abstraction of a secondary hydrogen atom by an alkoxy radical is much easier than abstraction of a primary hydrogen, although in certain cases attack at the primary hydrogens may be facilitated by favourable geometrical factors. Thus 2-hexyl nitrite afforded a 30 per cent yield of nitroso dimer on photolysis, but with 2-pentyl nitrite, where formation of a six-membered cyclic transition state requires abstraction of a primary hydrogen atom, the yield fell to 6 per cent. Presumably a tertiary hydrogen would react even more easily than a secondary, but apparently no comparative studies have been made.

Intramolecular hydrogen abstraction by alkoxy radicals is always accompanied to a greater or lesser extent by disproportionation, radical decomposition and intermolecular reactions. In the case of oxy radicals derived from tertiary nitrites, as in (17), reaction follows the normal course if tertiary or secondary hydrogen atoms are available for abstraction. But if only primary hydrogens are available the Barton reaction is superseded by alkoxy radical decomposition (4.20). With primary and secondary alkoxy radicals, disproportionation to form an alcohol and a carbonyl compound is often favoured over decomposition and always takes place to an appreciable extent, and particularly when the geometric and structural requirements of the Barton reaction are not met.

$$CH_3-\underset{\underset{ONO}{|}}{\overset{\overset{CH_3}{|}}{C}}-(CH_2)_2CH_3 \xrightarrow{h\nu} (CH_3)_2C{=}O + ON(CH_2)_2CH_3 \qquad (4.20)$$

(17)

The most important synthetic applications of the Barton reaction have been in the steroid series, particularly in the functionalisation of the two non-activated C-18 and C-19 angular methyl groups by photolysis of the nitrites of suitably disposed hydroxyl groups. In principle C-18 can be attacked by an alkoxy radical at C-8, C-11, C-15 or C-20, and C-19 by an alkoxy radical at C-2, C-4, C-6 and C-11 (4.21). Most of these

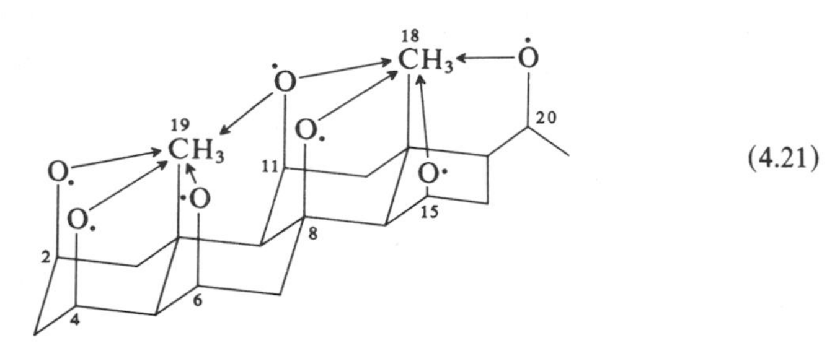

(4.21)

approaches have been realised in practice, either through the Barton reaction or by the related reactions described below (Nussbaum and Robinson, 1962; Heusler and Kalvoda, 1964; Kalvoda and Heusler, 1971). The reactions are facilitated by the conformational rigidity of the steroid framework and by the 1,3-diaxial relationship of the interacting groups, which allows easy formation of conformationally favoured six-membered cyclic transition states. Because of this, attack on the primary hydrogen atoms of the methyl groups is much easier than in the aliphatic series, and good yields of nitroso monomers or dimers are often obtained.

The reaction was used by Barton and his co-workers to effect the key step in their elegant synthesis of aldosterone (20), (R = H), (4.22), a biologically important hormone of the adrenal cortex (Barton, Beaton, Geller and Pechet, 1961), by photolysis of the 11β-nitrite (18) in toluene solution. The oxime (19) separated from the solution in 21 per cent yield and on hydrolysis with nitrous acid afforded aldosterone-21-acetate (20), (R = CH_3CO) directly.

It may happen that more than one site in a molecule is favourably situated for attack by the newly formed alkoxy radical, and in such cases a mixture of products may result. Thus, in the reaction (4.22), attack at C-19 instead of C-18 led to (21).

In another application photolysis of the nitrite (22) formed an important step in Corey's synthesis of perhydrohistrionicotoxin (Corey, Arnett and Widiger, 1975). The oxime (23), formed in 20 per cent yield, gave the spiro bicyclic lactam (24) on Beckmann rearrangement.

A modified procedure leads to the introduction of a hydroxyl group at the site of an unactivated C—H bond, a reaction common in nature but not easily effected in the laboratory. If photolysis of the nitrite is effected in the presence of oxygen, the product of rearrangement is not

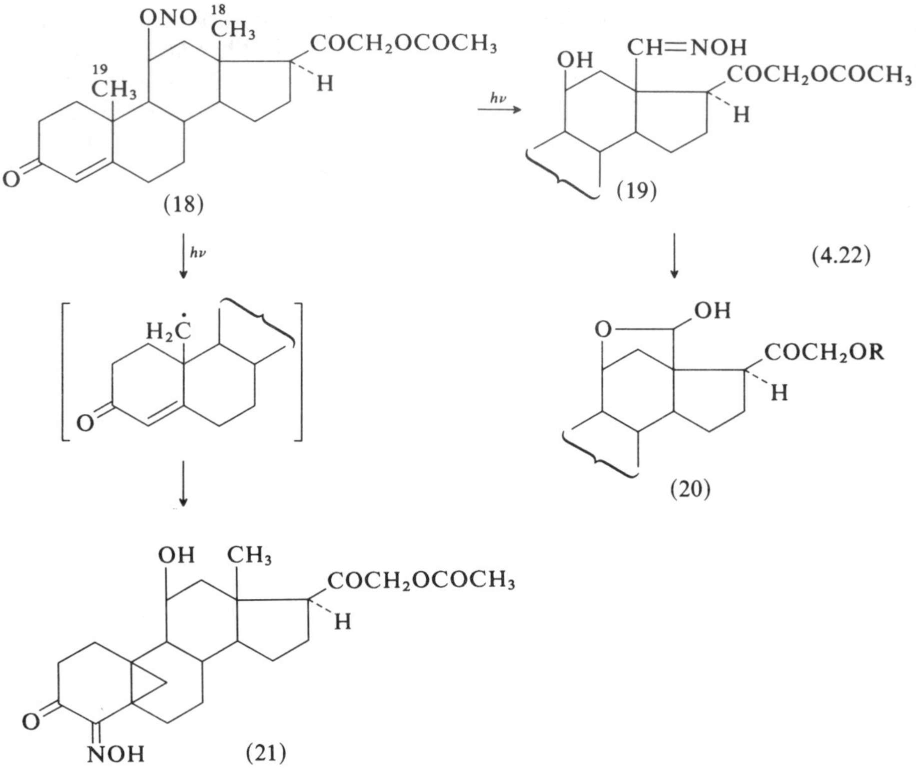

(18) (19) (20) (21) (4.22)

a nitroso compound or an oxime but a *nitrate*, which is easily converted into the corresponding alcohol by mild reduction (Allen *et al.*, 1973). The reaction is believed to take a pathway in which the initial carbon radical is captured by oxygen instead of nitric oxide.

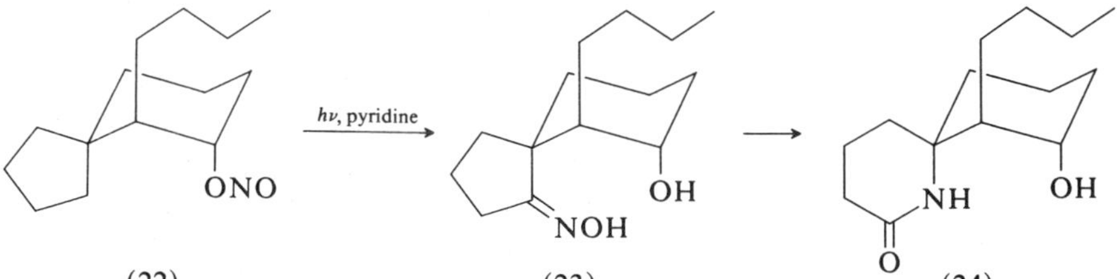

(22) (23) (24) (4.23)

Photolysis of N-nitrosoamides

N-Nitrosoamides also rearrange on photolysis, with abstraction of hydrogen on a δ-carbon atom by an amido radical through a cyclic six-membered transition state, and formation of a C-nitroso derivative.

Thus, the *N*-nitrosodehydroabietylamide (25) was converted into the 6α-nitroso compound (26) in 40 per cent yield on irradiation in benzene solution. In presence of oxygen the corresponding 6α-nitrato derivative was obtained and gave the 6α-hydroxy compound on reduction with lithium aluminium hydride.

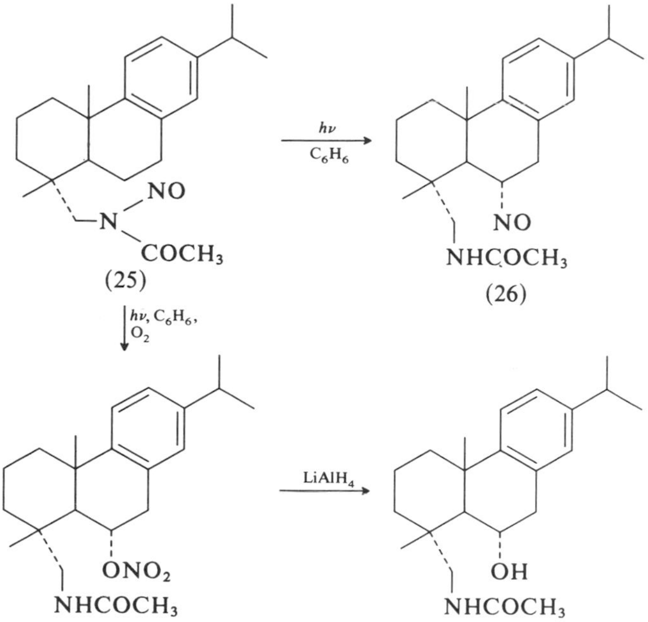

NO
N
COCH$_3$
(25)
$h\nu$
C_6H_6
NO
NHCOCH$_3$
(26)
(4.24)
$h\nu$, C_6H_6, O_2
LiAlH$_4$
ONO$_2$
NHCOCH$_3$
OH
NHCOCH$_3$

Photolysis of hypohalites

The generation of alkoxyl radicals which can undergo intramolecular hydrogen abstraction can also be achieved by photolysis of hypochlorites. The initial products of the reaction are 1,4-chlorohydrins, formed again by preferential abstraction of hydrogen attached to the δ-carbon atom. The 1,4-chlorohydrins produced are easily converted into tetrahydrofurans by treatment with alkali and the sequence provides a convenient route to these compounds. The reactions proceed through a six-membered cyclic transition state as in the photolysis of nitrites (Walling and Padwa, 1963; Akhtar and Barton, 1964) (4.25).

A competing reaction is β-cleavage of the oxy radical to form, in the case of a tertiary hypochlorite, a ketone and an alkyl chloride (4.26). The extent of this reaction varies with the structure of the hypochlorite. It becomes the predominant reaction if the carbon chain is too short to permit 1,5-abstraction of hydrogen, or if the geometrical requirements for a six-membered transition state are not met.

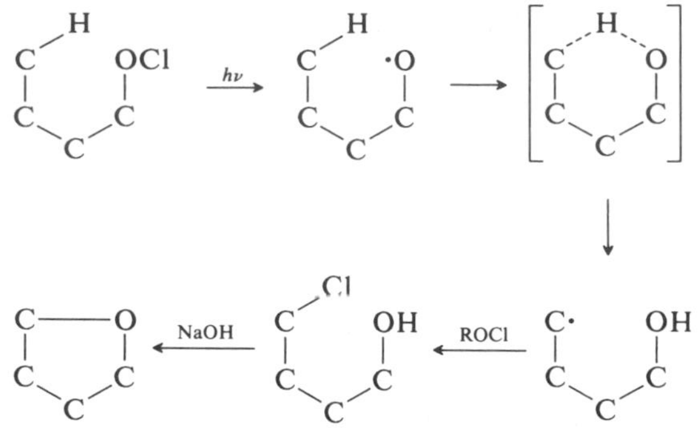

$$\text{(4.25)}$$

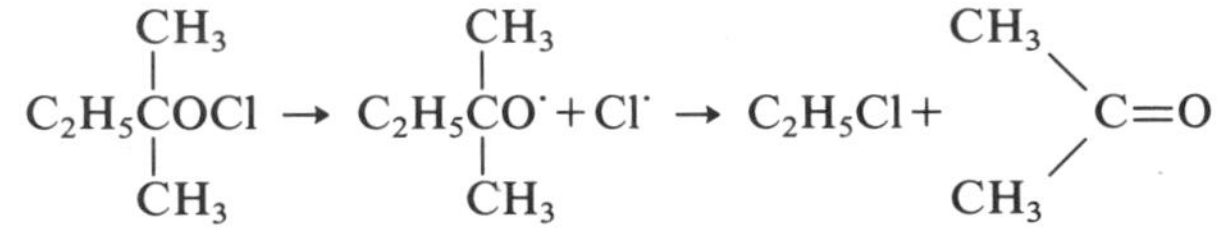

$$C_2H_5\underset{CH_3}{\overset{CH_3}{C}}OCl \rightarrow C_2H_5\underset{CH_3}{\overset{CH_3}{C}}O^{\cdot} + Cl^{\cdot} \rightarrow C_2H_5Cl + \begin{matrix} CH_3 \\ CH_3 \end{matrix}\!\!>C{=}O \qquad (4.26)$$

Hypobromites have also been employed in a limited number of cases. Thus, the trimethylcyclohexyl hypobromite (27), on irradiation and treatment with alkali, was converted into the cyclic ether (28) in 60 per cent yield, presumably by way of the 1,4-bromohydrin (4.27).

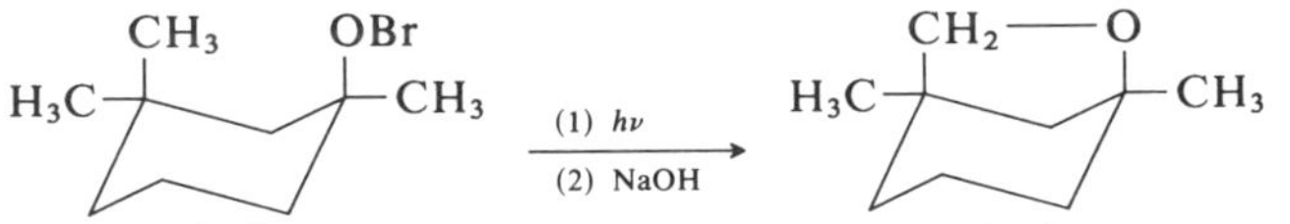

(27) →(1) hν (2) NaOH→ (28) (4.27)

In practice, a more convenient method for generating oxy radicals for intramolecular hydrogen abstractions is by cleavage of hypoiodites prepared *in situ* from the corresponding alcohol. A number of methods have been employed, including treatment of the alcohol with lead tetraacetate and iodine, with or without irradiation, and irradiation of a solution of the alcohol in presence of iodine and mercuryII oxide. The products of the reaction may be either five-membered ring oxides or acetals (Akhtar and Barton, 1964; Heusler and Kalvoda, 1964; Mihailović, Gojković and Konstantinović, 1973). More recently high yields of five-membered cyclic ethers have been obtained by irradiation of alcohols in the presence of iodosobenzene diacetate and iodine (Concepcion *et al.*, 1984).

Most of the applications of the hypoiodite method have been in the steroid series. The diacetate (29), for example, on irradiation in carbon tetrachloride solution in presence of lead tetra-acetate and iodine, gave the cyclic ether (30) in 90 per cent yield (4.28).

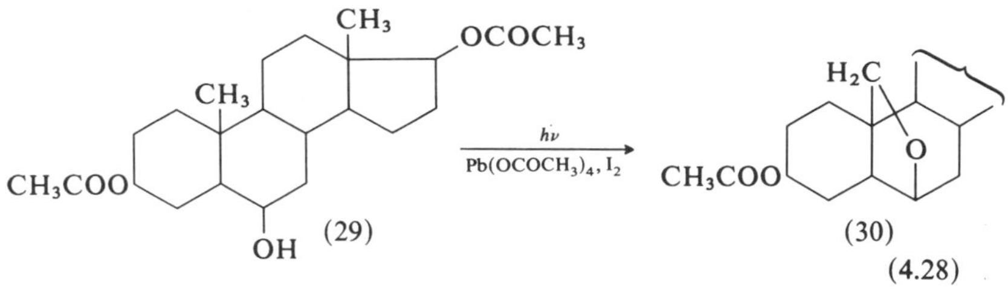

(4.28)

Under certain conditions, especially with lead tetra-acetate and iodine, a second substitution of the group being attacked can take place, and hemiacetals are produced. These can be oxidised to lactones (4.29).

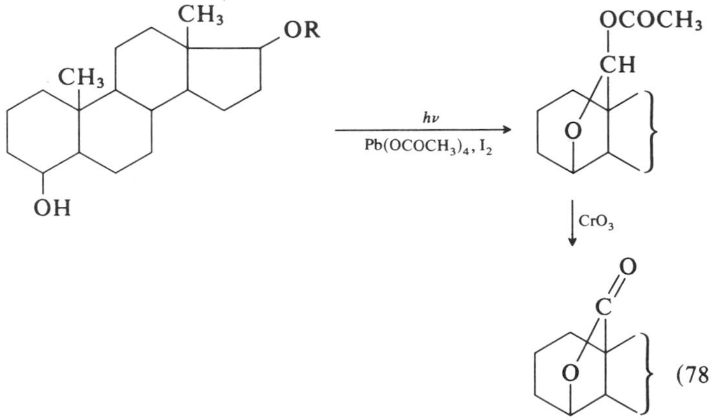

(4.29)

Outside the steroid series the reaction has recently been used to effect a key step in the synthesis of the naturally occurring spiro-acetal (±)-talaromycin, as shown in (4.30) (Kay and Bartholomew, 1984).

All these reactions are thought to take place by initial formation of the unstable hypoiodite which is converted into the 1,4-iodohydrin in a Barton-type transformation (4.31). In subsequent steps which are influenced by steric factors, the iodohydrin may be converted into the tetrahydrofuran, or, by further reaction with the reagent, into an iodo derivative from which the hemiacetal is derived.

A useful extension of the hypoiodite reaction, again involving attack on an unactivated C—H bond, is the formation of γ-lactones from carboxylic acids by photolysis of the corresponding *N*-iodoamides. The iodoamides are prepared *in situ* from the amides with lead tetra-acetate and iodine or t-butyl hypochlorite and iodine, and lactonisation is conveniently effected by photolysis of the amide in presence of the iodinating agent, followed by hydrolysis of the product. Under these conditions stearamide forms γ-stearolactone together with a smaller amount of the

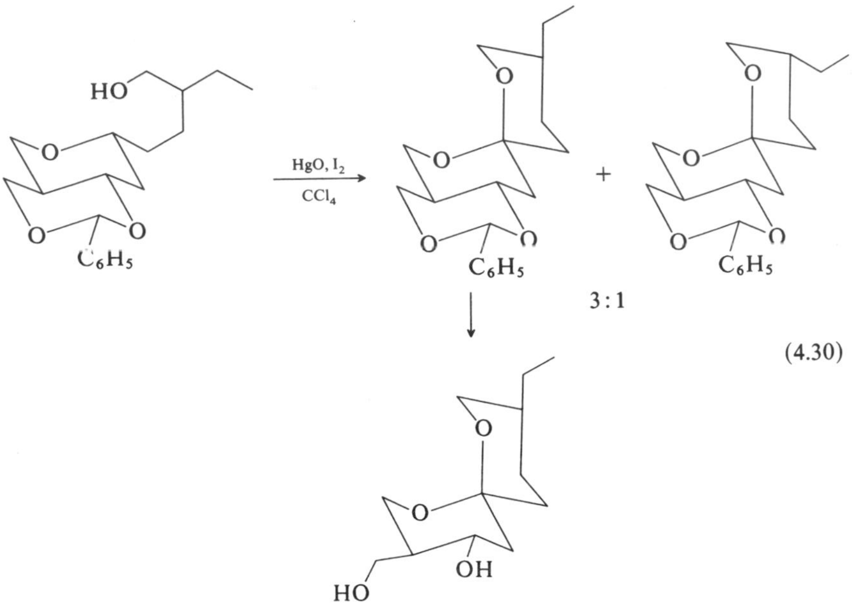

HO O O O C_6H_5 $\xrightarrow[CCl_4]{HgO, I_2}$ O O O O C_6H_5 + O O O O C_6H_5 3:1

(4.30)

O O HO OH

Talaromycin

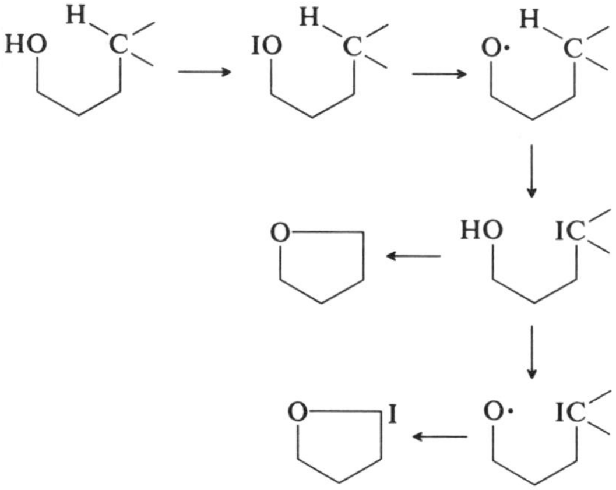

(4.31)

δ-lactone, and 4-phenylbutyramide is converted into phenylbutyrolactone.

The reaction is thought to proceed by homolysis of the N—I bond and intramolecular hydrogen abstraction by the nitrogen radical through a six-membered cyclic transition state, followed by formation of the γ-iodoamide. Intramolecular substitution to an imino lactone and hydrolysis then leads to the γ-lactone (4.32). In agreement with the proposed mechanism, the isolation of a derivative of the postulated imino lactone was achieved in the reaction with γ-phenylbutyramide.

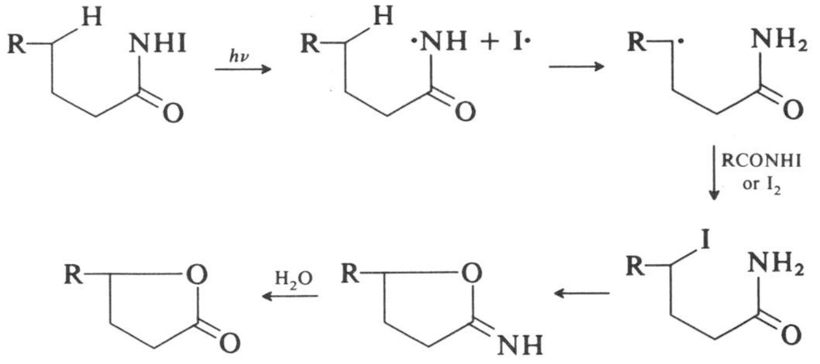

(4.32)

γ-Lactones can also be conveniently obtained from carboxylic acids by photolysis of the *N*-chloroamides, and this method offers certain advantages, particularly with the lower aliphatic acids where the iodoamide method gives poor yields. Chloroamides are readily obtained from the amides by action of t-butyl hypochlorite.

4.4. Reaction of monohydric alcohols with lead tetra-acetate

Another method which has been extensively used for the preparation of tetrahydrofurans, particularly in the steroid series, is oxidation of monohydric alcohols with lead tetra-acetate (see Heusler and Kalvoda, 1964). Like the hypoiodite reaction, this method has the advantage over the photolysis of nitrites and hypochlorites that the unstable reactive intermediate below does not have to be prepared and isolated in a separate step. It is produced from the alcohol *in situ* and converted directly into the alkoxy radical.

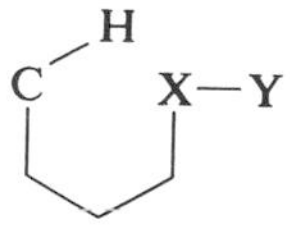

It was originally suggested that the lead tetra-acetate reaction proceeded via an electron-deficient oxonium ion intermediate, but it is now believed that it represents yet another example of an alkoxyl radical rearrangement. The initial reaction is formation of the triacetoxy lead alkoxide which is easily split thermally or photolytically to give the corresponding alkoxyl radical which then attacks the δH as before. Formation of the lead alkoxide is dependent on steric effects in the alcohol, and tertiary or strongly hindered secondary alcohols react only slowly.

$$R{-}OH + Pb(OCOCH_3)_4 \rightleftharpoons RO{-}Pb(OCOCH_3)_3 + HO_2CCH_3 \quad (4.33)$$

This reaction has been extensively applied in the steroid series, and it also provides a useful method for the preration of simple alkyltetrahydrofurans from primary and secondary aliphatic alcohols. Thus, 1-heptanol, treated with lead tetra-acetate in boiling cyclohexane gives 2-propyltetrahydrofuran in 50 per cent yield, accompanied by a small amount of 2-ethyltetrahydropyran and 2-heptanol is converted into a mixture of *cis-* and *trans-*2-ethyl-5-methyltetrahydrofuran. In the steroid series the products, in general, are similar to those obtained in the hypoiodite reaction (4.34). In all the reactions involving oxy radicals

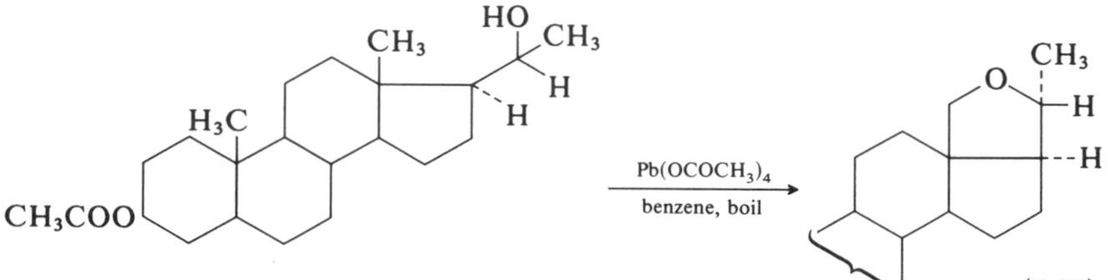

CH_3COO ... H_3C ... CH_3 ... HO ... CH_3 ... H ... H $\xrightarrow[\text{benzene, boil}]{Pb(OCOCH_3)_4}$ ... O ... CH_3 ... H ... H (4.34)

which have been discussed carbonyl-forming fragmentation reactions may accompany hydrogen abstraction. In particular, fragmentations which give rise to allyl or benzyl radicals, or to a radical adjacent to an oxygen function are strongly favoured, and in such cases hydrogen abstraction may be completely suppressed (4.35).

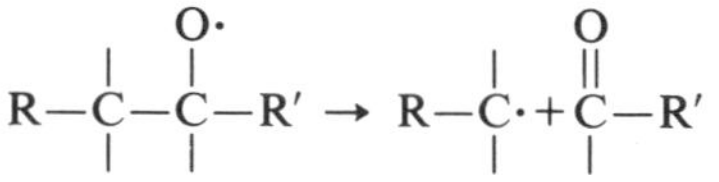

$$R-\overset{|}{\underset{|}{C}}-\overset{O\cdot}{\overset{|}{\underset{|}{C}}}-R' \rightarrow R-\overset{|}{\underset{|}{C}}\cdot+\overset{O}{\overset{\|}{\underset{|}{C}}}-R' \qquad (4.35)$$

4.5. Miscellaneous reactions

Unsaturated alcohols from hydroperoxides

Introduction of a carbon-carbon double bond into a saturated aliphatic hydrocarbon chain, although it occurs in nature, is not a reaction which is easily effected in the laboratory with ionic reagents. It has recently been found, however, that alkyl hydroperoxides are conveniently converted into unsaturated alcohols by reduction with ironII ion in presence of copperII salts. (Čeković and Green, 1974). The key step in this reaction is once again a 1,5-intramolecular hydrogen transfer to the oxy radical produced by reduction of the hydroperoxide; subsequent oxidation of the resulting carbon radical by copperII ion with simultaneous expulsion of a proton leads to formation of the olefin. Thus, 2-methyl-2-hexyloxy radical generated by ferrous ion reduction of the hydroperoxide (31) undergoes intramolecular hydrogen transfer to give the carbon radical (32) which in presence of copperII ion is oxidised to a mixture of 2-methyl-4-hexen-2-ol and (mainly) 2-methyl-5-hexen-2-ol (4.36). Similarly, the hexyl hydroperoxide (33) affords the two alkenes (34) and (35). The same reaction can be effected by photolysis of organic nitrites in presence of copperII acetate (Čeković and Srnic, 1976).

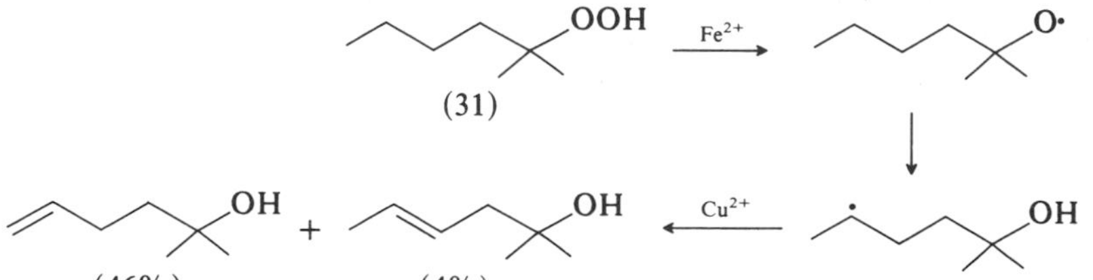

OOH $\xrightarrow{Fe^{2+}}$ O·

(31)

OH + OH $\xleftarrow{Cu^{2+}}$ OH

(46%) (4%) (32)

(4.36)

OOH ⟶ OH + OH

(76%) (<10%)

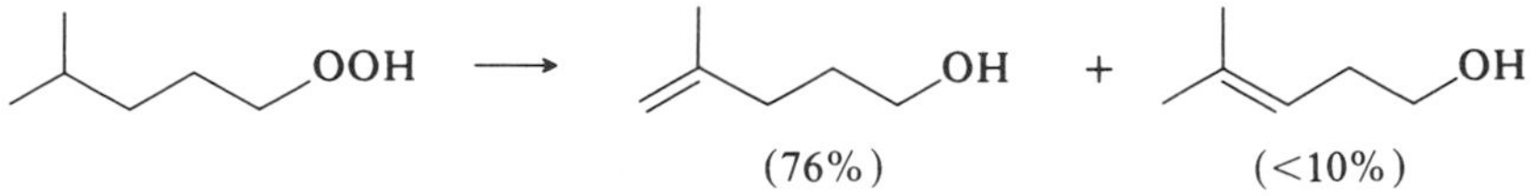

(33) (34) (35)

It is of interest that the alkene with the double bond in the 4,5-position to the hydroxyl group, and not necessarily the thermodynamically more stable alkene, is the main product. This is possibly due to a directive effect of the hydroxyl group exerted through a cyclic transition state of the form (36) which models show is much less strained than the form (37) leading to the γ-alkene, the minor product of the reactions (4.37).

Cu^{I} OH H (36)

OH H Cu^{I} (37)

(4.37)

Alcohols from aliphatic and alicyclic hydrocarbons

Another reaction which is common in nature but for which there is yet no convenient laboratory analogy entails the introduction of a hydroxyl group at the site of an unactivated aliphatic C—H bond. One method by which this transformation can be effected in the laboratory in certain cases, by photolysis of an organic nitrite in presence of oxygen, has been described on p. 273. A number of recent approaches involving oxidation of hydrocarbons with ozone (Beckwith and Duong, 1979) or peroxyacids (Mazur, 1975) or, more closely related to the reaction in nature, by direct oxygenation with molecular oxygen in the presence of an iron complex (Barton, Gastiger and Motherwell, 1983; Barton, Motherwell and Motherwell, 1983) have given promising results.

Cyclobutanols by photolysis of ketones

Intramolecular abstraction of hydrogen by an oxy radical is also involved in the formation of cyclobutanol derivatives by photolysis of aldehydes and ketones, and in the photolytic fragmentation of ketones to give alkenes and ketones of smaller carbon number by cleavage between the α- and β-carbon atoms. Thus, irradiation of methyl neopentyl ketone gives acetone and 2-methylpropene (4.38). A structure in which

$$(CH_3)_3CCH_2COCH_3 \xrightarrow{h\nu} (CH_3)_2C{=}CH_2 + CH_3COCH_3$$

$$\text{[cyclic six-membered transition state: } -\overset{|}{\underset{|}{C}}-C(O\cdot)\cdots H\cdots C{<}(\gamma),\ \alpha H_2C,\ \beta\ CH_2] \longrightarrow -\overset{|}{\underset{|}{C}}-C(OH){=}CH_2 + H_2C{=}C{<} \quad (4.38)$$

there is a hydrogen atom on the γ-carbon atom is necessary for those reactions to take place, and a mechanism involving hydrogen transfer through a diradical six-membered cyclic transition state has been proposed. In many cases these fragmentation reactions are accompanied by another reaction leading to cyclobutanol derivatives, and although the yields obtained are small the reaction provides a practicable route to otherwise difficulty accessible compounds (Orban *et al.*, 1963). Thus, 2-pentanone on irradiation in iso-octane solution affords acetone, ethylene and methylcyclobutanol (12 per cent) and 2-octanone gives 1-methyl-2-propylcyclobutanol in addition to acetone and 1-pentene (4.39).

$$CH_3CO(CH_2)_5CH_3 \xrightarrow[\text{iso-octane}]{h\nu} \text{1-methyl-2-propylcyclobutanol } (H_3C, OH, C_3H_7)\ (17\%) \quad (4.39)$$

The reaction has also been effected with cyclic ketones, and has been applied in the steroid series for reaction at the unactivated C-18 methyl group through attack by a suitably situated carbonyl group at C-20 (4.40).

$h\nu$ cyclohexane

OH H (64%) + OH H (11%)

CH_3 C=O CH_3 H_3C CH_3COO $h\nu$

CH_3 OH POCl$_3$ CH_3

(4.40)

These reactions are thought to take place by initial photoexcitation of the carbonyl group to a diradical, followed by 1,5-intramolecular hydrogen transfer to the oxy radical and intramolecular combination of the two carbon radicals with formation of the four-membered ring (4.41).

(4.41)

There is evidence, however, that the reaction may not be entirely a stepwise process involving discrete radical intermediates. Experiments using optically active ketones with an asymmetric carbon atom at the γ-position gave cyclobutanol derivatives which retained some optical activity, showing partial retention of configuration during cyclisation. For example, the aldehyde (38) on irradiation under the usual conditions gave the cyclobutanol (39) which had at least 24 per cent of the original optical activity (4.42). This can be explained by competitive participation of stepwise and concerted mechanisms, or by intervention of a short-lived diradical whose rates of racemisation and cyclisation are of the same order of magnitude.

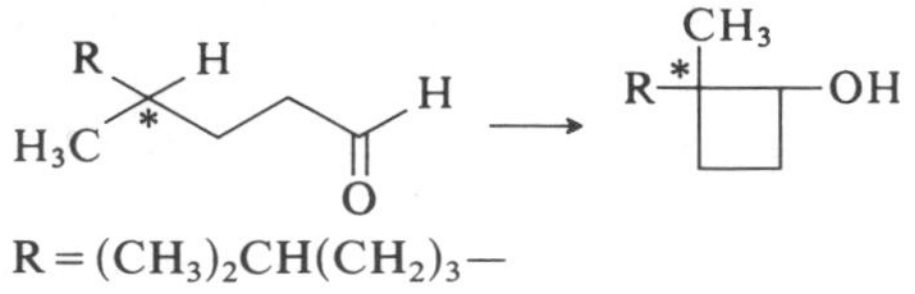

$$R = (CH_3)_2CH(CH_2)_3— \quad (38) \longrightarrow (39) \qquad (4.42)$$

In an interesting extension of this reaction, photochemical cyclisation of ketones has been used to effect oxidation of remote unactivated methylene groups in a series of esters derived from benzophenone-4-carboxylic acid and long-chain alcohols. The general scheme is shown in (4.43). It was found that hydrogen abstraction can occur over very large distances, much farther than the six atoms common in the reactions discussed above. Attack on the C—H bonds of the alkyl chain by the oxy radical is not completely indiscriminate. With the ester derived from hexadecanol, for example, there was considerable preference for reaction at C-14.

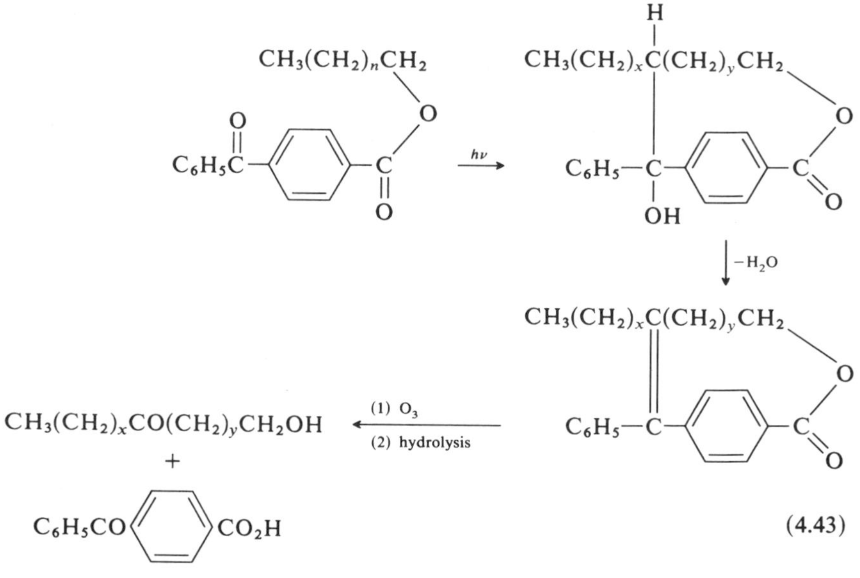

$$CH_3(CH_2)_nCH_2O-C(=O)-C_6H_4-C(=O)C_6H_5 \xrightarrow{h\nu} CH_3(CH_2)_xC(H)(CH_2)_yCH_2O-C(=O)-C_6H_4-C(OH)(C_6H_5)- \xrightarrow{-H_2O} CH_3(CH_2)_xC(CH_2)_yCH_2O-C(=O)-C_6H_4-C(C_6H_5)= \xrightarrow[(2)\ \text{hydrolysis}]{(1)\ O_3} CH_3(CH_2)_xCO(CH_2)_yCH_2OH + C_6H_5CO-C_6H_4-CO_2H \qquad (4.43)$$

This procedure has been used with conspicuous success to introduce functional groups at unactivated positions of the steroid nucleus. Here the rigidity of the molecules allows much greater selectivity than in the experiments with long-chain alkanes (Breslow, 1972; Breslow *et al.*, 1973). Thus, irradiation of the benzophenone-4-propionic ester of cholestan-3α-ol (40, $n = 2$) led specifically to attack at three α-oriented axial hydrogen atoms at C-7, C-12 and C-14 to give, after further manipulation of the initial products, the 8(14)- and 14(15)-cholestenols and the ketones (41) and (42). The point of attack on the steroid nucleus can be controlled to some extent by varying the length and the point of attachment of the ester side chain and by manipulation of the experimental conditions. Thus, with the acetate (40, $n = 1$) the 14(15)-cholestenol (48) was obtained specifically in 55 per cent yield.

An alternative procedure which leads to unsaturated products uses a free-radical chain reaction of aryliodine chlorides (Breslow *et al.*, 1977). Thus, the α-cholestanyl ester (43), on irradiation followed by treatment with methanolic potassium hydroxide, gave 9(11)-cholestenol (46) selectively in 43 per cent yield, by way of the carbon radical (45) formed by selective attack on the α-H at C-9 in (44). By using other templates or templates attached at other positions different centres of the steroid nucleus are accessible to attack. With the *p*-dichloroiodophenylacetate (47), for example, which can 'reach' further into the steroid nucleus, reaction took place selectively at C-14 to give, after reaction with potassium hydroxide, the 14(15)-cholestenol (48). These conversions were even more readily effected by irradiation of the appropriate iodoaryl ester in the presence of phenyliodine dichloride. Transfer of a chlorine atom from the phenyliodine dichloride to the iodo ester leads to the formation *in situ* of the species (as 44) which actually effects the hydrogen transfer.

An interesting development is the exclusive formation of 9(11)-cholestenol (48) in 83 per cent yield on irradiation of the tris-silyl ester

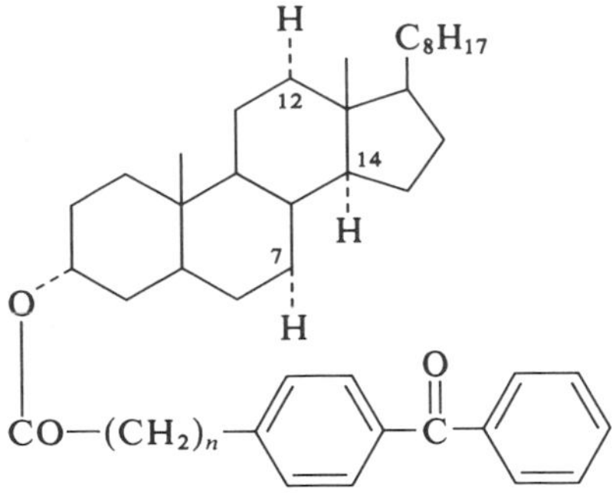

(4.44)

(40)

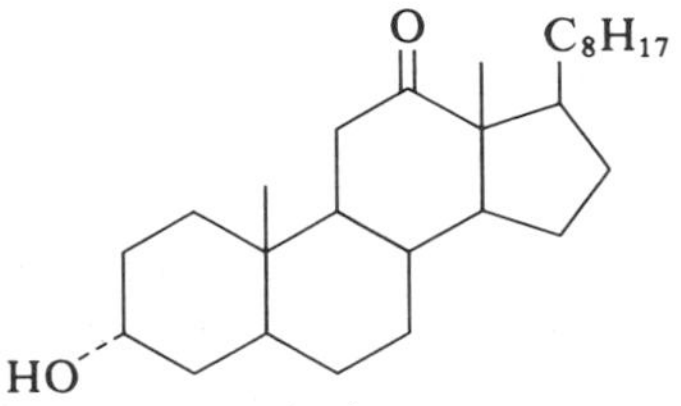

(41)

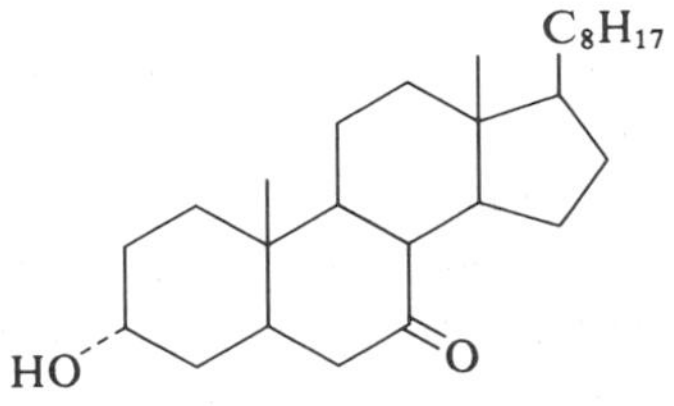

(42)

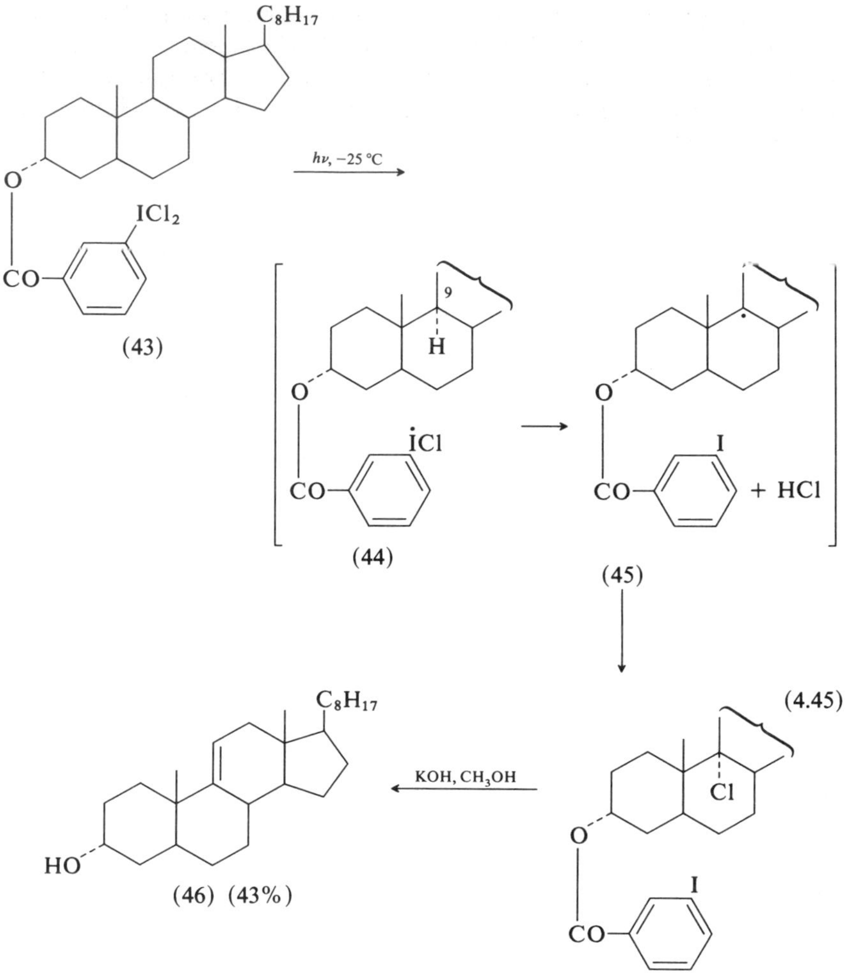

C_8H_{17}

O

ICl_2

CO

(43)

$h\nu$, −25 °C

9

H

$\dot{I}Cl$

CO

(44)

O

I

CO

+ HCl

(45)

(4.45)

C_8H_{17}

KOH, CH_3OH

Cl

HO

(46) (43%)

O

I

CO

C_8H_{17}

O

(1) $h\nu$, −25 °C

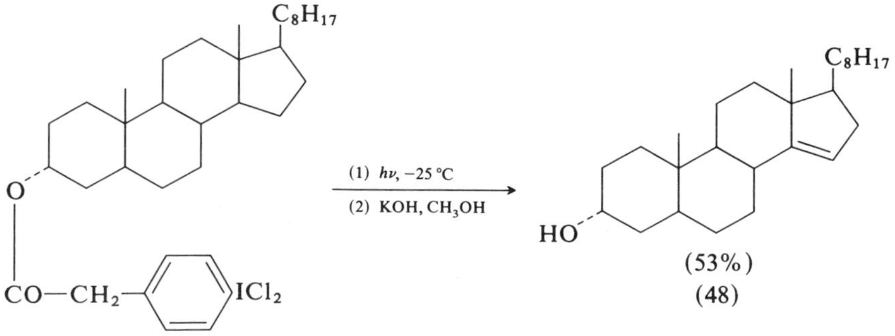

(2) KOH, CH_3OH

C_8H_{17}

HO

(53%)

(48)

$CO—CH_2$ ICl_2

(47)

(49) in the presence of sulphuryl chloride or phenyliodine dichloride, followed by dehydrochlorination with alkali. The high yield indicates that all three steroid nuclei in the ester (49) are being chlorinated, as the regenerated template directs a second and then a third selective functionalisation (Breslow and Heyer, 1982).

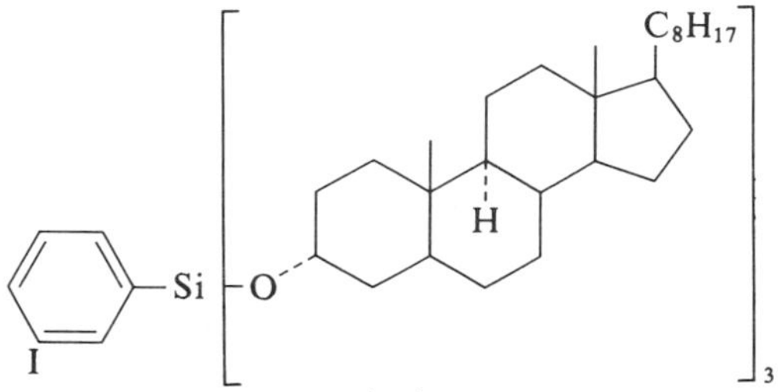

C_8H_{17}

H

Si

O

I

3

(49)

5 Synthetic applications of organoboranes and organosilanes

Organoboranes are obtained by addition of borane (which exists as the gaseous dimer diborane B_2H_6) to alkenes and alkynes. These versatile reagents undergo a wide variety of reactions many of which are of value in synthesis (Brown *et al.*, 1975; Cragg, 1973). In addition, borane itself and a number of its derivatives are useful reducing agents and reduce a variety of unsaturated groups (see p. 478).

Borane for reductions, or for the hydroboration reactions described below, can be prepared by reaction of boron trifluoride etherate with sodium borohydride, but solutions in tetrahydrofuran, where it is present as the complex BH_3.THF, are commercially available and this is a convenient source of the reagent for most purposes. Another useful source is the borane–dimethyl sulphide complex $BH_3.Me_2S$. It is more stable than BH_3.THF and has the additional advantage that it is soluble in a variety of organic solvents, such as ether and hexane (Lane, 1974; Brown, Mandal and Kulkarni, 1977).

5.1. Hydroboration

The most important synthetic application of borane in synthesis is in the preparation of alkyl- and alkenyl-boranes by addition to alkenes and alkynes, a process known as hydroboration (5.1). This reaction has been applied to a large number of alkenes of widely different structures. In nearly all cases the addition proceeds rapidly at room temperature, and only the most hindered alkenes do not react. With simple alkenes

$$\gt C{=}C\lt \; + H{-}B\lt \;\rightarrow\; H{-}\overset{|}{\underset{|}{C}}{-}\overset{|}{\underset{|}{C}}{-}B\lt$$

$$-C{\equiv}C- + H{-}B\lt \;\rightarrow\; \gt C{=}C\lt \text{ (H, B)} \;\rightarrow\; -CH_2{-}CH(B\lt)_2 \qquad (5.1)$$

(mono- and di-substituted ethylenes) a trialkylborane is produced but trisubstituted ethylenes normally give the dialkylborane and tetrasubstituted alkenes form only monoalkylboranes (5.2). This has been exploited in the preparation of a number of mono- and di-alkylboranes which are less reactive and more selective than borane itself (see p. 291).

$$(CH_3)_3CCH{=}CHCH_3 \xrightarrow[25\,^\circ C]{B_2H_6,\ diglyme} [(CH_3)_3CCH_2CH(CH_3)]_2BH$$

$$(CH_3)_2C{=}CHCH_3 \xrightarrow[25\,^\circ C]{B_2H_6,\ diglyme} \underset{(1)}{[(CH_3)_2CHCH(CH_3)]_2BH} \qquad (5.2)$$

$$(CH_3)_2C{=}C(CH_3)_2 \xrightarrow[25\,^\circ C]{B_2H_6,\ diglyme} \underset{(2)}{(CH_3)_2CHC(CH_3)_2{-}BH_2}$$

Particularly important in this respect are the so-called disiamylborane (1) and thexylborane (2), formed by addition of borane to 2-methyl-2-butene and 2,3-dimethyl-2-butene respectively. These partially alkylated boranes may themselves be used to hydroborate less hindered alkenes.

Hydroboration is readily effected with alkenes containing many types of functional groups, to give functionalised alkylboranes. Where the other functional group is not reduced by borane, hydroboration generally proceeds without difficulty and even some functional groups which react only slowly with borane may be tolerated, for example carboxylic esters. Easily reduced carbonyl groups of aldehydes and ketones, however, must be protected as their acetals, and carboxylic acids as esters.

Addition of borane to an unsymmetrical alkene could, of course, give rise to two different products by addition of the boron at either end of the double bond. It is found in practice, however, that in the absence of strongly polar neighbouring substituents, the reactions are highly selective and give predominantly the isomer in which boron is bound to the less highly substituted carbon atom (5.3). With disubstituted internal alkenes, however, there is little discrimination in reactions with borane itself (but see p. 291).

$$\underset{6\%\ \ 94\%}{CH_3(CH_2)_3CH{=}CH_2} \qquad \underset{2\%\ \ 98\%}{CH_3{-}C(CH_3){=}C(CH_3){-}H} \qquad \underset{58\%\ \ 42\%}{CH_3CH{=}CHC(CH_3)_3} \qquad (5.3)$$

All the available evidence suggests that hydroboration is a concerted process and takes place through a four-membered cyclic transition state

(3) formed by addition to the double bond of a polarised B—H bond in which the boron atom is the more positive. This is supported by the stereochemistry of the reactions (*syn* addition of B and H) and

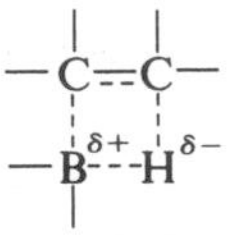

$$\begin{array}{cc} -\overset{|}{C} \cdots & \overset{|}{C}- \\ \vdots & \vdots \\ -\underset{|}{B}^{\delta+} \cdots & H^{\delta-} \end{array}$$

(3)

by the directive effect of polar substituents. Thus, in allyl derivatives and in nuclear-substituted styrenes the proportion of product formed by addition of boron to the α-carbon atom increases with the electronegativity of the substituent (5.4).

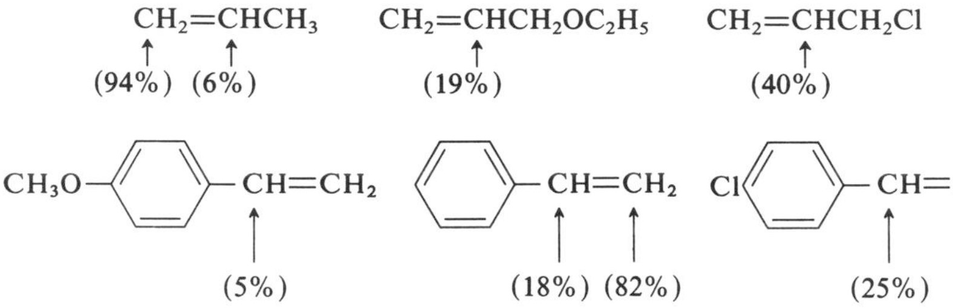

$CH_2{=}CHCH_3$ (94%) (6%) $CH_2{=}CHCH_2OC_2H_5$ (19%) $CH_2{=}CHCH_2Cl$ (40%)

CH_3O–C_6H_4–$CH{=}CH_2$ (5%) C_6H_5–$CH{=}CH_2$ (18%) (82%) Cl–C_6H_4–$CH{=}CH_2$ (25%) (5.4)

Hydroboration of alkenes and alkynes is highly stereoselective and takes place by *syn* addition to the less hindered side of the multiple bond. Thus, reduction of alkynes by hydroboration followed by protonolysis affords *Z*-alkenes (p. 294) and 1-alkylcycloalkenes, on hydroboration and oxidation yield *trans*-2-alkylcycloalkanols almost exclusively (5.5) (cf. p. 295). Both the oxidation of the boron–carbon bond to form an alcohol and protonolysis to a hydrocarbon are known to occur with retention of configuration, thus establishing the *syn* stereochemistry of the original addition of borane.

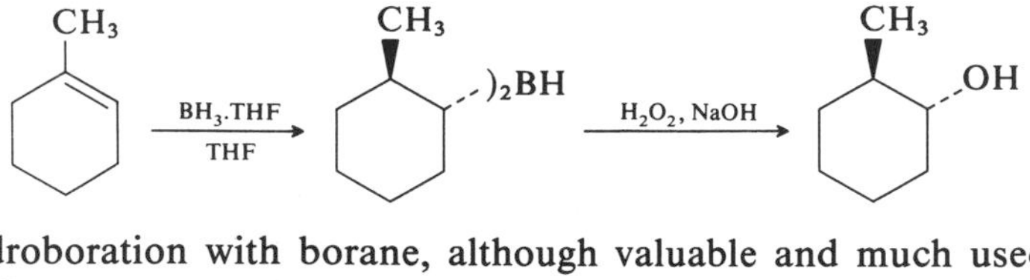

1-methylcyclohexene (CH₃) $\xrightarrow[\text{THF}]{BH_3.THF}$ (2-methylcyclohexyl)$_2$BH $\xrightarrow{H_2O_2,\ NaOH}$ *trans*-2-methylcyclohexanol (CH₃, OH) (5.5)

Hydroboration with borane, although valuable and much used, has limitations. For example, regioselectivity in the hydroboration of terminal alkenes, although high, is not complete and in 1,2-disubstituted alkenes there is little discrimination between the two termini of the double bond. Further, there is little difference in the rate of reaction of borane with differently substituted double bonds, so that it is rarely possible to achieve selective hydroboration of one double bond in the presence of others.

Another serious limitation is found in the hydroboration of terminal alkynes which proceeds past the desired alkenylborane by addition of a second molecule of borane to the 1,1-dibora-alkane. Many of these difficulties can be circumvented by hydroboration with a substituted borane rather than with borane itself (Brown, 1983). Three reagents frequently used are disiamylborane (1) and thexylborane (2) already referred to (p. 289), and the stable crystalline derivative 9-borabicyclo[3,3,1]nonane (4) (9-BBN) which is readily obtained by reaction of borane with 1,5-cyclo-octadiene (Soderquist and Brown, 1981).

$$[(CH_3)_2CHCH(CH_3)]_2BH \qquad (CH_3)_2CHCH(CH_3)_2BH_2$$

(1) (2)

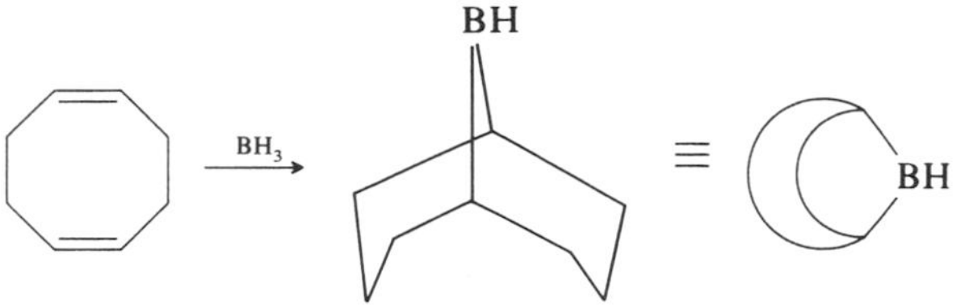

$$\xrightarrow{BH_3} \quad (4) \quad \equiv \quad BH \qquad (5.6)$$

These reagents are less reactive and more selective than borane. Disiamylborane, for example, hydroborates 1-hexene with 99 per cent attack at C-1 (borane gave 94 per cent) and the large steric requirement of the reagent results in a desirable regioselectivity between the two carbon atoms of the double bond in *cis*-4-methyl-2-pentene (5.7); borane itself was almost completely non-selective. Even greater discrimination is shown by 9-BBN.

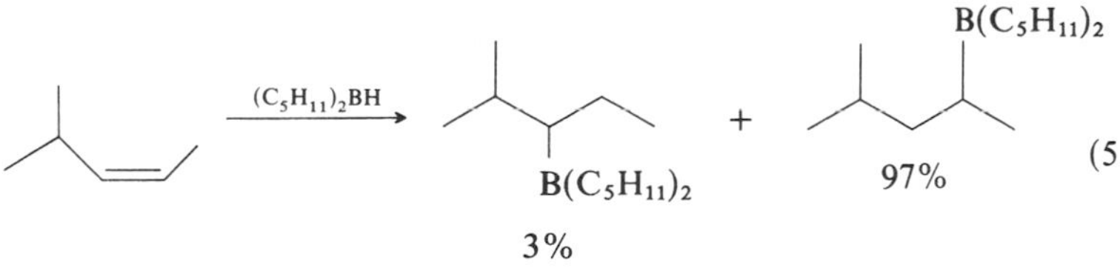

$$\xrightarrow{(C_5H_{11})_2BH} \quad B(C_5H_{11})_2 \; 3\% \quad + \quad B(C_5H_{11})_2 \; 97\% \qquad (5.7)$$

Disiamylborane is more sensitive to the structure of the alkene than is borane itself. Terminal alkenes react more rapidly than internal alkenes and *Z*-alkenes more rapidly than their *E*-isomers. This sometimes allows the selective hydroboration of one double bond in a compound which contains several. Vinyl cyclohexene, for example, was readily monohydroborated to (5), identified by oxidation to the alcohol (6) (p. 295). Similar results were obtained on hydroboration with 9-BBN. Disiamylborane also permits the monohydroboration of alkynes, thus making

$$\text{4-vinylcyclohexene} \xrightarrow{(C_5H_{11})_2BH} \underset{(5)}{\text{4-[2-}B(C_5H_{11})_2\text{-ethyl]cyclohexene}} \xrightarrow{[O]} \underset{(6)}{\text{4-(2-hydroxyethyl)cyclohexene (OH)}} \quad (5.8)$$

available the corresponding alkenylboron derivatives. Similar results are obtained with other substituted boranes such as catecholborane, dibromoborane or thexylmonochloroborane (p. 294). Oxidation of the products with alkaline hydrogen peroxide affords ketones (5.9). The same result is, of course, achieved by acid-catalysed hydration of alkynes but in some cases the route via hydroboration may be more convenient. Reaction of terminal alkynes with borane yields mainly dihydroboration products, but with disiamylborane high yields of monohydroboration products are obtained. These compounds are synthetically useful because in them the boron atom is attached to the end of the chain and on oxidation they afford aldehydes. The series of reactions, hydroboration followed by oxidation, thus complements the usual acid-catalysed hydration of terminal alkynes which leads to methyl ketones.

$$C_2H_5C{\equiv}CC_2H_5 \xrightarrow{(C_5H_{11})_2BH} C_2H_5\underset{B(C_5H_{11})_2}{\underset{|}{C}}{=}CHC_2H_5 \xrightarrow{H_2O_2,\ NaOH} \underset{(68\%)}{C_2H_5COC_3H_7}$$

$$CH_3(CH_2)_5C{\equiv}CH \xrightarrow{(C_5H_{11})_2BH} CH_3(CH_2)_5CH{=}CHB(C_5H_{11})_2 \xrightarrow{H_2O_2,\ NaOH} \underset{(70\%)}{CH_3(CH_2)_5CH_2CHO} \quad (5.9)$$

Other valuable selective hydroborating agents are thexylborane and the derived thexylchloroborane-methyl sulphide complex (7) (Brown, Sikorski, Kulkarni and Lee, 1982), catecholborane (8) and the mono- and di-halogenoborane-methyl sulphide complexes (9) and (10). The latter two reagents are readily available and are more stable than the

$$\underset{(7)}{(CH_3)_2CH{-}C(CH_3)_2{-}B(H)(Cl)\cdot(CH_3)_2S} \qquad \underset{(8)}{C_6H_4O_2BH} \quad (5.10)$$

$$\underset{(9)}{H_2BX.(CH_3)_2S} \qquad \underset{(10)}{HBX_2.(CH_3)_2S}$$

$$(X = Cl, Br)$$

etherates formerly employed. Both react selectively at the less substituted end of a carbon–carbon double bond and at C-1 of terminal alkynes (Brown, Ravindran and Kulkarni, 1979; Brown and Chandrasekharan, 1983). They do not always show the same reactivity as other borane derivatives. For example, the far higher reactivity of dibromoborane towards the group $-C(CH_3){=}CH_2$ as compared with $-CH{=}CH_2$, in contrast to disiamylborane, makes possible the selective hydroboration of the double bonds in 2-methyl-1,5-hexadiene (5.11).

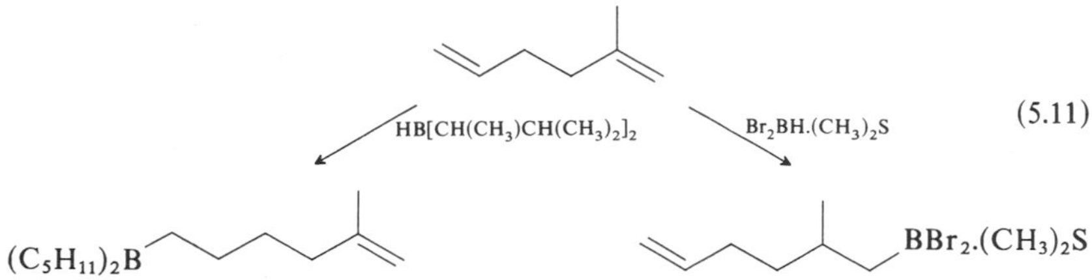

$HB[CH(CH_3)CH(CH_3)_2]_2$ $Br_2BH.(CH_3)_2S$ (5.11)

$(C_5H_{11})_2B$ $BBr_2.(CH_3)_2S$

Another valuable hydroborating agent is thexylborane, the most readily available of the monoalkylboranes. It is useful for the cyclic hydroboration of dienes and numerous other applications. Hydroboration of dienes with borane itself usually leads to the formation of polymers, but with thexylborane cyclic or bicyclic organoboranes are formed, some of which undergo synthetically useful transformations (see p. 305). 1,5-Hexadiene, for example, is converted mainly into the boracycloheptane (11). Thexylborane has also been used to make trialkylboranes containing

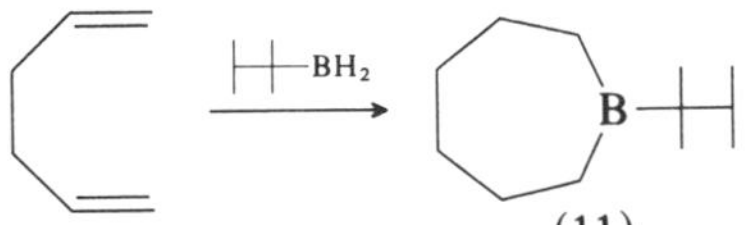

BH_2 B (5.12)

(11)

three different alkyl groups by stepwise addition to two different alkenes as in (5.13). The procedure is limited in scope, however, for the first alkene must be relatively unreactive (otherwise two groups would add to the borane at this stage) and so it cannot be used to make trialkylboranes containing two different primary alkyl groups. This difficulty

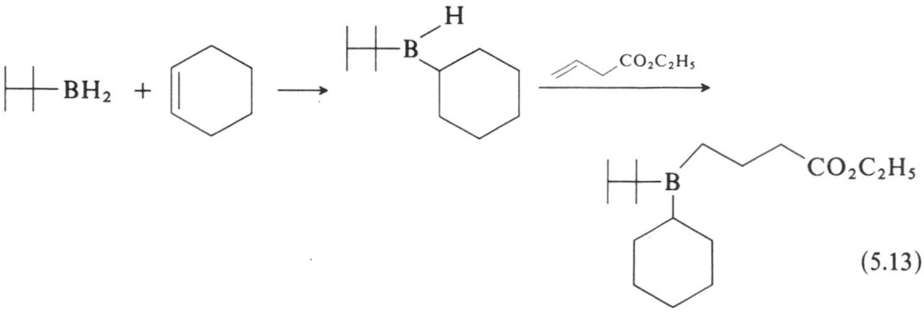

BH_2 + → H B → $CO_2C_2H_5$

B $CO_2C_2H_5$

(5.13)

can be overcome by the use of thexylchloroborane which can be prepared from thexylborane with ethereal hydrogen chloride (Zweifel and Pearson, 1980) or by monohydroboration of 2,3-dimethyl-2-butene with monochloroborane-dimethyl sulphide complex (Brown *et al.*, 1982*b*). Reaction with an alkene gives an alkylthexylchloroborane which may be converted into a dialkylthexylborane by reaction with one equivalent of a Grignard reagent or an alkyl-lithium or by hydridation with lithium aluminium hydride in the presence of an alkene (5.14). The dialkylthexylboranes are valuable reagents because on carbonylation (p. 303) they are converted into ketones. The insect pheromone (13), for example, was readily obtained from the dialkylthexylborane (12).

H, Cl; + $C_8H_{17}CH{=}CH_2$ ⟶ $C_{10}H_{21}$, Cl

C_5H_{11} MgCl

$C_{10}H_{21}$ O=C C_5H_{11} (74%) (13) ⟵ (1) NaCN (2) C_6H_5COCl (3) H_2O_2, NaOH — $C_{10}H_{21}$ B C_5H_{11} (12)

(5.14)

5.2. Reactions of organoboranes

The usefulness of the hydroboration reaction in synthesis arises from the fact that the alkylboranes formed can be converted by further reaction into a variety of other products. On hydrolysis (protonolysis), for example, the boron atom is replaced by hydrogen, and under the appropriate conditions boranes are readily oxidised to alcohols or carbonyl compounds. A practical advantage of these reactions is that it is often unnecessary to isolate the intermediate organoborane.

Protonolysis

Protonolysis is best effected with an organic carboxylic acid and provides a convenient method for the reduction of carbon–carbon multiple bonds. Boiling propionic acid is often used with alkylboranes but alkenylboranes are much more reactive and often undergo rapid protonolysis with acetic acid at room temperature. The reaction takes place with retention of configuration at the carbon atom concerned and in accordance with the proposed mechanism of hydroboration alkynes are cleanly converted into *Z*-alkenes. An advantage of the hydroboration–protonolysis procedure is that it can sometimes be used for the reduction

$$R_2B-R \cdots O=C(C_2H_5)-O-H \longrightarrow R_2BOCOC_2H_5 + RH \longrightarrow\!\!\!\!\rightarrow 3RH$$

$$C_4H_9CH{=}CH_2 \xrightarrow{BH_3.THF} (C_4H_9CH_2CH_2)_3B \text{ (not isolated)} \xrightarrow[\text{acid}]{\text{reflux, propionic}} C_4H_9CH_2CH_3 \quad (91\%) \qquad (5.15)$$

$$C_2H_5C{\equiv}CC_2H_5 \xrightarrow[\text{diglyme}]{(C_5H_{11})_2BH} (C_2H_5)HC{=}C(C_2H_5)B< \xrightarrow[25\,°C]{CH_3CO_2H} (C_2H_5)HC{=}CH(C_2H_5)$$

(68%; almost 100% Z)

of double or triple bonds in compounds which contain other easily reducible groups. Allyl methyl sulphide, for example, is converted into methyl propyl sulphide in 78 per cent yield.

Oxidation

Oxidation of organoboranes to alcohols is usually effected with alkaline hydrogen peroxide although other methods can be used (Kabalka and Hedgecock, 1975). The reaction is of wide applicability and many functional groups are unaffected by the reaction conditions, so that a variety of substituted alkenes can be converted into alcohols by this procedure (5.16). Several examples have already been given. A valuable feature of the reaction is that it results in overall anti-Markownikoff addition of water to the double or triple bond, and it thus complements the more usual acid-catalysed hydration. This follows from the fact that in the hydroboration step the boron atom adds to the less substituted carbon of the multiple bond. Terminal alkynes, for example, give aldehydes in contrast to the methyl ketones obtained by acid-catalysed hydration.

$$CH_2{=}CHCH_2CO_2C_2H_5 \xrightarrow{BH_3.THF} >BCH_2CH_2CH_2CO_2C_2H_5 \xrightarrow{H_2O_2,\ NaOH} HOCH_2CH_2CH_2CO_2C_2H_5$$

$$C_6H_{13}C{\equiv}CH \xrightarrow[\text{(2) } H_2O_2,\ NaOH]{\text{(1) } (C_5H_{11})_2BH \text{, diglyme, 0 °C}} C_6H_{13}CH_2CHO \quad (70\%) \qquad (5.16)$$

Another noteworthy feature of the reaction is that it leads to *cis* addition of the elements of water to the double bond. This is determined by the reaction pathway and not by the stability of the product, for the thermodynamically more stable product is not invariably formed. Thus, hydroboration-oxidation of 1-methylcyclohexene affords the more stable (E)-2-methylcyclohexanol, but 1,2-dimethylcyclohexene gives the less stable (E)-1,2-dimethylcyclohexanol. β-Pinene similarly gives the less stable *cis*-myrtanol (5.17). In each case *syn* addition of B—H (to the less hindered side of the double bond in (5.17)) is followed by oxidation of

$$\xrightarrow{B_2H_6} CH_2B\langle \;\; \xrightarrow{H_2O_2,\ NaOH} CH_2OH \qquad (5.17)$$

the carbon–boron bond with retention of configuration. A reaction path involving intramolecular transfer of an alkyl group from boron to carbon in an intermediate 'ate' compound has been proposed (5.18). Direct oxidation of primary trialkylboranes to aldehydes and of secondary trialkylboranes to ketones, without isolation of the alcohol, is possible with pyridinium chlorochromate or aqueous chromic acid (Rao, Kulkarni and Brown, 1979).

$$R_3B \xrightarrow{HOO^-} R_2\overset{R}{\overset{|}{B^-}}—O—OH \rightarrow R_2B—O—R \xrightarrow{OH^-} R_2\overset{OH}{\overset{|}{B^-}}—O—R \rightarrow R_2B—O^- + R—OH \rightarrow 3ROH + B(OH)_3 \qquad (5.18)$$

Hydroboration of alkenes with an optically active alkylborane followed by oxidation has been used in the asymmetric synthesis of optically active secondary alcohols. The best results so far have been obtained with diisopinocampheylborane (14) (Ipc_2BH) (Brown, Desai and Jadhav, 1982) and mono-isopinocampheylborane (15) ($IpcBH_2$) (Brown *et al.*,

$$)_2BH \quad (14) \qquad BH_2 \quad (15) \qquad (5.19)$$

1982*a*) which are readily prepared in either form by reaction of borane with (+)- or (−)-α-pinene under the appropriate conditions. Optically active secondary alcohols of high optical purity have been obtained from several disubstituted Z-alkenes by initial hydroboration with (14). Thus, reaction of (*Z*)-2-butene with (−)-diisopinocampheylborane, from (+)-α-pinene, followed by oxidation with alkaline hydrogen peroxide gave (*R*)-2-butanol of 87 per cent optical purity (5.20). With *E*-disubstituted alkenes the best results have been obtained using monoisopinocampheylborane, but as with the *Z*-alkenes, the success of the reactions is strongly dependent on the bulk of the alkyl substituents on the double bond of the alkene. (*E*)-2-butene, for example, gave (*S*)-2-butanol of 73 per cent optical purity, but with (*E*)-di-t-butylethene the corresponding alcohol obtained had 92 per cent optical purity. Even better results were obtained with the discovery that many alkyl isopinocampheylboranes can be obtained optically pure by crystallisation or by ageing the crystalline compound in tetrahydrofuran. Oxidation of the resultant dialkylborane then affords the corresponding optically pure secondary alcohol. 2-Methyl-2-butene, for example, gave enantiomerically pure (*S*)-3-methyl-2-butanol, after purification of the crystalline intermediate dialkylborane by ageing (Brown and Singaram, 1984).

$)_2BH$ → $)_2B$, H —H_2O_2, NaOH→ HO, H (5.20)

(*R*, 87% optically pure)

BH_2 → H, B, H —H_2O_2, NaOH→ HO, H

(*S*, 73% optically pure)

A related sequence leads to optically active acyclic ketones, based on the observation that trialkylboranes derived from mono-isopinocamphenylborane, on treatment with an aldehyde selectively eliminate α-pinene to give a chiral borinic ester. Thus, hydroboration of (*E*)-2-butene with mono-isopinocampheylborane and subsequent reaction with 1-pentene gave the trialkylborane (16), converted by acetaldehyde into the optically active borinic ester (17). Carbonylation of this product by reaction with dichloromethyl methyl ether, followed by oxidation (p. 307) then gave (*R*)-3-methyl-4-nonanone of 70 per cent optical purity (Brown, Jadhav and Desai, 1984).

C_5H_{11}, B, H (16) —CH_3CHO→ C_5H_{11}, B, C_2H_5O, H (17) → O, C_5H_{11}, H, CH_3 (5.21)

(70% optically pure)

Hydroboration followed by oxidation has also been used to bring about diastereoselective hydration of the double bond in acyclic alkenes which already contain a nearby chiral centre. Good levels of 1,3-asymmetric induction have been observed in terminal alkenes of the type (18, R_L and R_M are sterically large- and medium-sized substituent groups respectively). The alkene (19), for example, gave predominantly the diastereomer (20) on hydroboration with disiamylborane followed by oxidation with alkaline hydrogen peroxide. In reactions of this kind the stereochemistry of hydroboration appears to be controlled primarily by the size of the groups on the nearby chiral centre (Evans, Bartroli and Godel, 1982).

H_3C R_M H, R_L (18) —(1) R_2BH; (2) H_2O_2, NaOH→ HO … R_L, H_3C H R_M H (major) + HO … R_L, H CH_3 R_M H (minor)

$(CH_3)_3SiO$, O, OR, CH_3 CH_3 CH_3 (19) —(1) $(C_5H_{11})_2BH$; (2) H_2O_2, NaOH→ (5.22)

HO, $(CH_3)_3SiO$, O, OR, CH_3 CH_3 CH_3 (20) + HO, $(CH_3)_3SiO$, O, OR, CH_3 CH_3 CH_3 (ratio = 87 : 13)

In a related sequence acyclic secondary allylic alcohols (21) give the corresponding *threo*-1,3-diols with good diastereoselection on hydroboration with 9-BBN or thexylborane followed by oxidation (5.23). Thus, the allylic alcohol (22) gave mainly the *threo* diol (23) and, in a remarkable example which formed a step in an elegant synthesis of the ansa bridge of rifamycin, the double allylic alcohol (24) was converted selectively into (25), one prochiral centre controlling the formation of four other chiral centres. Reactions are believed to take place by attack of the borane on the less hindered side of the double bond in the conformation (26) (Still and Barrish, 1983). In an intramolecular example the 1,5-diene (27) was converted into the diol (30) with very high stereoselection at C4—C6, showing how cyclic hydroboration can be used to control acyclic stereoselection (Still and Shaw, 1981; see also Morgans, 1981). The strong preference of the C4—C5 bond in (28) for the conformation shown (least steric interference about the double bond) ensures preferential (>20 : 1)

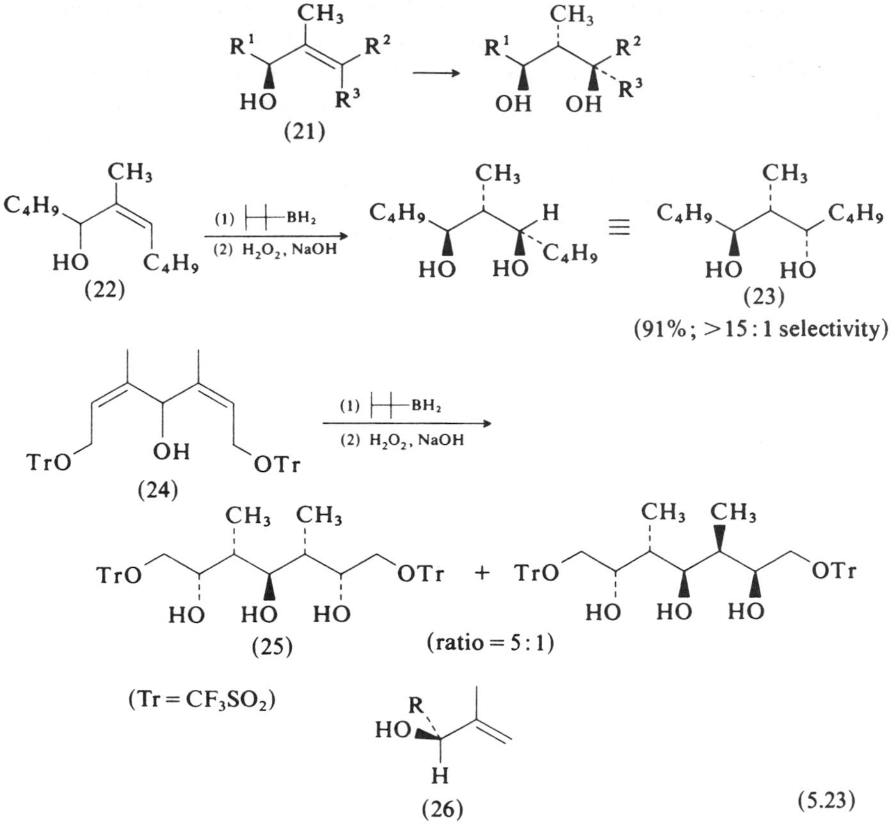

CH_3 R^1 R^2 HO R^3 → CH_3 R^1 R^2 OH OH R^3

(21)

CH_3 C_4H_9 HO C_4H_9 (22) (1) $\vdash\!\!\vdash BH_2$ (2) H_2O_2, NaOH → CH_3 C_4H_9 H HO HO C_4H_9 ≡ CH_3 C_4H_9 C_4H_9 HO HO

(23)

(91%; >15:1 selectivity)

TrO OH OTr (24) (1) $\vdash\!\!\vdash BH_2$ (2) H_2O_2, NaOH →

CH_3 CH_3 TrO OTr HO HO HO (25) + CH_3 CH_3 TrO OTr HO HO HO

(ratio = 5:1)

(Tr = CF_3SO_2)

R HO H

(26)

(5.23)

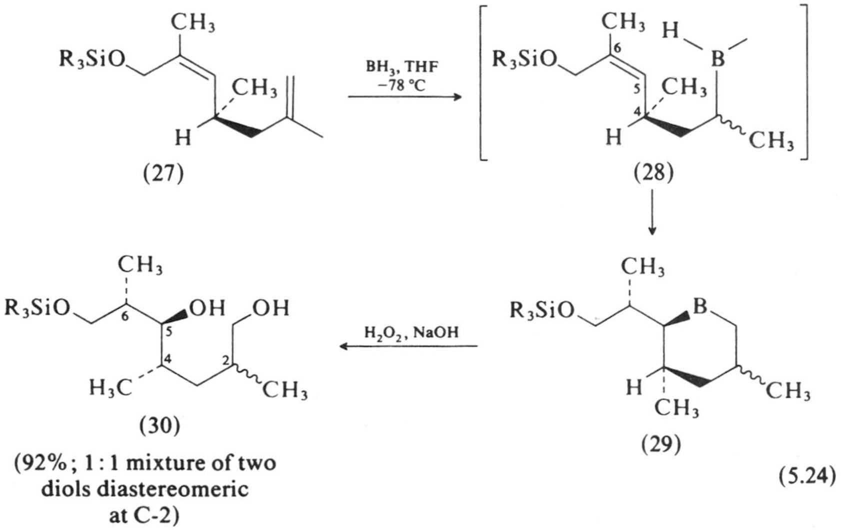

CH_3 R_3SiO CH_3 H (27) BH_3, THF −78 °C → [CH_3 H B R_3SiO 6 5 4 CH_3 H CH_3] (28) ↓

CH_3 R_3SiO B H CH_3 CH_3 (29)

H_2O_2, NaOH ←

CH_3 R_3SiO 6 5 OH OH 4 2 H_3C CH_3

(30)

(92%; 1:1 mixture of two diols diastereomeric at C-2)

(5.24)

β-attack on the double bond in the key intramolecular hydroboration (28 → 29).

The high 1,2-asymmetric induction in the above example is a consequence of the fact that in this intramolecular reaction the side chain bearing the monoalkylborane in (28) is sterically constrained to the β-face of the C5—C6 double bond by the asymmetric centre at C-4. Good asymmetric induction has also been achieved in other examples where there is no pre-existing chiral centre or where the asymmetric centre is so far removed from the double bond attacked in the second hydroboration that this particular steric restraint can no longer apply. Thus, hydroboration of the diene (31) with thexylborane and oxidation of the derived cyclic borane proceeded with substantial 1,4-asymmetric induction to give the diastereomers (32) and (33) in the ratio 6:1.

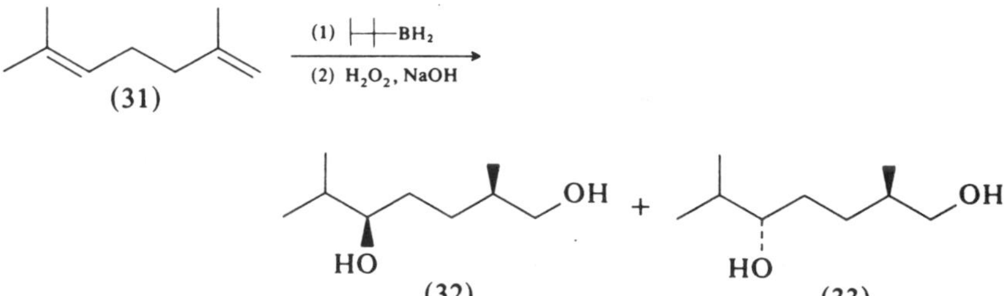

(31) —(1) thexyl-BH_2; (2) H_2O_2, NaOH→ (32) + (33)

(81%; 6:1) (5.25)

Stereocontrol here in the intramolecular hydroboration step is effected through the preferred conformation of the medium ring transition state leading to the boracycle. Formation of the main product (32) takes place by way of the boat-like transition state (34) rather than the chair (35). It so happens that the thexyl substituent is lost as tetramethylethene before cyclisation and the préference for the boat over the chair may be a consequence of the fact that the boron–hydrogen bond eclipses the π-system of the double bond in the former but not in the latter. The transition state for the intramolecular hydroboration would thus resemble the planar four-centre transition state postulated for intermolecular hydroboration of alkenes (p. 290) (Still and Darst, 1980).

(34) (35) (5.26)

Related reactions, giving 1,3-asymmetric induction proceed through five-membered transition states, and formation of *meso* diols with high 1,4- and 1,5-asymmetric induction has been achieved by way of seven- and eight-membered boracycles. The diene (36), for example, gave almost exclusively, the diol (37). In each case, formation of the major product may be rationalised in terms of a transition state having eclipsed B—H and C=C bonds and the least strained conformation of the connecting chain.

(1) $\vdash\!\!\vdash$BH$_2$
(2) H_2O_2, NaOH

(36)

HO ... OH + HO ... OH

(37)
(86%; 20:1) (5.27)

Other useful conversions of trialkylboranes which proceed in good yield include their oxygenation to alkylhydroperoxides, their conversion with hydroxylamine-O-sulphonic acid into primary amines and the reaction of dialkylchloroboranes with organic azides to give secondary amines. Primary organic bromides and iodides are also readily obtained by reaction of trialkylboranes derived from terminal alkenes with bromine or iodine in the presence of methanolic sodium methoxide (Brown *et al.*, 1975*b*).

Isomerisation of alkylboranes

A useful reaction of organoboranes is their ready isomerisation on gentle heating to compounds in which the boron atom is attached to the least hindered carbon of the chain. The reaction is catalysed by borane and other molecules containing boron–hydrogen bonds and proceeds by a succession of eliminations and additions of borane leading, eventually, to the most stable borane with the boron atom at the least substituted position. In line with this mechanism it is found that the boron atom can migrate along a chain of carbon atoms and past a single alkyl branch, but it cannot pass a completely substituted carbon atom. The reaction can be used to bring about 'contrathermodynamic isomerisation' of alkenes, for after rearrangement the isomerised alkene can be displaced from the organoborane by heating it with a higher-boiling alkene. Oxidation of the rearranged alkene provides a means of converting internal alkenes into primary alcohols, as illustrated in (5.28).

$$\text{1-ethylcyclohexene} \xrightarrow{B_2H_6} \text{(2-ethylcyclohexyl)}B\langle \xrightarrow{150\ ^\circ C} \text{(2-cyclohexylethyl)}CH_2CH_2B\langle\ (85\%) \xrightarrow{H_2O_2,\ NaOH} \text{cyclohexyl}CH_2CH_2OH \tag{5.28}$$

5.3. Formation of carbon–carbon bonds

One of the reasons why organoboranes are important in synthesis is the variety of ways in which they can be used to make carbon–carbon bonds. As will be seen, many of the reactions used involve the migration of a group from boron to carbon. Most of them proceed in high yield under mild conditions.

Carbonylation of organoboranes

One of the most useful reactions of organoboranes in synthesis is their reaction with carbon monoxide, which, under appropriate conditions, can be directed to give primary, secondary and tertiary alcohols, aldehydes and open chain, cyclic and polycyclic ketones (Brown, 1972). At a temperature of 100–125 °C in diglyme solution many organoboranes absorb one molecule of carbon monoxide at atmospheric pressure to form intermediates which are oxidised to tertiary alcohols by alkaline hydrogen peroxide in excellent yield (5.29). The reaction is of wide applicability, and for trialkylcarbinols containing bulky groups gives much higher yields than any other method. Tricyclohexylcarbinol, for example, is obtained from cyclohexene in 85 per cent yield, whereas the Grignard method gives only 7 per cent.

$$R_3B + CO \xrightarrow[\text{diglyme}]{100\text{–}125\ ^\circ C} R_3CBO \xrightarrow{H_2O_2,\ NaOH} R_3COH \tag{5.29}$$

The reaction obviously involves migration of alkyl groups from boron to the carbon atom of carbon monoxide, and this was shown to occur intramolecularly by the fact that carbonylation of an equimolecular mixture of triethylborane and tributylborane gave, after oxidation, only triethylcarbinol and tributylcarbinol; no 'mixed' carbinols were formed. Similarly, dicyclohexyloctylborane gave only dicyclohexyloctylcarbinol. The stepwise reaction pathway in (5.30), involving three successive

$$R_3B + CO \underset{}{\overset{a}{\rightleftharpoons}} R_3\bar{B}-\overset{+}{C}\equiv O \rightleftharpoons R_2B-\underset{\underset{O}{\|}}{C}-R$$

$$R_2B-\underset{\underset{O}{\|}}{C}-R \xrightarrow{b} RB\underset{O}{\diagdown\!\!\diagup}CR_2 \xrightarrow{c} OB-CR_3 \xrightarrow[NaOH]{H_2O_2} R_3C-OH \qquad (5.30)$$

intramolecular transfers has been proposed. If the carbonylation is conducted in the presence of a small amount of water, migration of the third alkyl group (step *c*) is inhibited. Oxidation of the hydrate produced then gives the dialkyl ketone instead of the trialkyl carbinol (5.31). Alkaline hydrolysis leads to the secondary alcohol. Yields obtained are generally high, and the sequence provides a very convenient synthetic route to ketones. 1-Octene, for example, was smoothly converted into dioctyl ketone in 80 per cent yield, and cyclopentene gave dicyclopentyl ketone in 90 per cent yield.

$$RB\underset{O}{\diagdown\!\!\diagup}CR_2 \xrightarrow[(fast)]{H_2O} \underset{OH}{\underset{|}{RB}}-\underset{OH}{\underset{|}{CR_2}} \begin{cases} \xrightarrow{^-OH} RCHOHR \\ \xrightarrow{[O]} R-\underset{\underset{O}{\|}}{C}-R \end{cases} \qquad (5.31)$$

The method can be extended to the synthesis of unsymmetrical ketones by using 'mixed' organoboranes prepared from thexylborane or thexylchloroborane (see pp. 292–294). The thexyl group shows an exceptionally low aptitude for migration, and carbonylation of trialkylboranes containing a thexyl group, in the presence of water, followed by oxidation,

$$\text{Thx}-BH_2 \xrightarrow{\text{alkene A}} \text{Thx}-BHR_A \xrightarrow{\text{alkene B}} \text{Thx}-B\begin{matrix} R_A \\ R_B \end{matrix} \qquad (5.32)$$

leads to high yields of the ketone R_ACOR_B. Because of the bulky nature of the thexyl group, carbonylation of these compounds requires more vigorous conditions than usual and generally has to be effected under pressure. Functional groups in the alkene do not interfere with the reaction, and the procedure can be used to synthesise a ketone from almost any two alkenes (5.33).

Dienes similarly yield cyclic ketones, and in a notable extension of the reaction bicyclic ketones have been prepared, as illustrated in (5.34) for the stereospecific conversion of cyclohexanone into the thermodynamically disfavoured *trans*-perhydroindanone. *Trans*-1-decalone is obtained similarly from 1-allylcyclohexene. The stereoselectivity of the

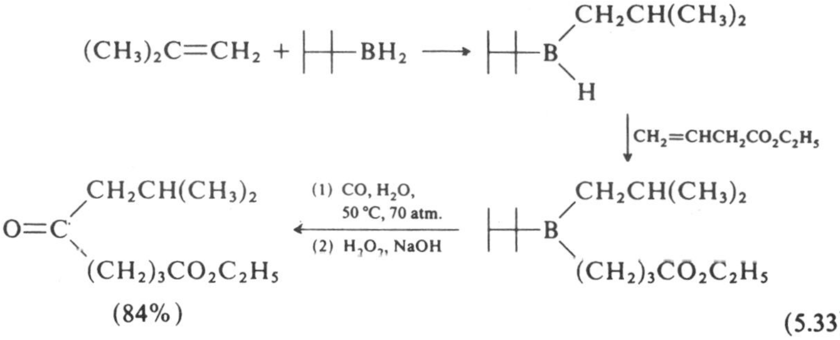

(5.33)

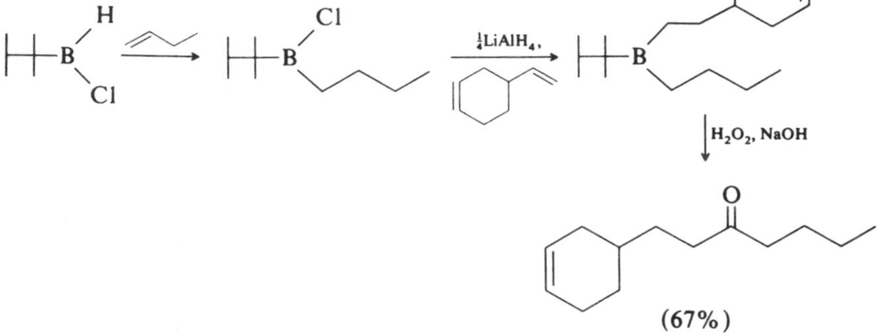

reactions, leading exclusively to the *trans* fused compounds, is a result of the mechanism of hydroboration which requires *syn* addition of the B—H group to the double bond of the alkene. An alternative procedure is described on p. 306.

The carbonylation reaction can be adapted to the preparation of aldehydes and primary alcohols. In the presence of certain hydride reducing agents, such as lithium hydridotrimethoxyaluminate, the rate of reaction of carbon monoxide with organoboranes is greatly increased and the products, on oxidation with buffered hydrogen peroxide afford aldehydes. Alkaline hydrolysis gives the corresponding primary alcohol containing one more carbon atom than the original alkene, and thus differing from the product obtained on direct oxidation of the original alkylborane. The reaction is believed to proceed by reduction of one of the intermediates R_3BCO or R_2BCOR (see p. 303) by the complex hydride, thus precluding migration of further alkyl groups (5.35). A disadvantage of this direct procedure is that only one of the three alkyl groups of the alkylborane is converted into the required derivative; the others are effectively wasted. This difficulty can be overcome by hydroboration of the alkene with 9-borabicyclo[3,3,1]nonane (9-BBN) (p. 291).

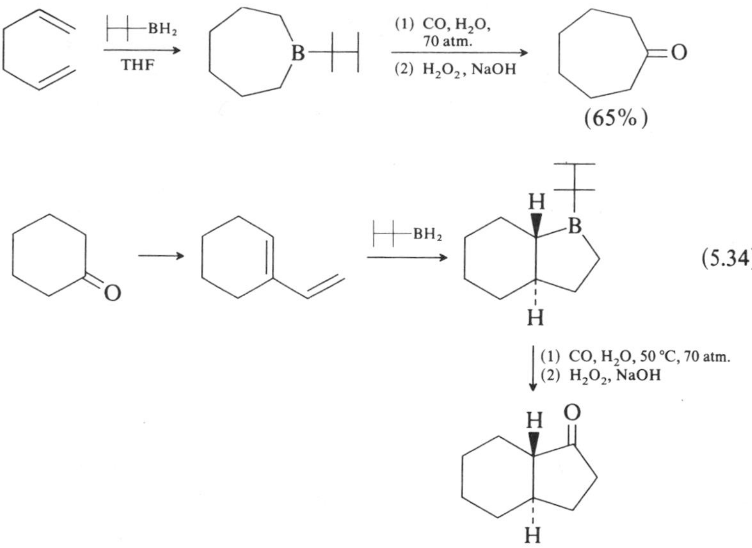

THF; (1) CO, H_2O, 70 atm. (2) H_2O_2, NaOH

(65%)

(5.34)

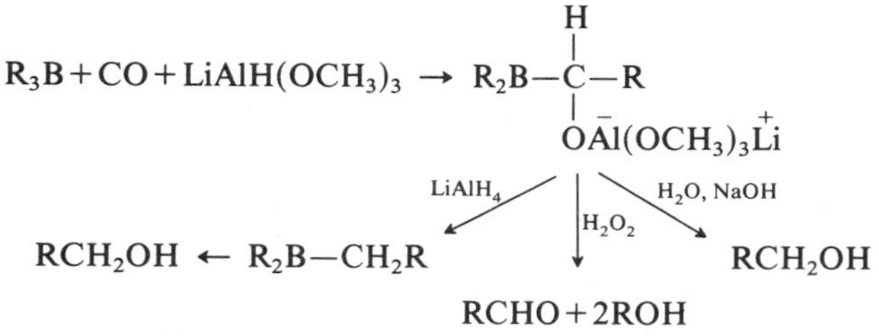

(1) CO, H_2O, 50 °C, 70 atm. (2) H_2O_2, NaOH

$$R_3B + CO + LiAlH(OCH_3)_3 \rightarrow R_2B{-}\underset{\underset{O\bar{A}l(OCH_3)_3\overset{+}{Li}}{|}}{\overset{\overset{H}{|}}{C}}{-}R \qquad (5.35)$$

$LiAlH_4$: $RCH_2OH \leftarrow R_2B{-}CH_2R$

H_2O_2: $RCHO + 2ROH$

H_2O, NaOH: RCH_2OH

Reaction of the resulting *B*-alkyl derivative with carbon monoxide and lithium hydridotrimethoxyaluminate takes place with preferential migration of the alkyl group, and high yields of aldehydes containing a variety of functional groups have been obtained from variously substituted

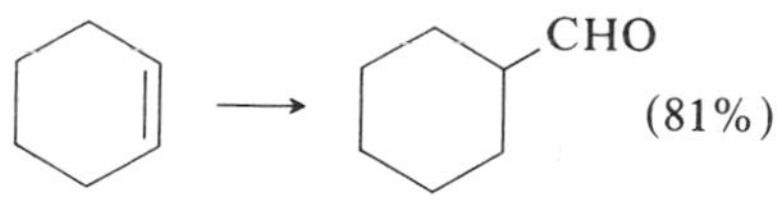

(81%)

(5.36)

$$CH_2{=}CHCH_2CO_2C_2H_5 \rightarrow OCH(CH_2)_3CO_2C_2H_5 \ (83\%)$$

alkenes by this method (5.36). A further advantage of the 9-BBN derivatives is that further reduction of the intermediate (38) with lithium aluminium hydride gives the corresponding homo-alkyl derivative providing another route to the homo primary alcohol (5.37) (Brown, Ford and Hubbard, 1980).

All these reactions are highly stereoselective, the original boron–carbon bond being replaced by a carbon–carbon bond with complete retention

$$CH_3(CH_2)_7{-}B\langle \xrightarrow{CO,\ LiAlH(OCH_3)_3} CH_3(CH_2)_7CHOH{-}B\langle\ \ (38)$$

$$\downarrow LiAlH_4 \qquad (5.37)$$

$$CH_3(CH_2)_7CH_2OH \xleftarrow{H_2O_2,\ NaOH} CH_3(CH_2)_7CH_2B\langle$$

of configuration. 1-Methylcyclopentene, for example, is converted entirely into *trans*-2-methylcyclopentylmethanol (5.38).

$$\text{1-methylcyclopentene}\ (-CH_3) \longrightarrow \text{cyclopentane with } CH_2OH \text{ and } CH_3\ (trans) \qquad (5.38)$$

There is a very useful alternative to the carbonylation route to ketones and trialkylmethanols from alkylboranes (Pelter *et al.*, 1975*b*; Pelter *et al.*, 1975*a*). In this sequence the key intermediate is a cyanoborate and the carbon atom which eventually becomes the carbonyl group of the ketone or the 'methanol' carbon of the trialkylmethanol originates not in carbon monoxide but in cyanide ion. If a solution of a trialkylborane is added to a suspension of sodium cyanide in tetrahydrofuran or diglyme, the cyanide dissolves and a trialkylcyanoborate is formed. Addition of one molar equivalent of benzoyl chloride or, better, trifluoroacetic anhydride induces two successive migrations of alkyl groups from boron to the adjacent carbon atom of the cyanide, forming the cyclic organoborane intermediate (39), oxidation of which without isolation affords the ketone in high yield (5.39). As in the carbonylation reaction, maximum utilisation of the alkyl groups on the borane is achieved by using thexylborane or thexylchloroborane (p. 292) as the hydroborating agent; the thexyl group

$$\text{Thx}{-}B(R^1)(R^2) \xrightarrow{NaCN} \text{Thx}{-}\overset{-}{B}(R^1)(R^2){-}C{\equiv}N \quad + \quad CF_3CO{-}O{-}COCF_3$$

$$\downarrow (CF_3CO)_2O,\ -78\,^\circ C \rightarrow 0\,^\circ C \qquad (5.39)$$

$$R^1{-}\overset{O}{\overset{\|}{C}}{-}R^2 \xleftarrow[NaOH]{H_2O_2} \text{Thx}{-}B\langle \text{ring: } C(R^1)(R^2){-}N{=}C(CF_3){-}O \rangle\ (39) \longleftarrow \left[\text{Thx}{-}B(R^2){-}C(R^1){=}N{-}C(CF_3){=}O\right]$$

does not migrate and unsymmetrical ketones are easily obtained from two different alkenes. If an excess of trifluoroacetic anhydride is employed at the rearrangement step, the reaction takes a different course; a third migration from boron to carbon ensues and oxidation of the product affords a trialkylmethanol (5.40).

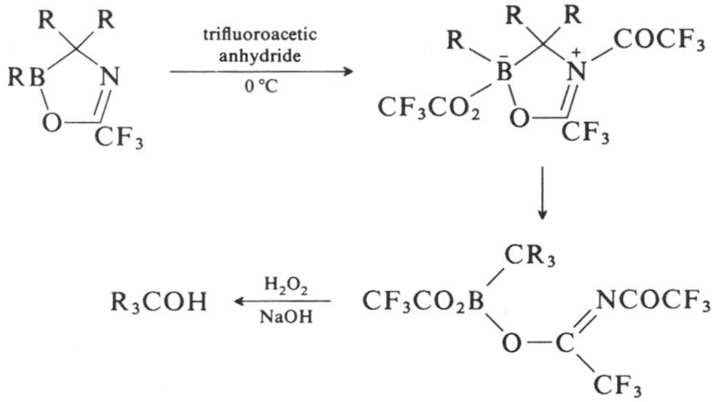

trifluoroacetic anhydride, 0 °C

$$R_3COH \xleftarrow[NaOH]{H_2O_2} CF_3CO_2B(CR_3)-O-C(CF_3)=NCOCF_3 \quad (5.40)$$

Both sequences give high yields of products with complete retention of configuration of the migrating groups. Some examples are given in (5.41).

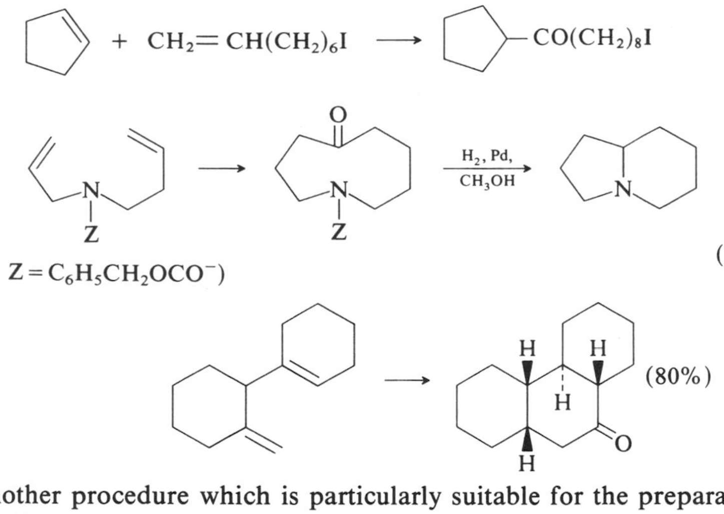

$$\text{cyclopentene} + CH_2{=}CH(CH_2)_6I \longrightarrow \text{cyclopentyl}{-}CO(CH_2)_8I$$

H_2, Pd, CH_3OH

$Z = C_6H_5CH_2OCO^-$)

(5.41)

(80%)

Another procedure which is particularly suitable for the preparation of trialkylmethanols containing tertiary or other hindered alkyl groups makes use of the reaction between trialkylboranes and dichloromethyl methyl ether in the presence of the strong hindered base lithium triethylcarboxide. All three groups of the alkylborane are transferred to a carbon atom of the ether, and oxidation of the product affords the corresponding tertiary alcohol. Tricyclopentylborane, for example, gave tricyclopentylmethanol in 97 per cent yield. It is not clear whether the reaction takes place via the α-halocarbanion or the corresponding carbene, but assuming the carbanion the following pathway has been proposed (5.42).

$$R_3B \xrightarrow[\text{THF}]{CH_3O\bar{C}Cl_2Li^+} R_2\bar{B}(R)-CCl(Cl)OCH_3Li^+ \xrightarrow{-LiCl} R_2B-C(R)(Cl)-OCH_3 \longrightarrow R-B(OCH_3)-CR_2Cl \longrightarrow Cl-B(OCH_3)-CR_3 \xrightarrow{[O]} R_3COH \quad (5.42)$$

This sequence has been adapted to provide a neat synthesis of ketones, starting not from a trialkylborane but from a dialkylborinic acid ester. The borinic acid esters themselves are readily obtained from an alkene and monobromoborane and reaction of the derived dialkylbromoborane with the appropriate alkoxide. 5-Nonanone, for example, was obtained from 1-butene in 99 per cent yield as shown in (5.43) (Brown, Ravindran

$$2\,CH_3CH_2CH{=}CH_2 + H_2BBr.(CH_3)_2S \longrightarrow (CH_3CH_2CH_2CH_2)_2BBr.(CH_3)_2S \xrightarrow{CH_3ONa,\ CH_3OH} (CH_3CH_2CH_2CH_2)_2B-OCH_3 \xrightarrow[(C_2H_5)_3COLi]{Cl_2CHOCH_3} CH_3(CH_2)_3CO(CH_2)_3CH_3\ (99\%) \quad (5.43)$$

and Kulkarni, 1979). This reaction has been used in an asymmetric synthesis of acyclic ketones via asymmetric hydroboration of a prochiral alkene with monoisopinocampheylborane (p. 296). In this sequence the borinic ester is obtained in a different way from that described above, by reaction of an aldehyde with a trialkylborane. Thus, successive alkylation with (E)-2-butene and 1-pentene of the monoisopinocamphenylborane derived from (+)-α-pinene gave the optically active mixed trialkylborane (40). Reaction of the latter with acetaldehyde then gave the borinic ester, (41) with elimination of α-pinene, and this on treatment with $\alpha\alpha$-dichloromethyl methyl ether and lithium triethylcarboxide, followed by oxidation, gave (R)-(−)-3-methyl-4-nonanone with 70 per cent optical purity. In this synthesis the high asymmetric induction realised in the hydroboration is retained in the carbon–carbon bond-forming reaction. An advantage of the method is that ketones of opposite configuration can be obtained by using monoisopinocampheylborane derived from either (+)- or (−)-α-pinene (Brown, Jadhav and Desai, 1982). A related route to optically active secondary alcohols uses a β-alkyl pinanediol boronic ester as the chiral auxiliary reagent. Here one of the

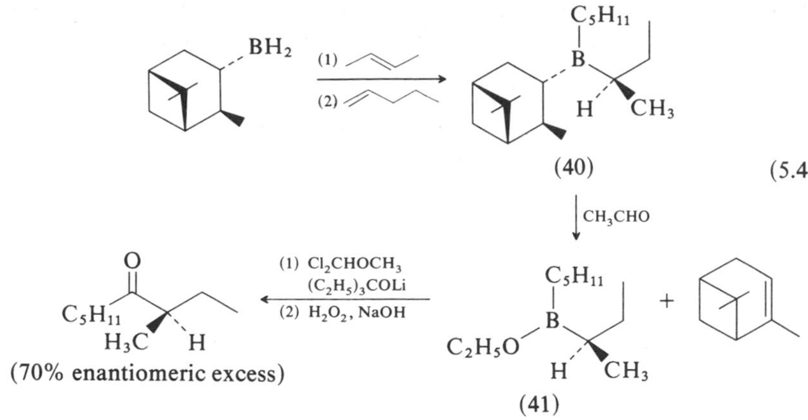

(1) propene, (2) but-2-ene → (40) (5.44)

(40) $\xrightarrow{CH_3CHO}$ (41) + α-pinene

(41) $\xrightarrow{(1)\ Cl_2CHOCH_3,\ (C_2H_5)_3COLi;\ (2)\ H_2O_2,\ NaOH}$ $C_5H_{11}COCH(CH_3)C_2H_5$

(70% enantiomeric excess)

alkyl groups of the dialkyl carbinol is derived from the β-alkyl borinic ester and the other from a Grignard reagent (Matteson and Sadhu, 1983).

Ketones and trialkylcarbinols can also be obtained by reaction of trialkylboranes with the readily available anion of tris(phenylthio)-methane. Two alkyl group migrate spontaneously from boron to carbon in the initial adduct (42) to give a product which can be oxidised to a ketone. A third migration can be induced by treatment with mercuric ion; oxidation of the product then affords a trialkylcarbinol (5.45) (Pelter and Rao, 1981).

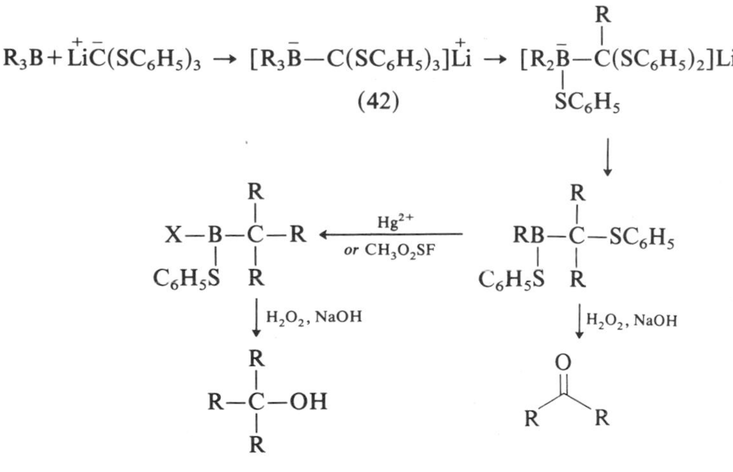

$$R_3B + \overset{+}{Li}\overset{-}{C}(SC_6H_5)_3 \rightarrow \underset{(42)}{[R_3\overset{-}{B}-C(SC_6H_5)_3]\overset{+}{Li}} \rightarrow [R_2\overset{-}{B}-CR(SC_6H_5)_2]Li^+ \quad (\text{B bearing } SC_6H_5)$$

$$\rightarrow RB(SC_6H_5)-CR_2-SC_6H_5 \xrightarrow{H_2O_2,\ NaOH} RCOR$$

$$RB(SC_6H_5)-CR_2-SC_6H_5 \xrightarrow{Hg^{2+}\ or\ CH_3O_2SF} X-B(SC_6H_5)-CR_3 \xrightarrow{H_2O_2,\ NaOH} R_3C-OH \quad (5.45)$$

Closely similar in conception is a useful route to aldehydes by reaction of *B*-alkylboronic esters with methoxy(phenylthio)methyl-lithium, LiCH(OMe)SPh. Treatment of the intermediate (44) with mercuric chloride readily induces transfer of the alkyl group from boron to carbon and subsequent oxidation provides the corresponding aldehyde in good yield (Brown and Imai, 1983). Thus, methylcyclopentene gave *trans*-2-methylcyclopentylcarboxaldehyde in 64 per cent yield (5.46). The 1,3,2-dioxaborinanes (43) were found to give the best results. They are conveniently obtained by reaction of the dibromoalkylborane methyl sulphide complex with 1,3-*bis*(trimethylsiloxy)propane. The migration step (44 → 45) proceeds with retention of configuration of the migrating

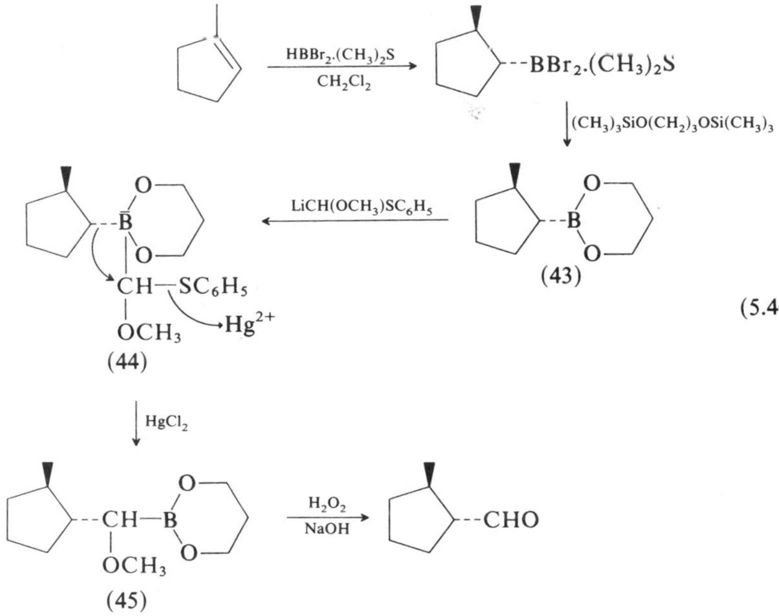

$HBBr_2.(CH_3)_2S$, CH_2Cl_2 → $BBr_2.(CH_3)_2S$

$(CH_3)_3SiO(CH_2)_3OSi(CH_3)_3$

$LiCH(OCH_3)SC_6H_5$

(43)

$CH—SC_6H_5$ Hg^{2+} OCH_3

(44)

(5.46)

$HgCl_2$

$CH—B$ OCH_3

(45)

H_2O_2, NaOH → CHO

carbon atom, as with other related 1,2-migrations. This, in conjunction with the stereospecificity of the hydroboration step, is turned to advantage in a valuable stereoselective synthesis of 2,3-disubstituted aldehydes illustrated in (5.47) for the synthesis of *threo*- and *erythro*-2,3-dimethylpentanal from (*E*)- and (*Z*)-3-methyl-2-pentene respectively.

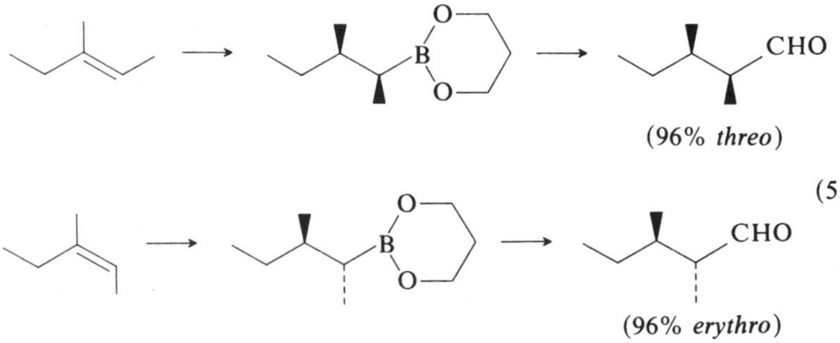

CHO

(96% *threo*)

(5.47)

CHO

(96% *erythro*)

Reaction with α-bromoketones and α-bromoesters

Organoboranes react readily with α-bromoketones and α-bromoesters in the presence of potassium t-butoxide or, better, the hindered base potassium 2,6-di-t-butylphenoxide, to give the corresponding α-alkyl or α-aryl derivatives in which the bromine atom has been replaced by an alkyl or aryl substituent from the borane (Brown, Nambu and Rogic, 1969). 2-Ethylcyclohexanone, for example, was readily obtained from triethylborane and 2-bromocyclohexanone. The reaction path (5.48) has been suggested; the key step again involves migration of

$$RCOCH_2Br + t\text{-}C_4H_9OK \rightarrow RCO\overset{-}{C}H(Br) \xrightarrow{BR'_3} RCOCH(Br)-\overset{-}{B}R'_3K^+$$
$$\downarrow \qquad (5.48)$$
$$RCOCH_2R' + R'_2BOC_4H_9\text{-}t \xleftarrow{t\text{-}C_4H_9OH} RCOCH(R')-BR'_2 + KBr$$

a group from boron to the adjacent carbon atom, with expulsion of bromide ion. The usefulness of the direct reaction using trialkylboranes is limited by the fact that organoboranes containing highly branched groups do not react and, in addition, only one of the three alkyl groups in the trialkylborane is used in the reaction. These difficulties can be circumvented by using an alkyl derivative of 9-borabicyclo[3,3,1]nonane (9-BBN) (p. 291) instead of the trialkylborane. The alkyl group is then fully utilised and the 9-BBN takes no part in the alkylation.

Mono-halogenoacetates undergo similar reactions, providing a useful alternative to the malonic ester route to substituted acetic acids. Again best yields are obtained with 9-aryl- or 9-alkyl-borabicyclononanes (5.49). The reaction can be extended to dibromoacetates and can be controlled to give either α-alkyl-α-bromoacetates or dialkylacetates. Dialkylation can be effected in two steps, allowing the introduction of two different groups into the acetic acid.

$$(CH_3)_2CHCH_2-B\langle\text{9-BBN}\rangle + BrCH_2CO_2C_2H_5 \xrightarrow[t\text{-}C_4H_9OH]{t\text{-}C_4H_9OK} (CH_3)_2CHCH_2CH_2CO_2C_2H_5$$

$$\text{1-methylcyclopentene} \xrightarrow{BH_3.THF} \text{(2-methylcyclopentyl)}--B\langle \xrightarrow[t\text{-}C_4H_9OK,\ t\text{-}C_4H_9OH]{BrCH_2CO_2C_2H_5} \text{(2-methylcyclopentyl)}--CH_2CO_2C_2H_5 \qquad (5.49)$$

Reaction with diazo compounds

Another route from organoboranes to ketones and esters proceeds by nucleophilic addition of a diazo compound to the organoborane.

Diazoketones, for example, react with organoboranes to form products which yield ketones on alkline hydrolysis. The reaction pathway (5.50) has been suggested in which migration of R from boron to carbon is facilitated by expulsion of nitrogen. Diazo-esters and -nitriles react similarly to give the corresponding substituted esters and nitriles (Hooz, Gunn and Kono, 1971). Better yields are often obtained by using a dialkylchloroborane instead of the corresponding trialkylborane (Brown, Midland and Levy, 1972).

$$R_3B + N{\equiv}\overset{+}{N}{-}\bar{C}HCOCH_3 \rightarrow R_3\bar{B}{-}CH(\overset{+}{N}{\equiv}N)COCH_3$$

$$\downarrow \tag{5.50}$$

$$RCH_2COCH_3 \xleftarrow{\text{hydrolysis}} R_2B{-}CHRCOCH_3 + N_2$$

5.4. Reactions of alkenylboranes and trialkylalkynylborates

Alkenylboranes and trialkylalkynylborates are used for the preparation of conjugated dienes and diynes, and of saturated and $\alpha\beta$-unsaturated ketones. These conversions involve the migration from boron to carbon of an alkenyl or alkynyl group instead of an alkyl group as in the reactions discussed above.

Alkenylboranes are readily prepared by the addition of borane or a substituted borane to alkynes (p. 292). They undergo a variety of useful synthetic reactions (Brown, 1981) and have been employed in the stereoselective synthesis of alkenes and of alkenyl halides (p. 152) and to make conjugated dienes. Both the *Z,E*- and *E,E*-isomers of the latter can be obtained by the appropriate choice of reaction. In each case the key step is the migration of an alkenyl group, with retention of configuration, from boron to an adjacent carbon atom. For the synthesis of *Z,E*-dienes the alkenylborane, obtained from an alkyne by appropriate hydroboration, is treated with iodine and sodium hydroxide when the diene is obtained directly. Thus, (*Z,E*)-4,5-diethyl-3,5-octadiene is obtained in high yield by way of the trialkenylborane (46) prepared by addition of borane to 3-hexyne. The reaction is believed to take the pathway shown in (5.51), in which iodide-assisted migration of the alkenyl group is followed by *anti* de-iodoboration.

E,E-Dienes are obtained from 1-chloro- or 1-bromo-alkynes. Hydroboration with thexylborane gives thexyl-1-chloroalkenylboranes which react further with alkynes to give thexyldialkenylboranes. Reaction of these with sodium methoxide forms intermediates which on protonolysis

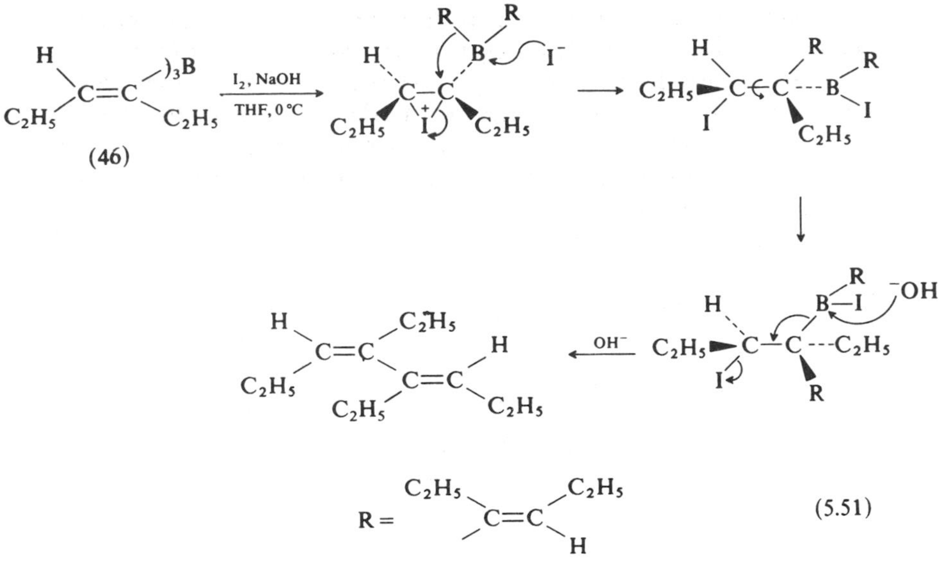

(5.51)

are converted into the corresponding *E,E*-dienes; oxidation with alkaline hydrogen peroxide on the other hand, gives α,β-unsaturated ketones (5.52) (Negishi and Yoshida, 1973).

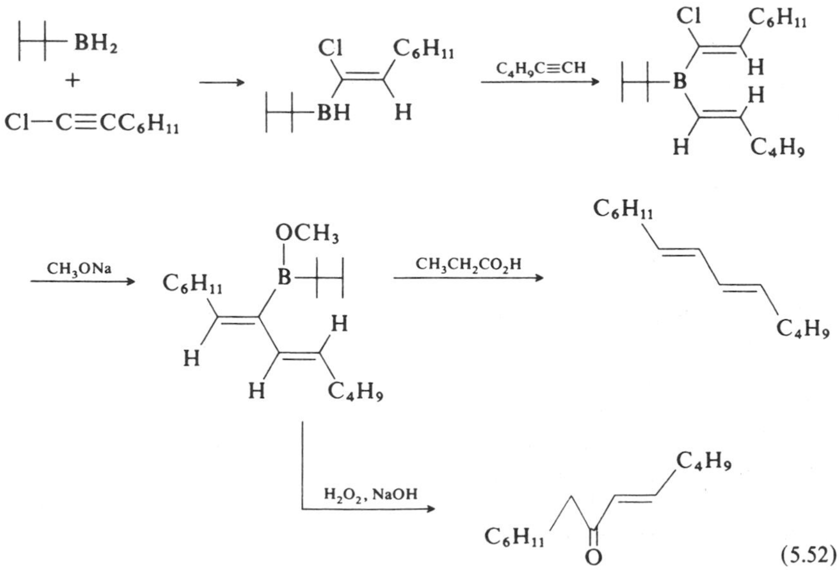

(5.52)

Sodium methoxyalkenyldialkylborates (as 47), which are obtained from alkenyldialkylboranes and sodium methoxide, react with copper(I) bromide–dimethyl sulphide complex at 0 °C to give symmetrical 1,3-dienes in high yield. The reaction can be adapted to give either the *Z,Z*- or the *E,E*-isomer. Apparently an alkenylcopper compound or a copper complex with boron is formed which undergoes thermal coupling at 0 °C (Campbell and Brown, 1980). At lower temperatures the complex is more stable and can be trapped by allyl bromides or iodides to give 1,4-dienes (5.53). Analogous reactions with 1-bromoalkynes give conjugated enynes which, of course, can be reduced to dienes (Brown and Molander, 1981).

$$HC{\equiv}C(CH_2)_2OCOCH_3 \xrightarrow[THF]{(C_6H_{11})_2BH} (C_6H_{11})_2B(H)C{=}C(H)(CH_2)_2OCOCH_3$$

$$\xrightarrow{CH_3ONa,\ THF} (C_6H_{11})_2\bar{B}(OCH_3)(H)C{=}C(H)(CH_2)_2OCOCH_3 \quad (47)$$

$$\xrightarrow[(2)\ CH_2{=}CHCH_2Br]{(1)\ CuBr.(CH_3)_2S,\ -15\,°C} CH_2{=}CHCH_2CH{=}CH(CH_2)_2OCOCH_3 \quad (76\%) \qquad (5.53)$$

Trialkylalkynylborates are obtained by reaction of alkynyl-lithiums with trialkylboranes. They are valuable intermediates for the synthesis of alkynes, ketones and conjugated dienes and diynes. The reactions again involve migration of a group from boron to carbon, this time induced by attack of an electrophile on the triple bond of the alkynyl substituent. Substituted alkynes, for example, are obtained in excellent yield by reaction of the borates with iodine (Midland, Sinclair and Brown, 1974). The same transformation can, of course, often be effected by direct alkylation of the alkali metal salt of the alkyne, but the present procedure can be used to introduce substituents bearing functional groups which might react with the alkynyl salt in the direct reaction (5.54). The mechan-

$$(C_4H_9)_3B + LiC{\equiv}CC_6H_5 \xrightarrow{THF} (C_4H_9)_2\bar{B}(C_4H_9){-}C{\equiv}CC_6H_5\ Li^+ \quad (I{-}I)$$

$$\xrightarrow{I_2,\ -78\,°C} (C_4H_9)_2B(C_4H_9)C{=}C(C_6H_5)I \longrightarrow C_4H_9C{\equiv}CC_6H_5 \quad (98\%) \qquad (5.54)$$

ism of the reaction is thought to involve electrophilic attack of iodine on the triple bond with migration of an alkyl group from boron to the adjacent carbon, and subsequent deiodoboration with regeneration of the triple bond.

The reaction has been extended to the coupling of alkynes by use of a dialkynylborate and it can be used to prepare unsymmetrical diynes containing two different alkyl substituents which are difficult to prepare by other methods (Sinclair and Brown, 1976; Pelter *et al.*, 1976*b*).

Terminal alkynes R—C≡CH may also be converted into ketones or $\alpha\beta$-unsaturated ketones via derived borates. Thus, alkylation or protonation of lithium alkynyltrialkylborates followed by oxidation gives ketones RR^1CHCOR^2 in good yield. The nature of the substituent groups R, R^1 and R^2 can be varied widely and the sequence provides a versatile route to α-substituted ketones (5.55) (Pelter *et al.*, 1976*a*).

$$(C_6H_{13})_3B + LiC{\equiv}CC_4H_9 \rightarrow [(C_6H_{13})_2\bar{B}(C_6H_{13}){-}C{\equiv}C{-}C_4H_9]Li^+ \quad CH_3{-}SO_4CH_3$$

$$\xrightarrow{(CH_3)_2SO_4,\ \text{diglyme},\ -78\,^\circ C} (C_6H_{13})_2B(C_6H_{13})C{=}C(C_4H_9)CH_3 \xrightarrow{H_2O_2,\ NaOH} C_6H_{13}COCH(C_4H_9)CH_3 \ (84\%) \qquad (5.55)$$

Carbanionic migration from boron to carbon-summary

All the reactions discussed so far take place by migration of a group from boron to an adjacent carbon or hetero-atom, induced by nucleophilic attack on the electron-deficient boron of an alkylborane or by electrophilic attack at an unsaturated substituent of a borate. They all proceed in high yield with retention of configuration of the migrating group. The main reactions are summarised in the table on page 316.

5.5. Free-radical reactions of organoboranes

There is another, smaller, group of reactions of organoboranes which proceed by a free-radical pathway, and some of these are useful in synthesis (Brown and Midland, 1972). One is the reaction leading to coupling of alkyl groups which takes place on treatment of trialkylboranes with alkaline silver nitrate. Tri-n-hexylborane, for example, gave dodecane in 70 per cent yield and, in an intramolecular example, 2,3-dimethylbutadiene was converted into *trans*-1,2-dimethylcyclobutane in

Table 5.1. *Reactions involving migration from boron to an adjacent atom*

$$R_3B + {}^{-}OOH \rightarrow R_2\bar{B}(R)\text{—}O\text{—}OH \rightarrow R_2B\text{—}OR \rightarrow ROH$$

$$R_3B + \bar{O}\text{—}\overset{+}{N}(CH_3)_3 \rightarrow R_2\bar{B}(R)\text{—}O\text{—}\overset{+}{N}(CH_3)_3 \rightarrow R_2B\text{—}OR \rightarrow ROH$$

$$R_3B + CO \rightarrow R_2\bar{B}(R)\text{—}\overset{+}{C}O \rightarrow R_2B\text{—}C(R){=}O \rightarrow RCOR,\ R_3COH$$

$$R_3B + CN^{-} \longrightarrow R_2\bar{B}(R)\text{—}C{\equiv}N^{\cdot} \xrightarrow{X^{+}} R_2B\text{—}C(R){=}NX \longrightarrow\!\!\!\rightarrow RCOR$$

$$R_3B + \bar{C}H(Br)CO_2C_2H_5 \rightarrow R_2\bar{B}(R)\text{—}CH(Br)CO_2C_2H_5 \rightarrow R_2B\text{—}CH(R)CO_2C_2H_5 \rightarrow RCH_2CO_2C_2H_5$$

$$\text{(thexyl)}B(R)CHR^1R^2 \xrightarrow{Br_2,\ h\nu} \text{(thexyl)}B(R)CR^1R^2Br\ (H_2O{:}) \rightarrow \text{(thexyl)}B(OH)\text{—}CRR^1R^2 \longrightarrow RR^1R^2COH$$

$$R_2\bar{B}(R)\text{—}C{\equiv}C\text{—}R^1 + R^2\text{—}X \rightarrow R_2B\text{—}C(R){=}C(R^1)R^2 \rightarrow R\text{—}C{\equiv}C\text{—}R^1,\ RCOCHR^1R^2$$

(for $R^2\text{—}X = I_2$)

$$CH_3O^{-} + \text{(thexyl)}B(CH{=}CHR)(C(Cl){=}CHR^1) \xrightarrow{^{-}OCH_3} \text{(thexyl)}B(OCH_3)\text{—}C(CH{=}CHR){=}CHR^1 \longrightarrow RCH{=}CH\text{—}CH{=}CHR^1$$

$$(HRC{=}CR)_3B \xrightarrow{I_2} R^1_2B\text{—}C(R)(R^1)\text{—}\overset{+}{I}\text{—}CHR\ (I^{-}) \longrightarrow R^1_2B\text{—}C(R)(R^1)\text{—}CH(R)I \longrightarrow HRC{=}C(R)\text{—}C(R){=}C(R)H$$

$$\left(R^1 = \begin{matrix} \\ R \end{matrix}\!\!\gt C{=}C\lt\!\!\begin{matrix} H \\ R \end{matrix}\right)$$

79 per cent yield. The reactions are believed to take place by way of alkyl free radicals formed by breakdown of intermediate silver alkyls.

$$H_3C(CH=CH_2) / H_3C(CH=CH_2) \xrightarrow[\text{(2) } AgNO_3,\ KOH,\ CH_3OH]{\text{(1) } BH_3.THF} \text{1,2-dimethylcyclobutane} \qquad (5.56)$$

Another useful free-radical reaction of organoboranes is the 1,4-addition to $\alpha\beta$-unsaturated aldehydes and ketones. Trialkylboranes do not attack isolated carbonyl groups but they do undergo fast addition to many $\alpha\beta$-unsaturated carbonyl compounds such as acrylic aldehyde, methyl vinyl ketone and 2-methylenecyclohexanone, to form aldehydes and ketones in which the alkyl chain has been extended by three or more carbon atoms (5.57). The reactions take place by free radical addition

$$R\cdot + CH_2{=}CHCHO \rightarrow RCH_2\dot{C}HCHO \longleftrightarrow RCH_2CH{=}CHO\cdot$$
$$\downarrow R_3B \qquad (5.57)$$
$$RCH_2CH_2CHO \longleftarrow RCH_2CH{=}CHOBR_2 + R\cdot$$

to the conjugated system to form the enol borinate, hydrolysis of which yields the aldehyde or ketone. In some cases, particularly with β-substituted $\alpha\beta$-unsaturated carbonyl compounds, the spontaneous reaction does not proceed readily, but it can be induced in the presence of catalytic amounts of diacetyl peroxide or of oxygen, or by irradiation of the reaction mixture with ultra-violet light, all of which favour the formation of radicals from the trialkylborane. The sequence provides a useful method for the synthesis of aldehydes and ketones from practically any combination of alkene and conjugated aldehyde or ketone (5.58).

$$(CH_3CH_2CH(CH_3))_3B \xrightarrow[\text{(2) } H_2O]{\text{(1) } CH_2{=}C(CH_3)CHO} CH_3CH_2CH(CH_3)CH_2CH(CH_3)CHO \quad (96\%)$$

$$(C_2H_5)_3B + CH_3CH{=}CHCOCH_3 \xrightarrow[\text{diglyme 25 °C}]{(CH_3CO_2)_2} C_2H_5CH(CH_3)CH_2COCH_3 \quad (88\%)$$

$$(C_2H_5)_3B + \text{cyclohex-2-enone} \xrightarrow[\text{THF, 25 °C}]{\text{cat. } O_2} \text{3-ethylcyclohexanone} \quad (68\%) \qquad (5.58)$$

5.6. Applications of organosilicon compounds in synthesis

Boron is not the only inorganic element whose derivatives are of value in organic synthesis. Reactions involving organosilicon compounds are also being used increasingly and a number of applications have already been given.

Silicon is electropositive with respect to carbon so that the silicon–carbon bond is polarised with the positive charge on silicon, making the bond susceptible to nucleophilic attack at silicon. This can be exploited to introduce silicon into organic compounds. For the same reason silicon forms comparatively strong bonds to electronegative elements and, in general, it is energetically profitable to replace a silicon–carbon bond by a bond to an electronegative element such as oxygen or, particularly, fluorine. This can often be exploited to remove silicon from the product of a reaction after it has performed its synthetic function.

Silicon can stabilise a negative charge on an adjacent carbon atom, and both this and its susceptibility to nucleophilic attack are ascribed, at least partly, to the presence of empty $3d$ orbitals in the silicon. They interact with $2p$ orbitals on the adjacent carbons and they may also aid nucleophilic substitution by allowing the nucleophile to attach itself to silicon before the leaving group departs. Another valuable feature of organosilicon compounds is that in many situations a carbon–silicon bond is able to stabilise an *adjacent carbonium ion* (that is on the carbon atom β to silicon). Because of the electropositive character of silicon, the orbitals of the silicon–carbon bond are favourably polarised and energetically suitable for effective overlap with an adjacent empty carbon p-orbital, provided that the geometrical disposition of the orbitals is suitable (p. 327).

Silicon is generally employed in organic synthesis in the form of a trialkylsilyl functional group, often the trimethylsilyl group, $(CH_3)_3Si$, and in this form it plays three main roles: (*a*) as a protecting group and in reactive intermediates such as silyl enol ethers, (*b*) for oxygen-capture with elimination to form alkenes (the Peterson reaction, p. 135) and (*c*) in the stabilisation of α-carbanions and β-carbonium ions.

Protection of functional groups; trimethylsilyl ethers

The trimethylsilyl group is well-known as a protecting group for amino, hydroxyl and terminal alkyne groups (Klebe, 1972). The derivatives are easily made by reaction of the alcohol or amine with chlorotrimethylsilane in the presence of a tertiary amine base such as pyridine, or from a suitable metallic derivative of the alkyne. Alkynes protected in this way have been used in coupling reactions to prepare polyalkynes (cf. Eastmond, Johnson and Walton, 1972). After reaction, the free terminal alkyne is readily liberated from the silyl derivative by treatment with methanol, aqueous sodium hydroxide or fluoride anion. A useful feature of these derivatives is that the large trialkylsilyl group screens the neighbouring triple bond, allowing selective partial hydrogenation of another triple bond elsewhere in the molecule (5.59).

$$\mathrm{BrMgC{\equiv}C{-}C{\equiv}CH} \xrightarrow[\text{THF, 0 °C}]{(C_2H_5)_3SiBr} \mathrm{(C_2H_5)_3Si(C{\equiv}C)_2H}$$

$$\downarrow O_2, Cu^+$$

$$\mathrm{H(C{\equiv}C)_4H} \xleftarrow[CH_3OH]{NaOH} \mathrm{(C_2H_5)_3Si(C{\equiv}C)_4Si(C_2H_5)_3} \quad (5.59)$$

OH, $C{\equiv}C{-}C{\equiv}CSi(CH_3)_3$ — (1) H_2, Pd-$BaSO_4$ quinoline; (2) F^-, THF → OH

Trialkylsilyl ethers of alcohols are very frequently employed in synthesis. In general the trimethylsilyl derivatives are too susceptible to solvolysis in protic media to be carried through a multi-step synthesis. Better results have been obtained with the t-butyldimethylsilyl ethers (Corey and Venkateswarlu, 1972). The t-butyldimethylsilyl derivatives are readily prepared by reaction of the alcohol with t-butylchlorodimethylsilane in the presence of imidazole, presumably by way of *N*-t-butyldimethylsilyl-imidazolyl ion (48), which would be a very reactive silylating agent (nucleophilic attack on silicon aided by a good leaving group) (5.60). These derivatives are not affected by conditions frequently encountered in synthetic transformations and are much more

ROH

$Si(CH_3)_2C_4H_9$-t

N, NH + t-$C_4H_9(CH_3)_2SiCl$ ⟶ N, $\overset{+}{N}$H (5.60)

$ROSi(CH_3)_2C_4H_9$-t

resistant to hydrolysis than the trimethylsilyl ethers. They can be reconverted into the alcohol by acid hydrolysis, under conditions comparable to those used for the cleavage of tetrahydropyranyl acetals, or selectively, by treatment with tetrabutyl-ammonium fluoride or aqueous hydrofluoric acid (Newton *et al.*, 1979; Nicolaou, Seitz and Pavia, 1981). (Cleavage of a silicon–oxygen bond by nucleophilic attack of fluoride ion and formation of an even stronger silicon–fluorine bond). Another advantage of the t-butyldimethylsilyl protecting group is that, because of its large size, it can sometimes be used to direct the stereochemical course of a reaction at a neighbouring functional group. Thus, in the conjugate addition of lithium dimethylcuprate to the cyclopentenone (49) attack is directed completely to the top face of the molecule at C-3 by the adjacent t-butyldimethyl ether.

CH_3 ... $OSi(CH_3)_2C_4H_9$-t (49) $\xrightarrow{(CH_3)_2CuLi}$ CH_3 ... $OSi(CH_3)_2C_4H_9$-t ... CH_3 (5.61)

Silyl *esters*, readily prepared from the acid and the appropriate trialkylsilyl chloride in the presence of a tertiary amine, are useful protecting groups for carboxylic acids. They are relatively stable to basic and oxidising conditions, but are readily cleaved on mild treatment with methanol or ethanol.

In addition to its direct use in the preparation of trimethylsilyl derivatives of alcohols and acids, chlorotrimethylsilane has found valuable application in trapping enolate anions as the corresponding trimethylsilyl enol ethers. It is a very useful adjunct in the acyloin condensation, for example. As is well known, this reaction provides a valuable and widely used method for the synthesis of medium-sized ring compounds from diesters. It is less effective for the synthesis of smaller rings or in intermolecular condensations, for in these cases side reactions induced by alkoxide ion become important and only poor yields of the acyloin are formed. Very much better results are obtained if the reactions are conducted in the presence of chlorotrimethylsilane; even four-membered rings can be prepared easily under these conditions. The chlorotrimethylsilane acts as a scavenger for the alkoxide ion liberated during the reaction, so that the reaction medium is kept neutral. At the same time the oxygen-sensitive ene-diol is protected as the *bis*-trimethylsilyl ether which may be isolated and purified before hydrolysis to the acyloin (5.62) (Ruhlmann, Seefluth and Becker, 1967; Bloomfield, 1968). Thus, diethyl suberate was converted into the seven-membered cyclic acyloin in 75 per cent yield and the four-membered ring compound (50) was obtained in 90 per cent yield; in the absence of chlorotrimethylsilane only poor yields were obtained in both reactions.

Silyl enol ethers

Trimethylsilyl enol ethers are valuable in synthesis in a number of ways: for the generation of isomerically pure enolate ions, in directed aldol condensations (p. 50), for the regiospecific formation of α-

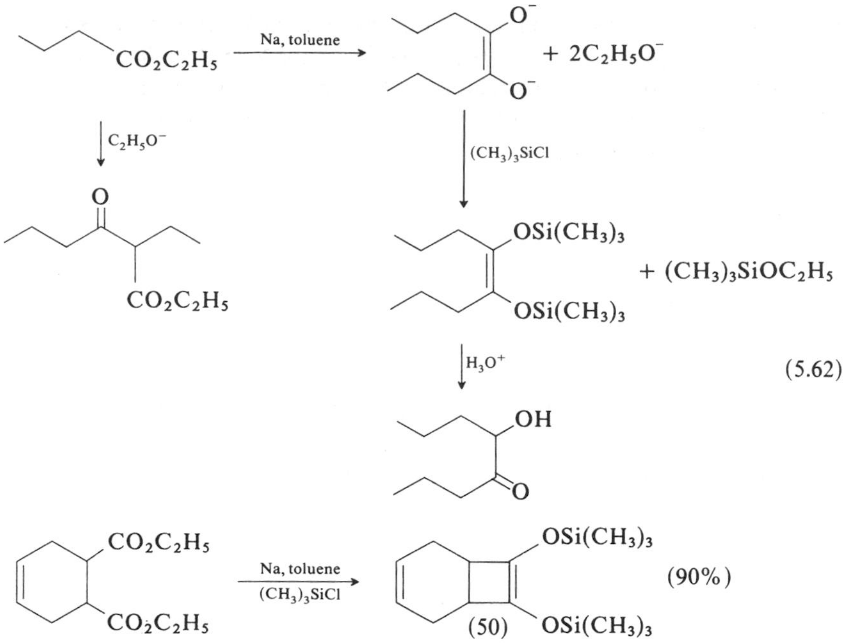

Na, toluene; $C_2H_5O^-$; $(CH_3)_3SiCl$; H_3O^+

$CO_2C_2H_5$ → $+ 2C_2H_5O^-$ → $OSi(CH_3)_3$, $OSi(CH_3)_3$ $+ (CH_3)_3SiOC_2H_5$ → OH, O

(5.62)

$CO_2C_2H_5$, $CO_2C_2H_5$ —Na, toluene / $(CH_3)_3SiCl$→ $OSi(CH_3)_3$, $OSi(CH_3)_3$ (50) (90%)

substituted carbonyl compounds, and in thermal and photochemical cycloaddition reactions (p. 200) They can be prepared from carbonyl compounds in a variety of ways (Brownbridge, 1983; Kuwajima, Nakamura and Hashimoto, 1983), but, in general, they are obtained by the action of base on the carbonyl compound and reaction of the enolate with chlorotrimethylsilane. In many cases the two possible ethers from an unsymmetrical ketone can be obtained selectively, depending on the method employed. Deprotonation with the hindered base lithium diisopropylamide and reaction with chlorotrimethylsilane gives mainly

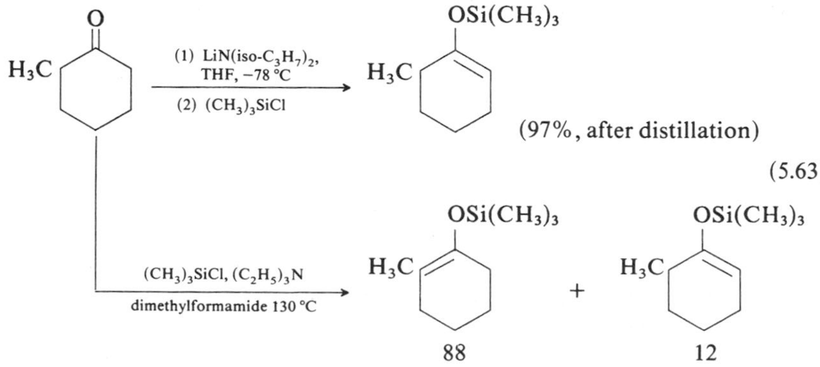

H_3C (ketone) —(1) $LiN(iso\text{-}C_3H_7)_2$, THF, −78 °C; (2) $(CH_3)_3SiCl$→ H_3C, $OSi(CH_3)_3$ (97%, after distillation)

(5.63)

—$(CH_3)_3SiCl$, $(C_2H_5)_3N$, dimethylformamide 130 °C→ H_3C, $OSi(CH_3)_3$ (88) + H_3C, $OSi(CH_3)_3$ (12)

the kinetic (less substituted) enol silyl ether (5.63) (Fleming and Paterson, 1979). The thermodynamic enol ether is obtained by equilibration of the mixture obtained with triethylamine and chlorotrimethylsilane or by the use of iodotrimethylsilane in conjunction with hexamethyldisilazane (Miller and McKean, 1979).

Carboxylic esters give the reactive silyl ketene acetals on reaction with lithium diisopropylamide and chlorotrimethylsilane (5.64).

OC_2H_5 (1) $LiN(iso\text{-}C_3H_7)_2$, THF (2) $(CH_3)_3SiCl$ OC_2H_5 $OSi(CH_3)_3$ (5.64)

Trimethylsilyl enol ethers and ketene acetals undergo a variety of synthetically useful reactions (Rasmussen, 1977; Brownbridge, 1983). They can be used with advantage for the generation of specific enolate anions from unsymmetrical ketones or, directly, as versatile carbon nucleophiles. Cleavage of the purified silyl ethers with methyl-lithium regenerates the corresponding lithium enolates which retain their structural integrity as long as the reaction medium is free of accessible protons which would make possible equilibration (see p. 17).

In the presence of Lewis acids ($TiCl_4$, $ZnBr_2$) trimethylsilyl enol ethers become very effective nucleophiles and are readily alkylated and acylated in reaction with alkyl or acyl halides (Mukaiyama, 1977). Experience shows that many alkylations of ketones and esters can be improved by changing from the basic conditions required for alkylation of enolates to the acidic conditions appropriate to the reactions of silyl enol ethers. The enolates and silyl enol ethers can have quite different reactivities and may thus complement each other in their synthetic applications. Thus, t-alkylation of carbonyl compounds cannot generally be achieved by reaction of the corresponding enolate with a tertiary halide, because of preferential elimination from the halide, but it can be effected with the silyl enol ether in the presence of catalytic titanium tetrachloride (5.65) (Reetz, Chatzüosifidis, Löwe and Maier, 1979; Paterson, 1979).

$OSi(CH_3)_3$ + $ClC(CH_3)_2CH_2CH_3$ $\xrightarrow[CH_2Cl_2,\ -40\ °C]{TiCl_4}$ O, $C(CH_3)_2C_2H_5$ (71%)

(5.65)

O → $(CH_3)_3SiO$ Cl $\xrightarrow{ZnBr_2}$ O (83%)

Allyl and benzyl halides are also effective alkylating agents under these conditions but unactivated secondary halides do not react. In an alternative procedure α-alkyl and α-alkylidene carbonyl compounds are obtained regiospecifically by acid catalysed alkylation of the appropriate enol silyl ether with an α-chloroalkyl phenyl sulphide followed by reductive or oxidative removal of sulphur, as illustrated in (5.66) (Patterson and Fleming, 1979).

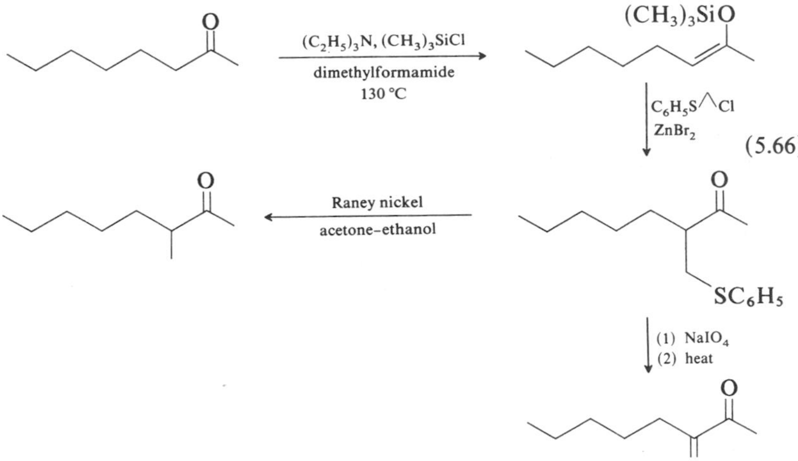

$(C_2H_5)_3N$, $(CH_3)_3SiCl$; dimethylformamide, 130 °C → $(CH_3)_3SiO$ enol ether; C_6H_5S∕∖Cl, $ZnBr_2$ → ketone with SC_6H_5; Raney nickel, acetone–ethanol; (1) $NaIO_4$ (2) heat

(5.66)

Lewis acid-catalysed reaction of enol ethers with acid chlorides similarly gives 1,3-dicarbonyl compounds (Fleming, Iqbal and Krebs, 1983) and with acetals β-alkoxyketones are obtained. Michael addition to $\alpha\beta$-unsaturated carbonyl compounds takes place readily, giving high yields of 1,5-dicarbonyl compounds without the concomitant 1,2-addition to the carbonyl group which often takes place in reactions using lithium enolates (Mukaiyama, 1977; Brownbridge, 1983) (5.67).

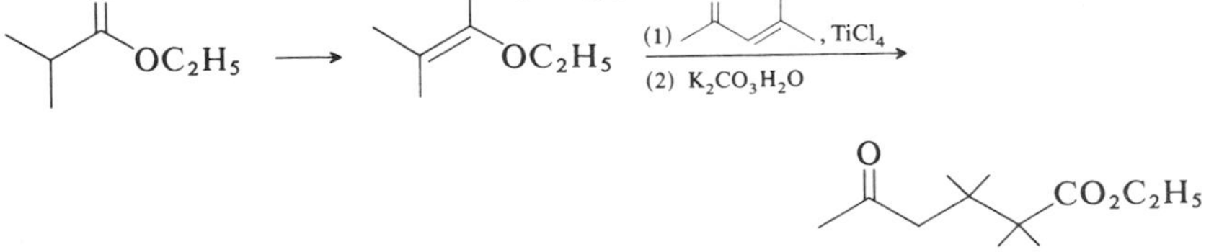

OC_2H_5 → $OSi(CH_3)_3$, OC_2H_5; (1) enone, $TiCl_4$ (2) K_2CO_3,H_2O → $CO_2C_2H_5$

(72%) (5.67)

There are numerous other synthetically useful reactions of silyl enol ethers (Brownbridge, 1983). The double bond is smoothly cleaved to give carbonyl compounds on ozonisation, and since enol ethers can be selectively produced under either thermodynamic or kinetic control, the sequence provides a useful method for the regiospecific α-cleavage of ketones. Alternatively, oxidation with dichlorodicyanobenzoquinone in the presence of collidine affords $\alpha\beta$-unsaturated ketones specifically (5.68) (Fleming and Paterson, 1979).

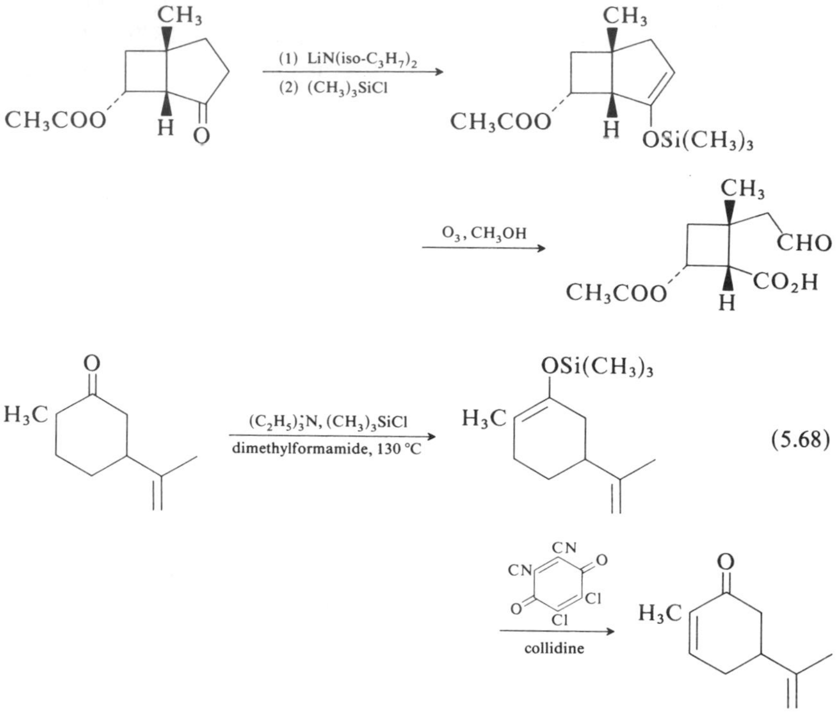

CH_3 CH_3COO H O

(1) $LiN(iso\text{-}C_3H_7)_2$
(2) $(CH_3)_3SiCl$

CH_3 CH_3COO H $OSi(CH_3)_3$

O_3, CH_3OH

CH_3 CHO CH_3COO H CO_2H

O H_3C

$(C_2H_5)_3N, (CH_3)_3SiCl$
dimethylformamide, 130 °C

$OSi(CH_3)_3$ H_3C

(5.68)

CN CN O O Cl Cl
collidine

O H_3C

Trimethylsilyl enol ethers of cyclic ketones react with dichlorocarbene to form 2-trimethylsiloxy-1,1-dichlorocyclopropane derivatives which undergo ring expansion on treatment with acid. (±)-Muscone, for example, was obtained as shown in (5.69). Another sequence leads from 1-silyloxy-1-vinylcyclopropanes to cyclobutanones or cyclopentanones, as illustrated in (5.70) (cf. Conia and Blanco, 1983). The required cyclopropanes are easily obtained from silyl enol ethers of $\alpha\beta$-unsaturated ketones by the Simmons–Smith reaction (p. 95).

5.7. Alkenylsilanes and allylsilanes

The stabilisation of carbonium ions by an adjacent carbon–silicon bond has important consequences for the reactions of organosilicon compounds and forms the basis for three synthetically useful reactions, namely the electrophilic substitution of alkenylsilanes and allylsilanes and the controlled rearrangement or cyclisation of carbonium ions.

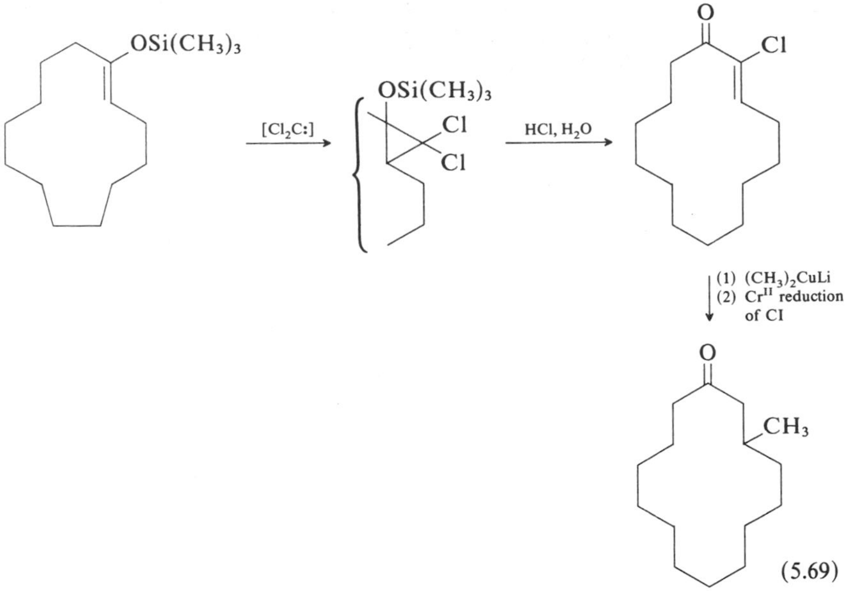

$OSi(CH_3)_3$ $\xrightarrow{[Cl_2C:]}$ $OSi(CH_3)_3$, Cl, Cl $\xrightarrow{HCl, H_2O}$ O, Cl

(1) $(CH_3)_2CuLi$
(2) Cr^{II} reduction of Cl

O, CH_3

(5.69)

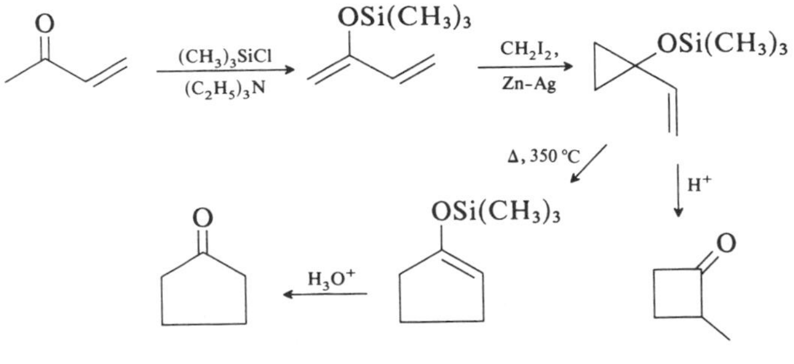

O $\xrightarrow[(C_2H_5)_3N]{(CH_3)_3SiCl}$ $OSi(CH_3)_3$ $\xrightarrow[Zn\text{-}Ag]{CH_2I_2,}$ $OSi(CH_3)_3$

Δ, 350 °C

H^+

(5.70)

O $\xleftarrow{H_3O^+}$ $OSi(CH_3)_3$ O

Alkenylsilanes are useful in synthesis because they often react with electrophiles both regio and stereospecifically, and their epoxides serve as masked aldehydes or ketones. Electrophilic substitution of alkenyltrialkylsilanes takes place readily to give, in general, substituted alkenes with replacement of the trialkylsilyl group, by way of the stabilised carbonium ion (51) (Chan and Fleming, 1979). Thus, the trimethylsilylcyclohexene derivative (52) gave the acetyl derivatives (53) on Friedel–Crafts acetylation, whereas the isomer (54) gave (55). Similarly vinyltrimethylsilane reacts with cyclopentenylcarboxylic acid chloride in the presence of tinIV chloride to give first the dienone (56) and thence in a Nazarov cyclisation, the bicyclo compound (57). In this reaction the vinylsilane acts as the synthetic equivalent of ethylene. It is a useful

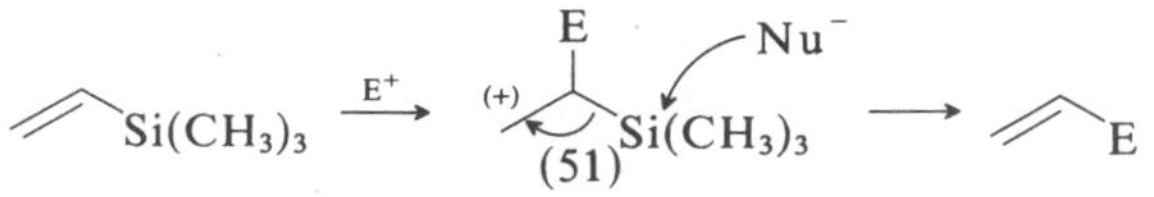

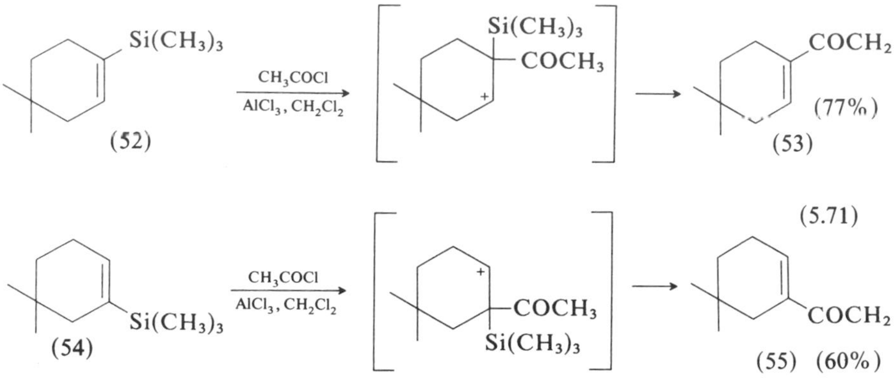

(51)

(52) (53) (77%)

(5.71)

(54) (55) (60%)

reagent for annelation of a five-membered ring at the site of an $\alpha\beta$-unsaturated acid (Cooke, Schwindeman and Magnus, 1979).

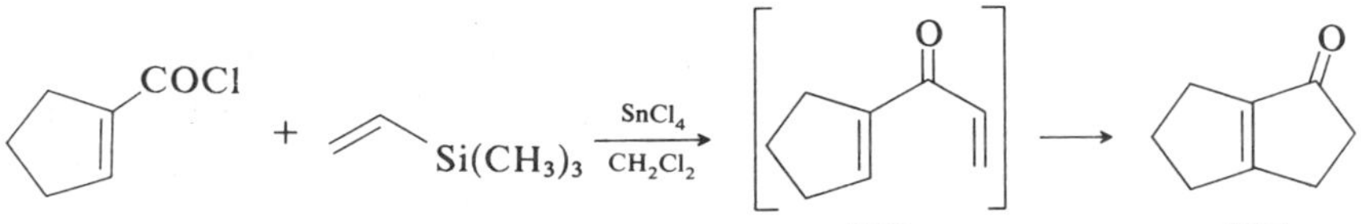

(56) (57)

(5.72)

With open-chain alkenylsilanes the reactions are often stereospecific as well as site specific and take place with retention of configuration of the double bond. Halogenation forms the main exception and may be accompanied by inversion (cf. Chan and Fleming, 1979). Thus, (E)-β-trimethylsilylstyrene on 'protonolysis' with deuterium chloride gives (E)-β-deuteriostyrene almost completely while (Z)-β-trimethylsilylstyrene yields only (Z)-β-deuteriostyrene. The stereospecificity is attributed to a bridging interaction between the silicon atom and the empty $2p_z$-orbital on the adjacent carbonium ion, or to an interaction of the p-orbital with the properly disposed carbon–silicon bond. On attack of the electrophile (D^+ in 5.73) on the double bond, rotation occurs about the developing carbon–carbon single bond in such a direction as to allow continuous interaction between the carbonium ion and the carbon–silicon bond; rotation in the opposite direction would bring the carbon–silicon bond into the nodal plane of the carbonium ion and thus interupt continuity.

Stabilisation of a carbonium ion by an adjacent carbon–silicon bond can only take place if the carbon–silicon bond lies in the same plane as

$$\xrightarrow[\text{AlCl}_3,\ \text{CH}_2\text{Cl}_2]{\text{CH}_3\text{COCl}} \quad (70\%)$$

Si(CH₃)₃ → COCH₃

(5.73)

the empty *p* orbital on the developing carbonium ion. In acyclic systems there is no difficulty since the molecule can generally adopt the required conformation, but with cyclic vinylsilanes, particularly in conformationally rigid systems, it may be geometrically difficult or impossible for the carbon–silicon bond to move into the same plane as the *p*-orbital carrying the positive charge. In such cases stabilisation of the carbonium ion will not be effected.

Another useful reaction of alkenylsilanes is their oxidation to epoxides with peroxy acids, for the resulting epoxides are readily converted into carbonyl compounds on treatment with acid under mild conditions. Reaction takes place through nucleophilic attack at the α-carbon atom by the pathway shown in (5.74) to give a product with the carbonyl group at the carbon atom which was originally attached to silicon. Methyl neopentyl ketone, for example, was obtained from the alkenylsilane (58) in 74 per cent yield.

$\alpha\beta$-Epoxysilanes undergo a variety of other useful reactions. With lithium organocuprates diasteriomerically pure β-hydroxysilanes are obtained as discussed on p. 136. β-Hydroxysilanes are also formed by reduction of the epoxides with lithium aluminium hydride by selective attack at the carbon bearing the trialkylsilyl group. With hydrobromic acid and acetic acid α-ring opening again occurs to give α-bromo- and α-acetoxy-β-hydroxysilanes which are smoothly converted into alkenyl bromides or enol acetates with high stereochemical purity, on treatment

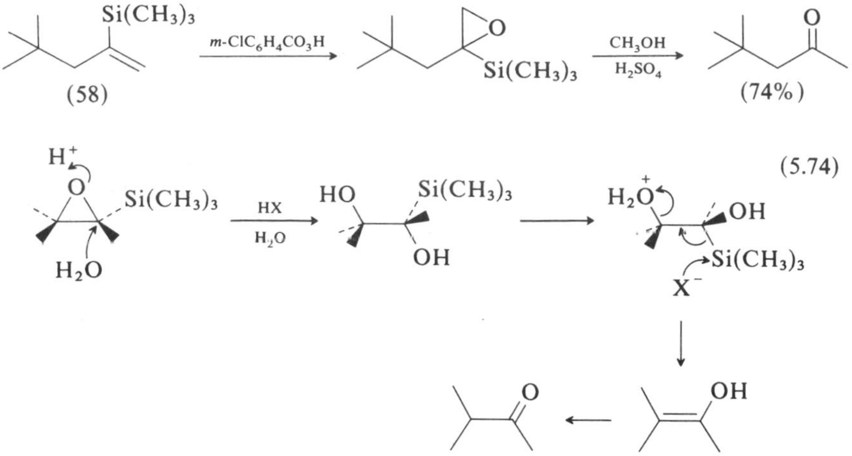

(5.74)

with boron trifluoride etherate. The overall stereochemistry is consistent with highly selective ring opening of the epoxide with inversion of configuration (5.75).

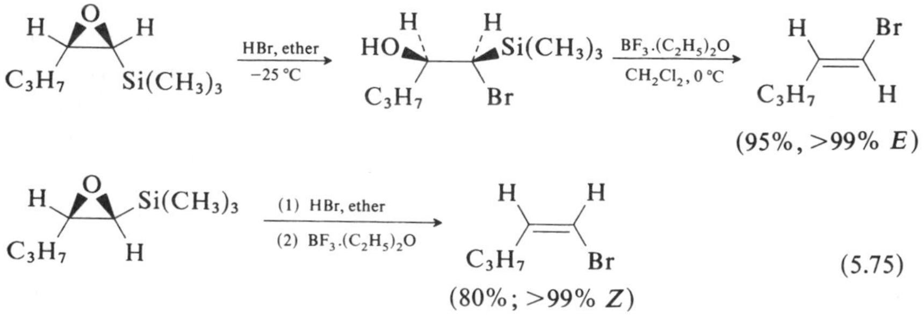

(5.75)

Two other useful synthetic sequences involving $\alpha\beta$-epoxysilanes which lead to 1,4- and 1,5-diketones are illustrated in (5.76). These reactions result, in effect, in the conjugate addition of an acyl anion, $RCO^{(-)}$, or an enolate anion, $RCOCH_2^{(-)}$, to an $\alpha\beta$-unsaturated carbonyl compound. They have the further practical advantage that the second carbonyl group is introduced in a masked form, so that the two carbonyl groups can be differentiated in subsequent reactions (Boeckman and Bruza, 1974).

The vinyltrimethylsilanes required for the above reactions are prepared in a number of ways (Chan and Fleming, 1979; Ager, 1982); alkynes, vinyl halides and ketones have all been used as starting materials. Two routes which exploit the regio and stereospecific addition of copper reagents to a terminal alkyne are illustrated in (5.77).

Alkynylsilanes, like alkenylsilanes, react with electrophiles to give, generally, substitution products (5.78).

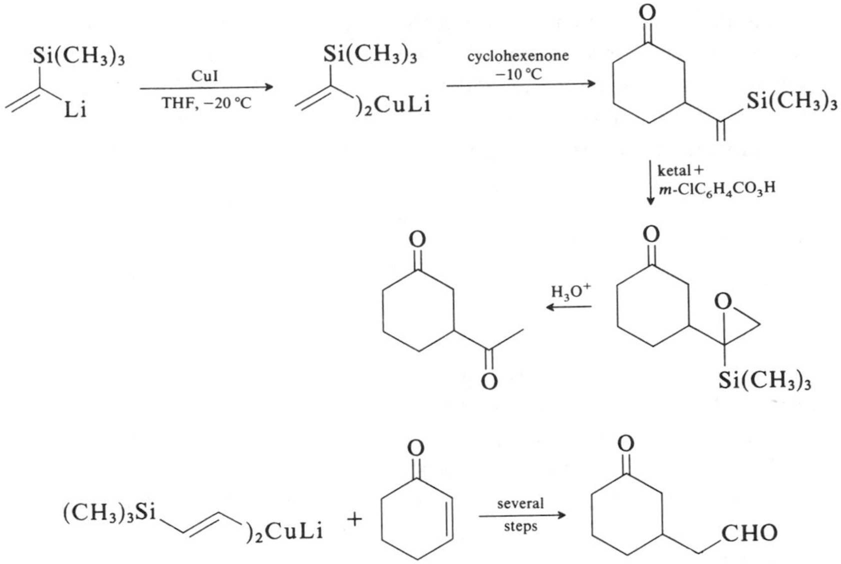

CuI, THF, −20 °C; cyclohexenone, −10 °C; ketal + m-$ClC_6H_4CO_3H$; H_3O^+

$Si(CH_3)_3$, Li, $)_2CuLi$, $Si(CH_3)_3$

$(CH_3)_3Si$ $)_2CuLi$ + several steps CHO

(5.76)

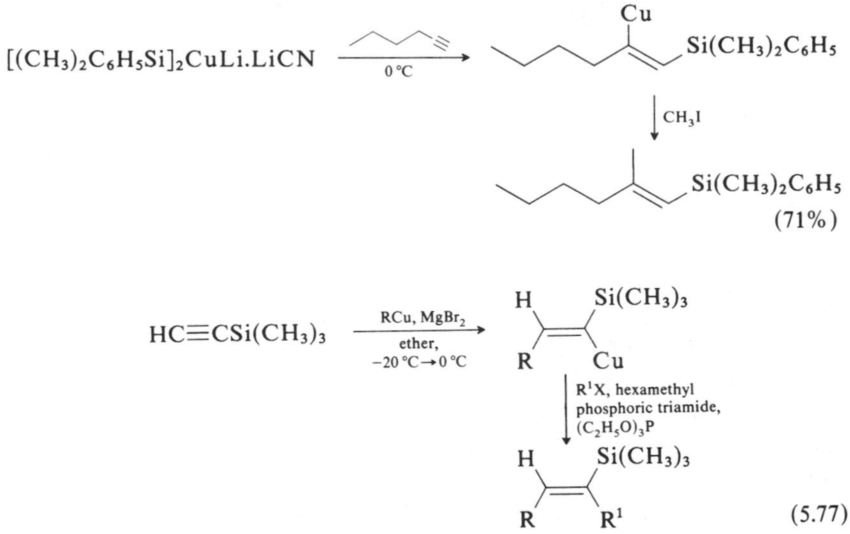

$[(CH_3)_2C_6H_5Si]_2CuLi.LiCN$ 0 °C Cu $Si(CH_3)_2C_6H_5$

CH_3I

$Si(CH_3)_2C_6H_5$ (71%)

$HC{\equiv}CSi(CH_3)_3$ RCu, $MgBr_2$, ether, −20 °C → 0 °C H $Si(CH_3)_3$ R Cu

R^1X, hexamethyl phosphoric triamide, $(C_2H_5O)_3P$

H $Si(CH_3)_3$ R R^1

(5.77)

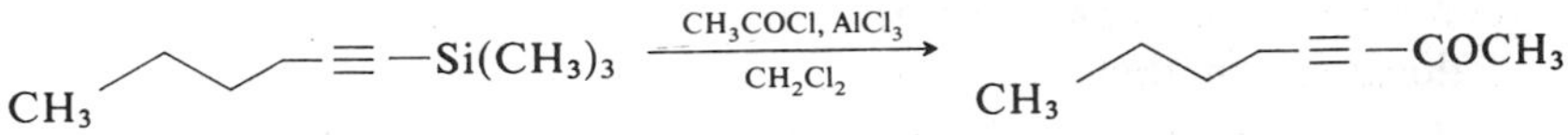

CH_3 $\equiv$ $-Si(CH_3)_3$ $CH_3COCl, AlCl_3$, CH_2Cl_2 CH_3 $\equiv$ $-COCH_3$

(5.78)

Allylsilanes also react readily with electrophiles. Here reaction takes place at the γ-carbon atom with migration of the double bond and displacement of silicon from the α-carbon (5.79). Again reaction is facilitated by stabilisation of a transient intermediate carbonium ion by the neighbouring carbon–silicon bond (Sakurai, 1982; Chan and Fleming,

$(CH_3)_3Si$ α β γ E^+ → $(CH_3)_3Si$ E Nu^- → E

(5.79)

1979). Thus, in a catalysed reaction with an acid chloride, artemisia ketone (60) was readily obtained from senecioyl chloride and trimethylisopentenylsilane (59), and the adduct (61), from Diels–Alder reaction of 1-trimethylsilylbutadiene and maleic anhydride, gave the anhydride (62) on protodesilylation, isomeric with the product obtained from maleic anhydride and butadiene itself (5.80). Alkylation can also be effected

$Si(CH_3)_3$ (59) —(1) O, Cl, $AlCl_3$, CH_2Cl_2, −60 °C (2) H_2O, NH_4Cl→ O (60) (90%)

$(CH_3)_3Si$ O O O (61) —p-$CH_3C_6H_4SO_3H$, C_6H_6, reflux→ O O O (62)

(5.80)

with t-butyl chloride and some other reactive alkyl halides, in conjunction with Lewis acids, and reaction with carbonyl compounds, triggered by Lewis acid or fluoride ion, affords homoallylic alcohols. An intramolecular example of the latter, which presumably takes place by way of the stabilised carbonium ion (63), is shown in (5.81). The first example in (5.81) shows how electrophilic substitution of the appropriate allyltrimethylsilane can be used to introduce two different alkyl substituents at the site of a carbonyl group. Conjugate addition of allylsilanes to $\alpha\beta$-unsaturated ketones takes place readily and sometimes gives better results than copper-catalysed addition of allyl Grignard reagents or addition of allylcuprates. Thus, in a recent synthesis of *Lycopodium* alkaloids the methylallyl group was introduced exclusively *trans* to the methyl substituent at C-5 in (64). Lithium bis(methylallyl)cuprate, in contrast, gave a mixture of products (Blumenkopf and Heathcock, 1983).

$(C_6H_5)_3P{=}CHCH_2Si(CH_3)_3$, THF, −78 °C → 25 °C; Si(CH₃)₃; TiCl₄, CH_2Cl_2, −78 °C (55%)

Si(CH₃)₃; CHO; $SnCl_4$, CH_2Cl_2, −78 °C or F^-, THF, 55 °C; (63); OH

(5.81)

H_3C; CN; Si(CH₃)₃; $TiCl_4$; (96%); (64) (5.82)

CH_3; $(C_4H_9)_4\overset{+}{N}\overset{-}{F}$ dimethylformamide; Si(CH₃)₃; (63%)

Again, intramolecular reactions are readily effected (Majetich, Casares, Chapman and Behnke, 1983).

An interesting new reagent is 2-bromo-3-(trimethylsilyl)propene (65), which can be used in the synthesis of α-methylene-γ-lactones from carbonyl compounds, and of methylene-five membered rings from $\alpha\beta$-unsaturated ketones (Trost and Coppola, 1982). In these reactions it is synthetically equivalent to the dianion (66). Lewis acid-catalysed addition to carbonyl groups, followed by carbonylation of the bromoalkene

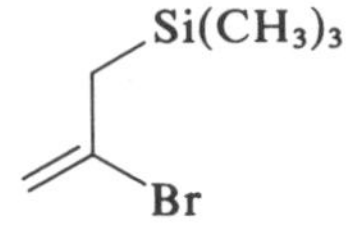

(65) 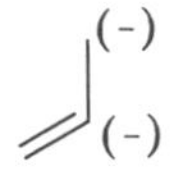 (66)

obtained with the nickel complex $(Ph_3P)_2Ni(CO)_2$, gives α-methylene-γ-butyrolactones in high yield as in (5.83). In an alternative sequence,

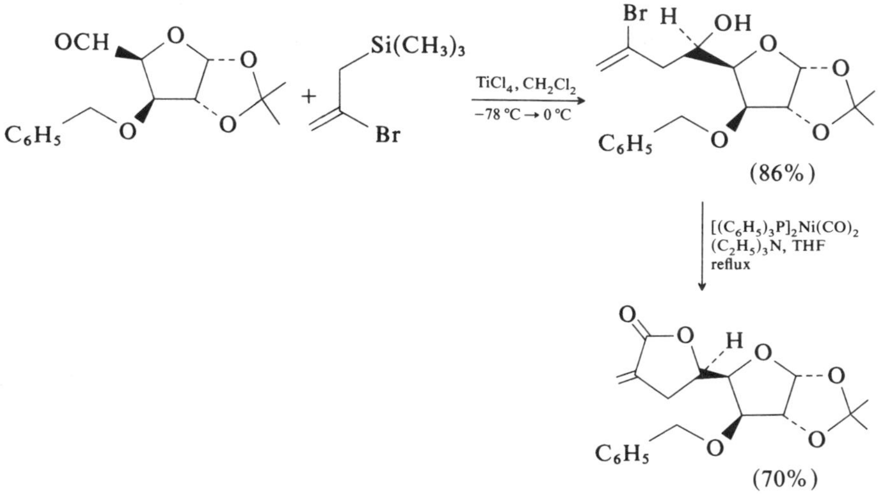

$TiCl_4$, CH_2Cl_2, −78 °C → 0 °C (86%)

$[(C_6H_5)_3P]_2Ni(CO)_2$, $(C_2H_5)_3N$, THF, reflux (70%)

(5.83)

which reverses the unmasking of the two pro-nucleophilic centres, conjugate addition of the Grignard reagent from (65) to an $\alpha\beta$-unsaturated ketone, followed by Lewis acid-catalysed addition of the resulting allylsilane to the carbonyl group, leads to methylenecyclopentanes (5.84).

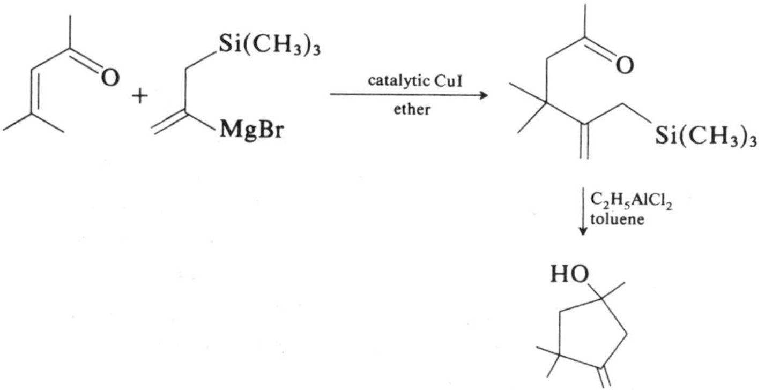

catalytic CuI, ether

$C_2H_5AlCl_2$, toluene

(5.84)

Electrophilic substitution of allylsilanes is a good method for the formation of carbon–carbon bonds, but its widespread application in synthesis is hampered by the lack of good general methods for making allylsilanes. Numerous routes to allylsilanes have been developed (Chan and Fleming, 1979), but few of them are of general application. Several more recent routes which appear to have some promise are exemplified in (5.85). Allylsilanes are obtained by conjugate addition of a silylcuprate reagent to allyl acetates or $\alpha\beta$-unsaturated esters (Fleming and Waterson, 1984; Fleming and Terrett, 1983) or by palladium-catalysed reaction of trimethylsilylmethylmagnesium chloride with iodoalkenes (Negishi, Luo and Rand, 1982). Excellent yields of allylsilanes have also been obtained by reaction of the anion of 1-benzenesulphonyl-2-trimethylsilylethane with aldehydes and ketones with subsequent reductive cleavage of the α-hydroxysulphone with sodium amalgam (p. 144) (Hsiao and Shechter, 1982).

CH_3COO ; $[(CH_3)_2C_6H_5Si]_2CuLi$, 0 °C ; $Si(CH_3)_2C_6H_5$ (95%)

C_6H_{13}, H, H_3C, I ; $(CH_3)_3SiCH_2MgCl$, $Pd[P(C_6H_5)_3]_4$, THF, 0 °C ; C_6H_{13}, H, H_3C, $CH_2Si(CH_3)_3$ (85%) (5.85)

$(CH_3)_3Si$ $SO_2C_6H_5$; (1) C_4H_9Li (2) O ; $(CH_3)_3Si$ $SO_2C_6H_5$ OH ; (1) CH_3SO_2Cl (2) Na–Hg ; $Si(CH_3)_3$ (92%)

5.8. Control of rearrangement of carbonium ions

The stabilisation of a carbonium ion centre by an adjacent carbon–silicon bond (sometimes called the β-effect) can be used to control the course of carbonium ion rearrangements and carbonium ion-induced cyclisations (Fleming, 1981). Thus, in a neat synthesis of the 7-hydroxynorbornene (69) rearrangement of the bromolactone (67), catalysed by silver ion, led smoothly to the lactone (68) and thence to

(69). The reaction did not proceed in the absence of the trimethylsilyl group. It facilitates the rearrangement by stabilisation of the positive charge developing at C-4 in (67), and the outcome is controlled by the ready loss of the silyl group (Fleming and Michael, 1981). In the cyclisation mode, the acetal (70) gave only the alkene (71) on treatment with stannic chloride. The corresponding methyl derivative (70, H for $SiMe_3$),

(5.86)

in contrast, gave a mixture of the three possible double bond isomers (Fleming and Pearce, 1981; Chow and Fleming, 1984). In the nitrogen series the imine (72) gave exclusively the bicyclic (73) (Overman and Burk, 1984). Another useful application is in the control of the position of the double bond in conjugated cyclopentenones produced in the

(5.87)

well-known Nazarov cyclisation of dienones. In the classical reaction the double bond normally resides in the thermodynamically more stable position with the highest degree of substitution. Judicious placement of a trimethylsilyl group directs introduction of the new double bond to the less substituted position, away from the ring fusion in bicyclic compounds (5.88) (Denmark and Jones, 1982). However, the 'β effect'

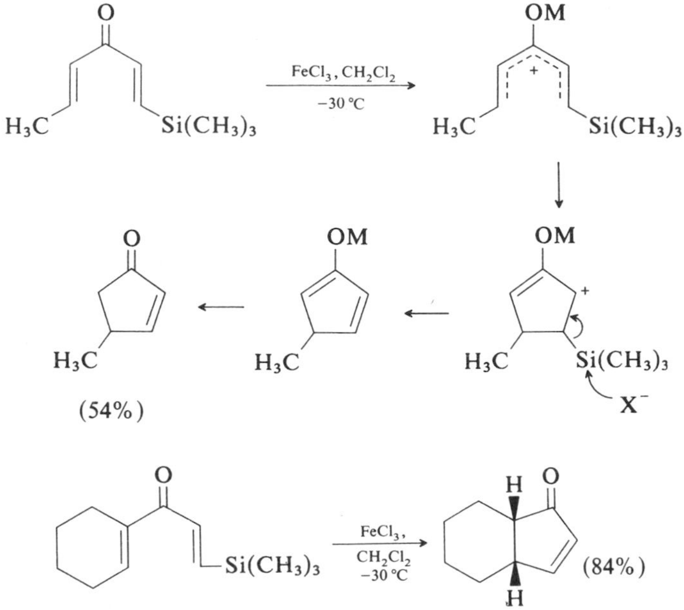

$FeCl_3$, CH_2Cl_2, −30 °C

(54%)

(5.88)

$FeCl_3$, CH_2Cl_2, −30 °C

(84%)

is a weak effect and is easily overridden. For example, Lewis acid-catalysed cyclisation of the allylsilane (74) affords not (77) as might have been expected, but the cyclohexenone (76), formed by way of the tertiary carbonium ion (75). Formation of (77) would require the primary carbonium ion (78) (5.89) (Mikami, Kishi and Nakai, 1983).

5.9. α-Silyl carbanions

Because of $(p-d)\pi$ back-bonding between silicon and carbon, silicon can stabilise an adjacent carbanion. Thus, tetramethylsilane can be deprotonated using n-butyl-lithium in tetrahydrofuran containing tetramethylethylenediamine to give α-lithiomethyltrimethylsilane. In practice the carbanion is usually flanked by another electron-withdrawing group, and α-silyl carbanions produced in this manner have been widely employed in the synthesis of heterosubstituted alkenes by the Peterson alkene synthesis (5.90) already discussed on p. 135.

COCl; $Si(CH_3)_3$; H_3C (74) $\xrightarrow[CH_2Cl_2]{AlCl_3}$

[H_3C, $Si(CH_3)_3$ (75) → H_3C $Si(CH_3)_3$ → CH_3] ↓

CH_3 (77) $\quad$ $Si(CH_3)_3$, CH_2^+, CH_3 (78) $\quad$ CH_3 (76) (76%)

(5.89)

$$R^1R^2C{=}O + Li{-}CH(Z)Si(CH_3)_3 \longrightarrow R^1R^2C(O^-Li^+){-}CH(Z)Si(CH_3)_3 \qquad (5.90)$$

$$\longrightarrow R^1R^2C{=}CHZ + (CH_3)_3SiOLi$$

(Z = e.g. C_6H_5, $Si(CH_3)_3$, SCH_3)

A useful reagent is α-chlorotrimethylsilylmethyl-lithium (79), which is readily obtained from α-chloromethyltrimethylsilane with s-butyl-lithium. It reacts with aldehydes and ketones to form αβ-epoxysilanes by way of the corresponding chlorohydrins which, surprisingly, do not form the chloro-alkene by elimination of trimethylsilanoxide as in (5.90) above. The αβ-epoxysilanes formed in this way are readily converted into aldehydes (p. 327) and the sequence thus provides a method for the homologation of aldehydes and ketones (Burford *et al.*, 1983). The related reagent α-methyl-α-chlorotrimethylsilylmethyl-lithium (80), can be used similarly to convert aldehydes and ketones into homologous methyl ketones (Cooke and Magnus, 1977).

A method for the construction of γ-lactones at the site of the carbonyl group of an aldehyde or ketone makes use of the (trimethylsilyl)allyl

$(CH_3)_3Si$ (79) ... R^1COR^2 ... $(CH_3)_3Si$ O^- R^1 Cl R^2 → $(CH_3)_3Si$ R^1 H O R^2

$(CH_3)_3Si$ ... LiCHClSi$(CH_3)_3$, THF, −55 °C; $HClO_4$, THF → CHO (>95%)

$(CH_3)_3SiC(Cl)(CH_3)$—Li (80); H_3C, $(CH_3)_3Si$; H_2SO_4, H_2O, CH_3OH → H_3C O (60%) (5.91)

anion, formed from allyltrimethylsilane with s-butyl-lithium. It reacts with aldehydes and ketones to give adducts formed by attack at the γ-carbon atom. Treatment of the derived epoxysilanes with methanol and boron trifluoride etherate gives the corresponding lactol methyl ethers which on Jones oxidation provide lactones (Ehlinger and Magnus, 1980). Thus, cyclopentanone gives the lactone (82) (5.92). In these reactions the (trimethylsilyl)allyl anion is synthetically equivalent to the homoenolate anion (81).

Si$(CH_3)_3$ — s-C_4H_9Li, THF, −76 °C → Li^+ Si$(CH_3)_3$ ≡ $^{(-)}$ CO_2H (81)

=O — Li^+ Si$(CH_3)_3$, THF, −30 °C → OH Si$(CH_3)_3$

m-$ClC_6H_4CO_3H$, CH_2Cl_2 → OH Si$(CH_3)_3$ O

CH_3OH, $BF_3.(C_2H_5)_2O$ → O OCH_3

Jones oxidation → O O (82) (5.92)

The ability of silicon to stabilise an adjacent carbanion has been exploited in an improved method for the annelation of ketones, that is for building a ring system on to an already existing carbon framework.

Annelation of 2-alkylcyclohexanones with methyl vinyl ketone and its homologues is an important route to fused polycyclic systems, for example in the steroid series (5.93). Reaction at the less substituted

R R R
+ → → (5.93)
O O O O O
(83)

α-carbon of the ketone can be effected using the enamine (p. 26), but the basic conditions required for reaction at the more substituted α-carbon generally result in polymerisation of the vinyl ketone, and low yields of the desired products are often obtained. Other difficulties are caused by the fact that the Michael adduct (83) and the original cyclohexanone have similar acid strengths and reactivities so that competitive reaction of the product with the vinyl ketone can ensue, and in reactions requiring the thermodynamically less stable enolate equilibration to the more stable isomer can take place by interaction with (83). The result is that, in addition to poor yields, mixtures of structural isomers may be obtained.

Many of these difficulties can be avoided by using α-silylated vinyl ketones instead of the vinyl ketones themselves. The silyl derivatives react readily with cyclohexanones under aprotic conditions to give high yields of condensation products. Thus, reaction of the lithium enolate of cyclohexanone with the silylated vinyl ketone (84) and cyclisation of the product with sodium methoxide in methanol, with simultaneous cleavage of the silyl group, gave the octalone (85) in 80 per cent yield (5.94). The corresponding reaction with methyl vinyl ketone itself gave less than 5 per cent of the Michael addition product. The trialkylsilyl group stabilises the anion of the Michael adduct relative to that of the starting ketone, thus facilitating the forward reaction and at the same

$Si(C_2H_5)_3$
+ $\xrightarrow[-78\,°C \to 0\,°C]{THF}$
OLi O
(84)

$Si(C_2H_5)_3$ $\xrightarrow{NaOCH_3,\ CH_3OH}$ (80%)
O O O
(85) (5.94)

time discourages equilibration of positionally unstable enolates. The steric effect of the large trialkylsilyl group may also play a part by preventing unwanted reactions at the α-carbon of the Michael adduct (Stork and Ganem, 1973; Stork and Singh, 1974; Boeckman, 1983).

5.10. β-Silylcarbonyl compounds

β-Silylcarbonyl compounds serve as masked enones, and the way in which they can be used in the synthesis of α-methylene-ketones and -lactones and for α-alkylation of $\alpha\beta$-unsaturated ketones is illustrated in (5.95). They can be prepared, among other methods, by conjugate addition of dimethylphenylsilyl-lithium to $\alpha\beta$-unsaturated

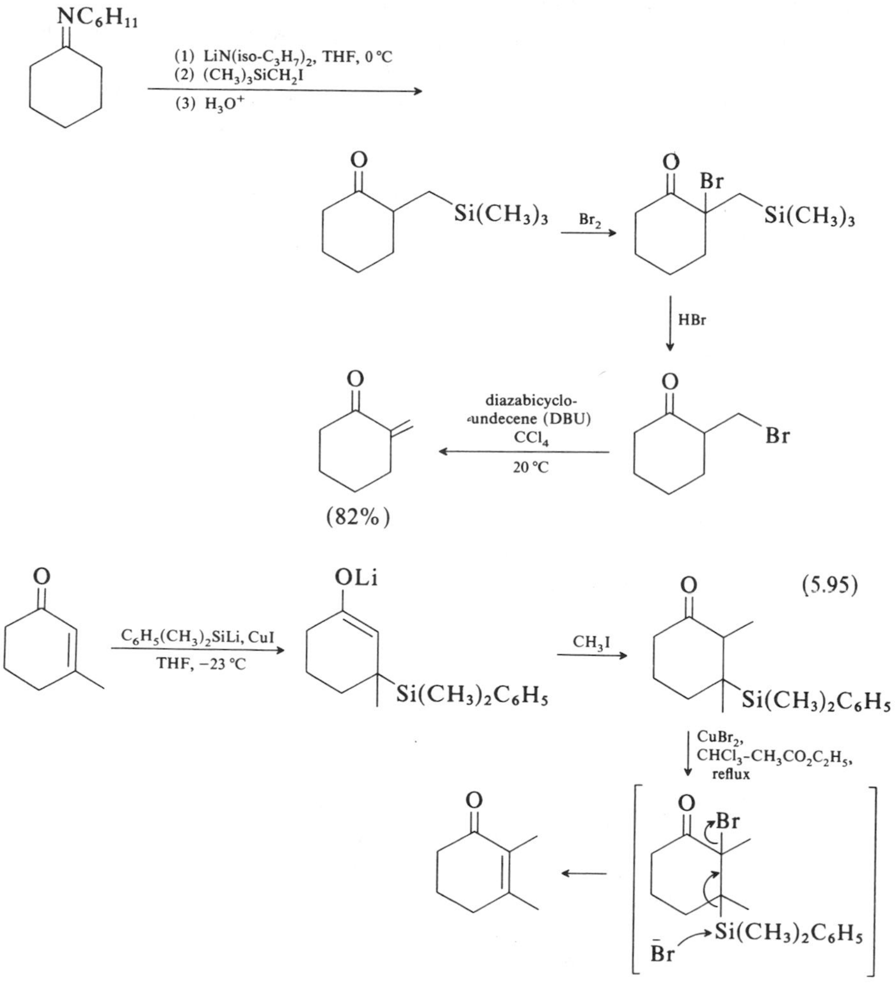

NC_6H_{11}

(1) $LiN(iso\text{-}C_3H_7)_2$, THF, 0 °C
(2) $(CH_3)_3SiCH_2I$
(3) H_3O^+

O $Si(CH_3)_3$ Br_2 → O Br $Si(CH_3)_3$

HBr

O Br

diazabicyclo-undecene (DBU) CCl_4 20 °C

O

(82%)

O $C_6H_5(CH_3)_2SiLi$, CuI THF, −23 °C → OLi $Si(CH_3)_2C_6H_5$ CH_3I → O $Si(CH_3)_2C_6H_5$ (5.95)

$CuBr_2$, $CHCl_3$-$CH_3CO_2C_2H_5$, reflux

O ← [O Br $Si(CH_3)_2C_6H_5$ $\bar{Br}$]

aldehydes and ketones in the presence of copper(I) iodide, or by alkylation of an enolate, or its equivalent, with trimethylsilylmethyl iodide. The enone system is unmasked by bromination–desilylbromination by reaction with bromine–hydrogen bromide and subsequent dehydrobromination or, in some cases, by reaction with copper(II) bromide (Ager and Fleming, 1978; Fleming and Goldhill, 1980).

α-Trialkylsilyl-aldehydes and -ketones are also valuable adjuncts in synthesis. In the Peterson alkene synthesis (p. 135) they are used to make alkenes through initial reaction with organolithium or Grignard reagents, or by reduction with lithium aluminium hydride and subsequent elimination of a trialkylsilanol (Hudrlik and Peterson, 1975; Utimoto *et al.*, 1976). In these reactions α-trimethylsilyl ketones serve effectively as tertiary vinyl cation equivalents (5.96), but attempts to use α-silyl aldehydes in the same way as secondary vinyl cation equivalents have been frustrated by lack of suitable methods for preparing α-silyl

$(CH_3)_3Si$–CH_2–C(=O)R $\xrightarrow{Nu^-}$ $(CH_3)_3Si$–CH_2–C(OH)(Nu)R $\longrightarrow$ CH_2=C(Nu)R

H_3C–CH=NC_6H_{11} $\xrightarrow[\text{(2) t-}C_4H_9(CH_3)_2SiCl]{\text{(1) LiN(iso-}C_3H_7)_2\text{, THF, }-78\,°C\rightarrow 25\,°C}$ t-$C_4H_9(CH_3)_2Si$–CH_2–CH=NC_6H_{11}

→ (1) $LiN(iso\text{-}C_3H_7)_2$ (2) $C_6H_{13}Br$ (3) H_2O, CH_3CO_2H → C_6H_{13}–CH($Si(CH_3)_2C_4H_9$-t)–CHO

→ $H_2O, CH_3CO_2H, CH_2Cl_2$ → t-$C_4H_9(CH_3)_2Si$–CH_2–CHO (86)

CH_3–C(=O)–OC_4H_9-t $\xrightarrow[\text{(2) t-}C_4H_9(CH_3)_2Si\text{–}CH_2\text{–CHO}]{\text{(1) LiN(iso-}C_3H_7)_2\text{, THF, }-78\,°C}$ t-$C_4H_9(CH_3)_2Si$–CH_2–CH(OH)–CH_2–C(=O)–OC_4H_9-t

$\xrightarrow{BF_3.(C_2H_5)_2O,\ CH_2Cl_2}$ CH_2=CH–CH_2–C(=O)–OC_4H_9-t (64%)

(5.96)

aldehydes. This limitation has been overcome by the finding that α-t-butyldimethylsilyl aldehydes are readily obtained by careful hydrolysis of the corresponding imines (5.96). α-t-Butyldimethylsilyl acetaldehyde (86) can be used to introduce a vinyl group in the α-position to the carbonyl group of both ketones and esters by reaction with the corresponding lithium enolates (Hudrlik and Kulkarni, 1981).

5.11. Trimethylsilyl cyanide

Trimethylsilyl cyanide is a useful silicon reagent which reacts readily with aldehydes and ketones in the presence of catalytic amounts of Lewis acids or of cyanide ion, to give the trimethylsilyl ethers of the corresponding cyanohydrins (Evans, Carroll and Truesdale, 1974). Even

$$^{-}CN + \rangle{=}O \xrightarrow{(CH_3)_3SiCN} \rangle C(CN)O^{-} \quad (CH_3)_3Si{-}CN \longrightarrow \rangle C(CN)OSi(CH_3)_3 + \bar{C}N \qquad (5.97)$$

normally unreactive ketones react readily with trimethylsilyl cyanide due to the formation of the strong Si—O bond which displaces the equilibrium in favour of the derivative. The reaction provides a valuable alternative to the base-catalysed addition of hydrogen cyanide to carbonyl compounds which often gives only poor yields. Tetralone, for example, is reported not to form a cyanohydrin, but it gives a trimethylsilyl derivative in excellent yield. The silylated cyanohydrins can be hydrolysed to α-hydroxy acids (Corey, Crouse and Anderson, 1975) and on reduction with lithium aluminium hydride they afford the corresponding β-amino alcohols in excellent yield. This sequence provides a better route to these valuable intermediates (they are used in the ring expansion of cycloalkanones) than the classical methods through reaction of hydrogen cyanide or nitromethane with the carbonyl compound. The derivatives from aromatic aldehydes are excellent acyl anion equivalents and have been used in 'umpolung' conversion of aldehydes into ketones and acyloins by reaction of the derived anions with alkyl halides and aldehydes or ketones (Deuchert *et al.*, 1979; Hünig and Wehner, 1979).

$$C_6H_5CH(OSi(CH_3)_3)CN \xrightarrow[THF,\ -78\,^{\circ}C]{LiN(iso\text{-}C_3H_7)_2} C_6H_5C^{(-)}(OSi(CH_3)_3)CN \xrightarrow[(2)\ H_2O]{(1)\ C_6H_5COCH_3} \underset{(85\%)}{C_6H_5COC(OH)(CH_3){-}C_6H_5} \qquad (5.98)$$

$\alpha\beta$-Unsaturated carbonyl compounds and p-quinones react exclusively at the carbonyl group; no 1,4-addition compounds are formed. This contrasts markedly with the base-catalysed addition of hydrogen cyanide to these compounds which generally gives the 1,4-addition products. In the reaction with quinones selective protection of one of the carbonyl groups can be achieved; subsequent transformations lead to p-quinols in yields much superior to those obtained by any other method (5.99) and to substituted quinones. The sequence has been exploited to prepare

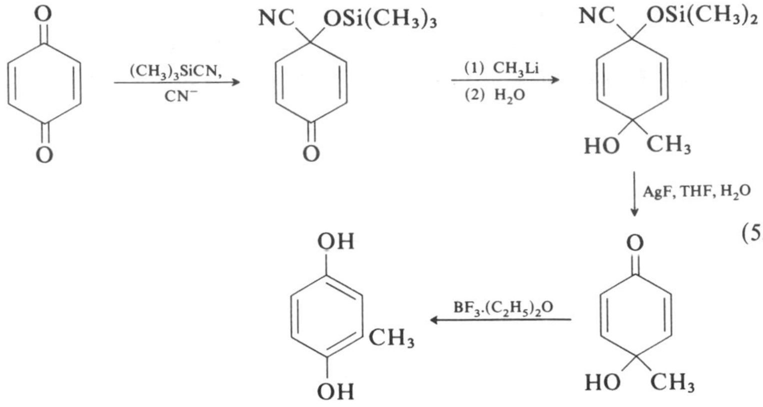

(5.99)

biologically important isoprenylquinones (Evans and Hoffmann, 1976; see also Hart, Cain and Evans, 1978).

5.12. Trimethylsilyl iodide and trimethylsilyl triflate

Two other useful silicon reagents are trimethylsilyl iodide and trimethylsilyl trifluoromethanesulphonate(triflate). The iodide is commercially available, but it is expensive and very sensitive to light and moisture and is probably best prepared *in situ* from reaction of chlorotrimethylsilane and sodium iodide in acetonitrile solution (Olah *et al.*, 1979). It is an excellent reagent for the cleavage of ethers, for the conversion of acetals and ketals into the corresponding carbonyl compounds and for the transesterification and cleavage of carboxylic esters under mild and neutral conditions. Sulphoxides are readily de-oxygenated to sulphides at room temperature and alcohols are converted into iodides in good yield (Schmidt, 1981). A convenient and inexpensive alternative to iodotrimethylsilane is provided by trichloromethylsilane and sodium iodide in acetonitrile. It is more selective than iodotrimethylsilane for the cleavage of ethers, esters and lactones (Olah *et al.*, 1983).

Trimethylsilyl triflate is a powerful silylating agent for organic compounds and acts as a catalyst for a variety of nucleophilic reactions in aprotic media (Noyori, Murata and Suzuki, 1981, Emde *et al.*, 1982). Excellent results in the silylation of carboxylic acids, alcohols, phenols and mercaptans have been obtained with a reagent prepared *in situ* from allyltrimethylsilane and trifluoromethanesulphonic acid (Olah *et al.*, 1981). Among catalysed nucleophilic reactions carbonyl compounds smoothly give acetals and thio-acetals by action of the corresponding trimethylsilyl ether or thioether, and epoxides give allylic alcohols (5.100). The latter transformation takes place by overall *trans* addition of trimethylsilyl triflate to the epoxide followed by *anti* elimination of trifluoromethanesulphonic acid away from the trimethylsiloxy group. Neighbouring double bonds may participate in the ring opening, as in (87) → (88).

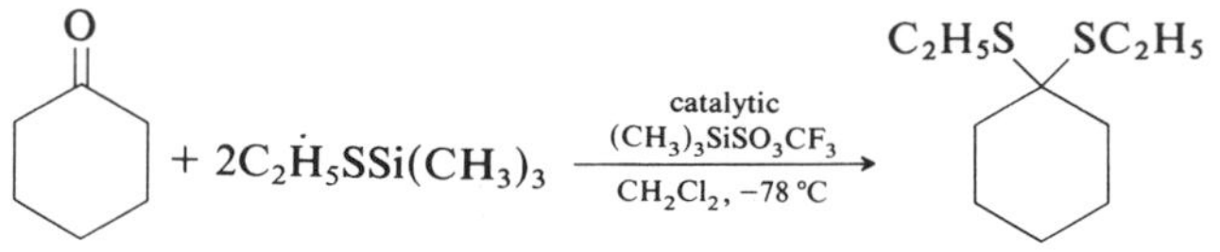

O

+ $2C_2H_5SSi(CH_3)_3$ —catalytic $(CH_3)_3SiSO_3CF_3$, CH_2Cl_2, −78 °C→ C_2H_5S SC_2H_5

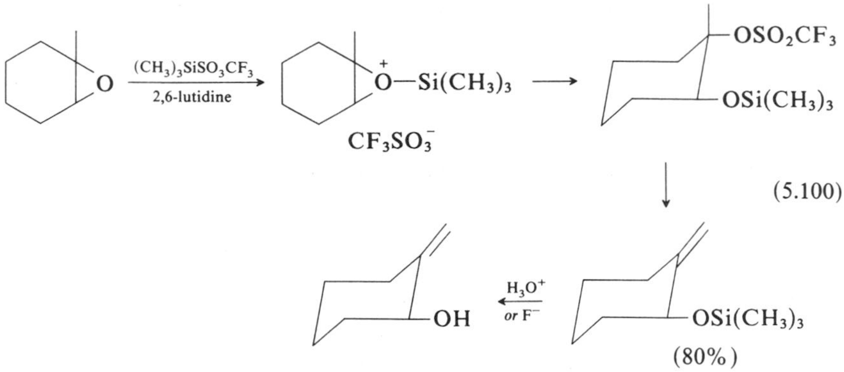

O —$(CH_3)_3SiSO_3CF_3$, 2,6-lutidine→ $\overset{+}{O}$—$Si(CH_3)_3$ $CF_3SO_3^-$ → OSO_2CF_3 $OSi(CH_3)_3$

(5.100)

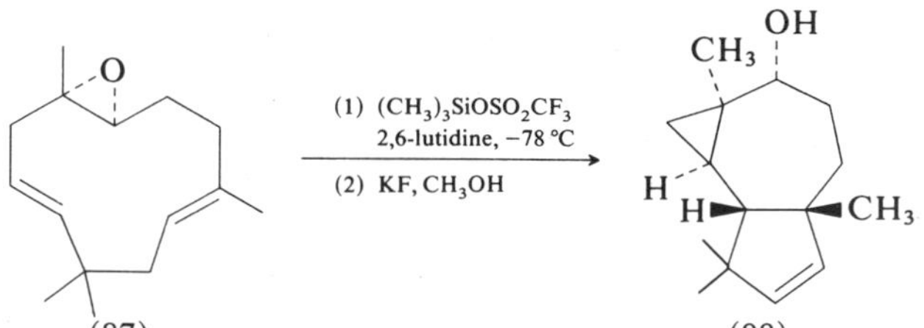

OH ←H_3O^+ or F^-— $OSi(CH_3)_3$ (80%)

O (87) —(1) $(CH_3)_3SiOSO_2CF_3$, 2,6-lutidine, −78 °C; (2) KF, CH_3OH→ (88)

OH, CH_3, H, H, CH_3

6 Oxidation

For practical purposes most organic chemists mean by 'oxidation' either addition of oxygen to the substrate, as in the conversion of ethylene into ethylene oxide, removal of hydrogen as in the oxidation of ethanol to acetaldehyde or removal of one electron as in the conversion of phenoxide anion to the phenoxy radical.

Of the wide variety of agents available for the oxidation of organic compounds, probably the most widely used are potassium permanganate and derivatives of hexavalent chromium. Permanganate, a derivative of heptavalent manganese, is a very powerful oxidant. Its reactivity depends to a great extent on whether it is used under acid, neutral or basic conditions. In acid solution it is reduced to the divalent manganeseII ion Mn^{2+} with net transfer of five electrons (MnVII → MnII), while in neutral or basic media manganese dioxide is usually formed, corresponding to a three electron change (MnVII → MnIV). Permanganate is generally used in aqueous solution and this restricts its usefulness since not many organic compounds are sufficiently soluble in water and only a few organic solvents are resistant to the oxidising action of the reagent. Solutions in acetic acid, t-butanol or dry acetone or pyridine can sometimes be employed.

Alternatively, oxidation with aqueous solutions of permanganate can be effected in the presence of a crown ether, dicyclohexano-18-crown-6 is often used, or of certain tetra-alkylammonium or phosphonium salts, known as phase-transfer reagents. Under these conditions the permanganate is soluble in benzene and the resulting solutions are excellent reagents for the oxidation of a variety of organic substrates. Thus, in the presence of dicyclohexano-18-crown-6, alkenes, alcohols, aldehydes and alkylbenzenes are rapidly oxidised to carboxylic acids in high yield at room temperature. Toluene gives benzoic acid in 78 per cent yield and α-pinene is converted into pinonic acid in 90 per cent yield. Similarly, heptanoic acid is obtained from 1-octene in 90 per cent yield with aqueous permanganate and benzene in the presence of cetyltrimethylammonium chloride and 1-eicosene gives nonadecanoic acid in 75 per cent yield

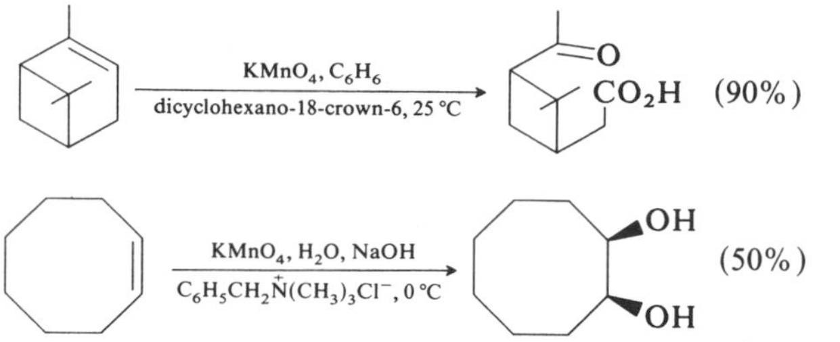

$$\xrightarrow[\text{dicyclohexano-18-crown-6, 25 °C}]{KMnO_4,\ C_6H_6} \quad CO_2H \ (90\%)$$

$$\xrightarrow[C_6H_5CH_2\overset{+}{N}(CH_3)_3Cl^-,\ 0\ °C]{KMnO_4,\ H_2O,\ NaOH} \quad OH,\ OH \ (50\%) \qquad (6.1)$$

(Lee, Lamb and Chang, 1981). The catalytic action of the quaternary salts is believed to be due to the ability of their organic-soluble cations to transfer anions (e.g. MnO_4^-) from the aqueous into the organic phase. With crown ethers an organic-soluble complex is formed. An alternative is to use permanganate salts which are soluble in organic solvents, such as tetra-n-butylammonium permanganate (Sala and Sargent, 1978).

Chromic acid, a derivative of hexavalent chromium, is one of the most versatile of the available oxidising agents, and reacts with almost all types of oxidisable groups. The reactions can often be controlled to yield largely one product, and for this reason chromic acid oxidation is a useful process in synthesis. In oxidations chromium is reduced from the hexavalent to the trivalent state (CrVI → CrIII) with production of a chromiumIII salt. The commonest reagents are chromiumVI oxide and sodium or potassium dichromateVI. ChromiumVI oxide is a polymer which dissolves in water with depolymerisation to form chromic acid. It

$$(CrO_3)_n + H_2O \rightarrow HO{-}\overset{\overset{\displaystyle O}{\|}}{\underset{\underset{\displaystyle O}{\|}}{Cr}}{-}OH$$

is commonly employed in solution in dilute sulphuric acid, sometimes containing acetic acid to aid dissolution of the substrate; a comparable solution is obtained by adding sodium or potassium dichromateVI to aqueous sulphuric acid. These solutions contain an equilibrating mixture of the acid chromateVI and the dichromateVI ions. ChromiumVI oxide

$$2HCrO_4^- \rightleftharpoons H_2O + Cr_2O_7^{2-}$$

may also be used in solution in acetic anhydride, t-butanol or in pyridine. In these solutions the reactive species present are chromyl acetate, t-butyl chromateVI and the pyridine–chromiumVI oxide complex, as shown (6.2).

The commonest lower oxidation state of chromium is CrIII and most oxidations with chromic acid lead to this state by a net transfer of three electrons. However, each stage in the oxidation of most organic

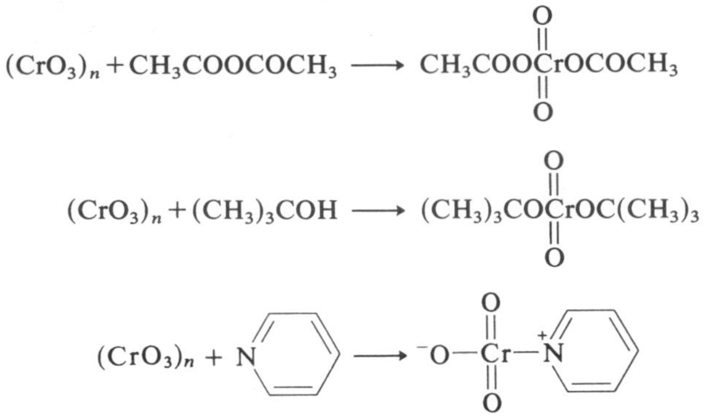

$$(CrO_3)_n + CH_3COOCOCH_3 \longrightarrow CH_3COO\overset{\overset{\displaystyle O}{\|}}{\underset{\underset{\displaystyle O}{\|}}{Cr}}OCOCH_3$$

$$(CrO_3)_n + (CH_3)_3COH \longrightarrow (CH_3)_3CO\overset{\overset{\displaystyle O}{\|}}{\underset{\underset{\displaystyle O}{\|}}{Cr}}OC(CH_3)_3 \qquad (6.2)$$

$$(CrO_3)_n + C_5H_5N \longrightarrow {}^{-}O-\overset{\overset{\displaystyle O}{\|}}{\underset{\underset{\displaystyle O}{\|}}{Cr}}-\overset{+}{N}C_5H_5$$

compounds involves transfer of only two electrons, and it is evident that most reactions must give rise to CrV or CrIV species as intermediates, as exemplified in the following scheme for the oxidation of a secondary alcohol. The CrIV and CrV ions are themselves powerful oxidants and may give rise to unwanted side reactions and to products different from those formed in the original oxidation by CrVI.

$$\mathrm{CrVI} + R_2CHOH \rightarrow R_2C{=}O + 2H^+ + \mathrm{CrIV}$$
$$\mathrm{CrIV} + \mathrm{CrVI} \rightarrow 2\mathrm{CrV}$$
$$\mathrm{CrV} + R_2CHOH \rightarrow R_2C{=}O + 2H^+ + \mathrm{CrIII}$$

6.1. Oxidation of hydrocarbons

Alkanes

Under vigorous conditions both chromic acid and permanganate attack alkanes, but the reaction is of little synthetic use for usually mixtures of products are obtained in low yield. The reaction is of importance in the Kuhn–Roth estimation of methyl groups. This depends on the fact that a methyl group is rarely attacked (the relative rates of oxidation of primary, secondary and tertiary C—H bonds are 1:110:7000) and is eventually converted into acetic acid. The usual method is to boil the substance with chromic acid in aqueous sulphuric acid and determine the amount of steam volatile acid formed (almost entirely acetic acid) by titration with alkali. The sesquiterpene hydrocarbon cadinene, for example, under these conditions gives three molecules of acetic acid, one from each of the ring methyl substituents and one other from the isopropyl group (6.3). Under less vigorous conditions intermediate oxidation products can sometimes be isolated, and this may be useful in degradative work. Controlled oxidation of 'unactivated' saturated CH_3, CH_2 and CH groups is common in nature under the influence of oxidising enzymes (cf. Ullrich, 1972) but there are very few methods for effecting controlled reactions of this kind in the laboratory (see p. 269).

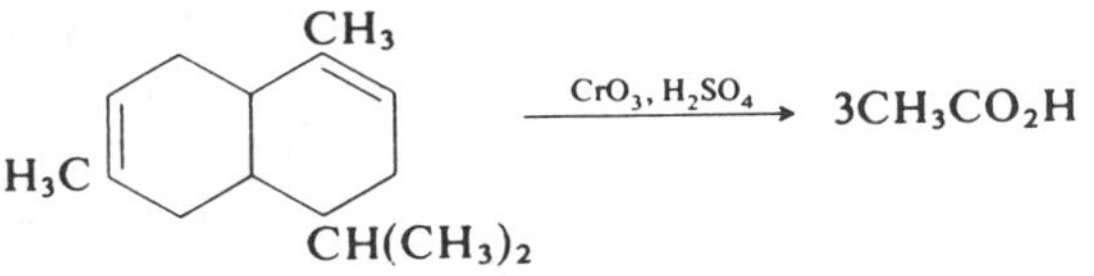

$$\text{(H}_3\text{C, CH}_3\text{, CH(CH}_3)_2\text{-substituted hexahydronaphthalene)} \xrightarrow{CrO_3,\ H_2SO_4} 3CH_3CO_2H \qquad (6.3)$$

Aromatic hydrocarbons

In the absence of activating hydroxyl or amino substituents benzene rings are only slowly attacked by chromic acid or permanganate, but alkyl side chains are degraded with formation of benzene carboxylic acids (6.4). This is a useful method for the preparation of benzene

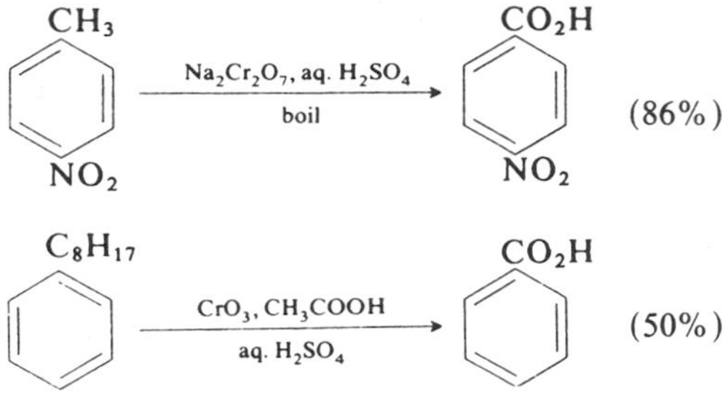

$$CH_3C_6H_4NO_2 \xrightarrow[\text{boil}]{Na_2Cr_2O_7,\ aq.\ H_2SO_4} HO_2CC_6H_4NO_2 \quad (86\%)$$

$$C_8H_{17}C_6H_5 \xrightarrow[aq.\ H_2SO_4]{CrO_3,\ CH_3COOH} C_6H_5CO_2H \quad (50\%) \qquad (6.4)$$

carboxylic acids, and also for determining the orientation pattern of unknown polyalkylbenzenes which are degraded to benzene polycarboxylic acids of known orientation. Nuclear hydroxyl or amino substituents, if present, must be converted into their methyl ethers or acetyl derivatives, for otherwise they activate the ring to attack, and quinones or, with excess of reagent, carbon dioxide and water are formed. With side chains longer than methyl, initial attack is thought to take place at the benzylic carbon atom. This is suggested by the fact that t-butylbenzene is very resistant to oxidation and ethylbenzene gives some acetophenone as well as benzoic acid. The rate-determining step in these chromic acid oxidations is known to be cleavage of the benzylic C—H bond, but it is uncertain at present whether attack on the benzylic hydrogen involves initial removal of a hydrogen atom to form a radical, or a concerted abstraction of hydride ion to give a chromate (6.5).

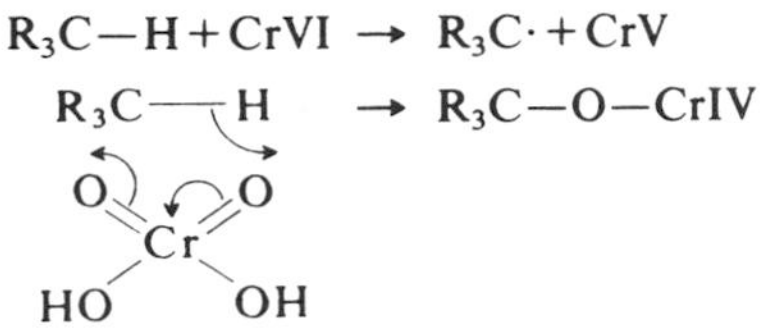

$$R_3C{-}H + CrVI \rightarrow R_3C\cdot + CrV$$

$$R_3C{-}H + (HO)_2CrO_2 \rightarrow R_3C{-}O{-}CrIV \qquad (6.5)$$

The conversion of a methyl group attached to a benzene ring into the formyl group can be achieved by oxidation with chromiumVI oxide in acetic anhydride in the presence of a strong acid, or with a solution of chromyl chloride in carbon disulphide or carbon tetrachloride (the Étard reaction) (6.6). The success of the first reaction is due to the initial formation of the di-acetate which protects the aldehyde group against further oxidation. In the Étard reaction a complex of composition 1 hydrocarbon: $2CrO_2Cl_2$ is first formed and is converted into the aldehyde by treatment with water. Ceric ion also readily oxidises aromatic methyl groups to aldehyde in acidic media. The aldehyde group is not oxidised further and, in a polymethyl compound only one methyl group is oxidised under normal conditions. Mesitylene, for example, gave 3,5-dimethylbenzaldehyde quantitatively, and acet-*p*-toluidine was converted into *p*-acetamidobenzaldehyde (94 per cent). Benzeneseleninic anhydride is another mild reagent for the conversion of some benzylic hydrocarbons into aldehydes or ketones. *para*-Xylene, for example, gave *para*-methylbenzaldehyde in 66 per cent yield (6.6) (Barton, Hui and Ley, 1982).

$$p\text{-}CH_3C_6H_4CH_3 \xrightarrow[H_2SO_4]{CrO_3,\ (CH_3CO)_2O} p\text{-}C_6H_4[CH(OCOCH_3)_2]_2 \xrightarrow[aq.\ H_2SO_4]{hydrolysis} p\text{-}C_6H_4(CHO)_2 \quad (52\%)$$

$$p\text{-}BrC_6H_4CH_3 \xrightarrow[CCl_4]{CrO_2Cl_2} [complex] \xrightarrow[H_2O]{hydrolysis} p\text{-}BrC_6H_4CHO \quad (80\%) \qquad (6.6)$$

$$p\text{-}CH_3C_6H_4CH_3 \xrightarrow[reflux]{(C_6H_5SeO)_2O,\ (0.33\ mol)} p\text{-}CH_3C_6H_4CHO \quad (66\%)$$

Alkenes

Oxidation of alkenes may take place both at the double bond and at the adjacent allylic positions and important synthetic reactions of each type are known involving oxidation with permanganate, peroxyacids, ozone and other reagents. With chromic acid mixtures of products are often formed and for this reason oxidation of alkenes with chromic acid is of limited value in synthesis.

Allylic oxidation of alkenes to give $\alpha\beta$-unsaturated carbonyl compounds can sometimes be effected with the chromium trioxide–pyridine complex (Collins' reagent; see p. 353) but selenium dioxide is more often used. With this reagent the immediate product may be an allylic alcohol, an $\alpha\beta$-unsaturated carbonyl compound or a mixture of the two depending on the experimental conditions (Jerussi, 1970). For example, with selenium dioxide in ethanol, 1-methylcyclohexene gave a mixture of 2-methyl-2-cyclohexenol and 2-methyl-2-cyclohexenone, but in water the ketone was formed exclusively. The allylic alcohols can be easily oxidised further to the $\alpha\beta$-unsaturated carbonyl compounds (see p. 354) if desired. For reaction in ethanol the order of reactivity of allylic groups is $CH_2 > CH_3 > CH$ but this may not hold for reaction under other conditions or for all types of alkene. Generally, reactions have been effected using stoichiometric amounts of selenium dioxide, but very good yields of more easily purified products are often obtained with catalytic amounts of selenium dioxide and t-butyl hydroperoxide, which serves to re-oxidise the spent catalyst (Umbreit and Sharpless, 1977).

The oxidations are believed to involve an ene reaction (p. 244) between the hydrated form of the dioxide and the alkene followed by a [2,3]-sigmatropic rearrangement of the resulting allylseleninic acid and final hydrolysis of the Se(II) ester to the allyl alcohol. Further oxidation of the alcohol gives the $\alpha\beta$-unsaturated carbonyl compound (Arigoni *et al.*, 1973; Woggon, Ruther and Egli, 1980). A useful application of this

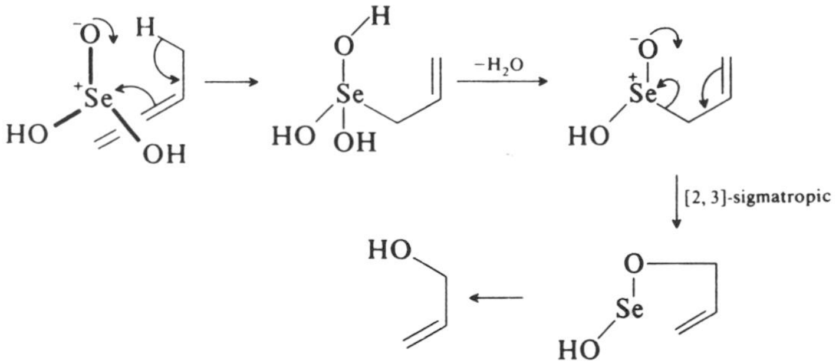

$-H_2O$ [2, 3]-sigmatropic (6.7)

reaction is in the oxidation of 1,1-dimethyl-alkenes to the corresponding *E*-allylic alcohols or aldehydes by selective attack on the *E*-methyl group (Bhalerao and Rapoport, 1971). Thus 2-methyl-2-heptene is converted almost exclusively into the (*E*)-allylic alcohol (1) and geranyl acetate gave a mixture of the *E*, *E*-alcohol (2) and the corresponding aldehyde (Umbreit and Sharpless, 1977). The high selectivity shown in these reactions is a consequence of the mechanism. The initial ene reaction proceeds through the favoured transition state (3), in which the substituent R is equatorial, to give the seleninic acid (4). The high

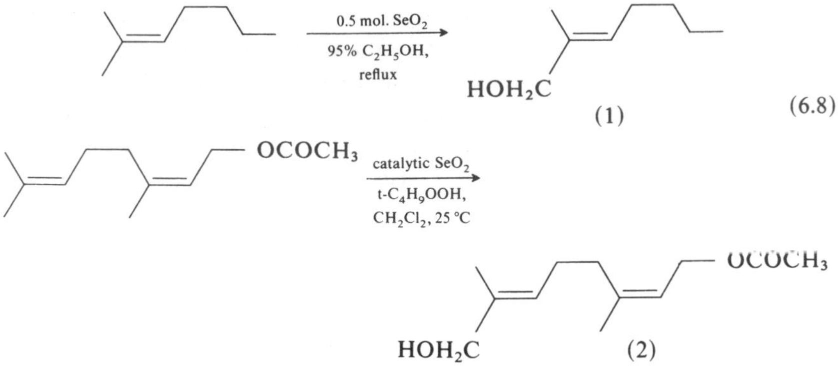

0.5 mol. SeO_2; 95% C_2H_5OH, reflux

HOH_2C (1)

(6.8)

$OCOCH_3$; catalytic SeO_2; t-C_4H_9OOH, CH_2Cl_2, 25 °C

$OCOCH_3$; HOH_2C (2)

E-selectivity is then established by the concerted [2,3]-sigmatropic rearrangement (cf. Woggon, Ruther and Egli, 1980).

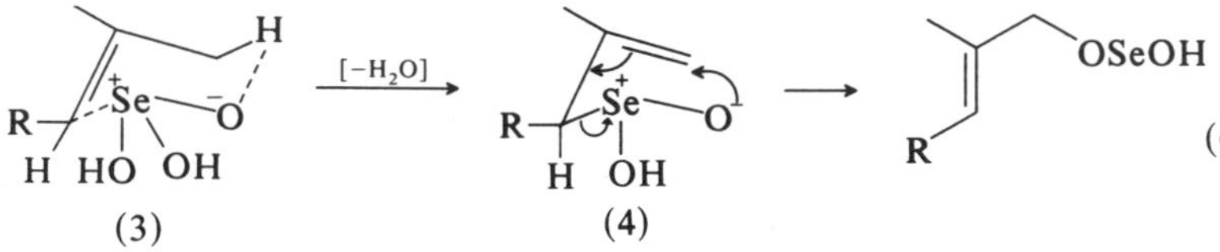

H; R; $\overset{+}{Se}$; $\overset{-}{O}$; H HO OH (3); [$-H_2O$]; R; $\overset{+}{Se}$; O; H OH (4); OSeOH; R

(6.9)

Allylic amination of alkenes can be effected by selenium or sulphur reagents of the type TosN=S=NTos (Sharpleess and Hori, 1976). Reactions take place easily at room temperature and follow the same sequence of ene reaction and [2,3]-sigmatropic rearrangement established for oxidations with selenium dioxide. 1-Hexene, for example, is smoothly converted into the 2-sulphonamido derivative by way of the sulphenamide (5). The sulphenamides are easily cleaved to the corresponding allylic sulphonamides with trimethyl phosphite in methanol.

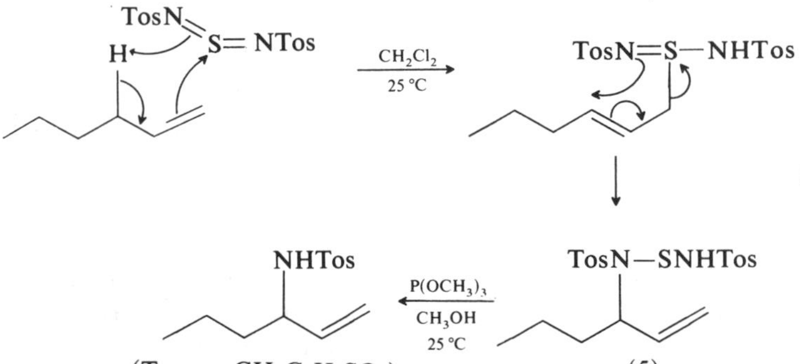

TosN; H; S= NTos; CH_2Cl_2, 25 °C; TosN=S—NHTos

(6.10)

NHTos; $P(OCH_3)_3$, CH_3OH, 25 °C; TosN—SNHTos

(Tos = p-$CH_3C_6H_4SO_2$) (5)

Another method for the conversion of an alkene into an allylic alcohol, but this time with a shift in the position of the double bond, proceeds from the corresponding β-hydroxyselenide. The latter are readily obtained from the alkene epoxide by reaction with phenylselenide anion or, better, directly from the alkene by addition of phenylselenenic acid or by acid-catalysed reaction with *N*-phenylseleno-phthalimide or -succinimide (Nicolaou *et al.*, 1979). The hydroxyselenide is not isolated but is oxidised directly with t-butyl hydroperoxide to the unstable selenoxide which spontaneously eliminates phenylselenenic acid to form the (*E*)-allylic alcohol (6.11) (cf. p. 121). Elimination always takes place

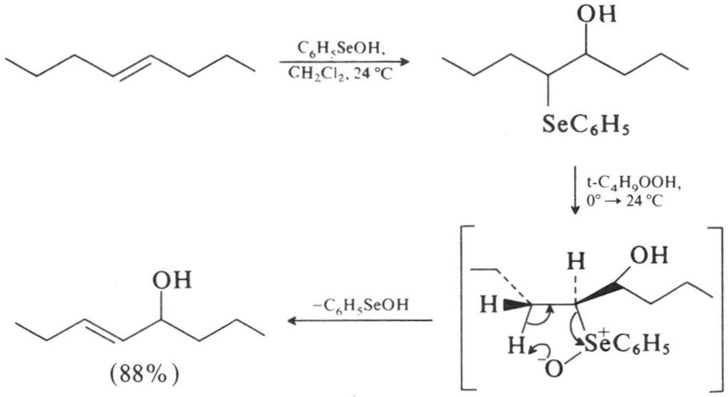

C_6H_5SeOH, CH_2Cl_2, 24 °C

OH

SeC_6H_5

t-C_4H_9OOH, 0° → 24 °C

H OH

H

H $\overset{+}{S}eC_6H_5$

O

$-C_6H_5SeOH$

OH

(88%)

(6.11)

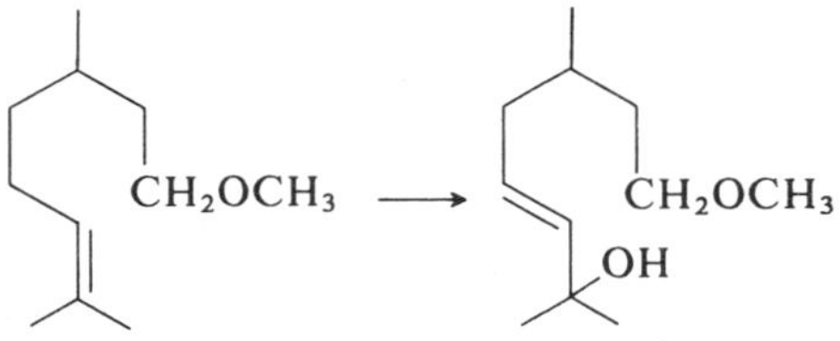

CH_2OCH_3 → CH_2OCH_3

OH

(6)

away from the hydroxyl group to give the allylic alcohol; no more than traces of the alternative ketonic products have ever been found in these reactions. Thus, 4-octene gave 5-octen-4-ol in 88 per cent yield. With trisubstituted alkenes addition of phenylselenenic acid is highly regioselective; citronellol methyl ether, for example, gave only (6) by initial Markovnikov addition of C_6H_5SeOH to the double bond (Hori and Sharpless, 1978).

6.2. Oxidation of alcohols

Chromic acid

One of the most important uses of chromic acid in synthesis is in the oxidation of alcohols, and particularly in the oxidation of secondary alcohols to ketones. This reaction is commonly effected with a solution of the alcohol and aqueous acidic chromic acid in acetic acid, or with aqueous acidic chromic acid in a heterogeneous mixture. If no complicating structural features are present high yields of ketone are usually obtained.

Oxidation of primary alcohols to aldehydes with acidic solutions of chromic acid is usually less satisfactory because the aldehyde is easily oxidised further to the carboxylic acid and, more importantly, because under the acidic conditions the aldehyde reacts with unchanged alcohol to form a hemiacetal which is rapidly oxidised to an ester (6.12). Satisfactory yields of aldehyde can be obtained in favourable cases by removing the aldehyde from the reaction medium by distillation as it is formed or, better, by use of the pyridine–chromiumVI oxide complex (p. 353).

$$C_3H_7CH_2OH \xrightarrow[H_2O,\ H_2SO_4,\ <20\,^{\circ}C]{Na_2Cr_2O_7} C_3H_7CHO \underset{C_3H_7CH_2OH}{\rightleftharpoons} C_3H_7CHOHOC_4H_9 \rightarrow C_3H_7COOC_4H_9 \qquad (6.12)$$

In general, tertiary alcohols are unaffected by chromic acid, but tertiary 1,2-diols are rapidly cleaved, provided they are sterically capable of forming cyclic chromate esters. *cis*-1,2-Dimethylcyclopentandiol, for example, is oxidised 17×10^3 times faster than the *trans* isomer (6.13).

$$\text{cis-1,2-dimethylcyclopentane-1,2-diol} \xrightarrow[HClO_4,\ H_2O]{Na_2Cr_2O_7} \text{cyclic chromate ester} \rightarrow CH_3CO(CH_2)_3COCH_3 \qquad (6.13)$$

Oxidation of alcohols by chromic acid is believed to take place by initial formation of a chromate ester, followed by breakdown of the ester, as shown for oxidation of isopropanol. Whether proton abstraction from the ester takes place by an intermolecular or intramolecular process is uncertain (6.14).

$$\text{B:} + (CH_3)_2CH{-}O{-}CrO_2{-}OH$$

or

$$(CH_3)_2CH{-}O{-}CrO_2{-}OH \text{ (intramolecular)} \longrightarrow (CH_3)_2C{=}O \qquad (6.14)$$

With unhindered alcohols the initial reaction to form the chromate ester is fast, and the subsequent decomposition of the ester is the rate-controlling step. Where formation of the ester results in steric overcrowding, ester decomposition is accelerated because steric strain is relieved in going from reactant to product. In extreme cases the initial esterification may become rate-determining. In the cyclohexane series it is found that axial hydroxyl groups are generally oxidised more rapidly than equatorial by a factor of about 3, presumably because of 1,3-diaxial interactions in the axial ester. This has been used in the determination of configurations of steroidal alcohols.

Oxidation with acid solutions of chromic acid is unsuitable for alcohols which contain acid-sensitive groups or other easily oxidisable groups such as carbon–carbon double bonds or allylic or benzylic C—H bonds elsewhere in the molecule. In such cases, and where the initial product of reaction is susceptible to further oxidation, it is often advantageous to effect reaction in the presence of an immiscible solvent such as benzene or ether, which serves to reduce contact between the organic compounds and the acidic oxidising solution. Another method which often allows selective oxidation of a hydroxyl group in such molecules is by dropwise addition of the stoichiometric amount of a solution of chromiumVI oxide in aqueous sulphuric acid (the Jones reagent) to a cooled (0–20 °C) solution of the alcohol in acetone. In many cases the 'end point' is easily observed by the persistence of the red colour of the chromic acid after addition of the theoretical amount of oxidant. Over-oxidation is thus lessened or prevented, and selective oxidation of unsaturated secondary alcohols to unsaturated ketones without appreciable oxidation or rearrangement of double bonds can often be achieved in good yield. Primary alcohols may give either aldehydes or carboxylic acids (6.15). Good yields of aldehydes and ketones have also been obtained from allylic and saturated primary and secondary alcohols with solutions of tetra-n-butylammonium chromate in chloroform (Cacchi, La Torre and Misiti, 1979).

$$CH_3CHOHC{\equiv}C(CH_2)_3CH_3 \xrightarrow[CH_3COCH_3]{CrO_3,\ H_2SO_4,} CH_3COC{\equiv}C(CH_2)_3CH_3 \quad (80\%)$$

$$CH{\equiv}C{-}CH{=}CHCH_2OH \rightarrow CH{\equiv}C{-}CH{=}CHCO_2H \quad (60\%) \qquad (6.15)$$

ChromiumVI oxide-pyridine complexes

Another useful mild reagent for the oxidation of alcohols that contain acid-sensitive functional groups is the chromiumVI oxide-pyridine complex $CrO_3.2C_5H_5N$, which is readily obtained by addition of chromiumVI oxide to pyridine. Reactions are best effected with a solution of the complex in dichloromethane – the so-called Collins' reagent – under anhydrous conditions (Collins, Hess and Fank, 1968). Primary and secondary alcohols are converted into the carbonyl compounds in good yield, and acid sensitive protecting groups are unaffected.

By this procedure, for example, 1-heptanol gave heptanal in 80 per cent yield and 5,9-dimethyl-5,9-decadienal was obtained in 90 per cent yield from the corresponding alcohol. Polyhydroxy compounds can sometimes be selectively oxidised at one position by protection of the other hydroxyl functions by acetal formation, as in the example in (6.16). Equally good results have been obtained in some cases simply by adding the alcohol to a solution of chromiumVI oxide in a mixture of pyridine and dichloromethane (Ratcliffe, 1976).

$$\xrightarrow{CH_3CHO,\ H^+} \qquad \xrightarrow[(2)\ H_3O^+]{(1)\ CrO_3,\ pyridine} \qquad (6.16)$$

A disadvantage of Collins' original procedure is that a considerable excess of reagent is usually required to ensure rapid and complete oxidation of the alcohol, and a number of modifications have been introduced to overcome this. Thus, excellent results have been obtained with pyridinium chlorochromate, $C_5H_5\overset{+}{N}HCrO_3Cl^-$, which is easily obtained by addition of pyridine to a solution of chromiumVI oxide in hydrochloric acid. When used in small excesss in solution in dichloromethane it gives good yields of aldehydes and ketones from the corresponding alcohols (Corey and Suggs, 1975*a*; Piancatelli, Scettri and D'Auria, 1982). The mildly acidic character of the chloro-chromate precludes its use with acid-sensitive compounds, however, and with such compounds Collins' reagent is superior. Another good reagent is pyridinium dichromate, $(C_5H_5N\overset{+}{H})_2Cr_2O_7$, which in dichloromethane solution oxidises primary alcohols to aldehydes in excellent yield (Corey and Schmidt, 1979) and allylic alcohols to $\alpha\beta$-unsaturated carbonyl compounds.

ManganeseIV oxide

Another useful mild reagent for the oxidation of primary and secondary alcohols to carbonyl compounds is manganeseIV oxide (Evans, 1959). The advantage of this reagent is that it is specific for allylic and benzylic hydroxyl groups, and reaction takes place under mild conditions (room temperature) in a neutral solvent (water, benzene, petroleum, chloroform). The general technique is simply to stir a solution of the alcohol in the solvent with the manganeseIV oxide for some hours. The manganeseIV oxide has to be specially prepared to obtain maximum activity. The best method appears to be by reaction of manganeseII sulphate with potassium permanganate in alkaline solution; the hydrated manganeseIV oxide obtained is highly active, but whether the actual oxidising agent is manganeseIV oxide itself or some other manganese compound adsorbed on the surface of the IV oxide is not clear at present.

Carbon–carbon double and triple bonds are unaffected by the reagent as illustrated in (6.17), and the reaction has been widely used for oxidation of polyunsaturated alcohols in the carotenoid and vitamin A series.

$$CH{\equiv}C{-}CH{=}\underset{}{\overset{CH_3}{\overset{|}{C}}}{-}CH_2OH \xrightarrow[CH_3COCH_3]{MnO_2} CH{\equiv}C{-}CH{=}\overset{CH_3}{\overset{|}{C}}{-}CHO \quad (35\%)$$

(6.17)

OH HO HO — MnO₂, CHCl₃, room temp. → OH HO O (62%)

Hydroxyl groups adjacent to triple bonds and cyclopropane rings are also easily oxidised, but under ordinary conditions saturated alcohols are not attacked (although they may be under more vigorous conditions), allowing selective oxidation of activated hydroxyl groups in appropriate cases. Potassium permanganate adsorbed on a solid support is reported to be a useful alternative to manganese dioxide in the oxidation of allylic alcohols (Noureldin and Lee, 1981). Triphenylbismuth carbonate is also very effective (p. 362).

In general oxidation of allylic primary alcohols with manganese dioxide takes place without significant further oxidation to carboxylic acids. In the presence of cyanide ions and an alcohol, however, high yields of the corresponding carboxylic esters can be obtained from $\alpha\beta$-unsaturated aldehydes (Corey, Gilman and Ganem, 1968). Thus, in methanol solution, cinnamaldehyde is converted into methyl cinnamate in 95 per cent yield and geranial gives methyl geranate in 85–95 per cent yield. Reaction is thought to proceed through the cyanhydrin (6.18). An important feature of the reaction is that oxidation takes place without any *cis–trans* isomerisation of the $\alpha\beta$-double bond. The traditional method of oxidising an aldehyde to a carboxylic acid using alkaline silver(I) oxide (Ag_2O) is relatively unsatisfactory for $\alpha\beta$-unsaturated aldehydes, since appreciable *cis–trans* isomerisation and other base-catalysed side reactions can occur.

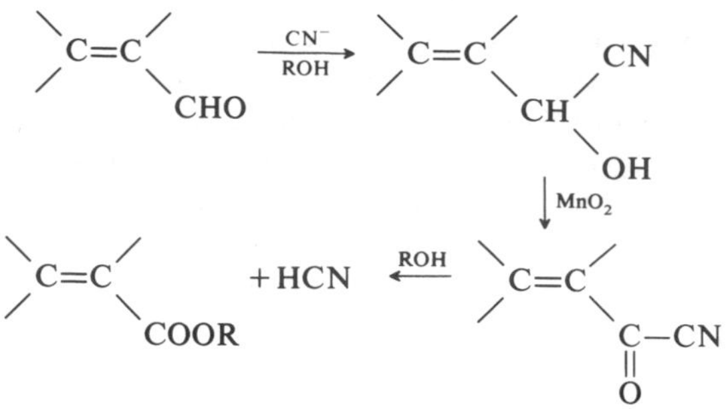

(6.18)

Silver carbonate

An excellent reagent for oxidising primary and secondary alcohols to aldehydes and ketones under mild and essentially neutral conditions is silver carbonate precipitated on celite (Fétizon and Golfier, 1968; McKillip and Young, 1979). The reaction is easily effected in boiling benzene and the product is recovered, usually in a high state of purity, by simply filtering off the spent reagent and evaporating off the solvent. Other functional groups are unaffected. Under these conditions nerol, for example, is converted into neral in 95 per cent yield. Highly hindered hydroxyl groups are not attacked, allowing selective oxidation in appropriate cases (6.19). Primary alcohols are oxidised more slowly than secondary which are themselves much less reactive than benzylic and allylic alcohols, and in acetone or methanol solution selective oxidation of benzylic or allylic hydroxyl groups is easily effected.

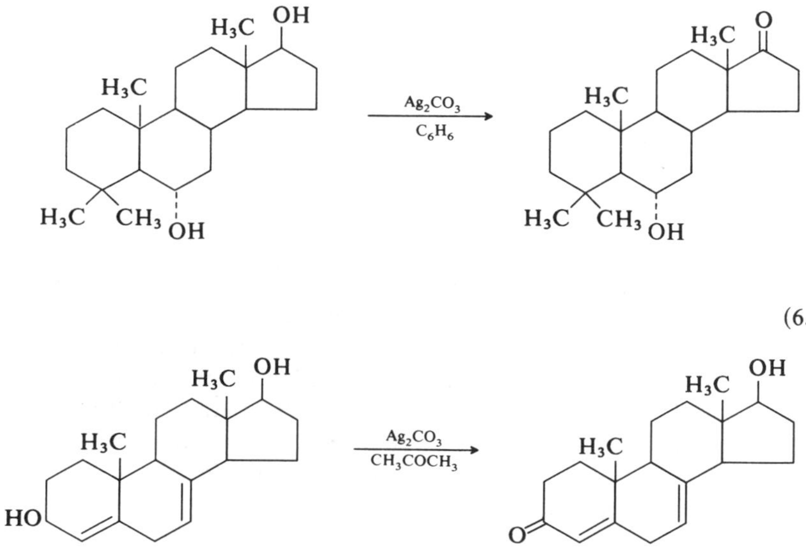

(6.19)

The behaviour of diols is different and depends on their structure. Generally only one of the hydroxyl groups is oxidised, allowing access to compounds not otherwise easily obtainable. Butan-1,4-diols, pentan-1,5-diols and hexan-1,6-diols, with two primary hydroxyl groups, are converted into the corresponding lactones, one of the alcohol groups being oxidised to the carboxylic acid. Thus, pentan-1,5-diol itself gave pentanolactone in 95 per cent yield and the diol (7) gave the lactone (8) exclusively, reflecting the greater reactivity of the allylic hydroxyl group (6.20). Other diols give hydroxy aldehydes or hydroxy ketones depending

H, CH_2OH, CH_2OH, H, O, O — Ag_2CO_3-celite, benzene → H, O, O, O, H

(6.20)

CH_2, OH, OH (7) — Ag_2CO_3-celite, benzene → CH_2, O, O (8) (80%)

on their structure. Thus, cyclohexan-1,2-diol gives 2-hydroxycyclohexanone and butan-1,3-diol forms 1-hydroxy-3-butanone, in line with the observation that secondary alcohols are more readily oxidised than primary alcohols with this reagent.

OH, OH — Ag_2CO_3-celite, benzene → O, OH (80%)

Oxidation via *alkoxysulphonium salts*

A number of methods for oxidising primary and secondary alcohols to aldehydes and ketones by action of a base on the derived alkoxysulphonium salts differ from each other mainly in the way in which the alkoxysulphonium salt is obtained from the alcohol (Mancuso and Swern, 1981). The conditions of reaction are mild and high yields of carbonyl compounds are generally obtained. One of the earliest procedures involved reaction of the alcohol with dimethyl sulphoxide and dicyclohexylcarbodiimide in the presence of a proton source (Epstein and Sweat, 1967). This method has been used to oxidise a number of sensitive compounds among natural products, including the 3′-*O*-acetylthymidine (9) (6.22).

$$\text{(9)} \xrightarrow[\text{CH}_3\text{SOCH}_3,\ \text{H}_3\text{PO}_4,\ 25\,^\circ\text{C}]{\text{dicyclohexylcarbodiimide,}} \text{aldehyde (90\%)} \tag{6.22}$$

(HOH$_2$C–, OCOCH$_3$ thymidine derivative (9) → HOC–, OCOCH$_3$ product, 90%)

The mechanism of the reaction has been elucidated by tracer experiments and is believed to involve initial formation of a sulphoxide-carbodiimide adduct which reacts with the alcohol to give the alkoxysulphonium ion. This then undergoes proton abstraction to form an ylid which collapses to the ketone and dimethyl sulphide by an intramolecular concerted process (6.23).

$$C_6H_{11}N{=}C{=}NC_6H_{11} + CH_3SOCH_3 \xrightarrow{H^+} C_6H_{11}NH{-}C({-}O{-}\overset{+}{S}(CH_3){-}CH_3){=}NC_6H_{11}$$

$$\xrightarrow{RCHOHR^1} (H_3C)(CH_3)\overset{+}{S}{-}O{-}CH(R)(R^1) + (C_6H_{11}NH)_2CO \tag{6.23}$$

$$\xrightarrow{\text{base}} (H_3C)(\overset{-}{C}H_2)\overset{+}{S}{-}O{-}CH(R)(R^1) \longrightarrow (H_3C)_2S + O{=}C(R)(R^1)$$

A disadvantage of this carbodiimide route is that the product has to be separated from the dicyclohexylurea formed in the reaction. To overcome this a number of other reagents have been used in conjunction with dimethyl sulphoxide including acetic anhydride, trifluoroacetic anhydride, thionyl chloride and oxalyl chloride. Best results appear to be obtained with oxalyl chloride as 'activator' (Mancuso, Huang and Swern, 1978). By reaction with dimethyl sulphoxide and oxalyl chloride followed by treatment of the resulting alkyloxysulphonium salt with a base, usually triethylamine, a wide variety of alcohols has been converted into the corresponding carbonyl compounds in high yield under mild conditions, and this must be counted one of the best methods for oxidising sensitive alcohols to the corresponding aldehydes or ketones (6.24).

$$\text{(geraniol)}\ \xrightarrow[\text{(2) }(C_2H_5)_3N]{\text{(1) }CH_3SOCH_3,\ (COCl)_2,\ CH_2Cl_2,\ -50\,^\circ C}\ \text{CHO}\ (95\%)$$

$$CH_2OH \longrightarrow CHO\ (80\%) \qquad (6.24)$$

The reaction is believed to proceed by way of the activated complex (11) formed by spontaneous loss of carbon dioxide and carbon monoxide from the oxysulphonium salt (10). Reaction then proceeds by way of the ylid (12) as in the dicyclohexylcarbodiimide route.

$$(CH_3)_2S{=}O + (COCl)_2 \xrightarrow[-60\,^\circ C]{CH_2Cl_2} \left[(CH_3)_2\overset{+}{S}{-}O{-}\overset{O}{\overset{\|}{C}}{-}\overset{O}{\overset{\|}{C}}{-}Cl\ \ Cl^-\right] (10) \xrightarrow{-CO,\ -CO_2} \left[(CH_3)_2\overset{+}{S}{-}Cl\right] Cl^- (11)$$

$$\downarrow RCHOHR^1$$

$$RR^1C{=}O + CH_3SCH_3 \longleftarrow CH_3(^-CH_2)S^+{-}O{-}CHRR^1\ (12) \xleftarrow{\text{base}} \left[(CH_3)_2S^+{-}O{-}CHRR^1\right] Cl^- \qquad (6.25)$$

A useful alternative approach makes use of the complexes formed from a methyl sulphide with chlorine or *N*-chlorosuccinimide (Corey and Kim, 1972). These complexes react readily with alcohols to form the corresponding alkoxysulphonium salts which decompose in the usual way in presence of triethylamine to give the carbonyl compound (6.26). Tracer experiments have shown that the transformation again involves an intermediate ylid. A valuable application is in the oxidation of secondary-tertiary-1,2-diols to α-ketols without rupture of the carbon–carbon bond (Corey and Kim, 1974). This unusual transformation depends on the fact that straightforward oxidation of the secondary alcohol function proceeds through a five-membered transition state and is thus favoured over glycol cleavage which requires a seven-membered transition state (6.26).

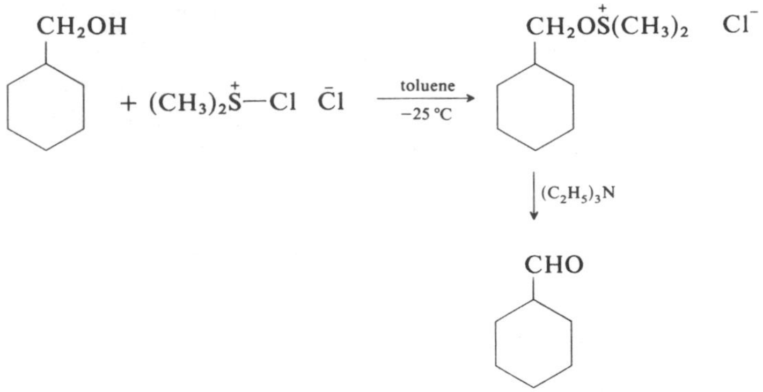

$$\text{C}_6\text{H}_{11}\text{CH}_2\text{OH} + (\text{CH}_3)_2\overset{+}{\text{S}}\text{—Cl}\ \overset{-}{\text{Cl}} \xrightarrow[-25\,^\circ\text{C}]{\text{toluene}} \text{C}_6\text{H}_{11}\text{CH}_2\text{O}\overset{+}{\text{S}}(\text{CH}_3)_2\ \text{Cl}^- \xrightarrow{(\text{C}_2\text{H}_5)_3\text{N}} \text{C}_6\text{H}_{11}\text{CHO}\ (93\%) \qquad (6.26)$$

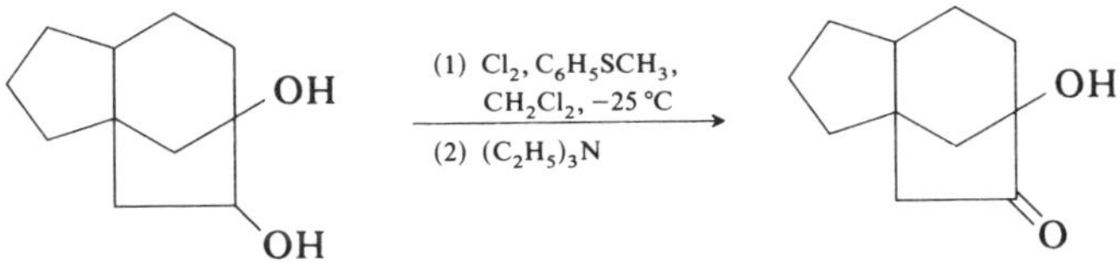

$$\xrightarrow[(2)\ (\text{C}_2\text{H}_5)_3\text{N}]{(1)\ \text{Cl}_2,\ \text{C}_6\text{H}_5\text{SCH}_3,\ \text{CH}_2\text{Cl}_2,\ -25\,^\circ\text{C}} \quad (80\%)$$

OH, OH → OH, O

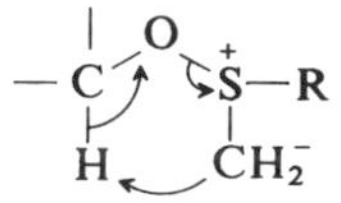

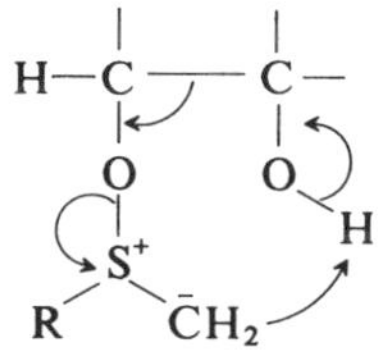

Related to these reactions is the oxidation of alkyl halides and toluene-*p*-sulphonates to carbonyl compounds with dimethyl sulphoxide. This useful reaction is simply effected by warming the halide or sulphonate in dimethyl sulphoxide, generally in presence of a proton acceptor such as sodium hydrogen carbonate or collidine. Oxidation never proceeds beyond the carbonyl stage and other functional groups are generally unaffected (Epstein and Sweat, 1967). The reaction has been applied to phenacyl halides, benzyl halides, primary sulphonates and iodides and a limited number of secondary sulphonates. Primary bromides and chlorides give only poor yields, but may be converted *in situ* into the corresponding sulphonates and oxidised without purification. Much improved yields of aldehydes from primary bromides can be obtained at room temperature by conducting the reaction in presence of silver ion and triethylamine (Ganem and Boeckman, 1974). With secondary sulphonates and halides elimination becomes an important side reaction and the reaction is less useful with such compounds. Where elimination is not possible good yields of ketones are obtained. It is thought that this reaction also probably proceeds through an ylid in most cases.

Other methods

Among a number of other useful methods for selective oxidation of primary and secondary alcohols to aldehydes and ketones under mild conditions are the Oppenauer oxidation, oxidation with lead tetra-acetate and catalytic oxidation with oxygen and platinum.

$$CH_3CHOHCH{=}CHCH{=}C(CH_3)CH{=}CH_3 \xrightarrow{\text{Al t-butoxide, acetone, boiling benzene}} CH_3COCH{=}CHCH{=}C(CH_3)CH{=}CH_2 \quad (80\%) \qquad (6.27)$$

$$\text{HO–steroid (H}_3\text{C)} \xrightarrow[CH_3COCH_3]{Al(OC_3H_7\text{-iso})_3} \text{O=steroid (H}_3\text{C)}$$

The Oppenauer oxidation (Djerassi, 1951) with aluminium alkoxides in acetone is the reverse of the Meerwein–Pondorff–Verley reduction discussed on p. 458. It has been widely used in the steroid series, particularly for the oxidation of allylic secondary hydroxyl groups to $\alpha\beta$-unsaturated ketones (6.27). $\beta\gamma$-Double bonds generally migrate into conjugation with the carbonyl group under the conditions of the reaction. The aluminium alkoxide serves only to form the aluminium alkoxide of the alcohol which is then oxidised through a cyclic transition state at the expense of the acetone. By use of excess of acetone the equilibrium is forced to the right (6.28).

$$CH_3-C(=O)-CH_3 \cdots H-C(R)(R)-O-Al \rightleftharpoons CH_3-CH(CH_3)-O-Al + R-C(=O)-R \qquad (6.28)$$

Lead tetra-acetate in refluxing benzene, hexane or chloroform is a good reagent for oxidation of primary and secondary alcohols to the corresponding aldehyde or ketone, provided there is no δ C—H group in the molecule (Butler, 1977). In this circumstance high yields of tetrahydrofuran derivatives are obtained (see p. 279). In pyridine solution lead tetra-acetate oxidises a variety of primary and secondary alcohols to the carbonyl compounds in good yield at room temperature whether they contain δ C—H groups or not. Both allylic and saturated alcohols are oxidised and cyclisation products appear not to be formed (6.29).

$$CH_3(CH_2)_3CH_2OH \xrightarrow[\text{pyridine}]{Pb(OCOCH_3)_4} CH_3(CH_2)_3CHO \quad (70\%)$$

$$CH_3CHOHCH_2CH_2CHOHCH_3 \longrightarrow CH_3COCH_2CH_2COCH_3 \quad (89\%)$$

$$C_6H_5CH{=}CHCH_2OH \longrightarrow C_6H_5CH{=}CHCHO \quad (91\%) \qquad (6.29)$$

Catalytic oxygenation with a platinum catalyst and molecular oxygen is another valuable method for oxidation of primary and secondary hydroxyl groups under mild conditions. With primary alcohols the reaction can be regulated to give aldehydes or acids. Double bonds, in general, are not affected and unsaturated alcohols such as tiglic alcohol (2-methyl-2-buten-1-ol) can be oxidised catalytically to the unsaturated aldehyde (Heyns and Paulsen, 1963) (6.30).

$$CH_3(CH_2)_{10}CH_2OH \xrightarrow[C_7H_{16},\ \frac{1}{2}h]{O_2,\ Pt} CH_3(CH_2)_{10}CHO \quad (77\%)$$

$$\xrightarrow{2\ h} CH_3(CH_2)_{10}COOH \quad (96\%) \qquad (6.30)$$

$$CH_3CH{=}C(CH_3){-}CH_2OH \xrightarrow[C_7H_{16}]{O_2,\ Pt} CH_3CH{=}C(CH_3){-}CHO$$

In general, primary hydroxyl groups are attacked before secondary, and in cyclic secondary alcohols axial groups appear to react before equatorial. The method has been widely used in the carbohydrate series to effect selective oxidation of specific hydroxyl groups. Thus, L-sorbose is oxidised at 30 °C to 2-keto-L-gulonic acid, an intermediate in a synthesis of ascorbic acid (6.31).

$$\begin{array}{c} CH_2OH \\ | \quad OH \\ C \\ | \\ HO-C-H \\ | \\ H-C-OH \quad O \\ | \\ HO-C-H \\ | \\ CH_2 \end{array} \xrightarrow{O_2,\ Pt} \begin{array}{c} COOH \\ | \quad OH \\ C \\ | \\ HO-CH \\ | \\ H-C-OH \quad O \\ | \\ HO-C-H \\ | \\ CH_2 \end{array} \qquad (6.31)$$

Triphenylbismuth carbonate, easily prepared by reaction of triphenylbismuth dichloride with potassium carbonate in aqueous acetone, is another good selective reagent for the oxidation of primary and secondary alcohols to aldehydes and ketones under mild conditions (Barton *et al.*, 1981). Many other functional groups such as thiol and amino groups do not interfere. The reagent is particularly effective for the oxidation of allylic and benzylic alcohols to the carbonyl compounds and has the advantage over manganeseIV oxide (p. 354) for this purpose that a large excess of reagent is not required; geraniol gave geranial in 95 per cent yield after two hours at 40 °C. Cleavage of 1,2-diols to the corresponding dicarbonyl compounds is also readily effected. Thus, cyclohexane-1,2-diol gave hexane-1,6-dial in quantitative yield.

SeCH$_3$ / HO
$(C_6H_5)_3Bi$(carbonate) ; CH_2Cl_2, 40 °C →
SeCH$_3$ / O (85%)

$$C_6H_5\overset{OH}{\overset{|}{C}}H.\underset{CH_3NCOCH_3}{\underset{|}{C}}HCH_3 \rightarrow C_6H_5\overset{O}{\overset{||}{C}}\underset{CH_3NCOCH_3}{\underset{|}{C}}HCH_3 \quad (75\%) \qquad (6.32)$$

Selective oxidation of a primary or a secondary hydroxyl group in the presence of the other is a useful transformation in synthesis. Few methods are available for such selective oxidation of primary hydroxyl groups (cf. Boeckman and Ganem, 1974) but secondary alcohols can be selectively oxidised in the presence of primary ones in several ways using silver carbonate (p. 356), sodium bromate in the presence of catalytic cerium ammonium nitrate (Tomioka, Oshima and Nozaki, 1982) or sodium hypochlorite in aqueous acetic acid (Stevens *et al.*, 1982). The latter inexpensive commercially available reagent is particularly suitable for large scale work; (−)-borneol gave (−)-camphor in 95 per cent yield and 2-ethyl-1,3-hexandiol afforded 2-ethyl-1-hydroxy-3-hexanone in 85 per cent yield.

OH ... OH → NaOCl ; CH_3CO_2H, H_2O, 25 °C → O ... OH (6.33)

One of the features of present day organic chemistry is the use of enzymes to bring about the specific transformation of functional groups to give enantiomerically pure compounds. A good example is provided by the oxidation of the *meso*-1,5-diol (13) by horse liver alcohol dehydrogenase to give the optically pure lactone (14) (Jakovac *et al.*, 1980; Ohta, Tetsukawa and Noto, 1982).

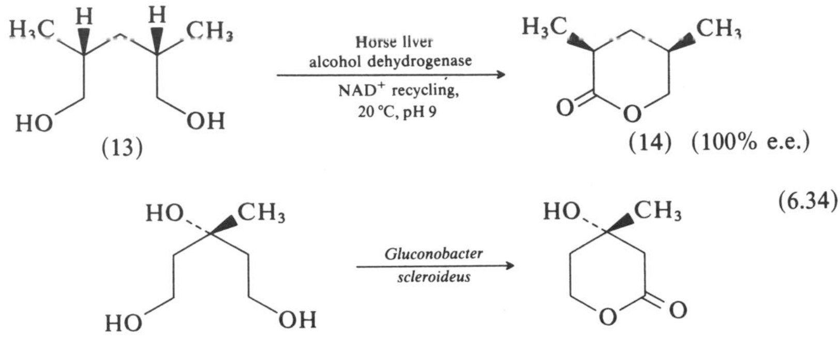

(13) $\xrightarrow[\text{NAD}^+ \text{ recycling, } 20\,°\text{C, pH 9}]{\text{Horse liver alcohol dehydrogenase}}$ (14) (100% e.e.)

$\xrightarrow[\textit{scleroideus}]{\textit{Gluconobacter}}$ (6.34)

6.3. Oxidation of carbon–carbon double bonds

Perhydroxylation

Perhydroxylation of carbon–carbon double bonds is useful in degradation and synthesis and can be effected stereospecifically with a number of different reagents.

$$\rangle C{=}C\langle \rightarrow HO{-}C{-}C{-}OH$$

For *cis-* (or *syn-*) perhydroxylation, in which the two hydroxyl groups are added to the same side of the double bond, the best methods are reaction with potassium permanganate, osmium tetroxide or iodine and moist silver acetate. The most important method of *trans-* (*anti-*) perhydroxylation is reaction with peroxy-acids, but the Prévost reaction, that is the action of iodine and silver acetate under anhydrous conditions is also useful.

The alkene may, of course, have either the *cis*-(*Z*-) or *trans*-(*E*-) configuration and by *cis-* (*syn-*) or *trans-* (*anti-*) perhydroxylation can give rise to isomeric diols which may be described by the terms *cis* or *trans*, *erythro* or *threo*, or *meso* or *racemic* depending on the nature of the other groups attached to the diol system. The relation between the alkenes and the diols produced is shown in Table 6.1. Thus, *trans*(*E*)-crotonic acid by *cis*-perhydroxylation gives the *threo*-dihydroxybutyric

Table 6.1. *Relation between alkenes of different configuration and diols produced by cis- and trans-perhydroxylation*

Alkene	Perhydroxylation	
	cis (*syn*)	*trans* (*anti*)
cis (*Z*)	*cis, meso, erythro*	*trans, racemic, threo*
trans (*E*)	*trans, racemic, threo*	*cis, meso, erythro*

acid (15) but by *trans*-perhydroxylation the *erythro*-isomer (16) is obtained. Similarly, maleic acid by *cis*-perhydroxylation gives *meso*-tartaric acid whereas fumaric acid affords racemic tartaric acid. Note that *cis*- and *trans*-perhydroxylation do not necessarily lead correspondingly to *cis*- and *trans*-diols; it depends on the configuration of the double bond. *Trans*-cyclo-octene, for example, by *cis*-perhydroxylation gives *trans*-1,2-dihydroxycyclo-octane. Perhaps for this reason the terms *syn* and *anti*-perhydroxylation would be less confusing than the presently employed *cis* and *trans*.

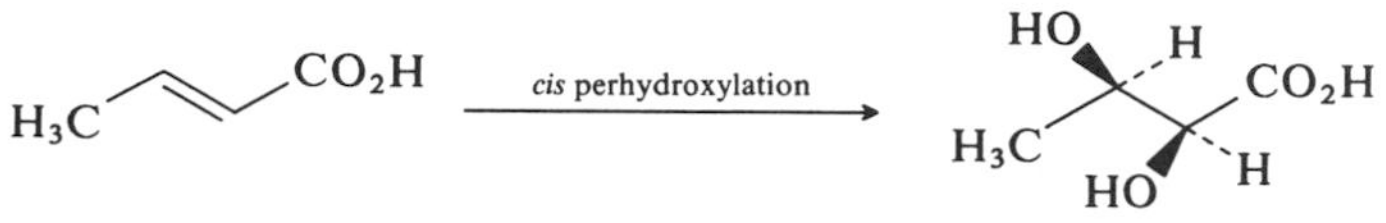

cis perhydroxylation → *threo* (15)

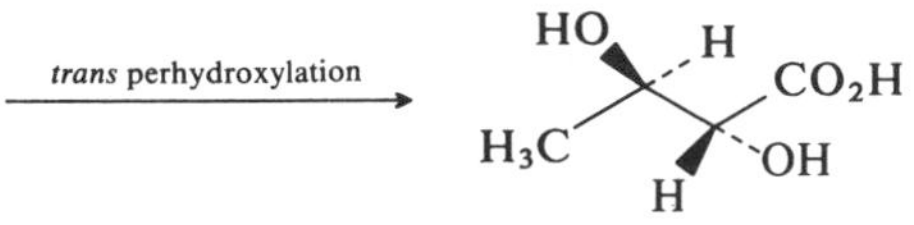

trans perhydroxylation → *erythro* (16) (6.35)

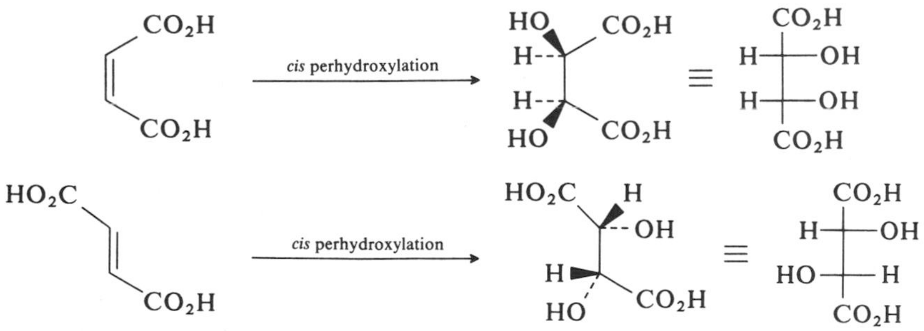

cis perhydroxylation

cis perhydroxylation

Potassium permanganate

Oxidation with potassium permanganate is a widely used method for *cis*-perhydroxylation of alkenes, but needs careful control to avoid over-oxidation. Best results are obtained in alkaline solution, using water or aqueous organic solvents (acetone, ethanol of t-butanol); in acid or neutral solution α-ketols or even cleavage products are formed. The method is particularly suitable for perhydroxylation of unsaturated acids, which dissolve in the alkaline solution; in many other cases only poor yields of the diols are obtained because of the insolubility of the substrate in the aqueous oxidising medium. Greatly improved yields in such cases can be obtained by effecting the oxidation in presence of a phase transfer catalyst, such as a quaternary ammonium salt, or a crown ether. Thus, oxidation of *cis*-cyclo-octene with aqueous alkaline permanganate in the presence of benzyltrimethylammonium chloride gave *cis*-cyclo-octan-1,2-diol in 50 per cent yield (equation 6.1); in absence of the catalyst the yield was only 7 per cent (cf. Lee, 1982).

These reactions are believed to proceed through the formation of cyclic manganese esters and it is this which controls the *cis* (*syn*) addition of the two hydroxyl groups. *Cis*-addition is shown by the conversion of maleic acid into *meso*-tartaric acid and of fumaric acid into (±)-tartaric acid, and the cyclic ester mechanism is supported by studies with ^{18}O permanganate which show that transfer of oxygen from permanganate to the substrate occurs. Competition between ring-opening of the cyclic ester by hydroxyl ion and further oxidation by permanganate accounts for the effect of pH on the distribution of products (Stewart, 1965; Lee, 1982). In acid solution cleavage of the double bond may occur.

$$\text{RHC=CHR} \xrightarrow[\text{(Mn(VII))}]{MnO_4^-} \text{cyclic ester } \bar{M}nO_2 \text{ (Mn(V))}$$

$$\text{cyclic ester (Mn(V))} \xrightarrow{\bar{O}H,\ H_2O} \text{R–CH(OH)–CH(OMnO}_3^{2-}\text{)–R (Mn(V))} \xrightarrow{H_2O} \text{R–CH(OH)–CH(OH)–R}$$

$$\text{cyclic ester (Mn(V))} \xrightarrow{MnO_4^-} \text{cyclic ester MnO}_2 \text{ (Mn(VI))} + H_2O \longrightarrow \text{R–CH(OMnO}_2^-\text{)–C(=O)–R (Mn(IV))} \xrightarrow{H_2O} \text{R–CH(OH)–C(=O)–R} \quad (6.36)$$

Osmium tetroxide

Reaction with osmium tetroxide is probably the best method for *cis*-perhydroxylation of alkenes, but the stoichiometric reaction is suitable only for small-scale work with valuable compounds because of the expense and toxicity of the reagent (Fieser and Fieser, 1967; Schröder, 1980).

Perhydroxylation of the alkene takes place by way of a cyclic osmium(VI) complex (17) which on reductive or oxidative hydrolysis yields the corresponding *cis*-diol. It is believed that (17) is itself formed by way of a cyclic four-membered ring species with an osmium–carbon bond (6.37) (Schröder and Constable, 1982; Hentges and Sharpless,

OsO_4 + || ⟶ [four-membered Os–C–C–O ring] $\xrightarrow[\text{(e.g. pyridine)}]{\text{Ligand}}$ L_2OsO_2(O–C–C–O) (6.37)

(17)

1980). Reaction is accelerated by tertiary bases, especially pyridine, and pyridine is often added to the reaction medium. Brightly coloured complexes (17), in which osmium is co-ordinated with two molecules of base, separate in almost quantitative yield. In an interesting development pyridine has been replaced by optically active bases to provide a direct route to optically active diols (Hentges and Sharpless, 1980).

Osmium tetroxide is frequently used catalytically in conjunction with other oxidising agents. Formerly chlorates and hydrogen peroxide were used but better results have been obtained with t-butyl hydroperoxide and tertiary amine oxides (Schröder, 1980). In these reactions the initial osmate ester is oxidatively hydrolysed by the oxidising agent with regeneration of osmium tetroxide with continues the reaction, so that a small amount suffices. A disadvantage of the earlier procedures using chlorates and hydrogen peroxide was that in some cases over-oxidation products were formed. This difficulty is largely avoided in the procedure using t-butyl hydroperoxide, which has the added advantage that it is effective for the oxidation of tri- and tetra-substituted double bonds which are often completely resistant to the chlorate and hydrogen peroxide reagents (6.38) (Akashi, Palermo and Sharpless, 1978). Because of the large steric requirements of the reagent, reactions with osmium tetroxide usually take place predominantly from the less hindered side of the double bond when there is a choice, as in the second example in (6.38).

Oxidation of allylic alcohols with osmium tetroxide provides a route to 1,2,3-triols, a structural feature found in some natural products. Furthermore the reaction with the alcohols and also the corresponding ethers, is highly stereoselective, giving preferentially the isomer in which the

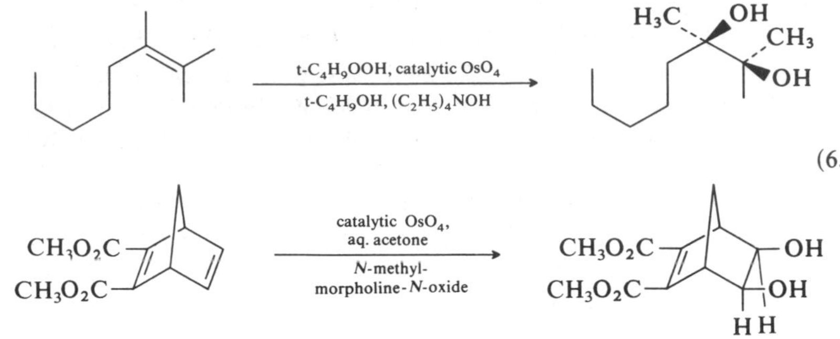

$$\xrightarrow[\text{t-}C_4H_9OH,\ (C_2H_5)_4NOH]{\text{t-}C_4H_9OOH,\ \text{catalytic } OsO_4}$$

H_3C OH CH_3 OH

(6.38)

CH_3O_2C CH_3O_2C

$$\xrightarrow[N\text{-methyl-morpholine-}N\text{-oxide}]{\text{catalytic } OsO_4,\ \text{aq. acetone}}$$

CH_3O_2C CH_3O_2C OH OH H H

original hydroxyl or alkoxyl group and the adjacent newly introduced hydroxyl group are in an *erythro* (*anti*; see p. 52) relationship. Thus, 2-cyclohexen-1-ol gave only the triol (18), and the open-chain allylic ether (19) gave the *erythro* (*anti*) and *threo* (*syn*) isomers (20) and (21) in the ratio 9:1 (Cha, Christ and Kishi, 1983). Reaction is believed to take place by preferential addition of osmium tetroxide to the allylic alcohol in the conformation (22) on the face of the double bond opposite to the hydroxyl or alkoxyl group (6.39).

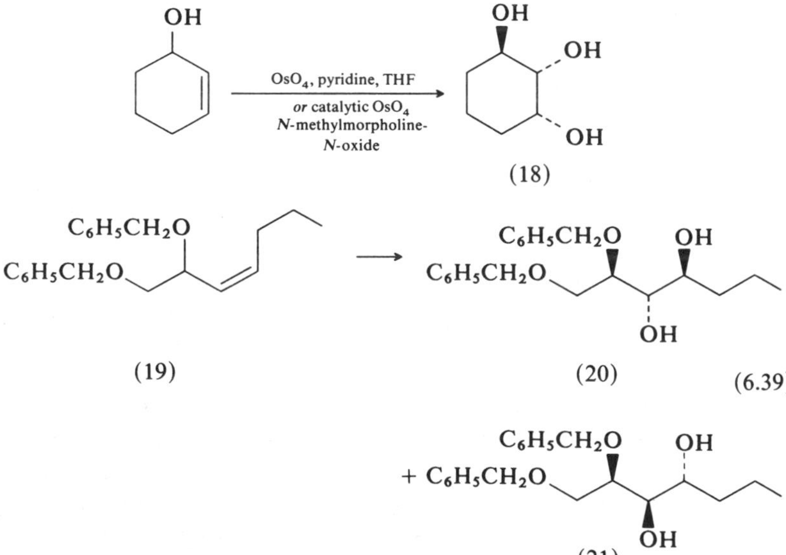

OH

$$\xrightarrow[\text{or catalytic } OsO_4\ N\text{-methylmorpholine-}N\text{-oxide}]{OsO_4,\ \text{pyridine, THF}}$$

OH OH OH

(18)

$C_6H_5CH_2O$ $C_6H_5CH_2O$ → $C_6H_5CH_2O$ OH $C_6H_5CH_2O$ OH

(19) (20) (6.39)

$C_6H_5CH_2O$ OH + $C_6H_5CH_2O$ OH

(21)

R^3 R^1 R^4 R^2O H

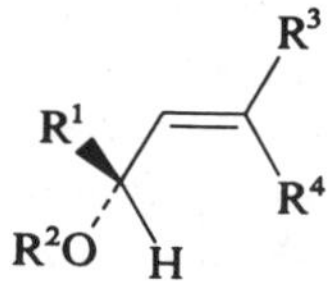

(22)

Osmium tetroxide has also been used in a procedure for converting alkenes into vicinal hydroxyamines. Reaction of the alkene with chloramine T in the presence of catalytic osmium tetroxide affords the corresponding vicinal hydroxy toluene-*p*-sulphonamide. The sulphonyl-imido osmium compound (23) is believed to be the effective reagent, and is continuously regenerated during the reaction (Sharpless, Chong and Oshima, 1976). The sulphonamides are readily converted into the *cis*-α-hydroxyamines by cleavage with sodium in liquid ammonia. The reagent (24) has been similarly used to convert alkenes into vicinal diamines (Chong, Oshima and Sharpless, 1977).

$$\text{TosNClNa} + \text{RCH=CHR} \xrightarrow[\text{t-}C_4H_9OH,\ 60\,^{\circ}C]{1\%\ OsO_4,} \text{RCH(OH)CH(NHTos)R} \xrightarrow{\text{Na, liq. }NH_3} \text{RCH(OH)CH(NH}_2\text{)R} \qquad (6.40)$$

(23) $O_3Os{=}NTos$

(24) $O_2Os(=NC_4H_9\text{-t})_2$

(Tos = p-$CH_3C_6H_4SO_2$)

Oxidation with iodine and silver carboxylates

Many of the difficulties attending the oxidation of alkenes to 1,2-glycols with other reagents can be avoided by using Prévost's reagent – a solution of iodine in carbon tetrachloride together with an equivalent of silver acetate or silver benzoate. Under anhydrous conditions this oxidant directly yields the diacyl derivative of the *trans*-glycol (Prévost conditions), while in presence of water the monoester of the *cis*-glycol is obtained (Woodward conditions). Thus, cyclohexene on treatment with iodine and silver benzoate in boiling carbon tetrachloride under anhydrous conditions gives the dibenzoate of *trans*-1,2-dihydroxycyclohexane. With iodine and silver acetate in moist acetic acid, however, the monoacetate of *cis*-1,2-dihydroxycyclohexane is formed. Similarly, oleic acid on oxidation under Prévost conditions gives *threo*-9,10-dihydroxystearic acid, while by the Woodward procedure the *erythro* isomer results.

The value of these reagents is due to their specificity and to the mildness of the reaction conditions; free iodine, under the conditions used, hardly affects other sensitive groups in the molecule. Reaction proceeds through formation of an iodonium cation which, in presence of acyloxy and silver ions, forms the resonance stabilised cation (25) (6.41). Attack on the cation by acetate ion in a bimolecular process gives the *trans*-diacyl compound. In the presence of water, however, a hydroxy acetal is formed which affords the *cis* hydroxy-acyloxy compound. With conformationally rigid molecules the *cis* diol obtained by the Woodward method may not have the same configuration as that obtained with osmium tetroxide. In an alternative procedure which avoids the use of expensive silver salts the *trans* iodoacetates are obtained in high yield from the alkene by reaction with iodine and thallium(I) acetate. Solvolysis in wet acetic acid under reflux affords the corresponding *cis* hydroxy-acetates, while in dry acetic acid the *trans* diacetate is obtained (Cambie *et al.*, 1974).

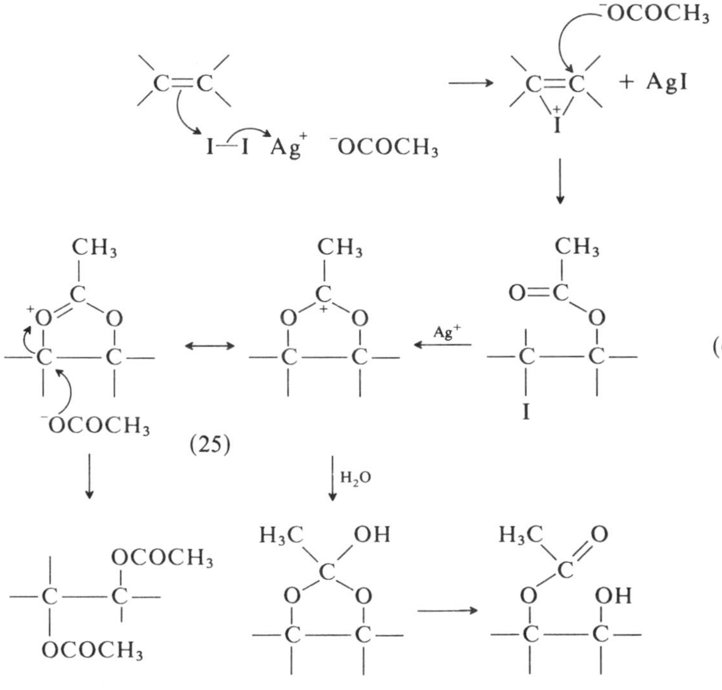

$I-I\ Ag^+\ {}^-OCOCH_3$; $+ AgI$; Ag^+ ; (25) ; H_2O

(6.41)

Oxidation with peroxy-acids

Oxidation of alkenes with peroxy-acids gives rise to epoxides (oxiranes) or to *trans*-1,2-diols, depending on the experimental conditions.

A number of peroxy-acids have been used in the past, including perbenzoic, performic and peracetic acid, but these have now been largely superseded, for the formation of epoxides at any rate, by *m*-chloroperbenzoic acid; it is commercially available and is an excellent reagent for the epoxidation of alkenes. It is more stable than the other peroxy-acids and has even been used at an elevated temperature (dichloroethane at 90 °C) to effect the epoxidation of unreactive alkenes (Kishi, *et al.*, 1972). The other peroxy-acids are rather unstable and generally have to be prepared freshly before use. Performic and peracetic acids, for example, are often prepared *in situ*, and not isolated, by action of hydrogen peroxide on the carboxylic acid. With acetic acid the equilibrium is attained only slowly, and sulphuric acid is usually added as a catalyst to hasten formation of the peroxy-acid. Solutions prepared in this way from acetic and formic acid are widely used for the preparation of *trans*-1,2-diols from alkenes, through ring opening of the initially formed epoxide (see page 381).

$$RCOOH + H_2O_2 \rightleftharpoons H_2O + RCO_3H$$

It is probable that reaction of all peroxy-acids with alkenes gives rise to the epoxide in the first place, but unless proper precautions are taken the epoxide may be converted directly into a mono-acyl derivative of the 1,2-diol. The best reagent for the preparation of the epoxide is *m*-chloroperbenzoic acid. Reaction is believed to take place by electrophilic attack of the peroxy-acid on the double bond as illustrated in (6.42) (Lee and Uff, 1967). In accordance with this mechanism the rate

(6.42)

of epoxidation is increased by electron-withdrawing groups in the peroxy-acid (trifluoroperacetic acid is more effective than peracetic acid) or electron-donating substituents on the double bond. Terminal mono-alkenes react only slowly with most peroxy-acids, and the rate of reaction increases with the degree of alkyl substitution. 1,2-Dimethyl-1,4-cyclohexadiene for example, reacts preferentially at the tetrasubstituted double bond (6.43). On the other hand, conjugation of the alkene double bond with other unsaturated groups reduces the rate of epoxidation because of delocalisation of the π electrons. $\alpha\beta$-Unsaturated acids and esters,

m-$ClC_6H_4CO_3H$, dioxan, 0 °C

(6.43)

m-$ClC_6H_4CO_3H$, $CHCl_3$ (80%)

for example, require the strong reagent trifluoroperacetic or *m*-chloroperbenzoic acid at an elevated temperature for successful oxidation. With $\alpha\beta$-unsaturated ketones reaction is complicated by competing Baeyer–Villiger oxidation at the carbonyl group (see p. 403). Epoxides of $\alpha\beta$-unsaturated aldehydes and ketones are best made by the action of nucleophilic reagents such as hydrogen peroxide or t-butyl hydroperoxide in alkaline solution. These reactions are believed to take the course shown in (6.44).

$$R_2C{=}CHCOR \xrightleftharpoons{^{-}OOH} R{-}\underset{HO-O}{\overset{R}{C}}{-}CH{=}\overset{O^{-}}{C}{-}R \longrightarrow R{-}\overset{R}{C}{-}CHCOR \text{ (epoxide, O bridging)} \qquad (6.44)$$

Epoxidations with peroxy acids are highly stereoselective and take place by *cis* addition to the double bond of the alkene. This follows from the results of numerous experiments and has been shown unequivocally by X-ray analysis of the products obtained by epoxidation of oleic and elaidic acid. Thus, oleic acid gave *cis*-9,10-epoxystearic acid whereas elaidic acid gave the isomeric *trans*-compound (6.45).

$$CH_3(CH_2)_7CH{=}CH(CH_2)_7CO_2H \ (cis) \xrightarrow{C_6H_5CO_3H} cis\text{-epoxide}$$

$$CH_3(CH_2)_7CH{=}CH(CH_2)_7CO_2H \ (trans) \xrightarrow{C_6H_5CO_3H} trans\text{-epoxide} \qquad (6.45)$$

With conformationally rigid cyclic alkenes the reagent usually approaches from the less hindered side of the double bond, as illustrated for the epoxidation of norbornene shown in (6.46), but with flexible

$$\text{norbornene} \xrightarrow[CHCl_3]{C_6H_5CO_3H} \text{exo-epoxide (94\%)} + \text{endo-epoxide (6\%)} \qquad (6.46)$$

molecules it may be more difficult to predict the stereochemical outcome (cf. Berti, 1973). Where there is a polar substituent in the allylic position this may influence the direction of attack by the peroxy-acid. Thus, whereas 2-cyclohexenyl acetate gives mainly the *trans*-epoxide as expected (attack from the less hindered side of the double bond), the free alcohol gives the *cis*-epoxide in 80 per cent yield under the same conditions (6.47). The rate of reaction is faster with the hydroxy compound

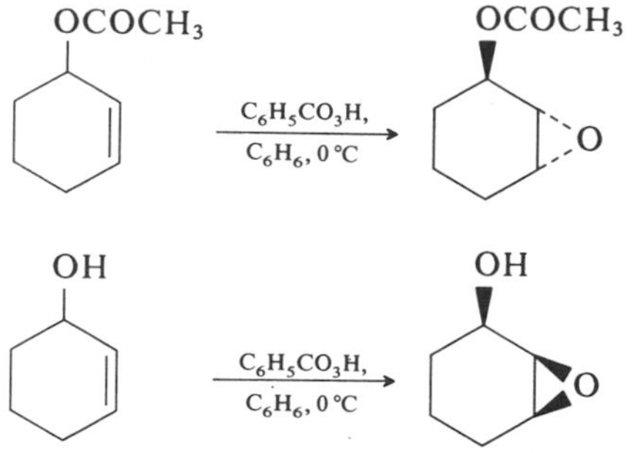

$OCOCH_3$ $\xrightarrow[C_6H_6,\ 0\,°C]{C_6H_5CO_3H,}$ $OCOCH_3$

(6.47)

OH $\xrightarrow[C_6H_6,\ 0\,°C]{C_6H_5CO_3H,}$ OH O

and it is believed that hydrogen bonding causes association of the reactants in an orientation favourable for *cis*-epoxidation. Highly stereoselective epoxidation of open-chain allylic alcohols with *m*-chloroperbenzoic acid has been observed in certain cases. Thus, the *cis*-allylic alcohol (26) gave the epoxide (27) exclusively, through attack on the preferred conformation (28); epoxidation is directed to the top face of the double bond by co-ordination of the reagent with the ethereal oxygen as well as the allylic hydroxyl group. With the corresponding *E*-alcohol the stereoselection was less good (Johnson, Nakata and Kishi, 1979).

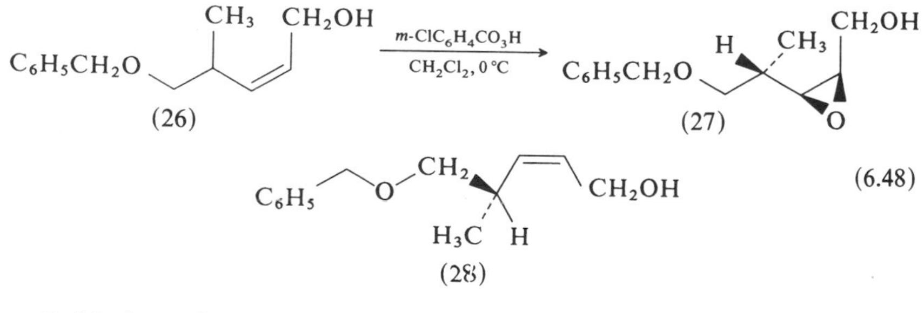

CH_3 CH_2OH $C_6H_5CH_2O$ (26) $\xrightarrow[CH_2Cl_2,\ 0\,°C]{m\text{-}ClC_6H_4CO_3H}$ $C_6H_5CH_2O$ H CH_3 CH_2OH O (27)

(6.48)

C_6H_5 O CH_2 CH_2OH H_3C H (28)

Oxidation with peroxy acids is not the only way to convert an alkene into an epoxide. It has recently been found that reaction of alkenes with t-butyl hydroperoxide in the presence of vanadium (V^{5+}) or molybdenum (Mo^{6+}) catalysts (vanadyl or molybdenyl acetylacetonates are frequently used) provides another excellent method for the preparation of epoxides (Sharpless and Verhoeven, 1979). It appears that the molybdenum catalysts are most effective for the epoxidation of isolated double bonds and the vanadium catalysts for allylic alcohols. Even terminal alkenes, which are among the most difficult to epoxidise with peroxy acids are readily oxidised under these conditions. 1-Decene, for example, was smoothly converted into its epoxide with t-butyl hydroperoxide and catalytic molybdenum hexacarbonyl in boiling 1,2-dichloroethane.

The molybdenum- and, in particular, the vanadium-t-butyl hydroperoxide reagents show remarkable reactivity towards the double bond of allylic alcohols, and this makes possible selective epoxidations not readily obtainable with other reagents (Sharpless and Michaelson, 1973; Rossiter, Verhoeven and Sharpless, 1979). Thus, reaction of geraniol with t-butyl hydroperoxide in boiling benzene in the presence of catalytic vanadium acetylacetonate gave the 2,3-epoxide (29) almost exclusively;

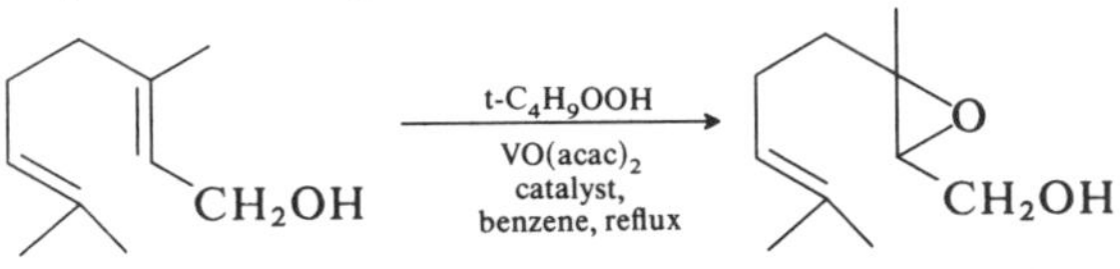

CH_2OH → (t-C_4H_9OOH; $VO(acac)_2$ catalyst, benzene, reflux) → O, CH_2OH (6.49)

(29) (93%)

with peroxy acids reaction takes place preferentially at the other double bond. Another advantage of these metal-catalysed reactions is that they are much more stereoselective than the reactions with peroxy acids. Thus, the acyclic allylic alcohol (30) with t-butyl hydroperoxide and vanadium acetylacetonate gave the *erythro*(*syn*) epoxide (31) almost exclusively;

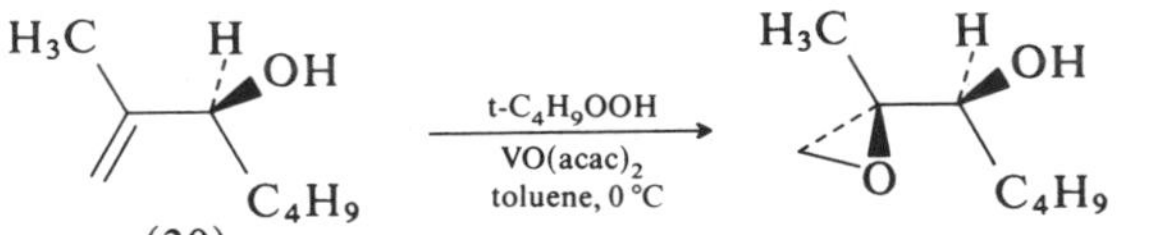

H_3C, H, OH, C_4H_9 (30) → (t-C_4H_9OOH; $VO(acac)_2$, toluene, 0 °C) → H_3C, H, OH, O, C_4H_9 (6.50)

(31) (>99% selectivity)

with *m*-chloroperbenzoic acid alone a mixture of the *erythro* and *threo* isomers was produced. Homoallylic and even bishomoallylic alcohols, in which the hydroxyl group is further removed from the double bond, show the effect (Mihelich, Daniels and Eickhoff, 1981; Fukuyama *et al.*, 1978). Thus, the homoallylic alcohol (32) gave mainly the epoxide (33) and with the cyclic alcohol (34) the *syn* directing effect of the hydroxyl group was even more marked than in the reaction with peroxy acids.

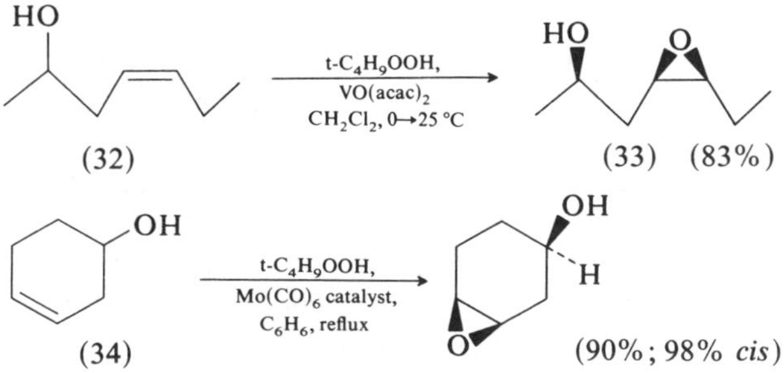

HO (32) → (t-C_4H_9OOH, $VO(acac)_2$, CH_2Cl_2, 0→25 °C) → HO, O (33) (83%)

OH (34) → (t-C_4H_9OOH, $Mo(CO)_6$ catalyst, C_6H_6, reflux) → OH, H, O (90%; 98% *cis*) (6.51)

The precise course of these reactions is not entirely clear, but the rate accelerations and high stereoselectivity observed suggest the formation of an intermediate in which the allylic hydroxyl group is co-ordinated to the metal. It seems to be generally agreed that transfer of oxygen to the double bond takes place in a vanadate ester, possibly of the form (35), in a conformation which minimises steric interactions among the various substituent groups.

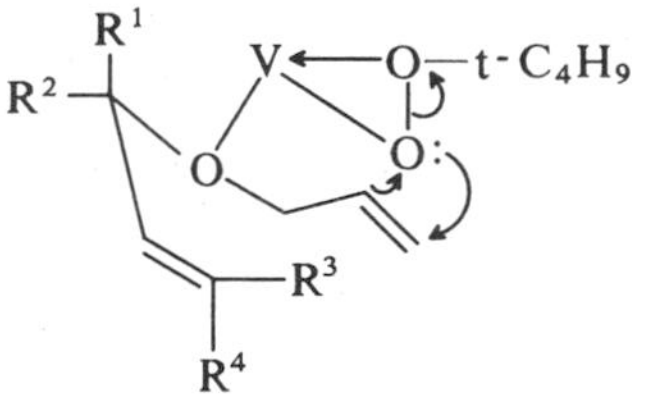

(35)

Related to the vanadium-catalysed reactions is an outstandingly useful new method for the *asymmetric* epoxidation of allylic alcohols, this time catalysed by a titanium complex. Oxidation of allylic alcohols with t-butyl hydroperoxide in the presence of either (+)- or (−)-diethyl tartrate and titanium tetraisopropoxide affords the corresponding asymmetric epoxide in high optical yield. The method is far more stereoselective than any other previously described for this transformation. The new chiral epoxidation system possesses two especially striking and useful features. It gives uniformly high asymmetric inductions throughout a range of substitution patterns in the allylic alcohol and, secondly, the absolute configuration of the epoxide produced can be predicted. Upon use of a given tartrate enantiomer the system delivers the epoxide oxygen to the same face of the double bond regardless of the substitution pattern. As shown in the diagram (36) if the allylic alcohol is drawn so that the hydroxymethyl group is at the lower right, oxygen is delivered at the bottom face in the presence of L-(+)-diethyl tartrate (the natural isomer) and from the top face in the presence of (R)-(−)-diethyl tartrate (Katsuki

D-(−)-diethyl tartrate (unnatural isomer)

"$:\ddot{O}:$"

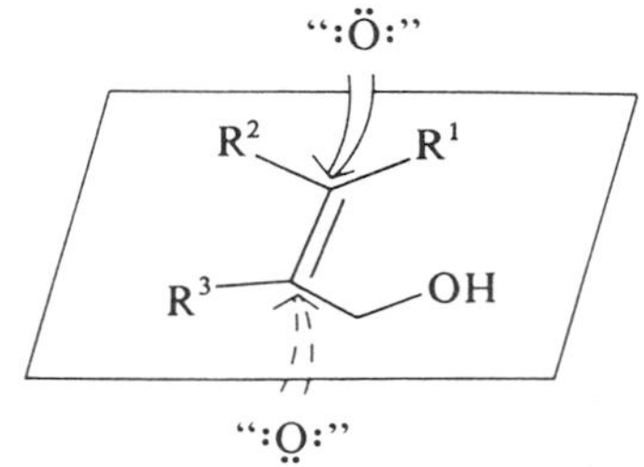

"$:\underset{..}{O}:$"

L-(+)-diethyl tartrate (natural isomer)

(36)

and Sharpless, 1980; Rossiter, Katsuki and Sharpless, 1981; Sharpless *et al.*, 1983). Numerous examples of highly enantioselective epoxidations of allylic alcohols by this procedure have now been recorded. Thus, the allylic alcohol (37) was selectively converted into the epimeric epoxides (38) and (39) in high yield and the octadienol (40), after acetylation, gave the epoxide (41) in greater than 95 per cent enantiomeric excess by selective attack on the allylic double bond.

C_9H_{19} OH

(37)

t-C_4H_9OOH, $Ti(OC_3H_7\text{-iso})_4$, (+)-diethyl tartrate, CH_2Cl_2, −20 °C

(38) (82%; 90% enantiomeric excess)

t-C_4H_9OOH, $Ti(OC_3H_7\text{-iso})_4$, (−)-diethyl tartrate CH_2Cl_2, −20 °C

(39) (80%; 90% enantiomeric excess)

(6.52)

(40)

(1) (+)-diethyl tartrate standard conditions
(2) $(CH_3CO)_2O$, pyridine

$OCOCH_3$

(41) (80%; >95% enantiomeric excess)

The titanium catalyst is sensitive to pre-existing chirality in the substrate, so that the epoxidation of racemic secondary allylic alcohols with a given tartrate–titanium isopropoxide combination proceeds rapidly with only one of the enantiomers, leaving the other slower-reacting enantiomer behind, produced, effectively, by a kinetic resolution (Martin *et al.*, 1981). Thus, in the oxidation of the racemic alcohol (42) using *L*-(+)-diisopropyl tartrate, the (*S*)-enantiomer reacts about a hundred times faster than the (*R*)-enantiomer, leading preponderantly to the *erythro*-epoxide (43). If the reaction is run to only 55 per cent completion the (*R*)-alcohol is recovered with greater than 96 per cent optical purity. In addition to being slower the reaction of the (*R*)-alcohol with the *L*-(+)-tartrate is much less stereoselective. The course of these kinetic resolutions can be predicted. If the secondary alcohol is written in the conventional way with the hydroxyl group at the lower right (see 36), the fast reacting enantiomer when using *L*-(+)-tartrate is the one in

(S)-enantiomer (42) $\xrightarrow[\text{Ti(OC}_3\text{H}_7\text{-iso)}_4\text{ (fast)}]{(L)\text{-(+)-diisopropyl tartrate}}$ erythro (43) + threo (98:2)

H, OH, C_6H_{11}, H

$$(6.53)$$

(R)-enantiomer $\xrightarrow[\text{(slow)}]{\text{some conditions}}$ (38:62)

H, OH, H, C_6H_{11}

which the substituent on the carbinol carbon (cyclohexyl in 42) is up. This new procedure appears to be general and in many cases will provide the best, if not the only, route to optically pure allylic alcohols. The actual catalyst is probably a dimeric titanium species. A reaction pathway invoking orbital-controlled approach of the carbon–carbon double bond to the peroxide oxygen O(1) in the direction of the axis O(1)–O(2) in a complex of the type (44) has been suggested (Sharpless, Woodward and Finn, 1983). More recently, modified catalyst systems have been found which effect epoxidation of allylic alcohols with an enantiofacial selectivity opposite to that obtained under the conditions described above (Lu, Johnson, Finn and Sharpless, 1984).

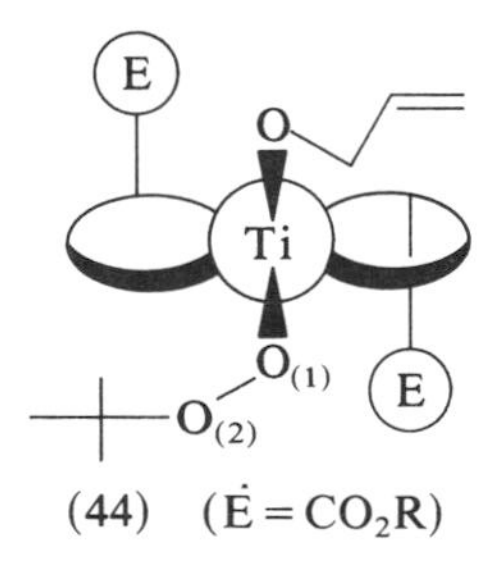

(44) ($\dot{E} = CO_2R$)

Epoxides may also be obtained by the action of base on bromohydrins, themselves conveniently prepared from alkenes by reaction with *N*-bromosuccinimide in aqueous dimethoxyethane. This method has found its most useful application in the selective epoxidation of terminal double bonds in acyclic polyalkenes (van Tamelen and Sharpless, 1967). Squalene, for example, was converted selectively into the 2,3-monoepoxide, and *Cecropea* juvenile hormone was obtained in 52 per cent yield from the corresponding triene (Corey, Katzenellenbogen, Gilman, Roman and Erickson, 1968). The selectivity shown in these reactions is highly dependent on the oxidising agent and on the solvent.

It is believed that in aqueous dimethoxyethane the polyalkene adopts a coiled conformation which protects the internal double bonds from attack. Free hypobromous acid is apparently not involved in the formation of the bromohydrins for it forms different products in reaction with squalene.

Intramolecular attack of alkoxide ion on an alkyl halide is a feature of useful methods for the stereoselective epoxidation of acyclic homoallylic alcohols and of $\gamma\delta$- and $\delta\varepsilon$-unsaturated carboxylic acids (cf. Bartlett, 1980). In the first sequence, the stereochemistry of the hydroxyl group in the homoallylic alcohol is cleverly transferred to the nearer carbon atom of the double bond by first converting the alcohol into its phosphate and thence, stereoselectively, into a cyclic iodophosphate by intramolecular attack on the double bond by the phosphate, triggered by reaction with iodine. Thus, treatment of the phosphate (45) derived from 4-penten-2-ol with iodine in acetonitrile gave the cyclic phosphate (47) in 87 per cent yield through Arbuzov rearrangement of the intermediate (46). Treatment of the cyclic phosphate with sodium ethoxide then gave the dialkyl phosphate of the *erythro* epoxide (48) with less than 2 per cent of the corresponding *threo* derivative (Bartlett and Jernstedt, 1977).

(45) 2 equivs. I_2, CH_3CN; 25 °C ⇌ (46)

(46) → $-C_2H_5I$ → (47)

(47) → $C_2H_5\bar{O}$, THF, 0 °C → (48)

(6.54)

The stereochemistry of the reaction is controlled by the reversible formation of the cyclic phosphonium ion (46), which strongly favours the *cis*-, di-equatorial, conformation shown. The phosphoryl group serves nicely to 'extend' the nucleophilic character of the hydroxyl group of the allylic alcohol, bringing its stereochemical influence within reach of the double bond, and at the same time, being a bulky non-chiral unit, it maximises the steric constraints in the cyclic intermediate phosphonium ion.

A carboxyl group can also be used to direct the epoxidation of an alkene by an 'oxidative cyclisation' process. Iodolactonisation of a $\gamma\delta$- or $\delta\varepsilon$-unsaturated acid under equilibrating conditions (I_2, CH_3CN) ensures thermodynamic control over the stereochemistry of the cyclic intermediate (as 49) and formation of the more stable *trans*-iodolactone. Alkaline methanolysis then leads to the epoxy ester with high asymmetric induction (Bartlett and Myerson, 1978; cf. also Chamberlin, Dezube and Dussault, 1981). Thus, the acid (48) gave predominantly the iodolactone (50) which was converted quantitatively into the epoxide (51) with sodium carbonate in methanol. Epoxides prepared stereoselectively in this way have been employed in the synthesis of a number of natural products (cf. Bartlett, 1980).

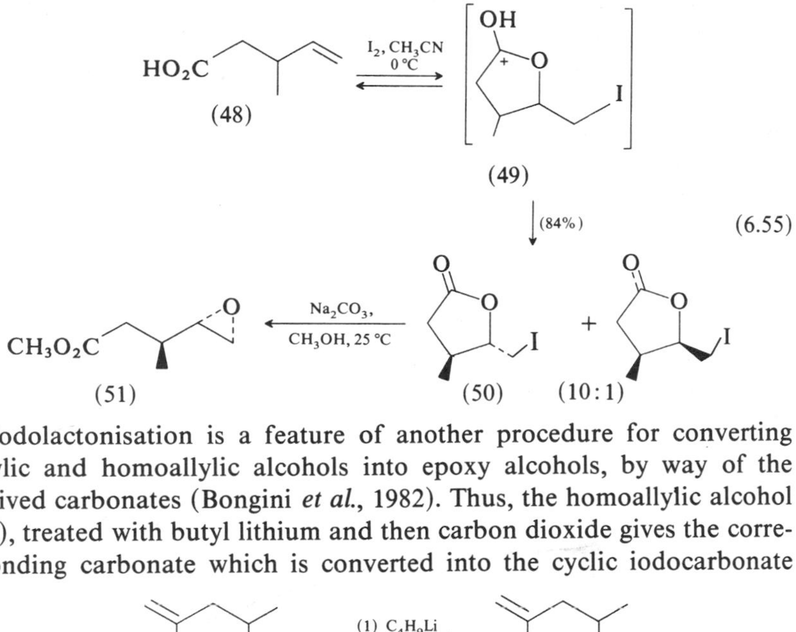

HO_2C (48) — I_2, CH_3CN, 0 °C ⇌ [OH, O, +, I] (49) — (84%) → (50) (10:1) — Na_2CO_3, CH_3OH, 25 °C → CH_3O_2C ... O (51)

(6.55)

Iodolactonisation is a feature of another procedure for converting allylic and homoallylic alcohols into epoxy alcohols, by way of the derived carbonates (Bongini *et al.*, 1982). Thus, the homoallylic alcohol (52), treated with butyl lithium and then carbon dioxide gives the corresponding carbonate which is converted into the cyclic iodocarbonate

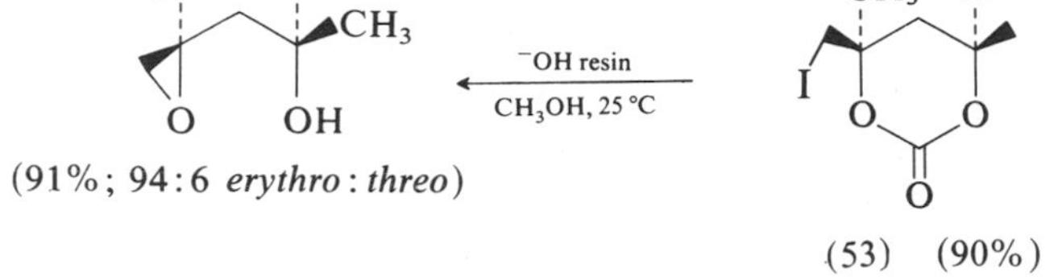

(52) OH — (1) C_4H_9Li (2) CO_2 → HO–C(=O)–O — I_2, THF → (53) (90%) — $^{-}$OH resin, CH_3OH, 25 °C → H_3C, H, CH_3, O, OH (91%; 94:6 *erythro* : *threo*)

(6.56)

(53) by reaction with iodine. The iodocarbonates are smoothly converted stereoselectively into the epoxy alcohols by treatment with a basic resin in methanol at room temperature.

Other useful reagents for the epoxidation of alkenes are the peroxyimidic acids,

$$\underset{\displaystyle \overset{\|}{NH}}{RC}-OOH, \qquad R = C_6H_5, CH_3, Cl_3C,$$

formed *in situ* by reaction of nitriles with hydrogen peroxide; they have the advantage that they can be used under mildly alkaline or neutral conditions (Bach and Knight, 1981; Arias, Adkins, Nagel and Bach, 1983). 2-Allylcyclohexanone was readily converted into the corresponding epoxide with the alkaline reagent, whereas with peroxyacetic acid Baeyer–Villiger ring expansion intervened. They may also show useful stereochemical preferences (cf. Woodward *et al.*, 1973). Peroxyarylseleninic acids, formed *in situ* from arylseleninic acids and hydrogen peroxide, have also been used. A useful feature of these reagents is that they preferentially attack isolated double bonds in the presence of allylic alcohol groups (Grieco *et al.*, 1977; Hori and Sharpless, 1978).

Epoxides can also be prepared from aldehydes and ketones by reaction with sulphur ylids (see p. 138), with arsonium ylids (Still and Novack, 1981) or with α-lithio derivatives of *N*-(*p*-tolylsulphonyl)sulphilimines (Johnson, Mori and Nakanishi, 1979). In an interesting recent development they have also been obtained from β-peroxy carbon free radicals, themselves formed by reduction of peroxy mercurials with sodium borohydride (Corey, Schmidt and Shimoji, 1983). Thus, 2-cyclohexen-1-ol was converted into *trans*-2,3-epoxycyclohexanol in 66 per cent yield by reaction with t-butyldimethylsilyl hydroperoxide and mercury trifluoroacetate followed by reduction of the resulting peroxy mercurial with sodium borohydride (6.57). Direct epoxidation of 2-cyclohexen-1-ol, of course, gives the *cis*-epoxide (p. 373).

OH
t-$C_4H_9(CH_3)_2SiOOH$, $Hg(OCOCF_3)_2$
CH_2Cl_2, −90 °C
OH
—OOR
—$HgOCOCF_3$
$NaBH_4$, THF, H_2O
(6.57)
OH
O
OH
—OOR

Epoxides are useful in synthesis because they react with a variety of nucleophiles with opening of the epoxide ring (Rao, Paknikar and Kirtane, 1983; Smith, 1984). The reactions take place under a wide range of conditions and provide one of the best methods for generating two contiguous stereochemically defined sp^3 carbon atoms. Good nucleophiles (e.g. RS^-, RSi^-) react with epoxides under neutral or alkaline conditions, and in an acidic medium, even weak nucleophiles like water react rapidly. With unsymmetrical epoxides, attack generally, but not always (cf. Behrens and Sharpless, 1983), takes place preferentially at the less substituted carbon atom, with inversion of configuration at the carbon atom attacked, giving a product formed, in effect, by *trans*-addition to the alkene from which the epoxide was prepared. Thus, hydrolysis of cyclohexene epoxide with dilute mineral acid affords *trans*-1,2-dihydroxycyclohexane (6.58), and the sequence of epoxidation followed by hydrolysis of the epoxide is widely used for the preparation of *trans*-1,2-diols from alkenes.

H O H; O; H^+; H_3O^+; OH; OH

(6.58)

C_4H_9 CH_3 → C_4H_9 O CH_3 $\xrightarrow{H_3O^+}$ HO H CH_3 C_4H_9 H OH

In conformationally rigid cyclohexene epoxides reactions leading to opening of the epoxide ring take place with formation of the *trans*-diaxial product in preponderant amount rather than the *trans*-diequatorial one. Thus, acid-catalysed hydrolysis of 2,3-epoxy-*trans*-decalin gives the *trans*-diaxial diol as the only isolated product (6.59). Numerous other examples are known in the steroid series. This propensity for diaxial ring opening determines the regiochemistry of attack on the epoxide by a nucleophile, that is which of two possible products of ring opening will be favoured. For example, reaction of $2\alpha,3\alpha$-epoxy-cholestane (as 54) with toluene-*p*-sulphonic acid gives specifically the 3α-hydroxy-2β-tosyloxy derivative (diaxial) (55), and not the alternative 2-hydroxy-3-tosyloxy compound, and reaction with hydrobromic acid leads to the diaxial bromohydrin (56).

Reduction of epoxides with lithium aluminium hydride affords alcohols. In general reaction takes place at the less substituted carbon of the epoxide to give the more substituted alcohol. With 2,3-epoxy

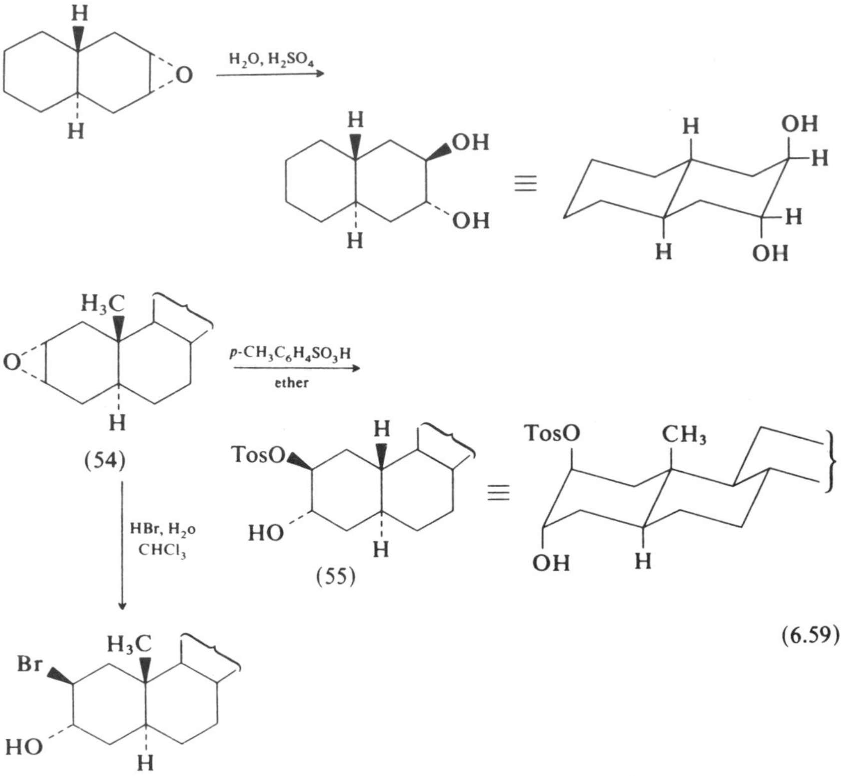

H_2O, H_2SO_4

(54)

p-$CH_3C_6H_4SO_3H$
ether

(55)

HBr, H_2O
$CHCl_3$

(56)

(6.59)

alcohols reduction could give 1,2- and 1,3-diols and in many cases mixtures are obtained, but it has been found recently that 2,3-epoxy alcohols bearing an ethereal oxygen on the α- or β-carbon of the substituent at the other end of the epoxide ring lead very selectively to 1,3-diols on reduction with 'sodium *bis*(methoxyethoxy)aluminium hydride [Red-al, $NaH_2Al(OCH_2CH_2OCH_3)_2$] (6.60). This transformation has already proved valuable in the synthesis of several natural products (Minami, Ko and Kishi, 1982; Ma, Martin, Masamune, Sharpless and Viti, 1982; Behrens and Sharpless, 1983). A mechanism involving intramolecular delivery of hydride ion has been proposed to account for the high selectivities observed.

Reaction of epoxides with carbon nucleophiles, for example Grignard reagents, leads to ring opening with formation of a new carbon–carbon bond. With some Grignard reagents only poor yields are obtained because of competing reactions. In such cases much better results are often obtained if the reaction is carried out in the presence of Cu(I) salts. Thus, cyclohexene oxide and phenylmagnesium bromide gave an 82 per

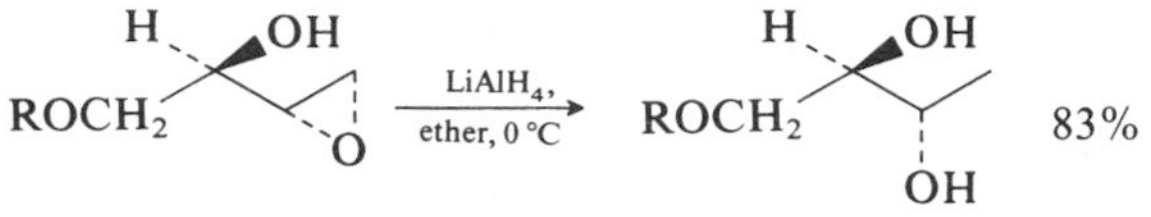

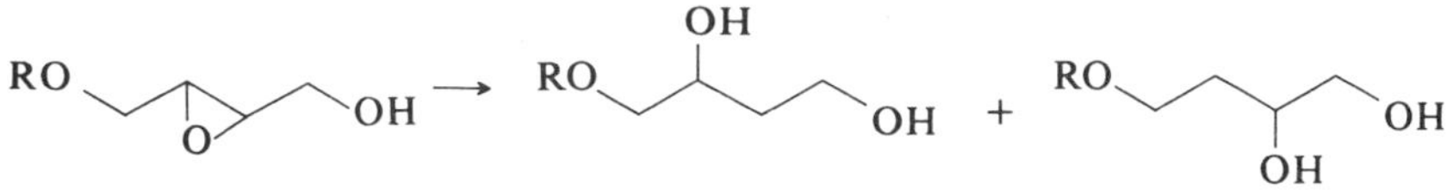

Red-al	100	1
$LiAlH_4$	11	1

(6.60)

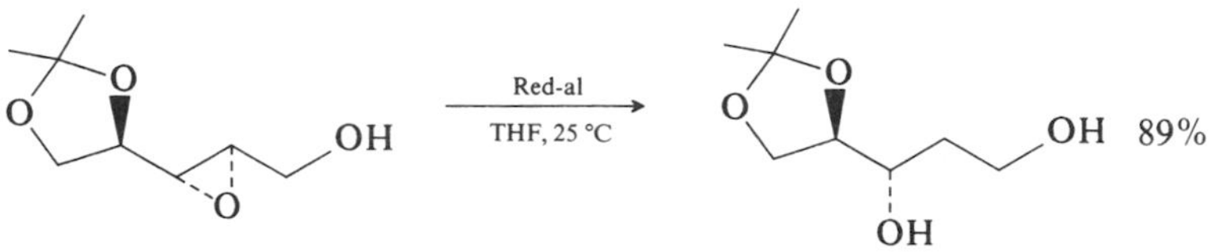

cent yield of *trans*-2-phenylcyclohexanol in the presence of catalytic copperI iodide; without the copper salt the yield was only 3 per cent (6.61) (Huynh, Derguini-Boumechal and Linstrumelle, 1979). Alternatively, the corresponding lithium dialkylcuprate may be used; these reagents react readily with epoxides (see p. 81).

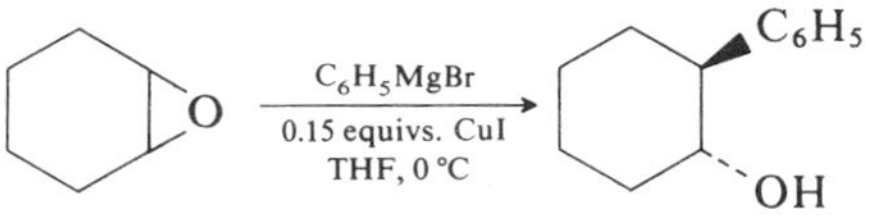

(6.61)

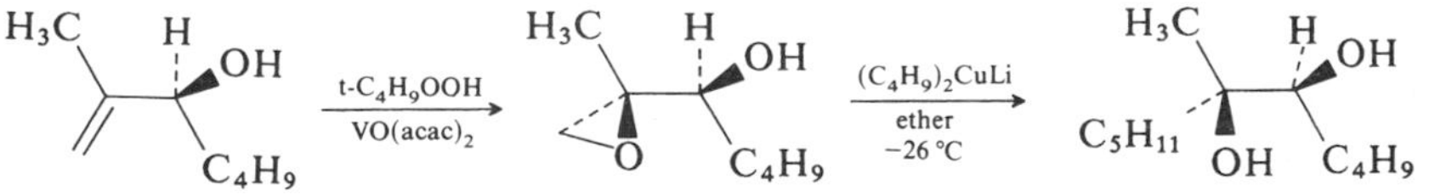

(88% overall)

Epoxides are readily attacked by enolate anions and both inter- and intramolecular reactions are employed in synthesis. Examples showing the formation of carbon rings and an oxygen heterocycle are shown in (6.62). An interesting feature of the reaction with the epoxynitrile is that cyclisation to form a four-membered ring is preferred to formation of a five-membered ring. This is a consequence of geometrical constraints in the transition state leading to the five-membered ring (Stork, Cama and Coulson, 1974). The position of attack on the epoxide is controlled by the line of approach of the enolate anion, and is most favoured when this is collinear with the breaking C—O bond in the epoxide. Models show that collinear attack is more easily achieved in the transition state

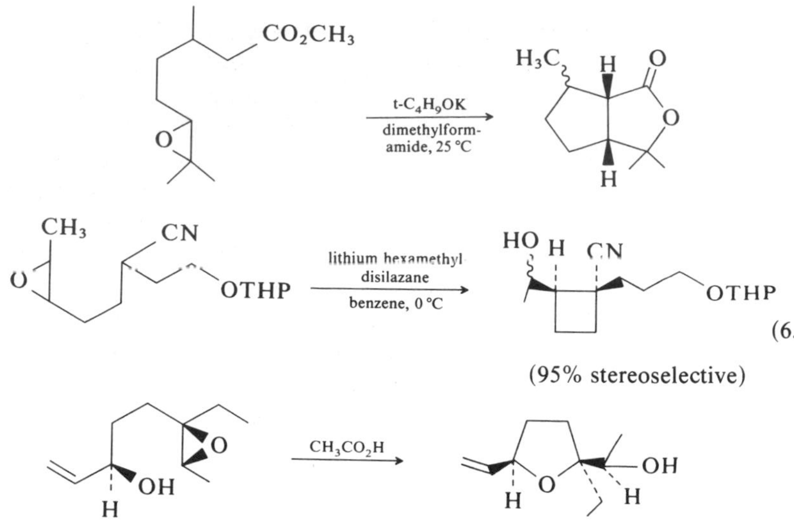

CO_2CH_3 → t-C_4H_9OK, dimethylformamide, 25 °C → H_3C, H, O, O, H

CH_3, O, CN, OTHP → lithium hexamethyldisilazane, benzene, 0 °C → HO, H, CN, OTHP

(6.62)

(95% stereoselective)

O, OH, H → CH_3CO_2H → O, OH, H, H

leading to a four-membered ring than in that which would give a five-membered ring.

De-oxygenation of epoxides to alkenes can be effected in a number of ways. Two of these involve selenium reagents. Reaction of the epoxide with triphenylphosphine selenide and trifluoroacetic acid (Clive and Denyer, 1973) or with potasssium selenocyanate (Behan, Johnstone and Wright, 1975) gives the corresponding alkene directly with retention of configuration of the substituent groups on the epoxide. Both conversions are believed to proceed by extrusion of selenium from the derived episelenide (6.63).

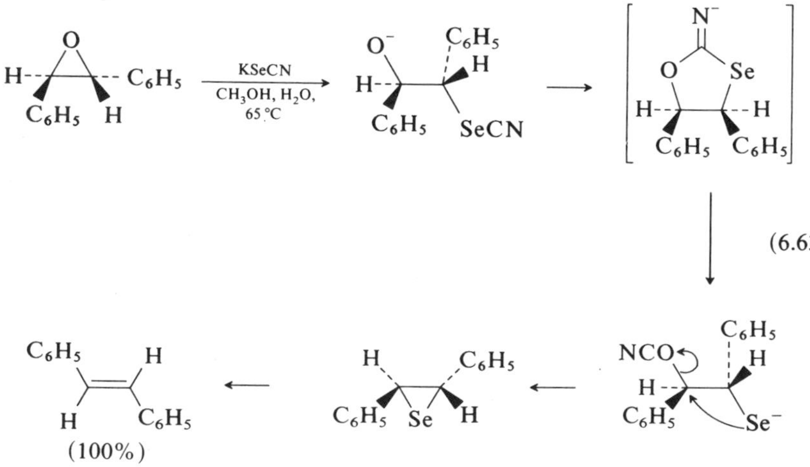

O, H, C_6H_5, C_6H_5, H → KSeCN, CH_3OH, H_2O, 65 °C → O^-, C_6H_5, H, H, C_6H_5, SeCN → [N^-, O, Se, H, H, C_6H_5, C_6H_5]

(6.63)

C_6H_5, H, H, C_6H_5 (100%) ← H, C_6H_5, C_6H_5, Se, H ← NCO, C_6H_5, H, H, C_6H_5, Se^-

Two other routes which also proceed with retention of configuration of substituent groups on the epoxide make use of organometallic reagents. In the first, the epoxide is deoxygenated in a rapid reaction at room temperature with lower-valent tungsten chlorides produced *in situ* by reaction of the hexachloride with an alkyl lithium. *trans*-Cyclodecene oxide, for example, is converted into *trans*-cyclodecene in 94 per cent yield with 95 per cent retention of stereochemistry (Sharpless *et al.*, 1972; Umbreit, and Sharpless, 1981). The second method uses sodium (cyclopentadienyl)-dicarbonylferrate as reagent, and is also highly stereoselective (Ciering, Rosenblum and Tancrede, 1972). Ti(II) species produced by reduction of dry $TiCl_3$ with lithium aluminium hydride reduce epoxides efficiently to alkenes but the reactions are not stereoselective (McMurry and Fleming, 1975). Smooth and quantitative conversion of epoxides into alkenes can also be achieved by reaction with iodotrimethylsilane, formed *in situ* from chlorotrimethylsilane and sodium iodide in acetonitrile, but insufficient examples have been reported to establish the stereoselectivity of the reaction. In one case (Z)-2,5-dimethyl-3-hexene oxide gave (Z)-2,5-dimethyl-3-hexene in 95 per cent yield (Caputo *et al.*, 1981).

$(CH_3)_3SiI$, CH_3CN, 25 °C; $(CH_3)_3SiO$; I^- (95%)

(6.64)

A number of methods are also available for the conversion of epoxides into alkenes with *inversion* of configuration. In one, reaction of the epoxide with lithium diphenylphosphide followed by hydrogen peroxide leads, by a single inversion, to a β-hydroxy-diphenylphosphine oxide. The latter, on treatment with sodium hydride in dimethylformamide undergoes a stereospecific *syn* elimination of diphenylphosphinate to give the alkene (6.65) (Bridges and Whitham, 1974; see also Newton and Whitham, 1979). Since alkenes can be epoxidised with retention of configuration the sequence provides an excellent method for the inversion of configuration of alkenes. *cis*-1-Methylcyclo-octene, for example, was converted into the *trans*-isomer in 63 per cent yield.

A useful synthetic reaction of epoxides is their acid-catalysed rearrangement to carbonyl compounds (Parker and Isaacs, 1959).

$$>C=C< \longrightarrow >C(O)C< \longrightarrow -\overset{|}{\underset{|}{C}}-\overset{|}{C}=O \qquad (6.66)$$

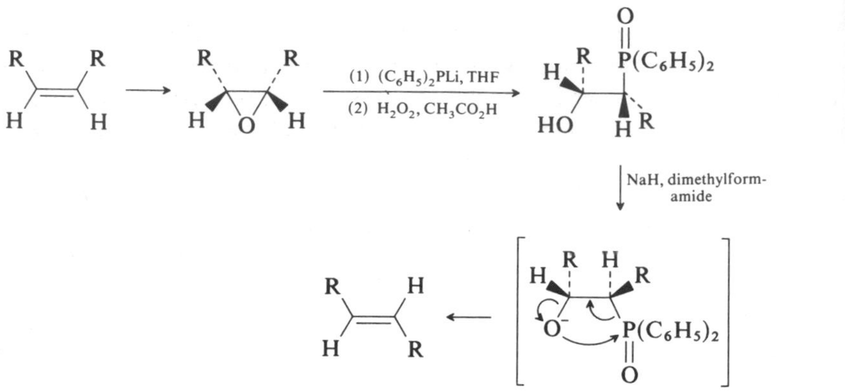

$$\text{RCH=CHR} \rightarrow \text{epoxide} \xrightarrow[\text{(2) } H_2O_2,\ CH_3CO_2H]{\text{(1) } (C_6H_5)_2PLi,\ THF} \text{RCH(OH)CH(R)P(O)(C_6H_5)_2} \xrightarrow{\text{NaH, dimethylformamide}} [\text{intermediate: } O^-,\ P(C_6H_5)_2=O] \rightarrow \text{RCH=CHR}$$

(6.65)

Mineral acids or Lewis acids such as boron trifluoride etherate or magnesium bromide are frequently used as catalysts. In some cases products formed by rearrangement of the carbon skeleton are obtained. Thus, 1-methyl-cyclohexene epoxide gives 2-methylcyclohexanone or 1-formyl-1-methylcyclopentane depending on the experimental conditions (6.67). The method has been used in synthesis for the conversion of cyclohexene derivatives into cyclohexanones, and also in a route to

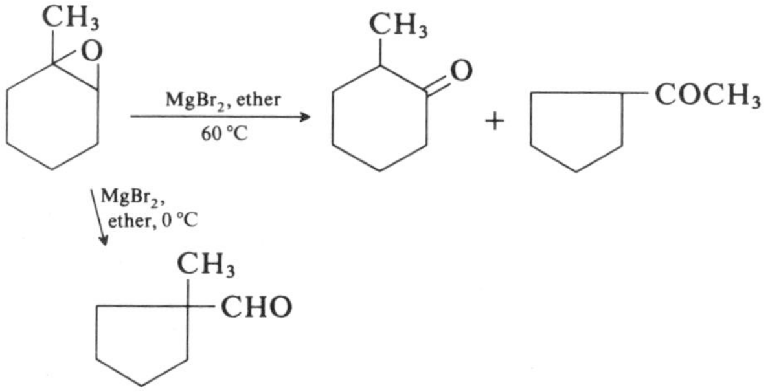

$$\text{1-methylcyclohexene epoxide} \xrightarrow[60\ °C]{MgBr_2,\ \text{ether}} \text{2-methylcyclohexanone} + C_5H_9COCH_3$$

$$\text{1-methylcyclohexene epoxide} \xrightarrow{MgBr_2,\ \text{ether, } 0\ °C} \text{1-}CH_3\text{-1-}CHO\text{-cyclopentane}$$

(6.67)

cyclobutanones by rearrangement of oxaspiropentanes (58). The latter compounds are themselves obtained by epoxidation of alkylidene cyclopropanes or, more conveniently, by reaction of the sulphur ylid (57) with carbonyl compounds (see p. 139).

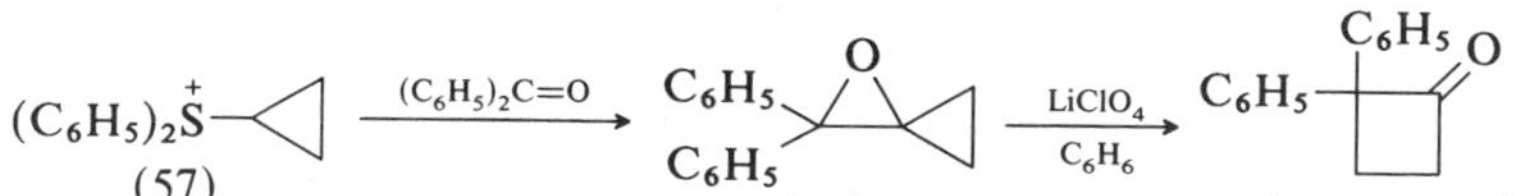

$$(C_6H_5)_2\overset{+}{S}\text{-}C_3H_4\ (57) \xrightarrow{(C_6H_5)_2C=O} (C_6H_5)_2C(O)C_3H_4\ (58) \xrightarrow[C_6H_6]{LiClO_4} \text{2,2-}(C_6H_5)_2\text{cyclobutanone}\ (90\%)$$

(6.68)

Epoxides also undergo a number of base-catalysed rearrangements (cf. Crandall and Luan-Ho Chang, 1967). The most useful synthetically is the rearrangement to allylic alcohols. A number of reagents have been used including lithium diisopropylamide (Rickborn and Thummel, 1969) and, more recently diethylaluminium 2,2,6,6-tetramethylpiperidide (Yasuda, Yamamoto and Nozaki, 1979) and the magnesium derivative of isopropylcyclohexylamine (Corey *et al.*, 1980). Reaction proceeds by abstraction of a proton from a carbon adjacent to the epoxide ring and with unsymmetrical epoxides can be highly selective. For example the epoxide (59) with diethylaluminium 2,2,6,6-tetramethylpiperidide gave the disubstituted allylic alcohol (60) in 96 per cent yield, while the isomeric epoxide (61), under the same conditions, give the trisubstituted *E*-allylic alcohol (62). Reaction is believed to take place by concerted

$LiN(C_2H_5)_2$, ether, reflux (68%)

$(C_2H_5)_2Al{-}N$, benzene, 0 °C: (59) → (60) (96%)

(6.69)

$(C_2H_5)_2Al{-}N$, benzene, 0 °C: (61) → (62) (83%)

syn elimination in a boat-like six-membered transition state (63) in which the substituents on the oxirane ring are arranged so as to minimise non-bonded interactions. Thus, for (61), transition state (63, $R^1 = Bu$) which leads to the *E*-alkene (62) actually obtained is preferred over (63, $R^2 = Bu$) which would have given the *Z*-alkene. *Syn* elimination in a cyclic transition state has been proposed for reactions with lithium diisopropylamide also (Thummel and Rickborn, 1970).

(63)

Ozonolysis

Ozonolysis, that is reaction of an alkene with ozone followed by splitting of the resulting ozonide, is a very convenient method for oxidative cleavage of carbon–carbon double bonds.

Recent physical measurements have shown that the ozone molecule is a resonance hybrid of the structures shown in (6.70). It is an electrophilic reagent and reacts with carbon–carbon double bonds to form

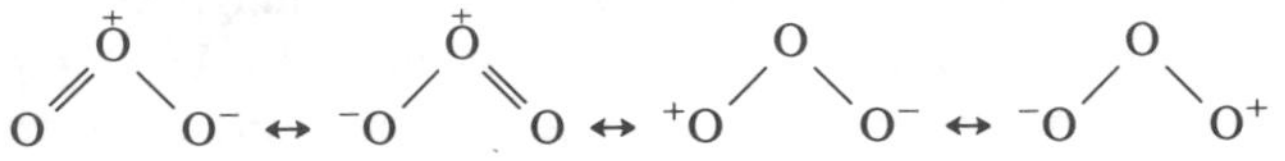

$$O{=}\overset{+}{O}{-}O^- \leftrightarrow {}^-O{-}\overset{+}{O}{=}O \leftrightarrow {}^+O{-}O{-}O^- \leftrightarrow {}^-O{-}O{-}O^+ \qquad (6.70)$$

ozonides which can be cleaved oxidatively or reductively to carboxylic acids, ketones or aldehydes; the nature of the products formed depends on the method used and on the structure of the alkene (Bailey, 1978, 1982). The reaction is usually carried out by passing a stream of oxygen containing 2–10 per cent of ozone into a solution or suspension of the compound in a suitable solvent, such as dichloromethane or methanol,

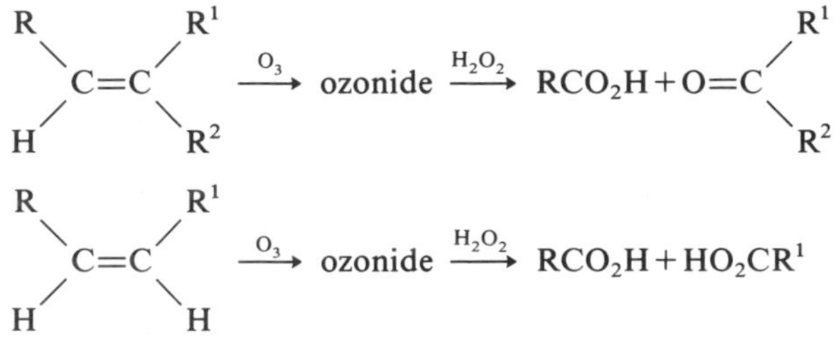

$$\begin{gathered} RHC{=}CR^1R^2 \xrightarrow{O_3} \text{ozonide} \xrightarrow{H_2O_2} RCO_2H + O{=}CR^1R^2 \\ RHC{=}CHR^1 \xrightarrow{O_3} \text{ozonide} \xrightarrow{H_2O_2} RCO_2H + HO_2CR^1 \end{gathered} \qquad (6.71)$$

at or below room temperature. Oxidation of the crude ozonisation product, without isolation, by hydrogen peroxide or other reagent, leads normally to carboxylic acids or ketones or both, depending on the degree of substitution of the alkene (6.71). Thus oleic acid gives azelaic and pelargonic acids (6.72).

$$CH_3(CH_2)_7CH{=}CH(CH_2)_7CO_2H \xrightarrow{(1)\ O_3;\ (2)\ \text{oxidation}} CH_3(CH_2)_7CO_2H + HO_2C(CH_2)_7CO_2H \qquad (6.72)$$

Reductive decomposition of the crude ozonide leads to aldehydes and ketones (6.73). Various methods of reduction have been used including catalytic hydrogenation and reduction with zinc and acids or with triethyl phosphite, but, in general, yields of aldehydes have not been high. Reaction with dimethyl sulphide in methanol has been found to give excellent results and this reagent appears to be superior to all others

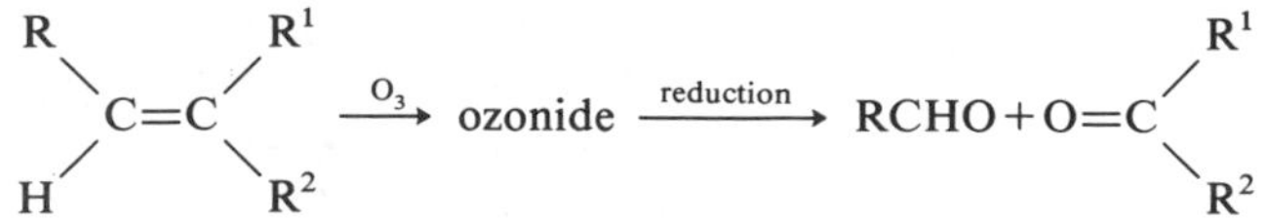

$$RHC{=}CR^1R^2 \xrightarrow{O_3} \text{ozonide} \xrightarrow{\text{reduction}} RCHO + O{=}CR^1R^2 \qquad (6.73)$$

previously used. Reaction takes place under neutral conditions and the reagent is highly selective; nitro and carbonyl groups, for example, elsewhere in the molecule are not affected (6.74). The procedure hinges

$$CH_2{=}CH(CH_2)_5CH_3 \xrightarrow{O_3} \text{ozonide} \xrightarrow{CH_3SCH_3,\ CH_3OH} \underset{(75\%)}{CHO(CH_2)_5CH_3 + CH_2O} \tag{6.74}$$

on the fact that hydroperoxides are rapidly and cleanly reduced to alcohols by sulphides (6.75). Ozonization of an alkene in methanol solution gives rise to a hydroperoxide which is reduced in the same way by

$$(CH_3)_3C{-}O{-}OH \xrightarrow{(CH_3)_2S} (CH_3)_3COH + (CH_3)_2SO \tag{6.75}$$

dimethyl sulphide to the hemi-acetal as shown (6.76). Equally good results have been obtained using odourless thiourea (Gupta, Soman and Dev, 1982).

$$>C(OCH_3)(O{-}OH) \xrightarrow{(CH_3)_2S} >C(OCH_3)(OH) \longrightarrow >C{=}O + CH_3OH + (CH_3)_2SO \tag{6.76}$$

Largely through the work of Criegee (1975) it now seems clear that most normal ozonolyses proceed by formation of a primary ozonide which decomposes to give a zwitterion and a carbonyl compound. The fate of the zwitterion depends on its structure, on the structure of the carbonyl compound and on the solvent. In an inert (non participating) solvent, if the carbonyl compound is reactive, it may react with the zwitterion to form an ozonide (64); otherwise the zwitterion may dimerise to the peroxide (65) or give ill-defined polymers. In nucleophilic solvents such as methanol or acetic acid, however, hydroperoxides of the type of (66) are formed (6.77). Strong evidence for the intermediacy of a zwitterion was found by Criegee in the ozonolysis of tetramethylethylene. In an inert solvent the cyclic peroxide and acetone were obtained. But when formaldehyde was added to the reaction mixture the known ozonide of isobutene was isolated. In the first case the intermediate zwitterion has dimerised; but in the second it has reacted preferentially with the highly reactive carbonyl compound (6.78). Criegee's general scheme is supported by much recent experimental work, although it appears that in some cases the exact mechanism is sensitive to a number of experimental factors (Murray, 1968; Kuczkowski, 1983).

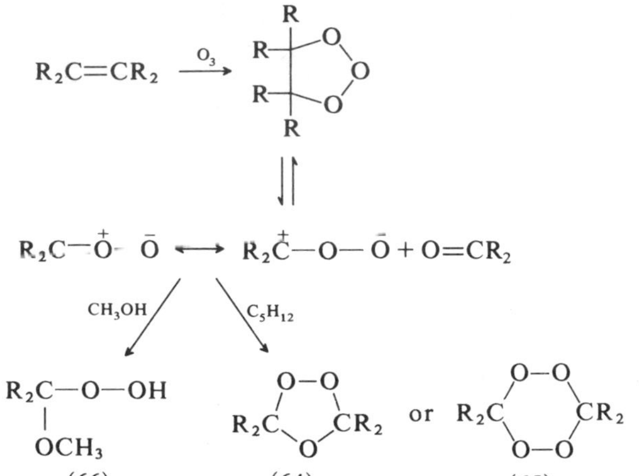

$$R_2C{=}CR_2 \xrightarrow{O_3} \text{(cyclic trioxolane)} \rightleftharpoons R_2C{-}\overset{+}{O}{-}\overset{-}{O} \longleftrightarrow R_2\overset{+}{C}{-}O{-}\overset{-}{O} + O{=}CR_2 \qquad (6.77)$$

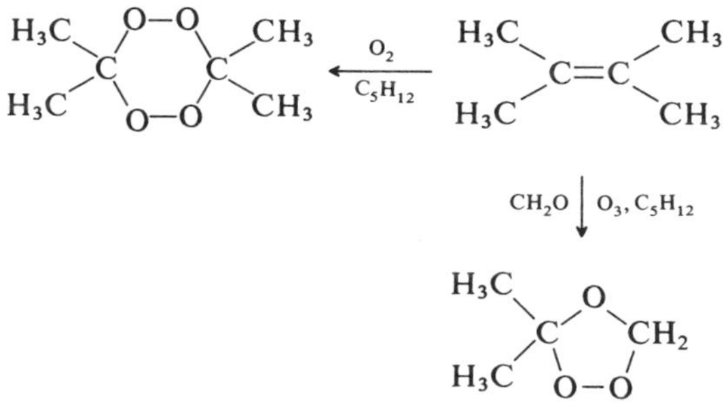

CH_3OH: $R_2C(OCH_3){-}O{-}OH$ (66); C_5H_{12}: (64) or (65)

$$\text{(6.78)} \quad O_2,\ C_5H_{12};\ CH_2O \mid O_3, C_5H_{12}$$

$\alpha\beta$-Unsaturated ketones or acids generally give products containing fewer than the expected number of carbon atoms. Thus, the tricyclic $\alpha\beta$-unsaturated ketone shown in (6.79) is converted into the keto acid

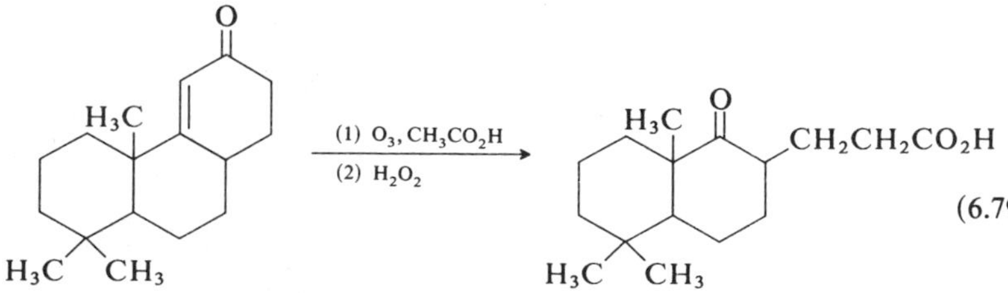

$$\xrightarrow[(2)\ H_2O_2]{(1)\ O_3,\ CH_3CO_2H} \quad CH_2CH_2CO_2H \qquad (6.79)$$

with loss of a carbon atom. The following general mechanism involving a release of electrons by O—H heterolysis has been suggested to account for these results (6.80). Abnormal results have also been observed in the oxonolysis of allylic alcohols, and amines.

Ozonolysis is widely used both in degradative work and in synthesis for the preparation of aldehydes, ketones and carboxylic acids. Reaction

$$R{-}CO{-}C(R^1){=}C\langle \xrightarrow[H_2O]{O_3} HO{-}C(R)(OH){-}C(R^1)\text{(ozonide)} \longrightarrow HO{-}C(R){=}O + R^1{-}C(=O){-}O^- + O{=}C\langle$$

$$\rangle C{=}C(R)CO_2H \xrightarrow{O_3} \text{ozonide} \longrightarrow \rangle{=}O + HO{-}C(R){=}O + CO_2 \qquad (6.80)$$

need not be confined to the straightforward ozonisation of alkenes. Cyclic ketones can serve as starting materials, and ozonisation of the derived enol ethers provides a convenient route to aliphatic compounds with differing functional groups at the ends of the chain (6.81).

$$\text{cyclohexanone} \longrightarrow \text{1-}OC_2H_5\text{-cyclohexene} \xrightarrow[(2)\ H_2,\ Pd]{(1)\ O_3} CO_2C_2H_5(CH_2)_4CHO \qquad (6.81)$$

A useful alternative to ozonolysis is oxidation with periodate in presence of a catalytic amount of potassium permanganate (6.82). With this reagent double bonds are cleaved to give ketones and aldehydes or, more usually, carboxylic acids formed by further oxidation of the aldehydes.

$$\text{(citronellal)}\ CHO \xrightarrow[H_2O,\ t\text{-}C_4H_9OH]{NaIO_4,\ KMnO_4} CO_2H\ \ CO_2H + CH_3COCH_3$$

$$CH_3(CH_2)_5CH(OCOCH_3)CH_2CH{=}CH(CH_2)_5CH_2OH \qquad (6.82)$$

$$\downarrow NaIO_4,\ KMnO_4,\ K_2CO_3;\ H_2O,\ t\text{-}C_4H_9OH$$

$$CH_3(CH_2)_5CH(OCOCH_3)CH_2CO_2H + HO_2C(CH_2)_5CH_2OH$$

Thus, citronellal is oxidised to acetone and 3-methyladipic acid in high yield, and the unsaturated alcohol shown (6.82) is cleaved without attack on the acetoxy and primary alcohol groups. Reaction is conducted at pH 7–8, and under these conditions the α-hydroxyketone or the 1,2-diol formed from the alkene with permanganate is cleaved by the periodate to carbonyl compounds. The permanganate is reduced only to the manganate stage at this pH and is re-oxidised by the periodate, which itself does not attack the double bond. Only catalytic amounts of permanganate are thus needed.

Similar results are obtained with a combination of periodate and osmium tetroxide, and this technique has the advantage that it does not proceed beyond the aldehyde stage. It produces the same result as ozonolysis followed by reductive cleavage of the ozonide. Cyclohexene, for example, is converted into adipaldehyde in 77 per cent yield. The osmium tetroxide oxidises the alkene to the diol, which is cleaved by the periodate. Catalytic amounts of osmium tetroxide are sufficient because the periodate oxidises the reduced osmium back to the tetroxide.

Another excellent method for cleaving carbon–carbon double bonds is by action of ruthenium tetroxide in combination with sodium periodate (see below). Carboxylic acids are usually produced from disubstituted olefins with this reagent.

6.4. Photosensitised oxidation of alkenes

Irradiation of dilute solutions of alkenes and conjugated dienes in presence of oxygen and a sensitiser gives rise to hydroperoxides and cyclic peroxides (6.83) (Denny and Nickon, 1973). The addition to dienes

$$\underset{\alpha}{\overset{H}{-C}}-\underset{\beta}{C}=\underset{\gamma}{C}- \xrightarrow[\text{sensitiser}]{O_2,\ h\nu} -\underset{\alpha}{C}=\underset{\beta}{C}-\underset{\gamma}{\overset{OOH}{C}}- \xrightarrow{\text{reduction}} -\underset{\alpha}{C}=\underset{\beta}{C}-\underset{\gamma}{\overset{OH}{C}}- \qquad (6.83)$$

$O_2,\ h\nu$ / sensitiser

is analogous to the Diels–Alder reaction and is discussed further on p. 196).

Photosensitised oxygenation of monoalkenes has been widely studied and provides a convenient method for introducing oxygen in a highly specific fashion into such compounds (Gollnick, 1968). The first products of the reaction are allylic hydroperoxides which may be isolated if desired or reduced directly to the corresponding allylic alcohols. The sequence thus provides a method for converting an alkene into an allylic alcohol with shift in the position of the double bond. For example, α-pinene is converted into *trans*-pinocarveyl hydroperoxide which on reduction gives *trans*-pinocarveol, and 2-methyl-2-pentene affords, after reduction of the hydroperoxides, a mixture of two methylpentenols corresponding to attack at each end of the double bond (6.84).

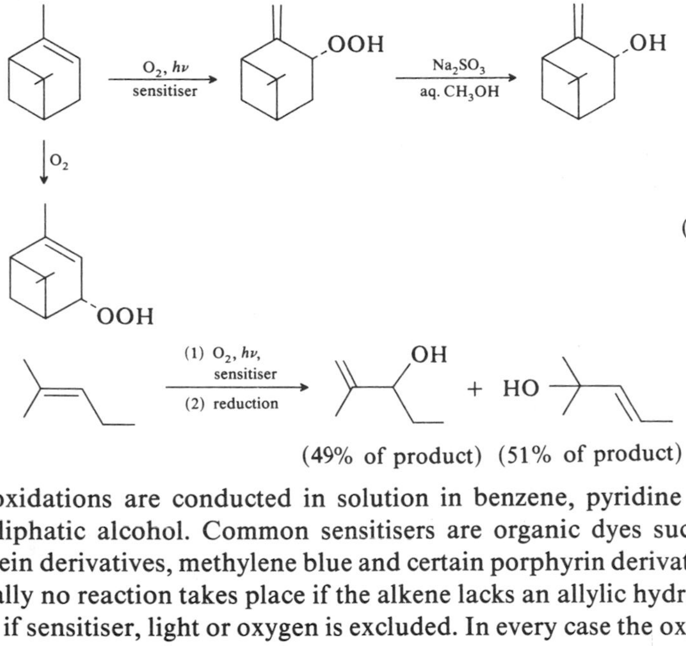

OOH
O_2, $h\nu$
sensitiser
OH
Na_2SO_3
aq. CH_3OH
O_2
OOH

(6.84)

(1) O_2, $h\nu$, sensitiser
(2) reduction
OH
+ HO

(49% of product) (51% of product)

The oxidations are conducted in solution in benzene, pyridine or a lower aliphatic alcohol. Common sensitisers are organic dyes such as fluorescein derivatives, methylene blue and certain porphyrin derivatives. Essentially no reaction takes place if the alkene lacks an allylic hydrogen atom or if sensitiser, light or oxygen is excluded. In every case the oxygen molecule adds to one carbon of the double bond and a hydrogen atom from the allylic position migrates to the oxygen with concomitant shift of the double bond. The reaction is generally held to proceed by a concerted cyclic mechanism, analogous to that postulated for the 'ene' reaction (p. 244) (6.85) (Foote, 1971; Yamaguchi, Yabushita, Fueno and

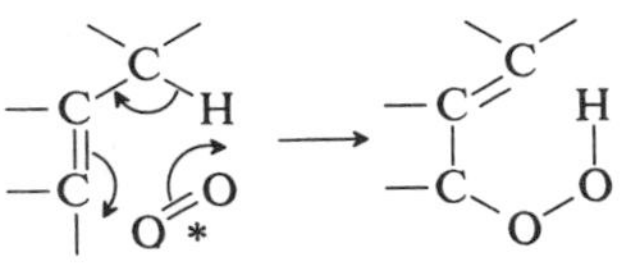

C H C C O O* → C H C C O O

(6.85)

Houk, 1981; Stephenson, Grdinon and Orfanipoulos, 1980) although other pathways involving biradials or an intermediate perepoxide have been considered.

The reaction should be distinguished from the familiar *radical* oxidation of alkenes by triplet oxygen which can also be initiated by sensitisers (such as benzophenone) and which may also give rise to allylic hydroperoxides. It is only in exceptional cases, however, that the products of these autoxidations are the same as those of photo-oxygenation. Autoxidation of α-pinene, for example, leads to verbenyl hydroperoxide and not to pinocarveyl hydroperoxide. The photosensitised reactions do not involve free-radical intermediates, as autoxidations do. This is shown, for example, by the fact that the alcohol (67) obtained as one product from photosensitised oxygenation of (+)-limonene is optically active (6.86). A (symmetrical) allylic free-radical intermediate (68) would have given a racemic product.

(1) O_2, $h\nu$, sensitiser
(2) reduction

HO

(6.86)

(67) optically pure (68)

Photosensitised oxygenation is useful in synthesis for the specific introduction of an oxygenated functional group at the site of a double bond, and has been employed in the synthesis of a number of natural products (cf. Wasserman and Ives, 1981). Thus, the compound (69), an analogue of the diterpene alkaloid garryfoline, was readily obtained by photo-oxygenation of an olefinic double bond precursor (6.87).

H_3C H CH_3 H H_3C CH_3

(1) O_2, $h\nu$, sensitiser
(2) $LiAlH_4$

H_3C H CH_3 OH H_3C CH_3

(6.87)

(69) (46%)
(two epimers formed)

Photo-oxygenations in many cases show a high degree of stereoselectivity. Experiments using steroidal alkenes have revealed that, in agreement with the proposed cyclic reaction path (p. 393), it is a *cis* reaction, that is the new C—O bond is formed on the same side of the molecule as the C—H bond which is broken. This is shown, for example, by photo-oxygenation of the deuterated cholesterols (6.88). The 7α-deutero compound gave, after reduction of the hydroperoxide, a diol which retained only 8 per cent of the original deuterium, whereas the diol from the 7β-deutero compound retained 95 per cent of the deuterium. In both these reactions approach of the activated oxygen takes place from the less hindered α-side of the steroid molecule, with migration of the *cis* ^{2}H or H from C-7 to oxygen.

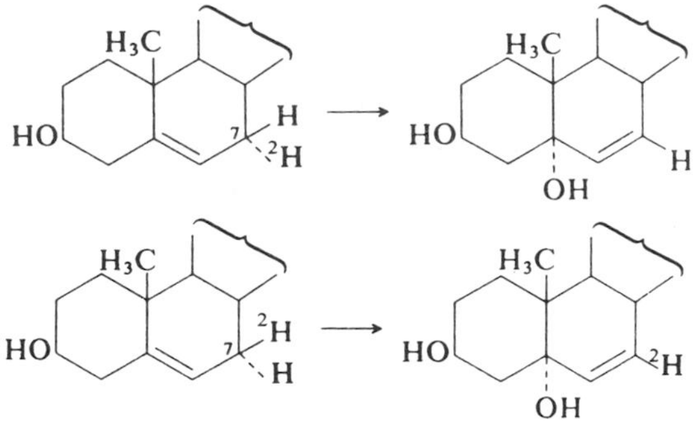

(6.88)

In all these sensitised photo-oxygenations attack of the activated oxygen on the π-orbital of the double bond takes place in a direction perpendicular to the olefinic plane, and the allylic hydrogen atom must be suitably oriented to allow transfer to oxygen. This requirement is best met when the C—H bond is perpendicular to the plane of the double bond, for this favours overlap of the developing empty *p*-orbital on the γ carbon atom with the π-orbital of the double bond (6.89). A consequence of this is that in cyclohexenoid systems, a *quasi* axial hydrogen

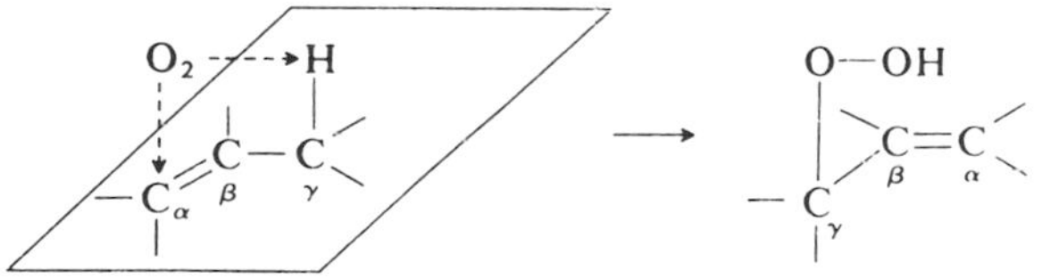

(6.89)

atom should be better disposed for reaction than a *quasi* equatorial one. In agreement, 5α-cholest-3-ene, with a *quasi* axial C-5—H bond, is more readily oxygenated than is 5β-cholest-3-ene, in which the C-5—H is *quasi* equatorial (6.90).

Consideration of the stereoelectronic course of the reaction also serves to explain some apparently anomalous results obtained with open chain alkenes. 2,4-Dimethyl-2-pentene, for example, affords the secondary hydroperoxide in more than 95 per cent yield; almost none of the tertiary hydroperoxide is formed. The most stable conformation of the alkene is

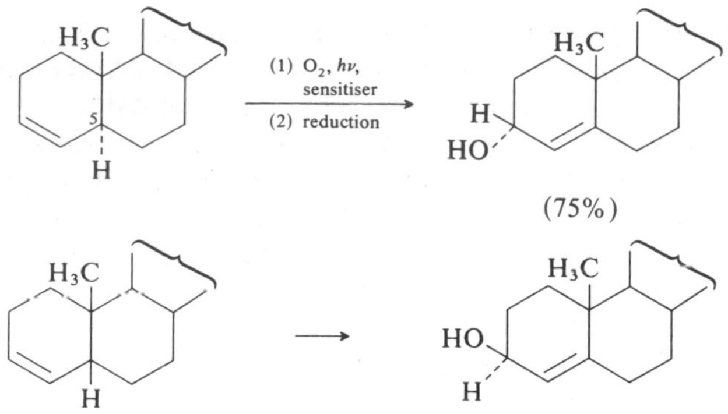

(1) O_2, $h\nu$, sensitiser

(2) reduction

(75%)

(6.90)

(40%)

(70) in which the hydrogen atom at C-3 almost eclipses the double bond. The alternative conformation (71) necessary for photo-oxygenation at C-1, in which the C-3—H bond is perpendicular to the double bond, is disfavoured by steric interaction between a C-3-methyl group and the dimethyl vinyl group. Consequently very little tertiary hydroperoxide is formed (6.91). In contrast, 1,1-dimethylbutene gives roughly equal

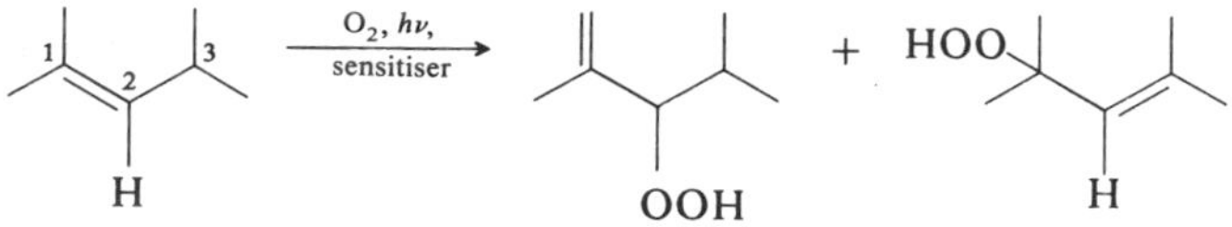

O_2, $h\nu$, sensitiser

(>95%) (<5%) (6.91)

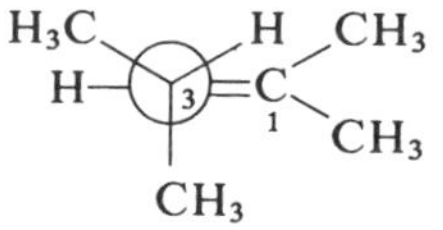

(70) 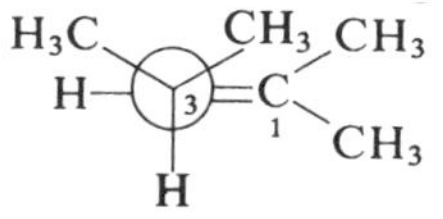 (71)

amounts of secondary and tertiary hydroperoxides because the conformation (72) appropriate to tertiary peroxide formation is no longer destabilised by steric factors (6.92).

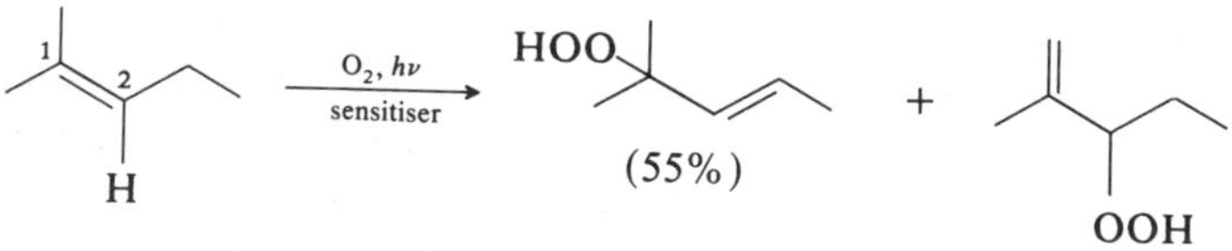

O_2, $h\nu$ sensitiser

(55%) (45%) (6.92)

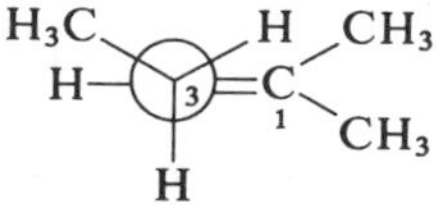

(72)

The reactive species in these reactions is believed to be singlet oxygen, formed by energy transfer from the light-energised triplet sensitiser to oxygen or an oxygen-sensitiser complex (cf. Wagner and Hammond, 1968). In line with the suggestion that singlet oxygen is involved it has been shown (Foote, 1968) that chemically generated singlet oxygen, obtained by reaction of hypochlorites with hydrogen peroxide, gives product distributions on oxidation of alkenes which are identical with those obtained in sensitised photo-oxygenations. The sensitisers used in the photo-oxidations are very bulky molecules and would be expected to exert a strong influence on the stereochemistry of the reaction intermediate if they were present in the transition state. No such effect is observed, and this would seem to tell against the suggestion that the active species is an oxygen–triplet sensitiser complex.

The synthesis of allylic alcohols by photo-oxygenation of alkenes is not without its difficulties, and is sometimes not very selective (cf. Adams, 1971). An alternative procedure makes use of selenium chemistry. Electrophilic addition to the alkene of phenylselenenic acid, PhSeOH, generated *in situ* from diphenyldiselenide and hydrogen peroxide, affords the 2-hydroxyphenylselenide. Oxidation with t-butyl hydroperoxide, without isolation, then gives the corresponding selenoxide which readily eliminates phenylselenenic acid (cf. p. 121) to give the allylic alcohol (6.93) (Hori and Sharpless, 1978).

SeC_6H_5

$C_6H_5SeSeC_6H_5$

H_2O_2, CH_2Cl_2
0 °C
[C_6H_5SeOH]

OH (6.93)

t-C_4H_9OOH, CH_2Cl_2,
0° → 20 °C

OH

6.5. Palladium-catalysed oxidation of alkenes

The oxidation of ethylene to acetaldehyde by oxygen and a solution of Pd(II) in aqueous hydrochloric acid is an important industrial process (the Wacker reaction). The palladium(II) is simultaneously reduced to the metal, but the reaction is made catalytic by addition of Cu(II) chloride in the presence of air or oxygen, whereby the palladium is continuously re-oxidised to Pd(II)

$$CH_2{=}CH_2 + PdCl_2 + H_2O \rightarrow CH_3CHO + Pd + 2HCl$$

$$Pd + 2CuCl_2 \rightarrow PdCl_2 + 2CuCl$$

$$2CuCl + \tfrac{1}{2}O_2 + 2HCl \rightarrow 2CuCl_2 + H_2O$$

The mechanism of this reaction has been very widely studied and it is now believed that it proceeds by an initial *trans* hydroxypalladation of the ethylene to form an unstable complex which then rapidly undergoes β-elimination with transfer of hydride ion from one carbon of the ethylene to the other, via the palladium. The hydride migration is required to explain the observation that when the reaction is conducted in deuterium oxide no deuterium is incorporated in the acetaldehyde produced. The

$$CH_3CH{=}O + H_2O + PdCl_2 \qquad (6.94)$$

reaction has been extended, with modifications, (Januszkiewicz and Alper, 1983; Mimoun *et al.*, 1980; Roussel and Mimoun, 1980) to provide a good laboratory method for the oxidation of terminal alkenes to methyl ketones. 1-Decene, for example, was converted into 2-decanone in 73 per cent yield with oxygen, palladium chloride and copperII chloride in aqueous benzene containing a quaternary ammonium phase transfer reagent. Internal alkenes are oxidised only slowly so that in the oxidation of dienes containing terminal and internal double bonds selective oxidation of the terminal double bond is possible (6.95).

$$\xrightarrow[H_2O,\ \text{dimethylformamide}]{PdCl_2,\ CuCl,\ O_2} \qquad (6.95)$$

6.6. Oxidation of ketones

Oxidation of ketones with chromic acid or permanganate under conditions vigorous enough to bring about reaction usually leads to rupture of the carbon chain adjacent to the carbonyl group with formation of carboxylic acids, and is of little value in synthesis. More important are controlled methods of oxidation leading to $\alpha\beta$-unsaturated ketones, α-ketols (acyloins) or lactones without disruption of the molecule.

$\alpha\beta$-Unsaturated ketones

Conversion of ketones into $\alpha\beta$-unsaturated ketones has been effected by bromination–dehydrobromination and, in certain cases, by oxidation with selenium dioxide or high potential quinones such as chloranil or 5,6-dichloro-2,3-dicyano-*p*-benzoquinone but none of these methods is entirely satisfactory. Yields are sometimes not good and the reactions lack selectivity.

A better method which gives excellent yields of $\alpha\beta$-unsaturated compounds under mild conditions, and which is of wide application proceeds from the α-phenylseleno carbonyl compound (as 73), which is itself readily obtained from the carbonyl compound by reaction with phenylselenyl chloride at room temperature, or from the corresponding enolate anion and the selenyl halide or diphenyl diselenide at −78 °C. By oxidation with hydrogen peroxide or sodium periodate the selenide is converted into the corresponding selenoxide which immediately undergoes *syn* β-elimination to form the *trans*-$\alpha\beta$-unsaturated ketone in high yield (cf. p. 121). Functional groups such as alcoholic hydroxyls, ester groups and carbon–carbon double bonds appear not to interfere. Thus, propiophenone is converted in 89 per cent yield into phenyl vinyl ketone, an alkene which is difficult to obtain by other means because of its ready polymerisation and susceptibility to nucleophilic attack, and 4-acetoxycyclohexanone gives 4-acetoxycyclohexenone. The procedure can also be used to make $\alpha\beta$-unsaturated esters and lactones from the saturated precursors, as shown in (6.96).

The sequence provides a method (*a*) for converting $\alpha\beta$-unsaturated ketones into β-alkyl derivatives by alkylation with an organocuprate and reaction of the intermediate copper enolate with phenylselenyl bromide, and (*b*) for the synthesis of enediones from 1,3-dicarbonyl compounds, a transformation which is difficult to bring about by other means (6.97) (Reich, Renga and Reich, 1974).

The reaction of enolates with phenylselenyl halides is very fast even at −78 °C and the kinetically generated enolates react without rearrangement to the more stable isomer. Unsymmetrical ketones may therefore be converted at will into one or other of the two alternative $\alpha\beta$-unsaturated compounds. 2-Methylcyclohexanone, for example, gave 2-methyl-2-cyclohexenone or 6-methyl-2-cyclohexenone selectively by way of the corresponding thermodynamic or kinetic enolate.

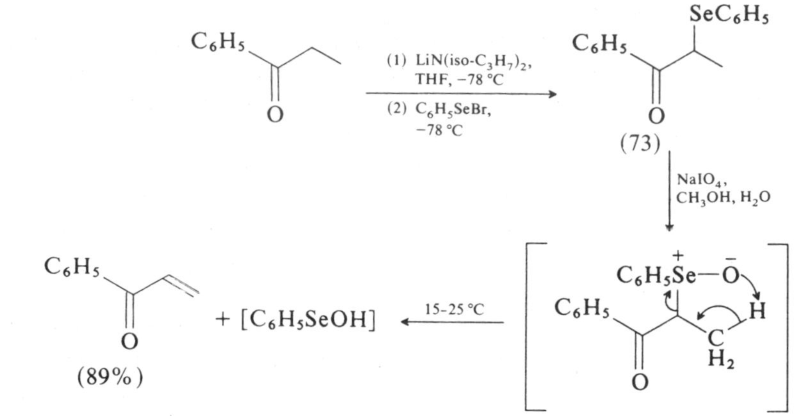

C_6H_5 O

(1) $LiN(iso\text{-}C_3H_7)_2$, THF, −78 °C
(2) C_6H_5SeBr, −78 °C

SeC_6H_5 C_6H_5 O

(73)

$NaIO_4$, CH_3OH, H_2O

C_6H_5 O

$+ [C_6H_5SeOH]$

15–25 °C

$C_6H_5\overset{+}{Se}-\overset{-}{O}$ H C_6H_5 C H_2 O

(89%)

(6.96)

O O

(1) $LiNR_2$, THF, −78 °C
(2) C_6H_5SeCl
(3) H_2O_2

O O

(56%)

CO_2CH_3

(1) $LiN(iso\text{-}C_3H_7)_2$, THF, −78 °C
(2) $C_6H_5SeSeC_6H_5$
(3) $NaIO_4$, CH_3OH, H_2O

CO_2CH_3

(80%)

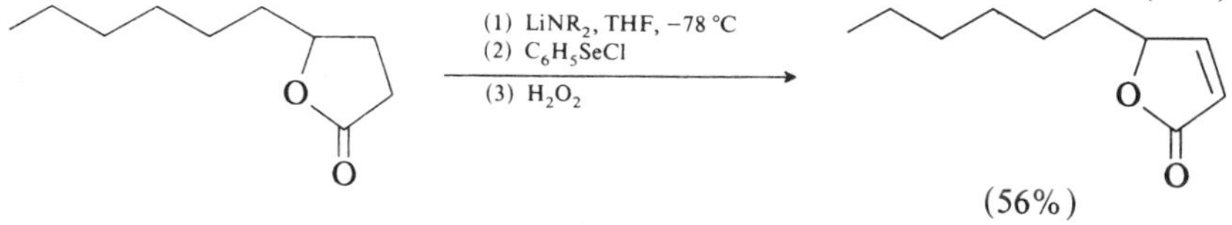

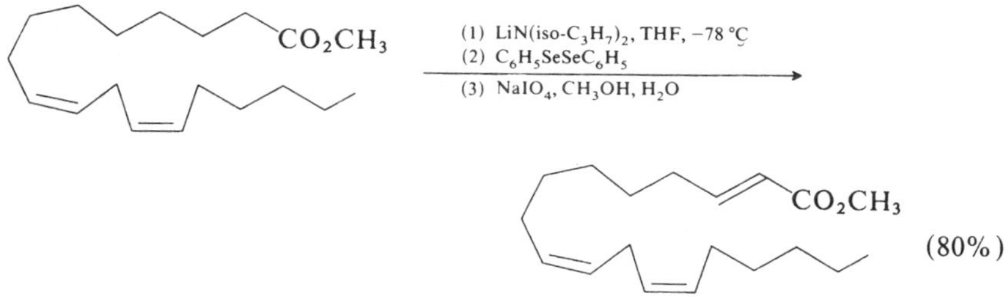

The same transformations can be effected by reaction of the corresponding trimethylsilyl enol ethers with palladium acetate in acetonitrile (Ito, Hirao and Saegusa, 1978).

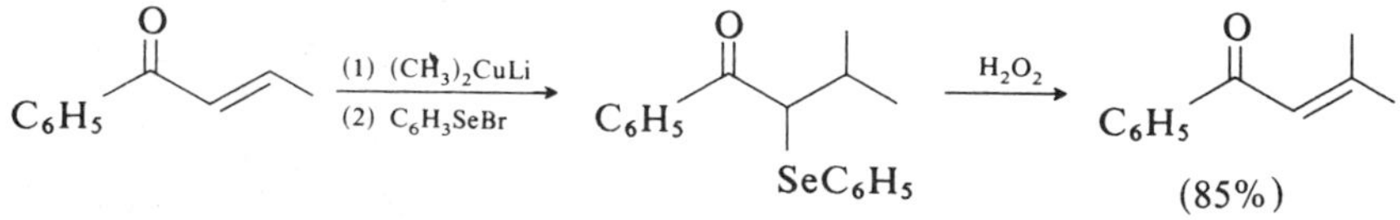

O C_6H_5

(1) $(CH_3)_2CuLi$
(2) C_6H_5SeBr

O C_6H_5 SeC_6H_5

H_2O_2

O C_6H_5

(85%)

(6.97)

O $CO_2C_2H_5$

(1) NaH, THF
(2) C_6H_5SeCl
(3) H_2O_2, CH_2Cl_2

O $CO_2C_2H_5$

(90%)

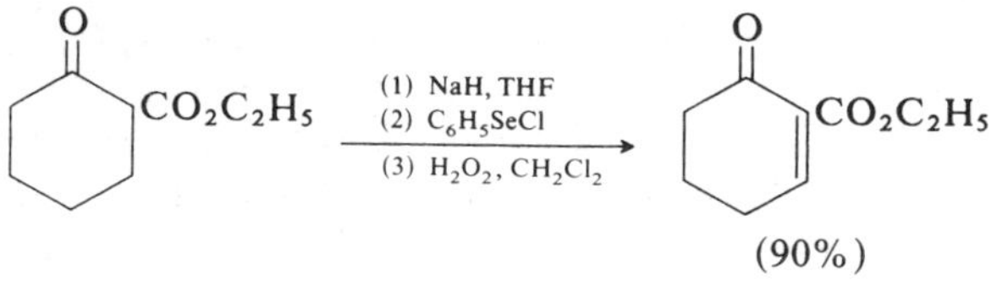

α-Ketols and 1,2-diketones

Oxidation of ketones at the α-carbon atom to give α-hydroxy-ketones (acyloins) or 1,2-diketones, is a synthetically useful transformation. Direct conversion of ketones into α-acetoxy derivatives can be effected in some cases by reaction with mercuric acetate or lead tetra-acetate, but these methods are not very selective and might well be inappropriate in cases where there were other functional groups in the molecule. Fortunately some other more selective methods are available.

In a strongly basic medium, such as a solution of potassium t-butoxide in t-butanol, ketones react rapidly with molecular oxygen to form α-hydroperoxides which, in some cases, can be reduced to the corresponding acyloin. Sometimes the hydroperoxide can be isolated and separately reduced with zinc and acetic acid, but better yields are often obtained by carrying out the oxidation in presence of triethyl phosphite which reduces the hydroperoxide directly to the acyloin. The method is useful for the introduction of a 17-α-hydroxyl substituent into 20-ketosteroids and has been used to introduce hydroxyl substituents at selected positions in the synthesis of a number of other natural products. The reactions appear to be highly stereoselective; in the examples which have been studied the hydroperoxy and derived hydroxyl substituents have the same stereochemistry as the hydrogen atoms which are replaced (Gardner, Carlon and Gnoj, 1968; see also Wasserman and Lipshutz, 1975).

$COCH_3$ CH_3 — (1) O_2, t-C_4H_9OH, t-C_4H_9OK; (2) Zn, CH_3CO_2H → $COCH_3$ OH CH_3 (78%)

(6.98)

RN H $COCH_3$ $CONH_2$ — (1) O_2, t-C_4H_9OH, t-C_4H_9OK; (2) $P(OC_2H_5)_3$, −20 °C → RN HO $COCH_3$ $CONH_2$ (96%)

This method, although useful, has the disadvantage that cleavage products are often formed, and where the α-hydroperoxide bears an α-hydrogen atom an α-diketone is likely to be produced by base-catalysed elimination, resulting in poor yields of the acyloin. An alternative procedure which avoids these difficulties uses the readily available molybdenum peroxide MoO_5.pyridine.hexmethylphosphoramide complex (Vedejs, Engler and Telschow, 1978). This reagent reacts readily with enolates at temperatures between −70 and −40 °C to form a MoVI ester which, after treatment with water, affords the α-hydroxy carbonyl compound in good yield without contamination by oxidative cleavage products. Ketones, esters and lactones with an enolisable methylene or methine group are all readily converted into α-hydroxy compounds by this route. With unsymmetrical ketones a single acyloin is formed with high selectivity by way of the kinetically generated enolate. 2-Phenylcyclohexanone, for example, is converted exclusively into *trans*-2-hydroxy-6-phenyl-cyclohexanone (6.99).

(1) $LiN(iso\text{-}C_3H_7)_2$, THF, −70 °C
(2) MoOPH, −70 → 0 °C
(3) H_2O

C_6H_5 → HO, C_6H_5 (70%)

C_4H_9 → HO, C_4H_9 (73%)

(6.99)

MoOPH ≡ MoO_5. Pyridine.hexamethylphosphoramide complex

Hydroxy ketones have also been obtained very conveniently by oxidation of silyl enol ethers, derived from ketones with either kinetic or thermodynamic control, with osmium tetroxide and *N*-methylmorpholine-*N*-oxide (McCormick, Tomasik and Johnson, 1981) or with chromyl chloride (Lee and Toczek, 1982) or by direct oxidation of ketones with *ortho*-iodosylbenzoic acid (Moriarty and Kwang Chung Hou, 1984). In the latter reaction the initial products are the α-hydroxy dimethylacetals, formed as in (6.100), which on hydrolysis afford the corresponding α-hydroxyketone. 3-Pentanone, for example, gave the α-hydroxy compound in 81 per cent yield.

Direct oxidation of aldehydes and ketones to 1,2-dicarbonyl compounds can be effected with selenium dioxide (Jerussi, 1970). Acetophenone, for example, is converted into phenylglyoxal in 70 per cent yield by warming with selenium dioxide in aqueous dioxan. The mechanism of this reaction has been in doubt, but Sharpless and Gordon (1976) have produced evidence that it proceeds by way of a β-ketoseleninic acid (74), formed by electrophilic attack of selenious acid on the enol, followed by a Pummerer-like rearrangement to a short-lived selenine. (6.101). The formation of $\alpha\beta$-unsaturated carbonyl compounds, which are often encountered in selenium dioxide oxidations of aldehydes and ketones, is readily explained by β-elimination from the ketoseleninic acid (cf. p. 121).

$$RCOCH_2R^1 \rightleftharpoons R{-}C(O^-){=}CHR^1 \xrightarrow{o\text{-}IOC_6H_4CO_2H} RCOCHR^1{-}I(OH){-}Ar$$

$$\downarrow CH_3O^-$$

$$R{-}C(OCH_3)_2{-}CHOHR^1 \xleftarrow{CH_3O^-} R{-}C(OCH_3)(O)CHR^1 + o\text{-}IC_6H_4CO_2^-$$

$$\downarrow H_3O^+$$

$$R{-}C(=O){-}CH(OH){-}R^1 \qquad (6.100)$$

$$RC(OH){=}CHR \xrightarrow[(SeO_2 + H_2O)]{H_2SeO_3} RCO{-}CH(Se(=O)OH){-}CH(H)R' \ (74) \xrightarrow{syn\ elimination} RCO{-}CH{=}CHR'$$

$$(74) \downarrow \left[RCO{-}C(R){=}Se{=}O\right] \rightarrow RCO{-}COR \qquad (6.101)$$

A more selective method which proceeds in high yield under mild conditions was developed by Woodward (1963) for use in his synthesis of colchicine. It culminates in the hydrolysis of a spiro-α-dithiaketal (as 75) which is itself obtained by reaction of the α-formyl derivative of the ketone with propane-1,3-dithiol di-toluene-*p*-sulphonate in the presence of sodium acetate (Woodward, Pachter and Scheinbaum, 1974) (6.102).

Baeyer–Villiger oxidation of ketones

On oxidation with peroxy-acids, ketones are converted into esters or lactones. This reaction was discovered in 1899 by Baeyer and Villiger who found that reaction of a number of cyclic ketones with Caro's acid (permonosulphuric acid) led to the formation of lactones (6.103). Better yields are obtained with organic peroxy acids such as perbenzoic acid, peracetic acid and trifluoroperacetic acid, although in practice nowadays

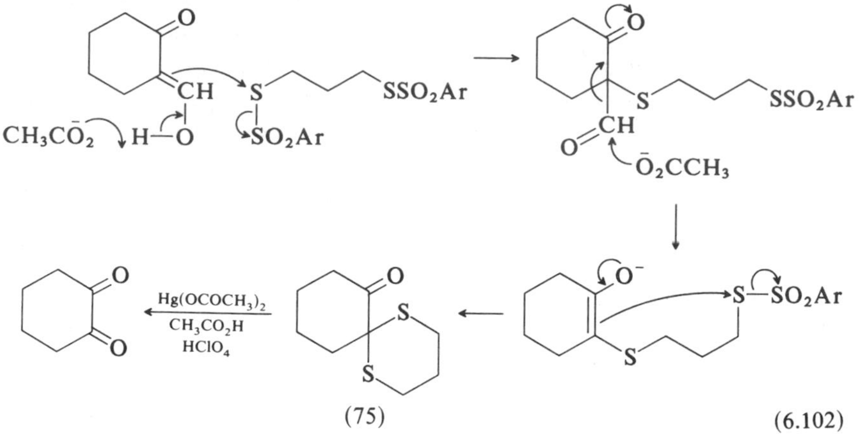

$CH_3CO_2^-$ H—O CH S SO_2Ar SSO_2Ar → S SSO_2Ar O=CH $^-O_2CCH_3$

↓

$Hg(OCOCH_3)_2$ / CH_3CO_2H / $HClO_4$ ← S S ← O^- S—SO_2Ar S

(75)

(6.102)

HO_2SO_3H

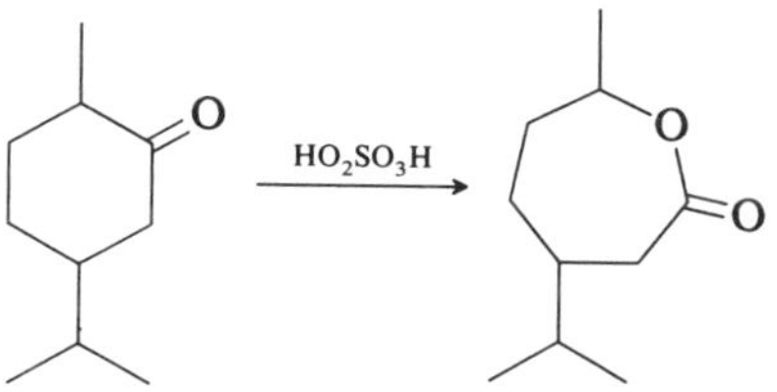

(6.103)

most reactions are effected with *m*-chloroperbenzoic acid. This is more stable than the other acids, which usually have to be prepared immediately before use, and is commercially available. The reaction occurs under mild conditions and has been widely used both in degradative work and in synthesis. It is applicable to open chain and cyclic ketones and to aromatic ketones, and has been used to prepare a variety of steroidal and terpenoid lacones, as well as medium and large-ring lactones which are otherwise difficult to obtain. It also provides a route to alcohols from ketones, through hydrolysis of the esters formed, and of hydroxy-acids from cyclic ketones by way of the lactones; lithium aluminium hydride reduction of the lactones gives diols with a defined disposition of the two hydroxyl groups (Lee and Uff, 1967; Hassall, 1957).

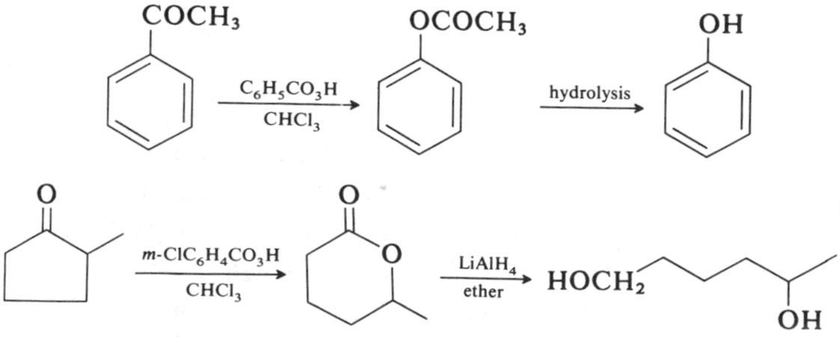

$COCH_3$ —$C_6H_5CO_3H$ / $CHCl_3$→ $OCOCH_3$ —hydrolysis→ OH

(6.104)

—m-$ClC_6H_4CO_3H$ / $CHCl_3$→ —$LiAlH_4$ / ether→ $HOCH_2$ OH

With unsaturated ketones mixtures of products sometimes result through competing attack at the carbon–carbon double bond. The new reagent bis(trimethylsilyl)peroxide, $Me_3SiOOSiMe_3$ provides a way round this difficulty. It behaves as a masked form of 100 per cent hydrogen peroxide and in the presence of catalytic trimethylsilyl trifluoromethanesulphonate it efficiently brings about Baeyer–Villiger oxidation of ketones but does not affect carbon–carbon double bonds. Thus, 3-cyclopentenone gave only 3-penten-5-olide, whereas with trifluoroperacetic acid 3,4-epoxycyclopentenone was obtained, and with 2-allylcyclohexanone reaction gave only 6-allyl-6-hexanolide (6.105) (Suzuki, Takada and Noyori, 1982).

$[(CH_3)_3SiO]_2$

Catalytic $(CH_3)_3SiOCOCl_3$,

CH_2Cl_2, −78 °C → −10 °C

(40%)

(6.105)

The reaction is thought to take place by a concerted intramolecular process involving migration of a group from carbon to electron-deficient oxygen, possibly by way of a cyclic transition state (Smith, 1963; Lee and Uff, 1967) (6.106). In the presence of a strong acid there may be addition of peroxy-acid to the protonated ketone, but additional acid is not needed and in its absence addition may take place to the ketone itself. The general mechanism is supported by the fact that the reaction is catalysed by acid and is accelerated by electron-releasing groups in the ketone and by electron-withdrawing groups in the acid. In an elegant experiment using [^{18}O]benzophenone Doering and Dorfman showed that the phenyl benzoate obtained had the same ^{18}O content as the ketone and that the ^{18}O was contained entirely in the carbonyl oxygen. The intramolecular concerted nature of the reaction is supported also by several demonstrations of complete retention of configuration in the migrating carbon atom. Thus optically active methyl α-phenethyl ketone was converted into α-phenethyl acetate with no loss of chirality.

$$RR^1C{=}O \xrightleftharpoons{R^2CO_3H,\ H^+} [\text{cyclic intermediate}] \longrightarrow RC(=O){-}OR^1 + R^2CO_2H \qquad (6.106)$$

An unsymmetrical ketone could obviously give rise to two different products in this reaction. Cyclohexyl phenyl ketone, for example, on reaction with peracetic acid gives both cyclohexyl benzoate and phenyl cyclohexanecarboxylate by migration of both the cyclohexyl and the phenyl group. It is found that the relative ease of migration of different groups in the reaction is in the order

t-alkyl > cyclohexyl ~ s-alkyl ~ benzyl ~ phenyl
> primary alkyl > methyl

That is, in the alkyl series migratory aptitudes are in the series tertiary > secondary > primary; among benzene derivatives migration is facilitated by electron-releasing *para* substituents, and hindered by electron-withdrawing ones. The methyl group shows the least tendency to migrate, so that methyl ketones always give acetates in the Baeyer–Villiger reaction. Phenyl *p*-nitrophenyl ketone gives only phenyl *p*-nitrobenzoate and s-butyl methyl ketone is converted into s-butyl acetate. Electronic factors evidently play a part here. It has been suggested that the ease of migration is related to the ability of the migrating group to accommodate a partial positive charge in the transition state, but it seems that in some cases steric effects may also be involved and the experimental conditions may influence the result as well. This is particularly noticeable in Baeyer–Villiger reactions of bridged bicyclic ketones (cf. Krow, 1981). Thus, while 1-methylnorcamphor (76) affords the expected lactone on oxidation with peracetic acid, the product obtained from camphor itself depends on the conditions and epicamphor (77) gives only the 'abnormal' product.

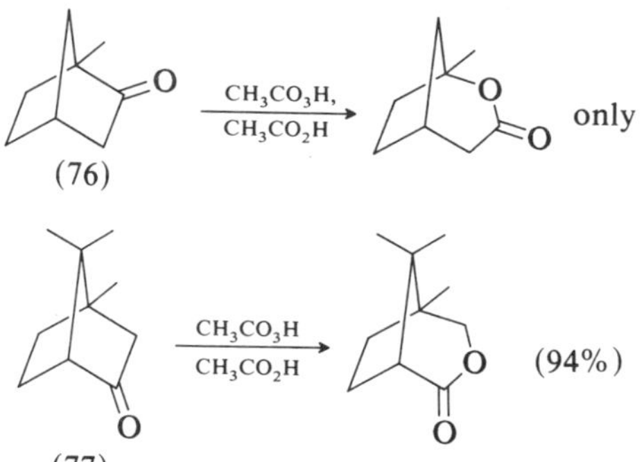

(76) $\xrightarrow[CH_3CO_2H]{CH_3CO_3H,}$ only

(77) $\xrightarrow[CH_3CO_2H]{CH_3CO_3H}$ (94%)

(6.107)

Baeyer–Villiger oxidation of bridged bicylic ketones is valuable in synthesis because it provides a method for preparing derivatives of cyclohexane and cyclopentane with control of the stereochemistry of the substituent groups, and numerous syntheses of natural products have exploited this possibility. Thus, the lactone (79), important in the synthesis of prostaglandins, was obtained in a sequence the key step of which was Baeyer–Villiger oxidation of the bridged bicyclic ketone (78).

Oxidation of aldehydes with peroxy-acids is not so synthetically useful as oxidation of ketones and generally gives either carboxylic acids or formate esters. But reaction of *o*- and *p*-hydroxy-benzaldehydes or

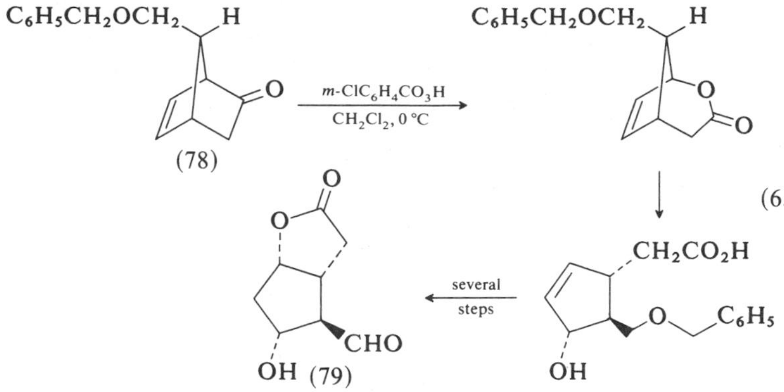

$C_6H_5CH_2OCH_2$ H (78) m-$ClC_6H_4CO_3H$, CH_2Cl_2, 0 °C → $C_6H_5CH_2OCH_2$ H → CH_2CO_2H, O C_6H_5, OH → several steps → CHO, OH (79)

(6.108)

-acetophenones with alkaline hydrogen peroxide is a useful method for making catechols and quinols (the Dakin reaction). With benzaldehyde itself only benzoic acid is formed, but salicylaldehyde gives catechol almost quantitatively and 3,4-dimethylcatechol was obtained by oxidation of 2-hydroxy-3,4-dimethylacetophenone, possibly by way of a cyclic transition state similar to that postulated for the Baeyer–Villiger reaction (Lee and Uff, 1967) (6.109).

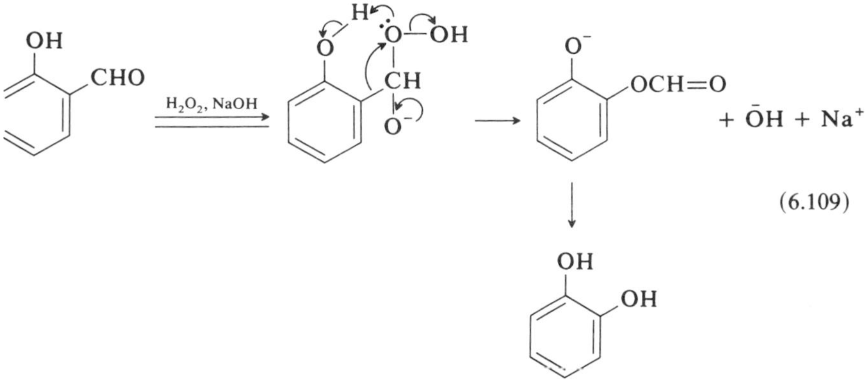

OH CHO $\xrightarrow{H_2O_2,\ NaOH}$ H O O—OH CH O^- → O^- OCH=O + $\bar{O}H$ + Na^+ → OH OH

(6.109)

6.7. Oxidations with ruthenium tetroxide

Ruthenium tetroxide is a powerful oxidising agent which oxidises a variety of functional groups at room temperature. Reactions are best effected in solution in carbon tetrachloride-acetonitrile (Carlson *et al.*, 1981) either with the pure tetroxide, which is expensive or, better, with a catalytic amount of tetroxide in the presence of sodium periodate which serves to oxidise the reduced ruthenium back to the active tetroxide (Lee and van den Engh, 1973).

Unlike osmium tetroxide, ruthenium tetroxide does not convert alkenes into diols. Instead, cleavage of the double bond takes place to give carboxylic acids or ketones, depending on the degree of substitution of the double bonds, and oxidation with ruthenium tetroxide and sodium periodate is a good alternative to ozonolysis. (*E*)-5-Decene, for example, gave pentanoic acid in 88 per cent yield. Primary alcohols give carboxylic acids (6.110) and secondary alcohols are oxidised smoothly to ketones

$$\text{C}_6\text{H}_5\text{-epoxide-CH}_2\text{OH} \xrightarrow[\text{CCl}_4,\ \text{CH}_3\text{CN},\ \text{H}_2\text{O},\ 25\,^\circ\text{C},\ 2\ \text{h}]{\text{catalytic RuCl}_3,\ \text{IO}_4^-} \text{C}_6\text{H}_5\text{-epoxide-CO}_2\text{H} \quad (75\%) \tag{6.110}$$

$$\underset{(80)}{\text{C}_6\text{H}_5\text{CH(CH}_3\text{)CH(OH)CH}_2\text{OH}} \longrightarrow \underset{(81)}{\text{C}_6\text{H}_5\text{CH(CH}_3\text{)CO}_2\text{H}} \quad (92\%)$$

in excellent yield at room temperature. Ketones are stable to further oxidation but aldehydes are rapidly converted into the carboxylic acid. The mildness of the reaction conditions is illustrated by the ready conversion of (80) into (81) without loss of optical activity (see also Still and Ohmizu, 1981).

A remarkable reaction is the smooth conversion of ethers into esters or lactones (see also Smith and Scarborough, 1980). Thus, dibutyl ether gave butyl butyrate in quantitiative yield and tetrahydrofuran similarly yielded butyrolactone. Benzene rings also are smoothly degraded to carboxylic acids. Phenylcyclohexane, for example, gave cyclohexanecarboxylic acid in 94 per cent yield. This is a better way of degrading aromatic rings than ozonisation (see Kasai and Ziffer, 1983).

$$\text{C}_6\text{H}_{11}\text{-C}_6\text{H}_5 \xrightarrow[\text{CH}_3\text{CN},\ \text{CCl}_4,\ \text{H}_2\text{O},\ 25\,^\circ\text{C},\ 24\ \text{h}]{\text{catalytic RuCl}_3,\ \text{IO}_4^-} \text{C}_6\text{H}_{11}\text{-CO}_2\text{H} \quad (94\%) \tag{6.111}$$

6.8. Oxidations with thallium(III) nitrate

Some useful oxidations of alkenes and alkynes can be effected with thallium(III) nitrate (McKillop and Taylor, 1973). The effectiveness of this reagent in oxidations is due to the energetically favorable conversion $\text{Tl}^{3+} + 2\text{e} \rightarrow \text{Tl}^{+}$ (−1.25 V), and to the weakness of the carbon–thallium bond (105–125 kJ mol^{-1}) which is easily cleaved heterolytically to form carbonium ion intermediates. Furthermore, nitrate ion is not very nucleophilic so that solvent may participate selectively as the nucleophile in the reactions.

Thallium(III) nitrate reacts with alkenes in methanol solution to give carbonyl compounds or 1,2-glycol dimethyl ethers. Particularly useful is the reaction with cyclic alkenes which leads by an oxidative rearrangement to ring-contracted products (McKillop *et al.*, 1973*a*). Thus, cyclohexene reacts almost instantaneously at room temperature in methanol solution to give the dimethylacetal of cyclopentanecarboxaldehyde, and cycloheptene is similarly converted into the dimethylacetal of cyclohexanecarboxaldehyde. The reaction is believed to take the course shown in (6.112). This reaction takes place most easily with six- and seven-membered rings in which the migrating ring bond, at C-2–C-3, is, or can easily become, *trans* anti-parallel to the carbon–thallium bond. With aliphatic alkenes and cyclic alkenes in which a similar conformationally favourable pathway is not possible, rearrangement does not take place to any appreciable extent and glycol dimethyl ethers are the main products (6.112).

$$\text{cyclohexene} \xrightarrow[CH_3OH]{Tl(ONO_2)_3} \text{(cation)}\ Tl(ONO_2)_2 \xrightarrow{CH_3OH} OCH_3,\ Tl(ONO_2)_2$$

$$\equiv \ \text{(2-}OCH_3\text{, 3; Tl}(ONO_2)_2\text{)} \longrightarrow \left[\ {}^+CHOCH_3\ \right] \xrightarrow{CH_3OH} CH(OCH_3)_2$$

$$CH_3(CH_2)_7CH{=}CH_2 \xrightarrow[CH_3OH]{Tl(ONO_2)_3} CH_3(CH_2)_7COCH_3 \quad (28\%)$$

$$+ \quad CH_3(CH_2)_7\underset{|}{\overset{OCH_3}{C}}H\!-\!\!-\!\overset{OCH_3}{C}H_2 \quad (52\%) \qquad (6.112)$$

Styrenes also form 1,2-dimethoxy derivatives when treated with thallium(III) nitrate in methanol, but in dilute nitric acid reaction takes a different course and high yields of arylacetaldehydes are formed. The transformation again involves a rearrangement (6.113).

With alkynes the product obtained depends on the structure of the alkyne (McKillop *et al.*, 1973*b*). Diarylalkynes react with two equivalents of thallium(III) nitrate in acid solution or in methanol to give benzils in high yield, but with dialkylalkynes reaction stops at the acyloin. Alkylarylalkynes in acid solution give mixtures of products, but in methanol a smooth rearrangement takes place to give methyl esters of α-alkylarylacetic acids, and the reaction provides a very effective route to these valuable intermediates (6.114). Arylacetic esters are also obtained

$$CH_3O\text{-}C_6H_4\text{-}CH{=}CH_2 \xrightarrow[HNO_3,\ H_2O]{Tl(ONO_2)_3} CH_3O\text{-}C_6H_4\text{-}CH(OH)\text{-}CH_2\text{-}Tl(ONO_2)_2 \longrightarrow CH_3O\text{-}C_6H_4\text{-}CH_2CHO \quad (75\%) \qquad (6.113)$$

by oxidation of acetophenones with thallium(III) nitrate in acidified methanol, and this procedure is a convenient alternative to the Willgerodt reaction (McKillip, Swann and Taylor, 1973).

$$C_6H_5\text{—}C{\equiv}C\text{—}CH_3 \xrightarrow[CH_3OH]{Tl(ONO_2)_3} C_6H_5(CH_3O)C{=}C(CH_3)Tl(ONO_2)_2 \xrightarrow{CH_3OH} (CH_3O)_2C{=}C(C_6H_5)CH_3 \xrightarrow{H_2O} C_6H_5\text{—}CH(CH_3)\text{—}CO_2CH_3 \ (80\%) \qquad (6.114)$$

With monoalkylacetylenes yet another reaction takes place. They are oxidised with two equivalents of thallium(III) nitrate in acid solution to carboxylic acids with loss of the terminal carbon atom. 1-Octyne, for example, gave heptanoic acid in 80 per cent yield. This novel reaction is believed to go through the α-ketol which is degraded further by the second molecule of the nitrate.

7 Reduction

There must be few organic syntheses of any complexity which do not involve a reduction at some stage. Reduction is here used in the sense of addition of hydrogen to an unsaturated group such as a carbon–carbon double bond, a carbonyl group or an aromatic nucleus, or addition of hydrogen with concomitant fission of a bond between two atoms, as in the reduction of a disulphide to a thiol or of an alkyl halide to a hydrocarbon.

Reductions are generally effected either chemically or by catalytic hydrogenation, that is by the addition of molecular hydrogen to the compound under the influence of a catalyst. Each method has its advantages. In many reductions either method may be used equally well. Complete reduction of an unsaturated compound can generally be achieved without undue difficulty, but the aim is often selective reduction of one group in a molecule in the presence of other unsaturated groups. Both catalytic and chemical methods of reduction offer considerable scope in this direction, and the method of choice in a particular case will often depend on the selectivity required and on the stereochemistry of the desired product.

7.1. Catalytic hydrogenation

Of the many methods available for reduction of organic compounds catalytic hydrogenation is one of the most convenient. Reaction is easily effected simply by stirring or shaking the substrate with the catalyst in a suitable solvent, or without a solvent if the substance being reduced is a liquid, in an atmosphere of hydrogen in an apparatus which is arranged so that the uptake of hydrogen can be measured, At the end of the reaction the catalyst is filtered off and the product is recovered from the filtrate, often in a high state of purity. The method is easily adapted for work on a micro scale, or on a large, even industrial, scale. In many cases reaction proceeds smoothly at or near room temperature and at atmospheric or slightly elevated pressure. In other cases high temperatures (100–200 °C) and pressures (100–300 atmospheres) are necessary, requiring special high pressure equipment. Detailed descriptions of equipment suitable for hydrogenations under different conditions are given by Schiller (1955) and by Augustine (1965).

Catalytic hydrogenation may result simply in the addition of hydrogen to one or more unsaturated groups in the molecule, or it may be accompanied by fission of a bond between atoms. The latter process is known as hydrogenolysis.

Most of the common unsaturated groups in organic chemistry, such as $>C{=}C<$, $-C{\equiv}C-$, $>C{=}O$, $-COOR$, $-C{\equiv}N$, $-NO_2$ and aromatic and heterocyclic nuclei can be reduced catalytically under appropriate conditions, although they are not all reduced with equal ease. Certain groups, notably allylic and benzylic hydroxyl and amino groups and carbon–halogen and carbon–sulphur single bonds, readily undergo hydrogenolysis, resulting in cleavage of the bond between carbon and the hetero-atom. This may be advantageous in certain circumstances. Thus, much of the usefulness of the benzyloxycarbonyl protecting group in peptide chemistry is due to the ease with which it can be removed by hydrogenolysis over a palladium catalyst and hydrogenolysis of the derived enol phosphates is a useful method for the reduction of ketonic carbonyl groups to methylene (e.g. Coates, Shah and Mason, 1982) (7.1).

(1) $LiN[Si(CH_3)_3]_2$, THF, −30 °C; (2) $(C_2H_5O)_2POCl$ → $OP(O)(OC_2H_5)_2$; H_2, Pt-C (7.1)

An alternative procedure which is sometimes advantageous is 'catalytic transfer hydrogenation' in which hydrogen is transferred to the substrate from another organic compound, usually a hydroaromatic compound or an alcohol. Reaction is easily effected simply by warming the substrate and hydrogen donor together in presence of a catalyst, usually palladium. The actual mechanism of transfer hydrogenation is not clear, but the method is useful on occasion for it sometimes shows different selectivity towards functional groups from that shown in catalytic reduction with molecular hydrogen (Johnstone and Wilby, 1981; Johnstone, Wilby and Entwistle, 1985).

The catalyst

Many different catalysts have been used for catalytic hydrogenations; they are mainly finely divided metals, metallic oxides or sulphides. The most commonly used in the laboratory are the platinum metals (platinum, palladium and, to a lesser extent rhodium and ruthenium), nickel and copper chromite. The catalysts are not specific and with the exception of copper chromite may be used for a variety of different reductions. The most widely used are the platinum metal catalysts and much of what can be accomplished by hydrogenation is best done with these catalysts (Rylander, 1967). They are exceptionally active catalysts and promote the reduction of most functional groups under mild conditions, with the notable exception of the carboxyl, carboxylic ester and amide groups. They are used either as the finely divided metal or, more commonly, supported on a suitable carrier such as asbestos, activated carbon, alumina or barium sulphate. In general, supported metal catalysts, since they have a larger surface area, are more active than the unsupported metal, but the activity is influenced strongly by the support and by the method of preparation, and this provides a means of preparing catalysts of varying activity. Platinum is very often used in the form of its oxide PtO_2, 'Adams' catalyst', which is reduced to metallic platinum by hydrogen in the reaction medium.

The platinum metal catalysts can be prepared by reduction of a metallic salt, in presence of the support if a supported catalyst is wanted. Detailed instructions are given by Wimmer (1955). It has been found that very effective catalysts can be obtained by reduction of various metal salts with either sodium borohydride (Brown and Brown, 1962) or a trialkylsilane (Eaborn *et al.*, 1969).

Most platinum metal catalysts, with the exception of Adams' catalyst, are stable and can be kept for many years without appreciable loss of activity, but they are deactivated by many substances, particularly by compounds of divalent sulphur. For best results in a hydrogenation, therefore, particularly with platinum catalysts, it is necessary to use pure materials and pure solvents. On the other hand, catalytic activity is often increased by addition of small amounts of promoters, usually platinum or palladium salts or mineral acid. The activity of platinum catalysts derived from platinum oxide is often markedly increased in presence of mineral acid; this may simply be due to neutralisation of alkaline impurities in the catalyst.

For hydrogenations at high pressure the most common catalysts are Raney nickel and copper chromite. Raney nickel is a porous, finely divided nickel obtained by treating a powdered nickel–aluminium alloy with sodium hydroxide. It is generally used at high temperatures and pressures, but with the more active catalysts many reductions can be effected at atmospheric pressure and normal temperature. Nearly all unsaturated groups can be reduced with Raney nickel but it is most frequently used for reduction of aromatic rings and hydrogenolysis of sulphur compounds (see p. 484). When freshly prepared it contains 25–100 ml adsorbed hydrogen per gram of nickel. The more adsorbed hydrogen the more active is the catalyst, and Raney nickel catalysts of graded activities can be obtained by variation of the preparative procedure (Augustine, 1965; Kimmer, 1955). Raney nickel catalysts are alkaline and may only be used for hydrogenations which are not adversely affected by basic conditions. They are deactivated by acids.

Copper chromite, $CuCr_2O_4$, is prepared by thermal decomposition of copper ammonium chromate; a more active catalyst is obtained by adding barium nitrate to the reaction mixture. It is a relatively inactive catalyst and is only effective at high temperatures (100–200 °C) and pressures (200–300 atm). It does not reduce aromatic rings under these conditions as Raney nickel does, and is principally used for reduction of esters to alcohols and of amides to amines (7.2). Hydrogenation using copper chromite is not so frequently used now because of the introduction of hydride reducing agents (p. 458) which can accomplish the same reactions under much milder conditions, but it is still useful for reactions on a large scale.

$$C_6H_5CH_2CO_2Et \xrightarrow[25\,°C,\ 200\ atm.]{H_2,\ CuCr_2O_4} C_6H_5CH_2CH_2OH$$

(7.2)

O, NH, CO_2Et $\xrightarrow[150\,°C,\ 300\ atm.]{H_2,\ CuCr_2O_4}$ O, NH, CH_2OH $\xrightarrow[250\,°C,\ 300\ atm.]{H_2,\ CuCr_2O_4}$ N, N

Selectivity of reduction

Many hydrogenations proceed satisfactorily under a wide range of conditions, but where a selective reduction is wanted, conditions may be more critical.

The choice of catalyst for a hydrogenation is governed by the activity and selectivity required. Selectivity is a property of the metal, but it also depends to some extent on the activity of the catalyst and on the reaction conditions. In general, the more active the catalyst the less discriminating it is in its action, and for greatest selectivity reactions should be run with the least active catalyst and under the mildest possible conditions consistent with a reasonable rate of reaction. The rate of a given hydrogenation may be increased by raising the temperature, by increasing the pressure or by an increase in the amount of catalyst used, but all these factors may result in a decrease in selectivity. For example, hydrogenation of ethyl benzoate with copper chromite catalyst under the appropriate conditions leads to benzyl alcohol by reduction of the ester group, while Raney nickel gives ethyl hexahydrobenzoate by selective attack on the benzene ring (7.3). At higher temperatures, however, the selective activity of the catalysts is lost and mixtures of the two products and toluene are obtained from both reactions.

$$C_6H_5CH_2OH \xleftarrow[160\,^\circ C,\ 250\ atm.]{H_2,\ CuCr_2O_4} C_6H_5CO_2C_2H_5 \xrightarrow[50\,^\circ C,\ 100\ atm.]{H_2,\ Raney\ Ni} C_6H_{11}CO_2C_2H_5 \quad (7.3)$$

Both the rate and, sometimes, the course of a hydrogenation may be influenced by the solvent used. The commonest solvents are methanol, ethanol and acetic acid. Many hydrogenations over platinum metal catalysts are favoured by strong acids. For example, reduction of β-nitrostyrene in acetic acid–sulphuric acid is rapid and affords 2-phenylethylamine in 90 per cent yield; but in the absence of sulphuric acid reduction is slow and the yield of amine poor.

Not all functional groups are reduced with equal ease. Table 7.1, due to House (1965), shows the approximate order of decreasing ease of catalytic hydrogenation of a number of common groups. This order is not invariable and is influenced to some extent by the structure of the compound being reduced and by the catalyst employed. In general, groups near the top of the list can be selectively reduced in the presence of groups near the bottom, but preferential reduction of groups at the bottom in presence of the more reactive groups at the top is more difficult. For example, reduction of an unsaturated ester or ketone to a saturated ester or ketone is, in most cases, readily accomplished by hydrogenation over palladium or platinum, but selective reduction of the carbonyl group to form an unsaturated alcohol is difficult to achieve by catalytic hydrogenation and is generally effected by chemical reduction. Similarly, nitrobenzene is easily converted into aniline, but selective reduction to nitrocyclohexane is not possible (but see p. 427).

Reduction of functional groups: alkenes

Hydrogenation of carbon–carbon double bonds takes place easily and in most cases can be effected under mild conditions. Only a few highly hindered alkenes are resistant to hydrogenation, and even these can generally be reduced under more vigorous conditions. Platinum

Table 7.1. *Approximate order of reactivity of functional groups in catalytic hydrogenation*

Functional group	Reduction product
$R-COCl$	$R-CHO$, $R-CH_2OH$
$R-NO_2$	$R-NH_2$
$R-C\equiv C-R$	$(H)(R)C=C(H)(R)$ (cis), RCH_2CH_2R
$R-CHO$	$R-CH_2OH$
$R-CH=CH-R$	$R-CH_2CH_2-R$
$R-CO-R$	$R-CHOH-R$, $R-CH_2-R$
$C_6H_5CH_2OR$	$C_6H_5CH_3+ROH$
$R-C\equiv N$	$R-CH_2NH_2$
Polycyclic aromatic hydrocarbons	Partially reduced products
$R-COOR'$	$R-CH_2OH+R'OH$
$R-CONHR'$	$R-CH_2NHR'$
benzene	cyclohexane
$R-CO_2^-Na^+$	inert

H. O. House, *Modern synthetic reactions*, copyright 1965, 1972, W. A. Benjamin, Inc., Menlo Park, California.

and palladium are the most frequently used catalysts. Both are very active and the preference is determined by the nature of other functional groups in the molecule and by the degree of selectivity required; platinum usually brings about a more exhaustive reduction. Raney nickel may also be used in certain cases. Thus cinnamyl alcohol is reduced to the dihydro compound with Raney nickel in ethanol at 20 °C, and 1,2-dimethylcyclohexene with hydrogen and platinum oxide in acetic acid is converted mainly into *cis*-1,2-dimethylcyclohexane (cf. p. 424) (7.4).

$$\text{1,2-dimethylcyclohexene} \xrightarrow[CH_3CO_2H]{H_2,\ PtO_2} \text{cis-1,2-dimethylcyclohexane } (82\%) + \text{trans-1,2-dimethylcyclohexane } (18\%) \quad (7.4)$$

Rhodium and ruthenium catalysts have not been much used in hydrogenation of alkenes, but they sometimes show useful selective properties. Rhodium is particularly useful for hydrogenation of alkenes when concomitant hydrogenolysis of an oxygen function is to be avoided. Thus, the plant toxin, toxol (3), on hydrogenation over rhodium-alumina in ethanol was smoothly converted into the dihydro compound; with platinum and palladium catalysts, on the other hand, extensive hydrogenolysis took place and a mixture of products was formed (7.5) (see also Evans and Morrissey, 1984).

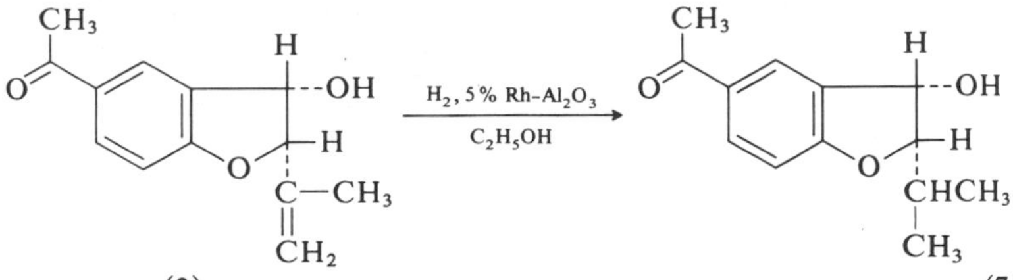

CH_3 H --OH H $C-CH_3$ CH_2 (3) $\xrightarrow[C_2H_5OH]{H_2, 5\% \ Rh-Al_2O_3}$ CH_3 H --OH H $CHCH_3$ CH_3 (7.5)

The ease of reduction of an alkene decreases with the degree of substitution of the double bond, and this sometimes allows selective reduction of one double bond in a molecule which contains several. Thus, limonene can be converted into *p*-menthene in almost quantitative yield by hydrogenation over platinum oxide if the reaction is stopped after absorption of one molecule of hydrogen. In contrast, the isomeric $\Delta^{1(7),8}$-*p*-menthadiene (4), in which both double bonds are disubstituted, gives only the completely reduced product (7.6).

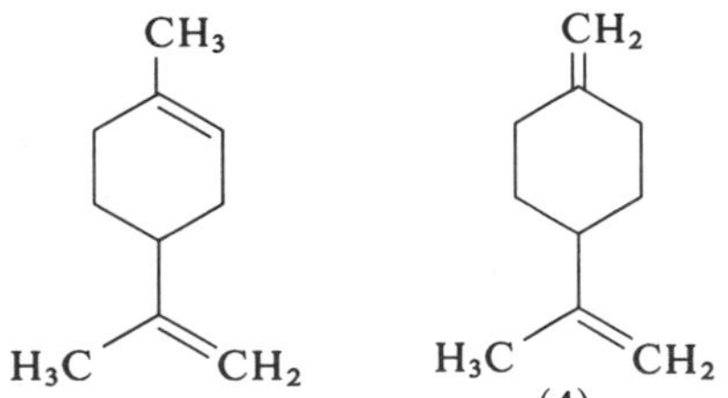

CH_3 H_3C CH_2 CH_2 H_3C CH_2 (4) (7.6)

Catalytic hydrogenation of alkenes over platinum metal catalysts is often accompanied by migration of the double bond, but unless tracers are used or special products result, or the new bond is resistant to hydrogenation, no evidence of migration remains on completion of the reduction. A not uncommon result in suitable structures is formation of a tetrasubstituted double bond which resists further reduction. Thus, the tetracyclic triterpene derivative methyl dihydromasticadienol acetate, with platinum oxide and deuterium in deuteroacetic acid gave the isomer (5); the location of the original double bond was deduced from the position of the deuterium atom in this product (7.7).

A further indication that catalytic hydrogenation of an alkene need not occur by straightforward addition of two hydrogen atoms at the site of the original double bond is found in the fact that catalytic deuteration generally results in the formation of products containing more and fewer

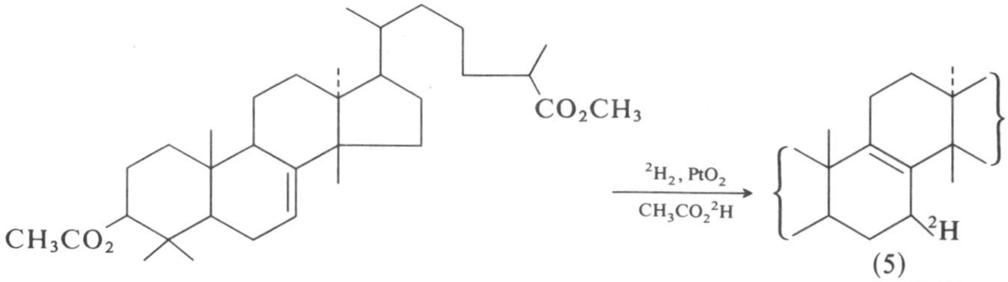

CO_2CH_3

CH_3CO_2

$^2H_2$, PtO_2

$CH_3CO_2{}^2H$

^2H

(5)

(7.7)

than two atoms of deuterium per molecule. Deuteration of 1-hexene, for example, over platinum oxide gave a mixture of products containing from one to six deuterium atoms per molecule. Heterogeneous catalytic deuteration of alkenes cannot therefore be safely used to prepare deuterated compounds for mechanistic or biosynthetic studies.

Selective reduction of carbon–carbon double bonds in compounds containing other unsaturated groups can usually be accomplished, except in the presence of triple bonds, aromatic nitro groups and acyl halides. Palladium is usually the best catalyst. Thus, 2-benzylidenecyclopentanone is readily converted into 2-benzylcyclopentanone with hydrogen and palladium in methanol; with a platinum catalyst, benzylcyclopentanol is formed. Unsaturated nitriles and aliphatic nitro compounds are also reduced selectively at the double bond with a palladium catalyst.

With palladium and platinum catalysts hydrogenation of allylic or vinylic alcohols, ethers and esters, is often accompanied by hydrogenolysis of the oxygen function, and this reaction is facilitated by acids. Hydrogenolysis can often be avoided with rhodium or ruthenium catalysts.

Hydrogenation of alkynes

Catalytic hydrogenation of alkynes takes place in a stepwise manner, and both the alkene and the alkane can be isolated. Complete reduction of alkynes to the saturated compound is easily accomplished over platinum, palladium or Raney nickel. A complication which sometimes arises, particularly with platinum catalysts, is the hydrogenolysis of propargylic hydroxyl groups (7.8).

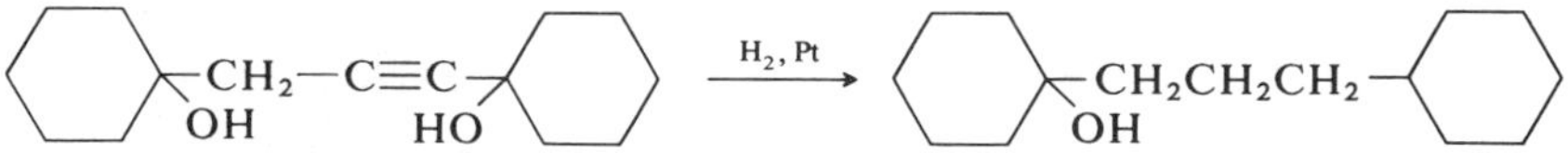

$CH_2—C\equiv C$ OH HO $\xrightarrow{H_2,\ Pt}$ $CH_2CH_2CH_2$ OH

(7.8)

More useful from a synthetic point of view is the partial hydrogenation of alkynes to *Z*-alkenes. This reaction can be effected in high yield with a palladium–calcium carbonate catalyst which has been partially deactivated by addition of lead acetate (Lindlar's catalyst) or quinoline. It is aided by the fact that the more electrophilic acetylenic compounds are adsorbed on the electron-rich catalyst surface more strongly than the corresponding alkenes. An important feature of these reductions is their high stereoselectivity. In most cases the product consists very largely of the thermodynamically less stable *Z*-alkene and partial catalytic hydrogenation of alkynes provides one of the most convenient routes to *Z*-1,2-disubstituted alkenes. Thus stearolic acid on reduction over Lindlar's catalyst in ethyl acetate solution affords a product containing 95 per cent of oleic acid (7.9). Partial reduction of alkynes with Lindlar's catalyst has been invaluable in the synthesis of carotenoids and many other natural products with *Z*-disubstituted double bonds.

$$CH_3(CH_2)_7C\equiv C(CH_2)_7CO_2H \xrightarrow{H_2,\ \text{Lindlar catalyst}} \underset{H}{\overset{CH_3(CH_2)_7}{}}\!\!>C=C<\!\!\underset{H}{\overset{(CH_2)_7CO_2H}{}} \qquad (7.9)$$

Hydrogenation of aromatic compounds

Reduction of aromatic rings by catalytic hydrogenation is more difficult than that of most other functional groups, and selective reduction is not easy. The commonest catalysts are platinum and rhodium, which can be used at ordinary temperatures, and Raney nickel or ruthenium which require high temperatures and pressures.

Benzene itself can be reduced to cyclohexane with platinum oxide in acetic acid solution. Derivatives of benzene such as benzoic acid, phenol or aniline, are reduced more easily. For large-scale work the most convenient method is hydrogenation over Raney nickel at 150–200 °C and 100–200 atm. Hydrogenation of phenols, followed by oxidation of the resulting cyclohexanols is a convenient method for the large-scale preparation of substituted cyclohexanones (7.10).

$$\text{2,5-dimethylphenol (OH, } CH_3\text{, } CH_3) \xrightarrow[(2)\ \text{Chromic acid}]{(1)\ H_2,\ \text{Raney Ni}} \text{2,5-dimethylcyclohexanone (O, } CH_3\text{, } CH_3) \qquad (7.10)$$

Reduction of benzene derivatives carrying oxygen or nitrogen functions in benzylic positions is complicated by the easy hydrogenolysis of such groups, particularly over palladium catalysts. Preferential reduction of the benzene ring in these compounds is best achieved with ruthenium, or preferably with rhodium, catalysts which can be used under mild conditions. Thus mandelic acid is readily converted into hexahydromandelic acid over rhodium-alumina in methanol solution, whereas, with palladium, hydrogenolysis to phenylacetic acid is the main reaction (7.11).

$$C_6H_5CHOHCO_2H \begin{cases} \xrightarrow{H_2,\ Rh\text{-}Al_2O_3} C_6H_{11}CHOHCO_2H \\ \xrightarrow{H_2,\ Pd\text{-}C} C_6H_5CH_2CO_2H \end{cases} \qquad (7.11)$$

With polycyclic aromatic compounds it is often possible, by varying the conditions, to obtain either partially or completely reduced products. Thus, naphthalene can be converted into the tetrahydro or decahydro-compound over Raney nickel depending on the temperature. With anthracene and phenanthrene the 9,10-dihydro compounds are obtained by hydrogenation over copper chromite, although, in general, aromatic rings are not reduced with this catalyst. To obtain more fully hydrogenated compounds more active catalysts must be used.

Hydrogenation of aldehydes and ketones

Hydrogenation of the carbonyl group of aldehydes and ketones is easier than that of aromatic rings but not so easy as that of most carbon–carbon double bonds.

For aliphatic aldehydes and ketones reduction to the alcohol is usually effected under mild conditions over platinum or the more active forms of Raney nickel. Ruthenium is also an excellent catalyst for reduction of aliphatic aldehydes and can be used to advantage with aqueous solutions. Palladium is not very active for hydrogenation of aliphatic carbonyl compounds but is effective for the reduction of aromatic aldehydes and ketones; excellent yields of the alcohols can be obtained if the reaction is interrupted after absorption of one mole of hydrogen. Prolonged reaction, particularly at elevated temperatures or in presence of acid, leads to hydrogenolysis with formation of the methyl or methylene compound and hydrogenation over a palladium catalyst in presence of acid is a convenient method for the reduction of aromatic ketones to methylene compounds.

Selective hydrogenation of a carbonyl group in the presence of carbon–carbon double bonds is difficult and in most cases is best effected with hydride reducing agents or by the Meerwein–Pondorff method (see p. 458). If the double bonds are highly substituted, selective reduction of the carbonyl group can sometimes be achieved catalytically.

Nitriles, oximes and nitro compounds

Functional groups with multiple bonds to nitrogen are also readily reduced by catalytic hydrogenation. Nitriles, oximes, azides and nitro compounds, for example, are all smoothly converted into primary amines. Reduction of nitro compounds takes place very easily and is generally faster than reduction of olefinic bonds or carbonyl groups. Raney nickel or any of the platinum metals can be used as catalyst, and the choice will be governed by the nature of other functional groups in the molecule. Thus β-phenylethylamines, useful for the synthesis of isoquinolines, are conveniently obtained by catalytic reduction of αβ-unsaturated nitro compounds (7.12).

$$C_6H_5CH{=}CHNO_2 \xrightarrow[C_2H_5OH,\ 25\,°C,\ H_2SO_4]{H_2,\ Pd\text{-}C} C_6H_5CH_2CH_2NH_2 \qquad (7.12)$$

Nitriles are reduced with platinum or palladium at room temperature, or with Raney nickel under pressure. Unless precautions are taken, however, large amounts of secondary amines may be formed in a side reaction of the amine with the intermediate imine (7.13). With the

$$R{-}C{\equiv}N \xrightarrow{H_2} RCH{=}NH \xrightarrow{H_2} RCH_2NH_2 \qquad (7.13)$$

$$RCH_2NH_2 \xrightleftharpoons{RCH=NH} RCH_2NHCHR(NH_2) \xrightleftharpoons{-NH_3} RCH_2N{=}CHR$$

$$RCH_2N{=}CHR \xrightarrow{H_2} RCH_2NHCH_2R$$

platinum metal catalysts this reaction can be suppressed by conducting the hydrogenation in acid solution or in acetic anhydride, which removes the amine from the equilibrium as its salt or as its acetate. For reactions with Raney nickel, where acid cannot be used, secondary amine formation is prevented by addition of ammonia. Hydrogenation of nitriles containing other functional groups may lead to cyclic compounds. Thus indolizidine and quinolizidine derivatives have been obtained as shown in (7.14).

[Scheme (7.14): 2-pyridyl-CH_2CH_2CN $\xrightarrow[C_2H_5OH]{H_2,\ PtO_2}$ 2-piperidyl-CH_2CH_2CN (NH) → bicyclic cyclic amidine (NH, N–H) → indolizidine (N) + NH_3] (7.14)

Reduction of oximes to primary amines takes place under conditions similar to those used for nitriles, with palladium or platinum in acid solution, or with Raney nickel under pressure.

Stereochemistry and mechanism

Hydrogenation of an unsaturated compound takes place by adsorption of the compound on to the catalyst surface, followed by transfer of hydrogen from the catalyst to the side of the molecule which is adsorbed on it. Adsorption onto the catalyst is largely controlled by steric factors, and it is found in general that hydrogenation takes place by *cis* addition of hydrogen atoms to the less hindered side of the unsaturated centre. However, it is not always easy to decide which is the less hindered side and in such cases it may be difficult to predict what the steric course of a hydrogenation will be.

Thus, hydrogenation of the *E*-stilbene derivative (6) forms the (±)-dihydro compound (7) by *cis* addition of hydrogen, while the (*Z*)-stilbene (8) gives the *meso* isomer (9) (7.15). Hydrogenation of the pinene

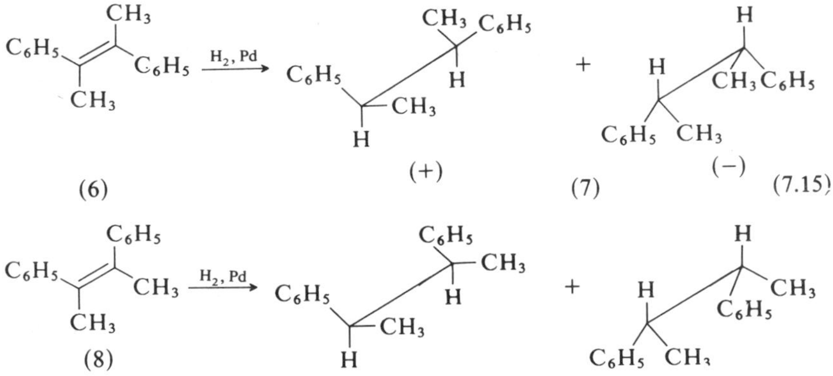

C_6H_5, CH_3, C_6H_5, CH_3 (6) $\xrightarrow{H_2, Pd}$ (+) + (−) (7)

C_6H_5, C_6H_5, CH_3, CH_3 (8) $\xrightarrow{H_2, Pd}$ *meso* (9)

(7.15)

derivative (10) and of the trimethylcyclohexanone (11) gave products formed by *cis* addition of hydrogen to the more accessible side of the double bonds (7.16). In all these examples the molecule possesses a certain degree of rigidity and it is clear which is the less hindered face of the double bond. With more flexible molecules it may be more difficult to decide on which side the molecule will be more easily adsorbed on the catalyst and to predict the steric course of a hydrogenation which, in such cases, may be influenced by the experimental conditions or by functional groups far removed from the centre of unsaturation. Where

CH_3 CH_3 CO_2H (10) $\xrightarrow[CH_3CO_2H]{H_2,\ PtO_2}$ CH_3 CH_3 CO_2H H (90%) (7.16)

CH_3 H O CH_3 CH_3 (11) $\xrightarrow[CH_3CO_2H]{H_2,\ PtO_2}$ CH_3 H OH H CH_3 CH_3 (83%) + CH_3 H H OH CH_3 CH_3 (17%)

there is no marked difference in the ease of approach to the two faces of the unsaturated centre mixtures of isomers are generally formed on hydrogenation. In some cases the affinity of a particular substituent group for the catalyst surface may induce addition of hydrogen from its own side of the molecule, irrespective of steric effects; the $—CH_2OH$ group is particularly effective in this respect. Thus, in the hydrogenation of the tetrahydrofluorene derivative in (7.17) there was 95 per cent *cis* addition of hydrogen when $R=CH_2OH$ but only 10 per cent *cis* addition when $R=CONH_2$ (see also Fujimoto, Kishi and Blount, 1980).

CH_3O R O O $\xrightarrow[CH_3OCH_2CH_2OH,\ 25\ °C]{H_2,\ Pd}$ CH_3O H R O O

$R=CH_2OH$ 90%
$R=CONH_2$ 10% (7.17)

The hydrogenation of substituted cyclic alkenes is anomalous in many cases in that substantial amounts of *trans*-addition products are formed, particularly with palladium catalysts. Thus, the octa-hydronaphthalene (12) on hydrogenation over palladium in acetic acid affords mainly *trans*-decalin, and 1,2-dimethylcyclohexene gives variable mixtures of *cis*- and *trans*-1,2-dimethylcyclohexane depending on the conditions (7.18). Similarly, in the hydrogenation of the isomeric xylenes over platinum oxide, the *cis*-dimethylcyclohexanes are the main products, but some *trans* isomer is always produced. The reason for the formation of the *trans* products is not completely clear, but it has been suggested that it may be due to migration of the methyl substituted double bond in a partially hydrogenated product on the catalyst surface accompanied by stereorandom allylic migration of hydrogen, followed by desorption and random readsorption.

(12) $\xrightarrow[CH_3CO_2H]{H_2,\ Pd\text{-}C}$ (21%) + (79%)

$\xrightarrow[CH_3CO_2H]{H_2,\ Pd}$ (16%) + (46%) (7.18)

$\xrightarrow[CH_3CO_2H]{H_2,\ PtO_2}$ (82%) (18%)

The rate and steric course of many hydrogenations are influenced by the presence or absence of acid or base in the reaction medium. For example, hydrogenation of 4-cholestene in neutral solution affords coprostane while in acid solution cholestane is obtained, and for the unsaturated ketone (13) the proportion of *cis*- and *trans*-decalones obtained on reduction varied with the solvent (7.19).

(13) $\xrightarrow{H_2,\ Pd\text{-}C}$ + (7.19)

Solvent	Product composition (%)	
C_2H_5OH	53	47
$C_2H_5OH + 10\%\ HCl$	93	7

The stereochemical course of the hydrogenation of the carbonyl group of cyclohexanones is also influenced by the nature of the solvent. The von Auwers–Skita rule, as modified by Barton (1953) is often used as a guide, and predicts that hydrogenation in strongly acid solution, which is rapid, will lead to formation of the axial alcohol, whereas in alkaline or neutral medium the equatorial alcohol will predominate if the ketone is not hindered; strongly hindered ketones will again give the axial alcohol. Thus, hydrogenations with Raney nickel, which is alkaline, or with Adams' catalyst in a neutral solvent, often leads to the equatorial alcohol, while with Adams' catalyst in acid solution the axial alcohol is formed (7.20). However, the rule must be used with caution, and the effect of acid and alkali is not always as straightforward as the von Auwers–Skita rule suggests.

CH_3 O $CO_2C_2H_5$ H CH_3 $\xrightarrow[CH_3COOH]{H_2, PtO_2}$ OH CH_3 H $CO_2C_2H_5$ H CH_3 (95%)

Raney Ni, H_2, methanol

H CH_3 OH $CO_2C_2H_5$ H CH_3 (95%)

(7.20)

A satisfactory mechanism for catalytic hydrogenation must explain not only the observed *cis* addition of hydrogen, but also the fact that alkenes are isomerised by hydrogenation catalysts, and that catalytic deuteration of an alkene leads to products containing more and fewer than two atoms of deuterium per molecule as well as the unexpected formation of *trans* addition products in some hydrogenations. These results can be rationalised on the basis of a mechanism in which transfer of hydrogen atoms from the catalyst to the adsorbed substrate is supposed to take place in a stepwise manner. The process is thought to involve equilibria between π-bonded forms A and B and a half hydrogenated form C which can either take up another atom of hydrogen or revert to starting material or to an isomeric alkene D (7.21). Another mechanism involving the π-allyl intermediate E has been postulated (7.22). π-Allyl intermediates have also been invoked to explain the hydrogenolysis of benzyl alcohols and their derivatives.

Homogeneous hydrogenation

Catalysts for heterogeneous hydrogenation of the types discussed above, although useful have some disadvantages. They may show lack of selectivity when more than one unsaturated centre is present, they may cause double-bond migration and, in reactions with deuterium, they usually bring about allylic interchanges with deuterium. This, in conjunction with double-bond migration, results in unspecific labelling with, in many cases, introduction of more than two deuterium atoms. Again, a number of functional groups suffer easy hydrogenolysis over heterogeneous catalysts and this sometimes leads to complications. The stereochemistry of reduction, despite a number of rules, is difficult to predict since it depends on chemisorption and not on reactions between catalyst surface

A

C

B

D

(7.21)

E

(7.22)

molecules. Some of these difficulties have been overcome by the introduction of soluble catalysts which allow hydrogenation in homogeneous solution (Harmon, Gupta and Brown, 1973; McQuillin, 1973; Parshall, 1980).

A number of soluble catalyst systems have been used, but the most effective found so far are the rhodium and ruthenium complexes tris(triphenylphosphine)chlororhodium $[(C_6H_5)_3P]_3RhCl$, and hydridochlorotris(triphenylphosphine)ruthenium $[(C_6H_3)_3P]_3RuClH$.

The rhodium complex is easily made by reaction of rhodium chloride with excess of triphenylphosphine in boiling ethanol (7.23). It is an extremely efficient catalyst for the homogeneous hydrogenation of non-conjugated alkenes and alkynes at ordinary temperature and pressure in benzene or similar solvents. Functional groups such as oxo, cyano, nitro,

$$RhCl_3.3H_2O \xrightarrow[\text{boiling } C_2H_5OH]{\text{excess of } (C_6H_5)_3P} [(C_6H_5)_3P]_3RhCl \qquad (7.23)$$

chloro, azo are not reduced under these conditions. Mono- and disubstituted double bonds are reduced much more rapidly than tri- or tetrasubstituted ones, permitting the partial hydrogenation of compounds containing different kinds of double bonds (Biellmann, 1968). Thus, in the reduction of linalool (14) addition of hydrogen occurred selectively at the vinyl group, giving the dihydro compound in 90 per cent yield;

$$\text{(14)} \xrightarrow[C_6H_6]{H_2,[(C_2H_5)_3P]_3RhCl} \text{dihydro compound}$$

$$\text{(15)} \xrightarrow[\text{benzene}]{H_2,[(C_6H_5)_3P]_3RhCl} \text{carvotanacetone} \qquad (7.24)$$

carvone (15) was similarly converted into carvotanacetone. The selectivity of the catalyst is shown further by the remarkable reduction of ω-nitrostyrene to phenylnitroethane (7.25).

$$C_6H_5CH{=}CHNO_2 \xrightarrow[C_6H_6]{H_2,[(C_6H_5)_3P]_3RhCl} C_6H_5CH_2CH_2NO_2 \qquad (7.25)$$

Hydrogenations take place by *cis* addition to the double bond. This was shown by the catalysed reaction of deuterium with maleic acid to form *meso*-dideuterosuccinic acid, while fumaric acid gave the (±)-compound. Likewise, interrupted reduction of 2-hexyne gave a mixture of hexane and *cis*-2-hexene.

An important practical advantage of this catalyst in addition to its selectivity is that deuterium is introduced without scrambling; that is, only two deuterium atoms are added, at the site of the original double bond. Thus methyl oleate is converted into methyl 9,10-dideuterostearate in quantitative yield, and 1,4-androstadien-3,17-dione (16) gave the dideutero compound (17) by *cis* addition to the α-face of the disubstituted double bond (7.26).

$^2H_2$, $[(C_6H_5)_3P]_3RhCl$ → C_6H_6-C_2H_5OH (7.26)

(16) (85%) (17)

Another very valuable feature of this catalyst is that it does not bring about hydrogenolysis, thus allowing the selective hydrogenation of carbon–carbon double bonds without hydrogenolysis of other susceptible groups in the molecule. Thus, benzyl cinnamate is converted smoothly into the dihydro compound without attack on the benzyl ester group (7.27), and allyl phenyl sulphide is reduced to phenyl propyl sulphide in 93 per cent yield.

$$CH_2{=}CHCH_2{-}S{-}C_6H_5 \rightarrow C_3H_7{-}S{-}C_6H_5$$
$$C_6H_5CH{=}CHCO_2CH_2C_6H_5 \rightarrow C_6H_5CH_2CH_2CO_2CH_2C_6H_5 \qquad (7.27)$$

Because of the strong affinity of $[(C_6H_5)_3P]_3RhCl$ for carbon monoxide it decarbonylates aldehydes, and olefinic compounds containing aldehyde groups cannot be hydrogenated with this catalyst under the usual conditions. Thus cinnamaldehyde is converted into styrene in 65 per cent yield, and benzoyl chloride gives chlorobenzene in 90 per cent yield.

$[(C_6H_5)_3P]_3RhCl$ dissociates in solvents (S) to give the solvated species $[(C_6H_5)_3P]_2Rh(S)Cl$ which in presence of hydrogen is in equilibrium with the dihydrido complex $[(C_6H_5)_3P]_2Rh(S)ClH_2$ in which the hydrogen atoms are directly attached to the metal (7.28). Reduction is believed to take place by displacement of solvent by the alkene followed by step-wise stereospecific *cis* transfer of the two hydrogen atoms from the metal to the loosely co-ordinated alkene by way of an intermediate with a carbon–metal bond. Diffusion of the saturated substrate away from the transfer site leaves the complex again ready to combine with dissolved hydrogen and continue the reduction.

$$[(C_6H_5)_3P]_2Rh(S)Cl \underset{}{\overset{H_2}{\rightleftharpoons}} [(C_6H_5)_3P]_2Rh(S)ClH_2 \xrightleftharpoons{RCH=CHR'} \tag{7.28}$$
$$[(C_6H_5)_3P]_2RhCl(RCH=CHR')H_2 \rightarrow RCH_2CH_2R' + [(C_6H_5)_3P]_2Rh(S)Cl$$

The ruthenium complex $[(C_6H_5)_3P]_3RuClH$, formed *in situ* from $[(C_6H_5)_3P]_3RuCl_2$ and molecular hydrogen in benzene in presence of a base such as triethylamine, is an even more efficient catalyst which is specific for the hydrogenation of monosubstituted alkenes $RCH=CH_2$. Rates of reduction for other types of alkenes are slower by a factor of at least 2×10^3. Thus 1-heptene was rapidly converted into heptane with molecular hydrogen in benzene solution but 3-heptene was unaffected. Some isomerisation of alkenes is observed with this catalyst but the rate is slow compared with the rate of hydrogenation. Very similar behaviour is shown by the rhodium complex hydridocarbonyl-tris-(triphenylphosphine)rhodium(I) $[(C_6H_5)_3P]_3RhH(CO)$. Terminal alkynes apparently react with the catalyst, but disubstituted alkynes are converted into *Z*-alkenes. Thus stearolic acid gave oleic acid and diphenylacetylene formed *Z*-stilbene (7.29).

$$CH_3(CH_2)_7C\equiv C(CH_2)_7CO_2H \xrightarrow[C_6H_6]{H_2,\,[(C_6H_5)_3P]_3RuCl_2} \underset{\textstyle H\qquad\qquad H}{\overset{\textstyle CH_3(CH_2)_7\qquad (CH_2)_7CO_2H}{C=C}} \tag{7.29}$$

Hydrogenations with $[(C_6H_5)_3P]_3RuClH$ and $[(C_6H_5)_3P]_3RhH(CO)$ are two-step processes which proceed by the reversible formation of a metal–alkyl intermediate. The actual catalyst in the former case is thought to be the square monomer $[(C_6H_5)_3P]_2RuClH$ with *trans* $P(C_6H_5)_3$ groups, formed by dissociation in solution (7.30). The high selectivity for reduction of terminal double bonds is attributed to steric hindrance by the bulky $P(C_6H_5)_3$ groups to the formation of the metal–alkyl intermediate with other types of alkenes.

$$[(C_6H_5)_3P]_2Ru\langle{}^{H}_{Cl} \xrightarrow{RCH=CH_2} \left[[(C_6H_5)_3P]_2Ru\langle{}^{H}_{Cl} \longleftarrow \begin{matrix} R \\ | \\ CH \\ \| \\ CH_2 \end{matrix} \right] \tag{7.30}$$
$$\rightarrow [(C_6H_5)_3P]_2\underset{\textstyle Cl}{\underset{|}{Ru}}-\underset{\textstyle R}{\underset{|}{C}}HCH_3 \xrightarrow{H_2} RCH_2CH_3 + [(C_6H_5)_3P]_2Ru\langle{}^{H}_{Cl}$$

Another useful catalyst is the iridium complex $[Ir(Cod)py(PCy_3)]PF_3$ (Cod = 1,5-cyclooctadiene; Cy = cyclohexyl) (Crabtree and Morris, 1977). It reduces tri- and tetra-substituted double bonds as well as mono- and di-substituted ones, although not so rapidly, and it appears to be unaffected by sulphur in the molecule. A valuable feature of this catalyst is the high degree of stereocontrol which can be achieved with it in the hydrogenation of cyclic allylic and homoallylic alcohols. Thus, the allylic alcohol (18) gave the saturated compound (19) almost exclusively (Corey and Engler, 1984; see also Stork and Kahne, 1983) and the homoallylic alcohols (20) and (22) gave the *cis*- and *trans*-indanols (21) and (23).

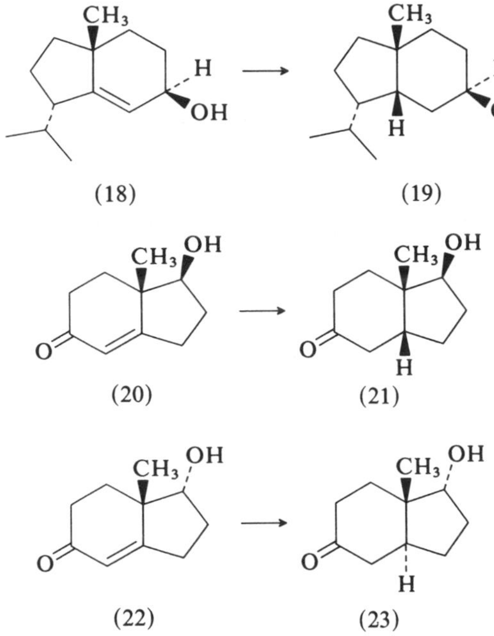

(18) (19)

(20) (21) (7.31)

(22) (23)

No stereoselectivity was observed in the hydrogenation of the corresponding acetates, or under heterogeneous conditions, and in these reactions the hydroxyl substituents exert the same kind of directing effect as they do in the Simmons–Smith reaction (p. 95) and the Sharpless epoxidation (p. 373).

Induced asymmetry via homogeneous hydrogenation

An interesting development has been the use of soluble metal catalysts to effect asymmetric hydrogenations (Bogdanović, 1973; ApSimon and Seguin, 1979; Čaplar *et al.*, 1981; Knowles, 1983). The catalysts used include a number of rhodium complexes containing as ligands chiral phosphines or amides. Trichlorotripyridylrhodium, treated with sodium borohydride in dimethylformamide, gives a complex [py_2.dmf.$RhCl_2(BH_4)$] which is highly active for the hydrogenation of alkenes. By using asymmetric ligands related to dimethylformamide catalysts are obtained which can bring about asymmetric hydrogenations. Thus, methyl 3-phenyl-2-butenoate is hydrogenated at a rhodium complex formed in (+)- or (−)-1-phenylethylformamide to give (+)- or (−)-methyl 3-phenylbutanoate in better than 50 per cent optical yield (7.32). Very high optical yields of α-amino acids have been obtained by

$$C_6H_5C(CH_3){=}CHCO_2CH_3 \rightarrow C_6H_5CH(CH_3)CH_2CO_2CH_3 \qquad (7.32)$$

hydrogenation of α-acylaminoacrylic acids using rhodium(I) complexes containing chiral phosphines (e.g. 24), or the cyclo-octadiene complex [Rh(1,5-cyclo-octadiene)Cl]$_2$. α-Acetylaminocinnamic acid for example, gave *N*-acetylphenylalanine in 95 per cent yield and 85 per cent optical purity and, using the laevorotatory form of the ligand (24) atropic acid was converted into (*S*)-hydratropic acid with 64 per cent optical purity (7.33) (Scott *et al.*, 1981; Bergstein and Martens, 1981).

H_3C, H_3C, O, O, H, H, $CH_2P(C_6H_5)_2$, $CH_2P(C_6H_5)_2$

(24)

$$C_6H_5CH{=}C(NHCOCH_3)CO_2H \xrightarrow[H_2,\ C_6H_6\text{-}C_2H_5OH,\ 25\,^\circ C]{RhCl[L^*]S} C_6H_5CH_2CH(NHCOCH_3)CO_2H$$

(85% optical purity)

$$CH_2{=}C(C_6H_5)CO_2H \xrightarrow[H_2,\ C_6H_6,\ 25\,^\circ C]{RhCl[L^*]S} CH_3CH(C_6H_5)CO_2H \qquad (7.33)$$

(64% optical purity)

7.2. Reduction by dissolving metals

General

Chemical methods of reduction are of two main types: those which take place by addition of electrons to the unsaturated compound followed or accompanied by transfer of protons; and those which take place by addition of hydride ion followed in a separate step by protonation (Augustine, 1968).

Reductions which follow the first path are generally effected by a metal, the source of the electrons, and a proton donor which may be water, an alcohol or an acid. They can result either in the addition of hydrogen atoms to a multiple bond or in fission of a single bond between atoms, usually, in practice, a single bond between carbon and a heteroatom. In these reactions an electron is transferred from the metal surface (or from the metal in solution) to the organic molecule being reduced, giving, in the case of addition to a multiple bond, an anion radical, which in many cases is immediately protonated. The resulting radical subsequently takes up another electron from the metal to form an anion which may be protonated immediately or remain as the anion until work-up. In the absence of a proton source dimerisation or polymerisation of the anion-radical may take place. In some cases a second electron may be added to the anion-radical to form a dianion, or two anions in the case of fission reactions (Birch, 1950; House, 1972). These transformations may be represented as shown in (7.34).

$$\mathrm{A{-}B} \xrightarrow{2e} \mathrm{A^- + B^-}$$
$$\mathrm{A{-}B} \xrightarrow{e} \begin{matrix}\mathrm{A^- + B^{\cdot}} \\ \mathrm{A^{\cdot} + B^-}\end{matrix} \xrightarrow{e} \mathrm{A^- + B^-}$$
$$\mathrm{A{=}B} \xrightarrow{e} \left\{\begin{matrix}\mathrm{A^-{-}B^{\cdot}} \\ \updownarrow \\ \mathrm{A^{\cdot}{-}B^-}\end{matrix}\right\} \xrightarrow{e} \mathrm{A^-{-}B^-} \tag{7.34}$$
$$\left\{\mathrm{A^-{-}B^{\cdot}} \leftrightarrow \mathrm{A^{\cdot}{-}B^-}\right\} \longrightarrow \begin{matrix}\mathrm{A{-}B^-} \\ | \\ \mathrm{A{-}B^-}\end{matrix}$$
$$\left\{\mathrm{A^-{-}B^{\cdot}} \leftrightarrow \mathrm{A^{\cdot}{-}B^-}\right\} \xrightarrow{H^+} \begin{matrix}\mathrm{AH{-}B^{\cdot}} \\ \text{or } \mathrm{A^{\cdot}{-}BH}\end{matrix} \xrightarrow{e} \begin{matrix}\mathrm{AH{-}B^-} \\ \text{or } \mathrm{A^-{-}BH}\end{matrix}$$

Thus, in the reduction of benzophenone with sodium in ether or liquid ammonia, the first product is the resonance-stabilised anion radical (25) which, in absence of a proton donor, dimerises to the pinacol. In presence of a proton source, however, protonation leads to the radical (26) which is subsequently converted into the anion and thence into benzhydrol (7.35). The presence in these anion radicals of an unpaired electron which interacts with the atoms in the conjugated system has been established by measurements of the electron spin resonance spectra of various anion radical solutions.

The metals commonly employed in these reductions include the alkali metals, calcium, zinc, magnesium, tin and iron. The alkali metals are often used in solution in liquid ammonia (Birch reduction, see p. 440) or as suspensions in inert solvents such as ether or toluene, frequently

$$\begin{array}{c}
C_6H_5\underset{\underset{O}{\|}}{C}C_6H_5 \xrightleftharpoons{\text{Na, liq. } NH_3} C_6H_5\dot{C}C_6H_5 \to (C_6H_5)_2\underset{O^-}{C}-\underset{O^-}{C}(C_6H_5)_2 \\
\quad :\ddot{O}:^- \updownarrow \qquad\qquad \downarrow H_3O^+ \\
C_6H_5\underset{:\ddot{O}^{\cdot}}{CH}C_6H_5 \xleftarrow{C_2H_5OH} C_6H_5\ddot{C}^-C_6H_5 \quad (C_6H_5)_2\underset{OH}{C}-\underset{OH}{C}(C_6H_5)_2 \\
\quad\; :\ddot{O}^{\cdot} \\
(26) \qquad\qquad (25) \\
\xrightarrow{Na} C_6H_5\underset{:\ddot{O}:^-}{CH}C_6H_5 \xrightarrow{H_3O^+} C_6H_5\underset{OH}{CH}C_6H_5
\end{array} \tag{7.35}$$

with addition of an alcohol or water to act as a proton source. Many reductions are also effected by direct addition of sodium or, particularly, zinc, tin or iron, to a solution of the compound being reduced in a hydroxylic solvent, such as ethanol, acetic acid or an aqueous mineral acid.

Reduction with metal and acid

In the well-known Clemmensen reduction of the carbonyl group of aldehydes and ketones to methyl or methylene, a mixture of the carbonyl compound and amalgamated zinc is boiled with hydrochloric acid, sometimes in the presence of a non-miscible solvent which serves to keep the concentration in the aqueous phase low and thus prevent bimolecular condensations at the metal surface. Amalgamation of the zinc raises its hydrogen overvoltage to the point where it survives as a reducing agent in the acid solution and is not consumed in reaction with the acid to give molecular hydrogen. The choice of acid is confined to the hydrogen halides, which appear to be the only strong acids whose anions are not reduced by zinc amalgam. The Clemmensen reaction as ordinarily effected employs rather vigorous conditions and is not suitable for the reduction of polyfunctional molecules such as 1,3- or 1,4-diketones or sensitive compounds (Buchanan and Woodgate, 1969) but for stable compounds it is an effective method of reduction. Stereophenone, for example, is converted into octadecylbenzene in 88 per cent yield. An alternative which proceeds under mild conditions uses zinc dust and solutions of hydrogen chloride gas in aprotic organic solvents such as ether (Vedejs, 1975). Other methods for converting carbonyl groups into methyl or methylene are described on pp. 476, 482.

The mechanism of the Clemmensen reaction is uncertain. Alcohols are not normally reduced under Clemmensen conditions and it would appear that they are not intermediates in the reduction of the carbonyl compounds. The reaction pathway in (7.36) involving transfer of electrons from the metal surface to the carbonyl carbon atom has been proposed (Nakabayashi, 1960).

$$\underset{\text{Zn}\ \ \text{Cl}^-}{\text{R—}\overset{\text{O}}{\overset{\|}{\text{C}}}\text{—R}'} \rightleftharpoons \text{R—}\underset{\text{ZnCl}}{\overset{\text{O}^-}{\text{C}}}\text{—R}' \overset{\text{H}^+}{\rightleftharpoons} \text{R—}\underset{\text{ZnCl}}{\overset{\text{OH}}{\text{C}}}\text{—R}' \xrightarrow[-\text{H}_2\text{O}]{\text{H}^+} \text{R—}\underset{\text{ZnCl}}{\overset{+}{\text{C}}}\text{—R}' \xrightarrow{2\text{Zn:}} \text{R—}\underset{\text{ZnCl}}{\overset{-}{\text{C}}}\text{—R}' + 2\text{Zn}^{\dot{+}} \xrightarrow{\text{H}^+} \text{R—}\underset{\text{ZnCl}}{\text{CH}}\text{—R}' \xrightarrow{\text{H}^+} \text{R—CH}_2\text{—R}' + \overset{+}{\text{Z}}\text{nCl} \qquad (7.36)$$

Reductive cleavage of α-substituted ketones, such as α-halo, α-acyloxy- and α-hydroxy-ketones to the unsubstituted ketone is commonly effected with zinc and acetic acid or dilute mineral acid. Metal-amine reducing agents (see p. 451) and chromium(II) salts (see p. 456) have also been used. The well-known reduction of an α-acetoxyketone with zinc and acetic acid can be represented as in (7.37). Transfer of two

$$\text{2-acetoxycyclohexanone} + \text{:Zn} \longrightarrow \text{zinc enolate (OZn)} \xrightarrow{\text{CH}_3\text{COOH}} \text{cyclohexanone}$$

$$\text{—COCHOH—} \xrightarrow[\text{CH}_3\text{COOH}]{\text{Zn,HCl}} \text{—COCH}_2\text{—} \qquad (7.37)$$

electrons to the carbonyl group, followed by departure of the substituent as the anion (the substituent must be a good leaving group for effective reaction) affords an enolate which is converted into the ketone by acid, or into the enol acetate if reaction is effected in presence of acetic anhydride. α-Ketols are similarly reduced to the ketone. The reductive cleavage of α-halo ketones may have a different mechanism.

Reductive eliminations of this type proceed most readily if the molecule can adopt a conformation where the bond to the group being displaced is orthogonal to the plane of the carbonyl group. Elimination of the substituent group is then eased by continuous overlap of the developing p-orbital at the α-carbon atom with the π-orbital system of the ion radical (7.38). For this reason cyclohexanone derivatives with axial α-substituents are reductively cleaved more readily than their equatorial isomers. For example, the (axial) hydroxyl group of the 20-keto-20a-α-hydroxy-20a-β-methyl-D-homo-steroid derivative (27) was cleaved in high yield with zinc and acetic acid, whereas the 20a-β-hydroxy epimer was unaffected (7.39).

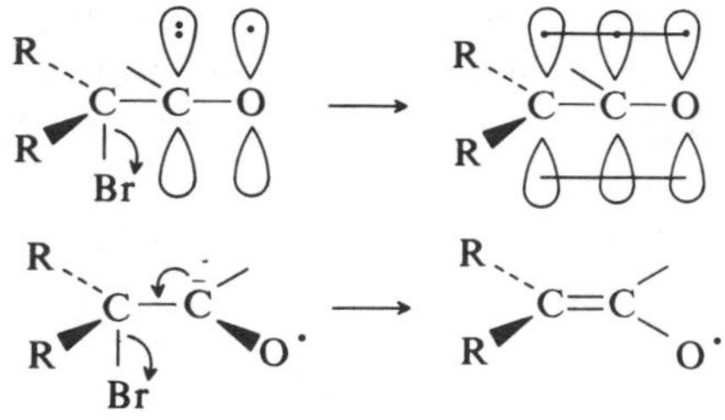

R, R, C–C–O, Br → R, R, C–C–O (7.38)

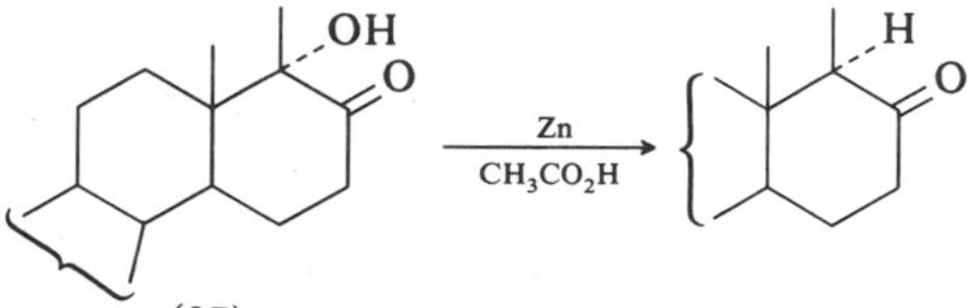

R, R, C–C, O·, Br → R, R, C=C, O·

OH, O, $\xrightarrow[\text{CH}_3\text{CO}_2\text{H}]{\text{Zn}}$ H, O (7.39)

(27)

Even carbon–carbon bonds may be broken in favourable circumstances. Thus, the cyclopropyl ketone (28) was converted into the bicyclic compound (29) with zinc–copper couple in acetic acid (7.40).

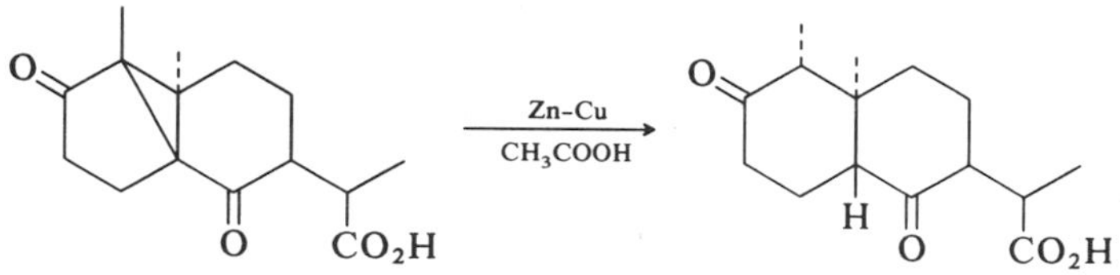

O, O, CO_2H $\xrightarrow[\text{CH}_3\text{COOH}]{\text{Zn-Cu}}$ O, H, O, CO_2H (7.40)

(28) (29)

Although alkyl halides are not normally cleaved by zinc and acid, benzylic halides are. For example, 1-chloromethylnaphthalene is readily converted into 1-methylnaphthalene. The more active of two halogen atoms can often be removed selectively, as in (7.41).

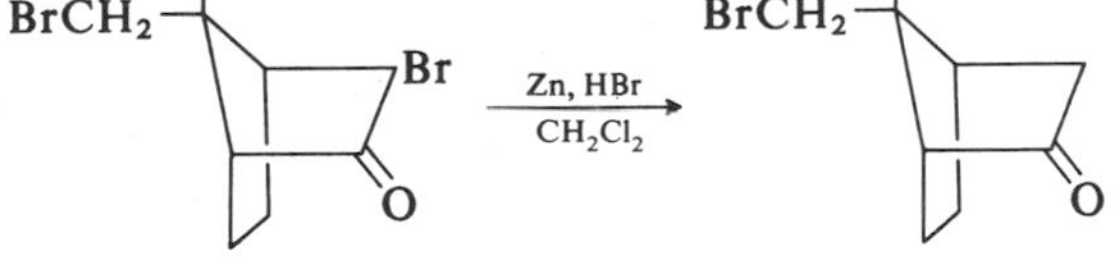

$BrCH_2$, Br, O $\xrightarrow[\text{CH}_2\text{Cl}_2]{\text{Zn, HBr}}$ $BrCH_2$, O (7.41)

β-Chloro-$\alpha\beta$-unsaturated ketones are reduced selectively to the halogen-free compounds with silver-promoted zinc dust in methanol (7.42). Since the β-chloroenones are readily obtained from the corresponding 1,3-diketones the sequence provides a method for converting 1,3-diketones into $\alpha\beta$-unsaturated ketones (Clark and Heathcock, 1973).

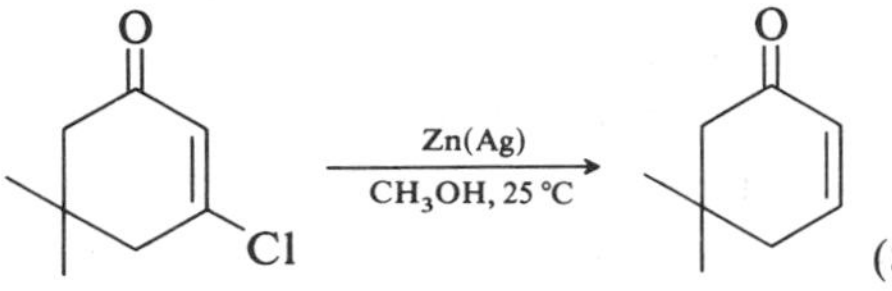

O, Cl $\xrightarrow[\text{CH}_3\text{OH, 25 °C}]{\text{Zn(Ag)}}$ O (81%) (7.42)

Reduction of carbonyl compounds with metal and an alcohol

Reduction of ketones to secondary alcohols can be effected catalytically (p. 420), with complex hydrides (p. 459) or with sodium (either as the free metal or as a solution in liquid ammonia) and an alcohol. The distinguishing feature of the sodium–alcohol method is that with cyclic ketones it gives rise in most cases, although not always, to the thermodynamically more stable alcohol either exclusively or in preponderant amount. Table 7.2 shows the proportions of more stable *trans*-(equatorial) alcohol formed by reduction of 2-methylcyclohexanone with different reducing agents. 4-t-Butylcyclohexanone similarly gives the more stable *trans*-4-t-butyl-cyclohexanol almost exclusively on reduction with lithium and propanol in liquid ammonia, and numerous other experiments confirm that in the vast majority of cases the more stable alcohol is the main product.

O CH$_3$ → OH CH$_3$ + OH CH$_3$ (7.43)

Table 7.2. *Proportion of trans-2-methylcyclohexanol by reduction of 2-methylcyclohexanone with different reagents*

Reagent	Proportion *trans* alcohol (%)
Na–alcohol	99
Lithium aluminium hydride	82
Sodium borohydride	69
Aluminium isopropylate	42
Catalyst and hydrogen	7–35

Two main hypotheses have been put forward to account for the high proportion of the more stable epimer formed in these reductions with sodium and an alcohol. In one due to Barton and Robinson (1954) it is supposed that a tetrahedral dianion is formed initially and adopts the more stable configuration with equatorial oxygen which, on protonation, affords the more stable arrangement of the alcohol. House (1972) on the other hand, considers that reaction proceeds by donation of one electron from the metal to the carbonyl group followed by protonation on oxygen to give a free-radical intermediate which adopts the more stable configuration with an equatorial hydroxyl group. Further reduction and protonation of the anionic intermediate takes place with retention of configuration at the carbon atom to give the observed equatorial alcohol (7.44).

(7.44)

It appears that neither of these mechanisms is universally applicable, however. There are some anomalies. In some cases the thermodynamically less stable isomer is the main product and in others the proportion of the two isomers produced depends on the experimental conditions. Reduction of norcamphor with lithium and ethanol, for example, gave mainly the less stable *endo* alcohol. Labelling experiments have shown also that in some cases reduction is accompanied by transfer of hydrogen from the α-carbon of the ketone to the carbinol carbon of the newly formed alcohol. This cannot be accounted for by either the Barton or the House mechanism (see Huffmann, 1983).

Aldehydes can be reduced to primary alcohols with sodium and ethanol but in many cases better yields are obtained by catalytic hydrogenation (p. 420) or with hydride reducing agents (p. 459).

Reduction of ketones with dissolving metals in absence of a proton donor leads to the formation of bimolecular products. The usual reagents are magnesium, magnesium amalgam or aluminium amalgam. Thus, reduction of acetone to pinacol follows the course shown in (7.45). Bimolecular reduction of this type is often a competing reaction in other reductions with dissolving metals.

$$(CH_3)_2C{=}O \xrightarrow[\text{boiling } C_6H_6]{\text{Mg-Hg}} (CH_3)_2\bar{C}{-}O^{\cdot}\ \tfrac{1}{2}Mg^{2+} \longleftrightarrow (CH_3)_2\dot{C}{-}O^{-}$$

$$\downarrow$$

$$\begin{matrix}(CH_3)_2C{-}OH \\ (CH_3)_3C{-}OH\end{matrix} \xleftarrow{H_2O} \begin{matrix}(CH_3)_2C{-}O^{-} \\ (CH_3)_2C{-}O^{-}\end{matrix} \qquad (7.45)$$

The pinacol condensation has not been widely employed in synthesis up to now, but its potentiality has now been greatly increased by the introduction of a number of new reagents derived from low-valent transition metal species which bring about both inter- and intra-molecular condensations in high yield. One of the most effective of these reagents is the Ti(II) species generated by reaction of titaniumIV chloride with amalgamated magnesium (Corey *et al.*, 1976). With this reagent cyclohexanone is converted into the corresponding pinacol in 93 per cent yield, and mixed pinacols are also readily obtained (7.46). Intramolecular

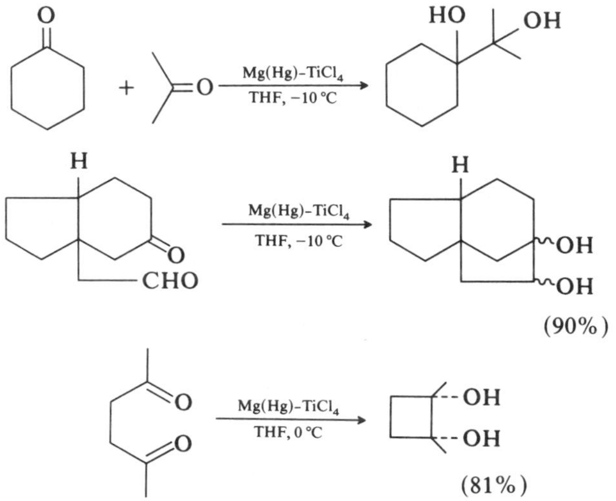

Mg(Hg)–$TiCl_4$, THF, −10 °C

Mg(Hg)–$TiCl_4$, THF, −10 °C (90%)

Mg(Hg)–$TiCl_4$, THF, 0 °C (81%)

(7.46)

condensations proceed equally well to give cyclic 1,2-diols, even cyclobutandiols being formed in remarkably high yield (7.59). The mechanism of these reactions is uncertain, but it seems clear that the species which supplies electrons to the carbonyl group (cf. equation 7.45) is some derivative of Ti(II).

The value of this improved route to pinacols lies not only in the provision of another method for the formation of carbon–carbon bonds, but also in the fact that the unsymmetrical pinacols which it makes readily available for the first time are themselves useful synthetic intermediates. Thus, treatment with methanesulphonyl chloride in triethylamine affords epoxides, and both the pinacols and the epoxides rearrange readily to carbonyl compounds under mild acidic conditions (7.47).

Esters are also reduced by sodium and alcohols to form primary alcohols. This, the Bouveault–Blanc reaction, is one of the oldest established methods of reduction used in organic chemistry, but has now been largely replaced by reduction with lithium aluminium hydride. It follows the same general course as the reduction of aldehydes and ketones. When

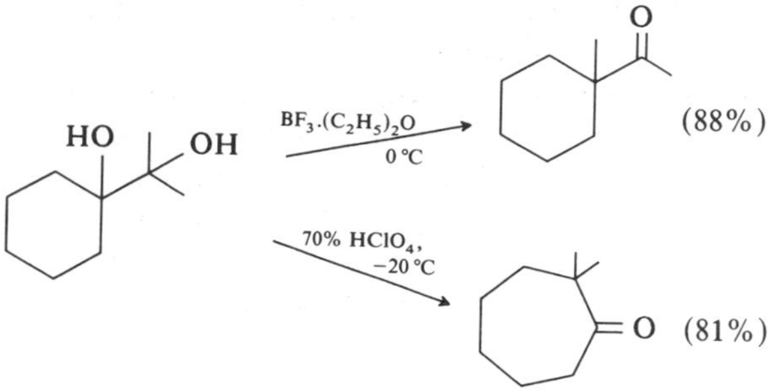

HO OH → $BF_3.(C_2H_5)_2O$, 0 °C → (88%)

70% $HClO_4$, −20 °C → O (81%)

(7.47)

the reaction is carried out in absence of a donor, for example with sodium in xylene or sodium in liquid ammonia, dimerisation, analogous to the formation of pinacols from ketones, takes place, and this is the basis of the well-known and synthetically useful acyloin condensation. Intramolecular reaction gives ring compounds, and the reaction has been extensively used for the synthesis of medium and large carbon rings as well as five- and six-membered rings (Finley, 1964). Thus, the dicarboxylic ester (30) gives the ten-membered ring ketol (31), and, in the steroid series, dimethyl marrianolate methyl ether (32) gives the ketol (33) which is readily converted into oestrone (7.48). Greatly improved yields are often obtained by carrying out the reactions in presence of chlorotrimethylsilane. This serves principally to remove alkoxide ion from the reaction medium, thus preventing wasteful base-catalysed side reactions such as β-eliminations and Claisen or Dieckmann condensations (see p. 320).

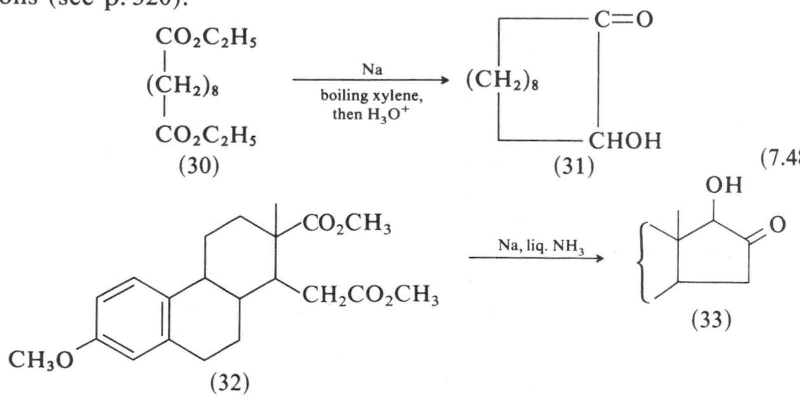

$CO_2C_2H_5$–$(CH_2)_8$–$CO_2C_2H_5$ (30) → Na, boiling xylene, then H_3O^+ → $(CH_2)_8$ C=O / CHOH (31)

(7.48)

CH_3O … CO_2CH_3, $CH_2CO_2CH_3$ (32) → Na, liq. NH_3 → OH, O (33)

The acyloin reaction is generally considered to proceed by dimerisation of the anion radicals formed by electron transfer from the metal to the carbonyl group of the ester (cf p. 433). Some doubt has been cast on this mechanism and an alternative which is said to account better for the anomalous products formed in some reactions has been proposed (Bloomfield *et al.*, 1975).

Reduction with metal and ammonia or an amine: conjugated systems

Isolated carbon–carbon double bonds are not normally reduced by dissolving metal reducing agents, because formation of the intermediate electron-addition products requires more energy than ordinary reagents can provide. Reduction is possible when the double bond is conjugated, because then the intermediate can be stabilised by electron delocalisation. By far the best reagent is a solution of an alkali metal in liquid ammonia, with or without addition of an alcohol – the so-called Birch reduction conditions. Under these conditions conjugated dienes, $\alpha\beta$-unsaturated ketones, styrene derivatives and even benzene rings can be reduced to dihydro derivatives (Birch and Smith, 1958; Birch and Subba Rao, 1972).

Birch reductions are usually effected with solutions of lithium, sodium or potassium in liquid ammonia. In some reactions an alcohol is added to act as a proton donor and as a buffer against the accumulation of the strongly basic amide ion; in other cases acidification with an alcohol or with ammonium chloride is effected at the end of the reaction. Solutions of alkali metals in liquid ammonia contain solvated metal cations and electrons (7.49), and part of the usefulness of these reagents arises from

$$\text{M} \xrightarrow{\text{liq. NH}_3} \text{M}^+(\text{NH}_3) \cdots e^-(\text{NH}_3) \qquad (7.49)$$

the small steric requirement of the electrons, which sometimes allows reactions which are difficult to achieve with other reducing agents, and in many cases leads to different stereochemical results. Wasteful reaction of the metal with ammonia to form hydrogen and amide ion is slow in the absence of catalysts; part of the superiority of lithium over sodium and potassium is due to the fact that its reaction with ammonia is catalysed to a lesser extent by colloidal iron present as an impurity in ordinary commercial liquid ammonia. Using distilled ammonia there is little to choose among the three metals. Reactions are usually carried out at the boiling point of ammonia (−33 °C), and since the solubility of many organic compounds in liquid ammonia is low at this temperature, co-solvents such as ether, tetrahydrofuran or dimethoxyethane are often added to aid solubility. A common technique is to add the metal to a stirred solution of the compound in liquid ammonia and the co-solvent, in presence of an alcohol where appropriate, but variations of this procedure are often used (Djerassi, 1963).

Isolated carbon–carbon double bonds are not normally reduced by metal–ammonia reagents alone, but in the presence of an alcohol terminal double bonds may be reduced. Conjugated dienes are readily reduced to the 1,4-dihydro derivatives with metal–ammonia reagents in the absence of added proton donors. Thus, isoprene is reduced to 2-methyl-2-butene by sodium in ammonia through the anion radical. The protons required to complete the reduction are supplied by the ammonia (7.50).

$$CH_2{=}\overset{\overset{\displaystyle CH_3}{|}}{C}{-}CH{=}CH_2 \xrightarrow[\text{liq. } NH_3]{Na} \underbrace{CH_2{\cdots}\overset{\overset{\displaystyle CH_3}{|}}{C}{\cdots}CH{\cdots}CH_2}_{\cdot\,-} \rightarrow CH_3{-}\overset{\overset{\displaystyle CH_3}{|}}{C}{=}CH{-}CH_3 \qquad (7.50)$$

Reduction of cyclic ketones with metal-ammonia-alcohol reagents leads predominantly to the alcohol with the thermodynamically more stable configuration. With $\alpha\beta$-unsaturated ketones the saturated ketone or the saturated alcohol can be obtained depending on the conditions. Reduction in the presence of not more than one equivalent of a proton donor followed by acidification with ethanol or ammonium chloride leads to the saturated ketone. The same transformation can, of course, be effected by catalytic hydrogenation, but metal-ammonia reduction is more stereoselective and in many cases leads to a product with different stereochemistry from that obtained by hydrogenation. A further advantage is that isolated double bonds elsewhere in the molecule are not reduced by the metal-ammonia procedure. Thus, the $\alpha\beta$-unsaturated ketone (34) on reduction with hydrogen and palladium affords the *cis*-fused ketone (35) while with lithium in liquid ammonia the *trans*-product (36) is obtained. Cyperone (37) with lithium in ammonia is converted into the ketone (38); hydrogenation reduces both double bonds. Reduction in the presence of ethanol as proton source gives the saturated alcohol (39).

The reduction of $\alpha\beta$-unsaturated ketones with alkali metals and liquid ammonia takes place in the following way. With cyperone, for example, the first step is formation of the anion radical (40) which subsequently abstracts a proton from the ammonia or from added alcohol to give (41) and thence, after addition of another electron, the enolate anion (42) (7.52). In the absence of a stronger acid this anion retains its negative charge and resists addition of another electron which would correspond to further reduction. Acidification with ammonium chloride then leads to the saturated *ketone*. The reaction of ammonium ions with solvated electrons apparently destroys the reducing system before further reduction of the ketone to the alcohol can take place.

In the presence of 'acids' (e.g. ethanol) sufficiently strong to protonate the enolate anion however, the ketone is generated in the reducing medium and is reduced further to the saturated alcohol (7.53). The formation of enolate anions such as (42) during metal-ammonia reduction of $\alpha\beta$-unsaturated ketones is shown by the ready formation of α-methyl ketones by treatment of the reaction mixture with methyl iodide before

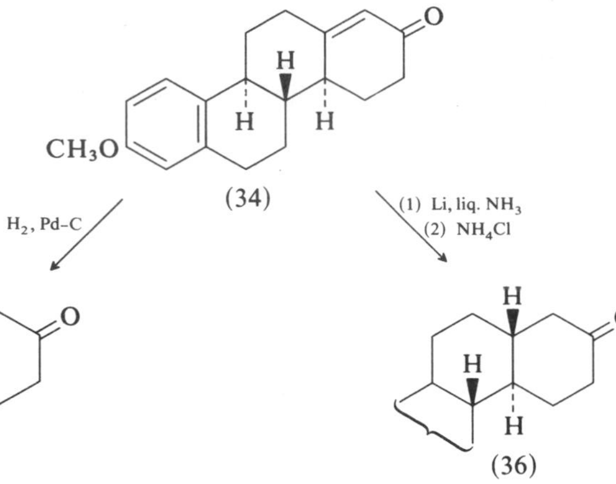

CH$_3$O

(34)

H_2, Pd–C

(1) Li, liq. NH_3
(2) NH_4Cl

(35)

(36)

(7.51)

(37)

(1) Li, NH_3 ether
(2) NH_4Cl

(38)

Li, NH_3, C_2H_5OH
dioxan

HO

(39)

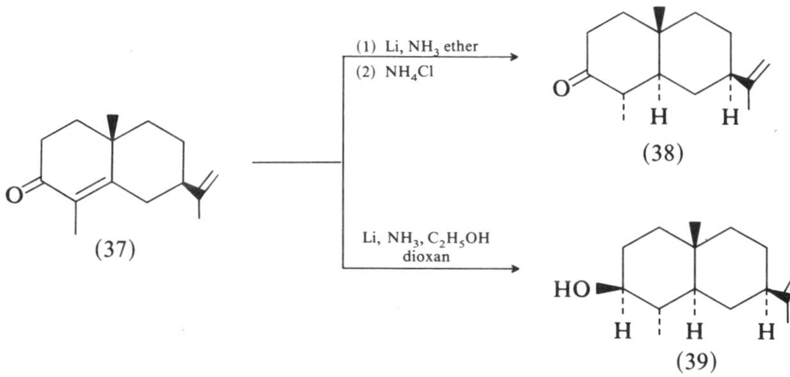

final acidification and this is, in fact, a good method for preparing specific α-alkyl derivatives of unsymmetrical ketones (cf. p. 21).

Reduction of cyclic $\alpha\beta$-unsaturated ketones in which there are substituents on the β- and γ-carbon atoms could apparently give rise to two stereoisomeric products (7.54), but in practice it is found that in many cases one isomer is formed predominantly, generally the more stable of the two. The guiding principle appears to be that protonation of the intermediate allylic anion (as 43) takes place axially, that is in a direction orthogonal to the plane of the double bond, and to the most stable conformation–configuration of the carbanion which allows the best overlap of the sp^3 orbital on the β-carbon atom with the π-orbital system of the double bond. Thus, reduction of 1(9)-octalin-2-one led almost exclusively to the *trans*-decalone (44) through axial protonation of the carbanion (45), where sp^3-π overlap is strong, rather than (46), where it is weak (7.55). However, this simple rule is not always followed. Other factors may be involved and the more stable isomer does not always result (see Caine, 1976).

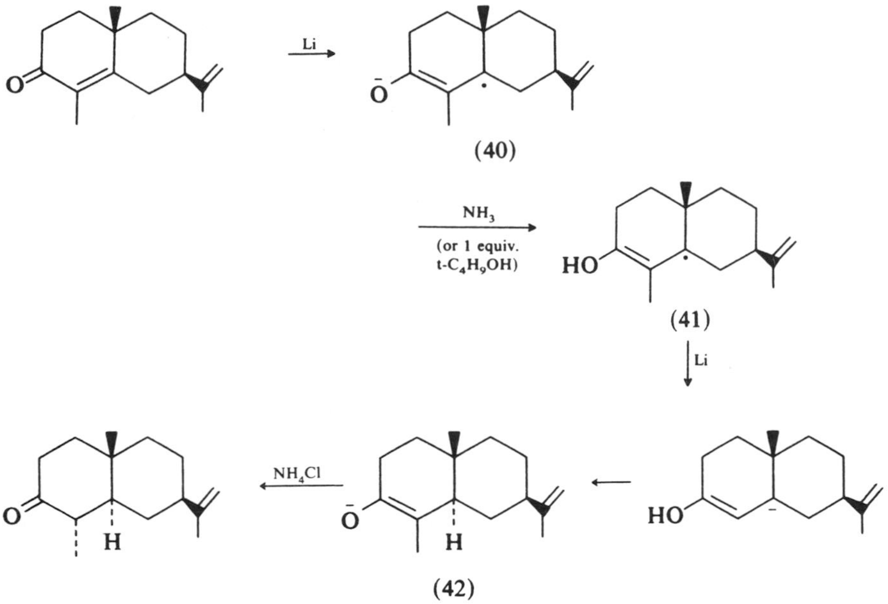

Li

(40)

NH_3
(or 1 equiv. t-C_4H_9OH)

HO

(41)

Li

NH_4Cl

H

(42)

HO

(7.52)

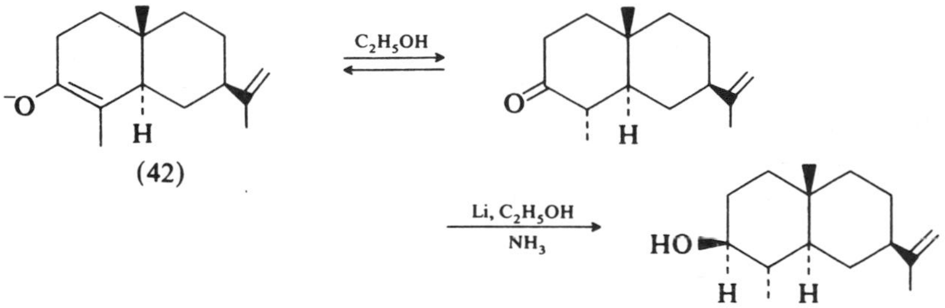

(42)

C_2H_5OH

H

Li, C_2H_5OH
NH_3

HO

H H

(7.53)

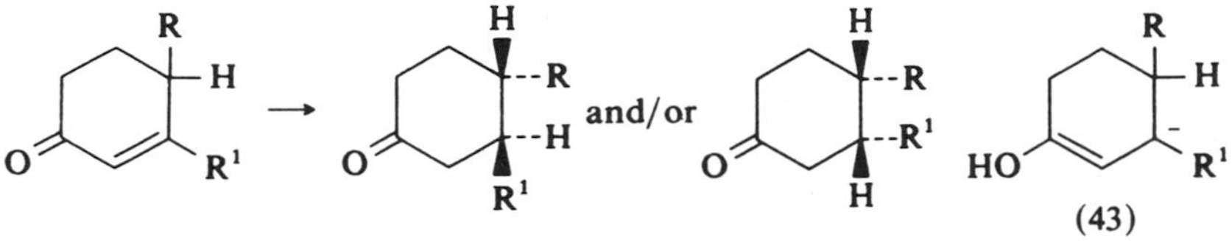

R
H
R^1

H
R
H
R^1

and/or

H
R
R^1
H

R
H
HO
R^1

(43)

(7.54)

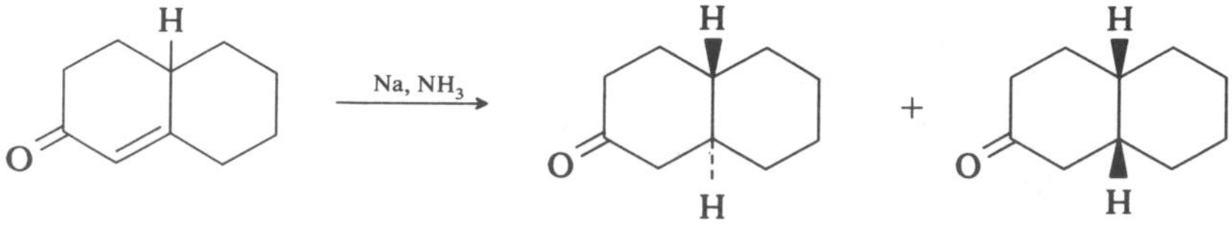

Na, NH_3

(44) (99%) (1%)

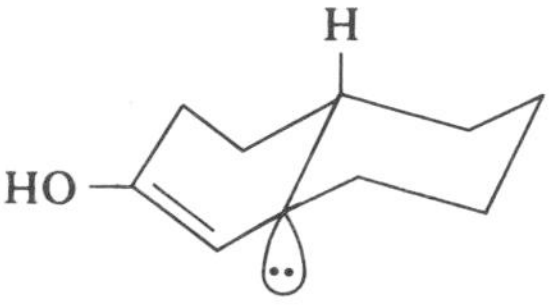

(45) 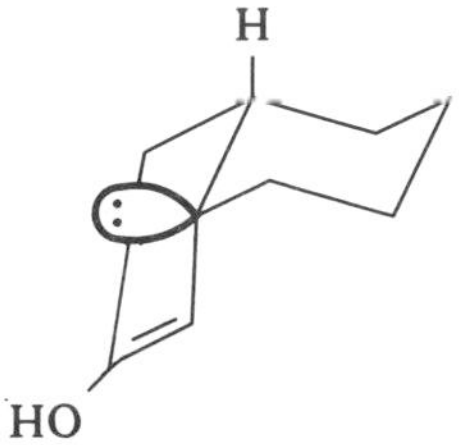 (46) (7.55)

After protonation at the β-carbon atom, the final product is obtained by protonation of the α-carbon atom (7.56). The stereochemistry of this process is not easy to predict. This protonation is kinetically controlled and usually takes place from the less hindered side of the enolate system

$$\underset{\beta}{-C}=\underset{\alpha}{C}-C=O \rightarrow -\dot{C}-C=C-\bar{O} \rightarrow -\dot{C}-C=C-OH$$
$$\downarrow \qquad (7.56)$$
$$-CH-CH-C=O \leftarrow -CH-C=C-\bar{O} \leftarrow \bar{C}-C=C-OH$$

to give a ketone which is not necessarily the thermodynamically more stable isomer. In many cases, however, the more stable isomer can be obtained by subsequent equilibration, and it is often formed directly by equilibration during isolation. For example, reduction of the indenone (47) with lithium in liquid ammonia followed by protonation with ammonium chloride affords a mixture of dihydro compounds in which *cis*-2-methyl-3-phenylindanone predominates, formed by addition of a proton to the less hindered side of the enolate anion. Protonation with ethanol, however, leads to the more stable *trans* isomer by equilibration of the initial product by ethoxide ion (7.57).

The presence of a hydroxyl substituent can sometimes affect the regio- and stereo-chemical course of reactions at a nearby carbon–carbon double bond (cf. pp. 95, 373). An interesting example is seen in the lithium–ammonia reduction of the dienone (48) which forms exclusively the product (49) in 89 per cent yield by selective attack on one of the double bonds. It is suggested that reaction takes place by intramolecular protonation by the hydroxyl proton in an intermediate (50) (Iwata *et al.*, 1981).

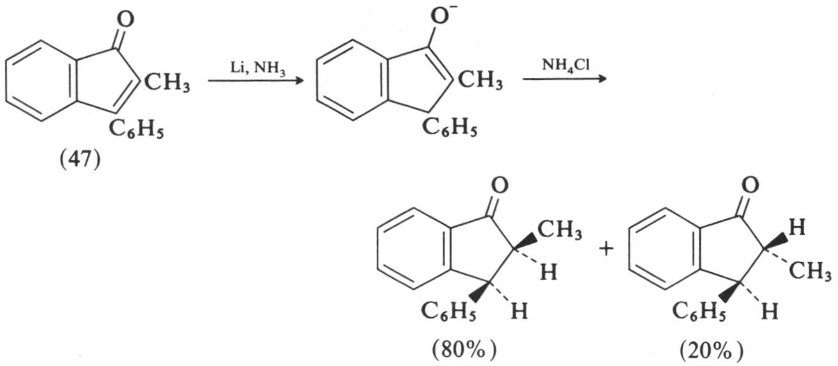

O, CH_3, C_6H_5 (47) $\xrightarrow{Li,\ NH_3}$ O^-, CH_3, C_6H_5 $\xrightarrow{NH_4Cl}$

O, CH_3, H, C_6H_5, H (80%) + O, H, CH_3, C_6H_5, H (20%)

(7.57)

O, CH_3, H, OH, H (48) $\xrightarrow{Li,\ NH_3}$ O, CH_3, H, OH, H (49) [O–H, OLi, CH_3, H] Li^+ (50)

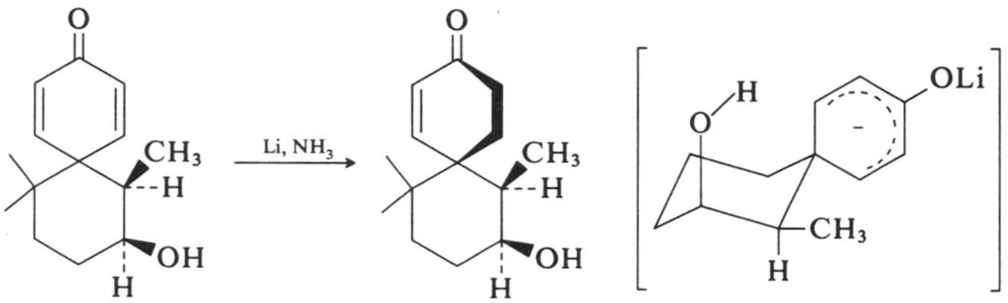

(7.58)

Aromatic compounds

One of the most useful synthetic applications of metal–ammonia–alcohol reducing agents is in the reduction of benzene rings to 1,4-dihydro derivatives. The reagents are powerful enough to reduce benzene rings, but specific enough to add only two hydrogen atoms. Benzene itself is reduced with lithium and ethanol in liquid ammonia to 1,4-dihydrobenzene by way of the anion radical (51) (Birch and Nasipuri, 1959; Akhrem, Reshetova and Titov, 1972) (7.59). The presence of an alcohol as a proton

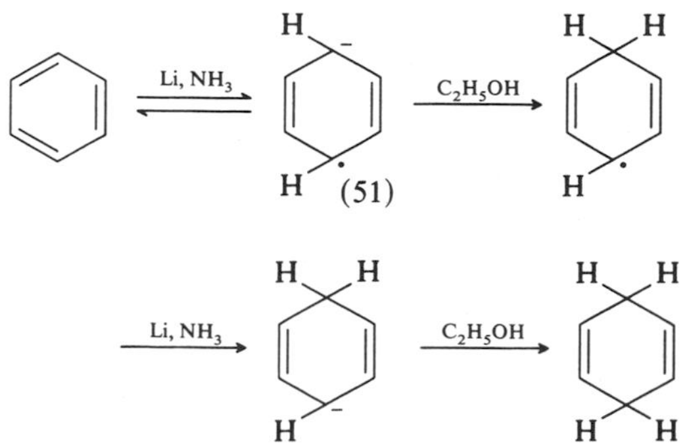

$\underset{}{\overset{Li,\ NH_3}{\rightleftharpoons}}$ H, H (51) $\xrightarrow{C_2H_5OH}$ H, H, H

(7.59)

$\xrightarrow{Li,\ NH_3}$ H, H, H $\xrightarrow{C_2H_5OH}$ H, H, H, H donor is necessary in these reactions, for the initial anion radical is an insufficiently strong base to abstract a proton from the ammonia. The alcohol also acts to prevent the accumulation of the strongly basic amide ion which might bring about isomerisation of the 1,4-dihydro compound to the conjugated 1,2-dihydro isomer which would be further reduced to tetrahydrobenzene.

Alkylbenzenes are reduced to 2-alkyl-1,4-dihydrobenzenes. Particularly useful synthetically is the reduction of methoxy- and aminobenzenes to dihydro compounds which are readily hydrolysed to cyclohexenones. Under mild conditions the $\beta\gamma$-unsaturated ketones are obtained, but these are readily isomerised to the conjugated $\alpha\beta$-unsaturated compounds (7.60). This is one of the best methods available for preparing substituted cyclohexenones. With substituted benzenes of these types, containing only electron-donating substituents addition of the two hydrogen atoms takes place at positions which are *para* to each other, avoiding, if possible, carbon atoms to which the electron-donating substituents are attached. On the other hand, reduction of benzene rings substituted by electron-withdrawing carbonyl or carboxamide groups (the only electron-withdrawing groups which are not reduced before the benzene ring) gives derivatives of 1,4-dihydrobenzoic acid. Usefully, the intermediate carbanions formed in these reductions can be alkylated in the α-position to the carboxyl group to give 1-alkyl-1,4-dihydrobenzoic acids which in many cases can be modified further to give other synthetically useful products. Thus, 2-heptyl-2-cyclohexenone was readily obtained from *ortho*-methoxybenzoic acid by way of the dianion (53). Alkylation followed by acid hydrolysis of the enol ether led to the

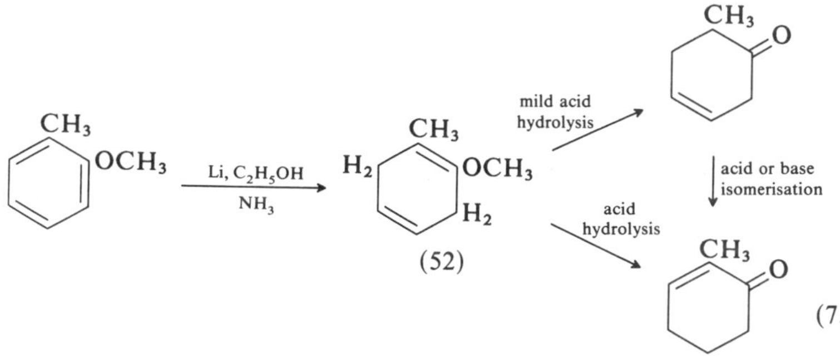

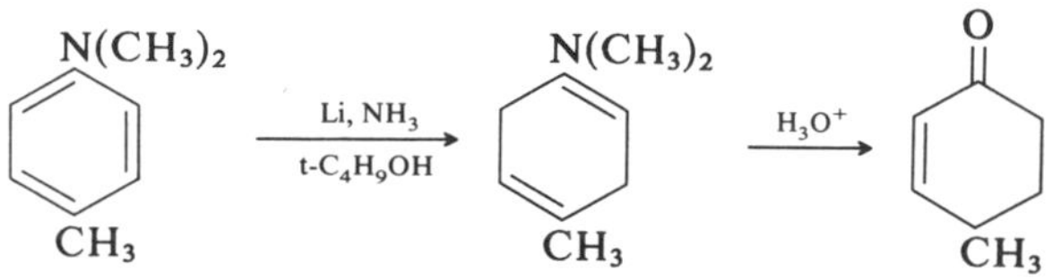

β-keto-acid (54) whence the heptylcyclohexenone was produced by decarboxylation (7.61). The intermediate dianion (53) here serves as a readily accessible operational equivalent of the anion of the β-keto-ester (55) (Birch and Slobbe, 1976*a*; Taber, Gunn and Ching Chiu, 1983, see also Cossey, Gunter and Mander, 1980). These orientation effects are in line with molecular orbital calculations of the sites of highest electron density in the anion radicals and with other theoretical aspects of the reactions (Birch, Hinde and Radom, 1980).

CO_2H CH_3 — 2 Na, liq. NH_3 → ^-O O^- CH_3 $2Na^+$ — H_3O^+ → H CO_2H CH_3 H_2

OCH_3 CO_2H — 2Na, NH_3 → OCH_3 O^- O^- $2Na^+$ (53) — $C_7H_{15}Br$ → CH_3O C_7H_{15} CO_2^- Na^+ (7.61)

↓ aq. HCl

O $CO_2C_2H_5$ (55) O C_7H_{15} ← [O C_7H_{15} CO_2H] (54)

Selective reduction of a benzene ring in the presence of another reducible group is possible if the other group can be protected or masked by reversible conversion into a saturated derivative, or into a derivative containing only isolated carbon–carbon double bonds or into an anion by salt formation. Ketones, for example, may be converted into acetals or enol ethers as in (7.62). Conversely, reduction of benzene rings takes place only slowly in the absence of a proton donor, and selective reduction of an $\alpha\beta$-unsaturated carbonyl system can be effected if no alcohol is present.

A spectacular demonstration of the high stereoselectivity shown by the metal–ammonia–alcohol reagents is found in the conversion of the compound (56) into the ketones (57) and (58) by simultaneous reduction of all the unsaturated centres. Although sixteen racemates of ketone (57) and sixty-four racemates of ketone (58) are possible, in fact only the two

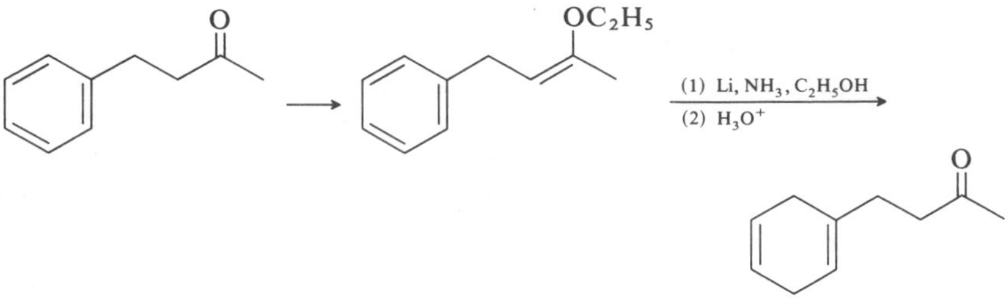

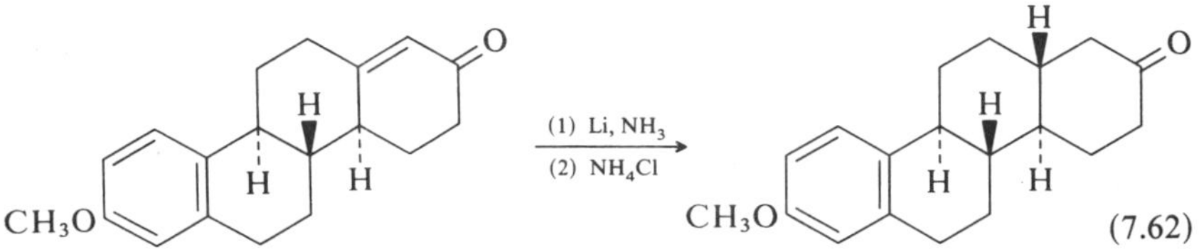

(1) Li, NH_3, C_2H_5OH
(2) H_3O^+

CH_3O (1) Li, NH_3 (2) NH_4Cl CH_3O (7.62)

racemates with the stereochemistry shown were obtained (Johnson *et al.*, 1953) (7.63).

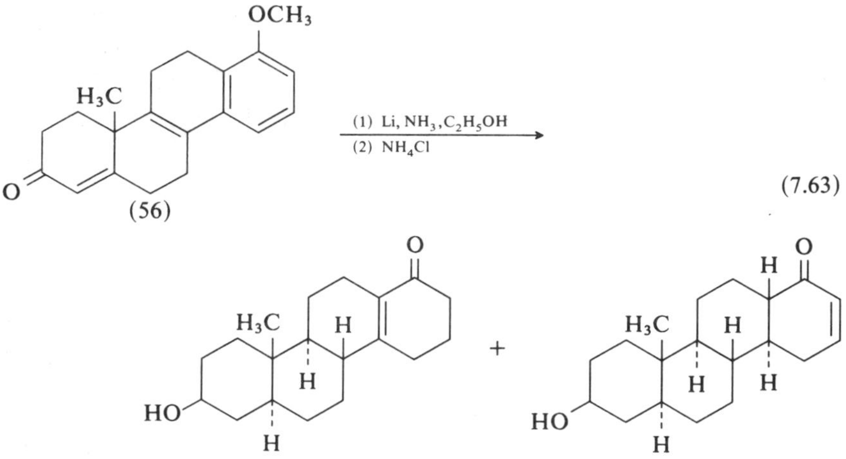

(56) (1) Li, NH_3, C_2H_5OH (2) NH_4Cl (7.63)

(57) + (58)

The applications in synthesis of the Birch reduction are manifold and go beyond the straightforward preparation of dihydrobenzenes or cyclohexenones. Equation (7.64) shows an example, taken from a synthesis of the alkaloid lycopine, in which a reduced anisole ring serves as a source of functionalised alkyl groups, and another in which oxidative decarboxylation of an (alkylated) dihydrobenzoic acid provides a method for alkylation of benzene derivatives with unusual orientation; in the example the pentyl group is introduced *meta* to the two methoxyl substituents (Birch and Slobbe, 1976*a*).

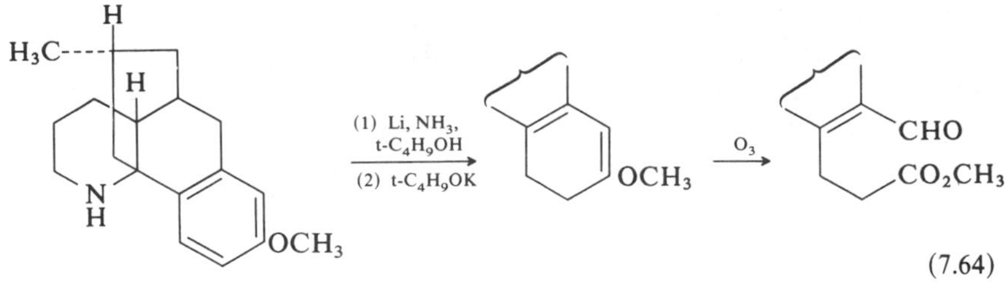

(1) Li, NH_3, t-C_4H_9OH; (2) t-C_4H_9OK → OCH_3 ; O_3 → CHO, CO_2CH_3

(7.64)

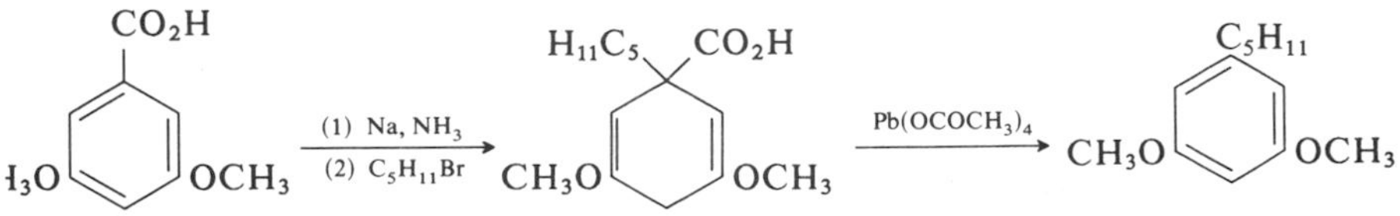

CO_2H, H_3CO, OCH_3 —(1) Na, NH_3; (2) $C_5H_{11}Br$→ $H_{11}C_5$, CO_2H, CH_3O, OCH_3 —$Pb(OCOCH_3)_4$→ C_5H_{11}, CH_3O, OCH_3

Polycyclic aromatic compounds also undergo Birch reduction. Naphthalene, for example, is converted into the 1,4,5,8-tetrahydro compound with sodium and alcohol in liquid ammonia. Some heterocyclic compounds have also been reduced; here the ease of reduction depends on the structure of the heterocycle and, sometimes, on the experimental conditions (Birch and Slobbe, 1976*b*). 1,4-Dihydropyridines, for example, are readily available by reduction of pyridines.

Reduction of alkynes with metal and liquid ammonia

Although carbon–carbon double bonds are not normally reduced by metal–ammonia reducing agents, the partial reduction of acetylenic bonds is very conveniently effected by these reagents. The procedure is highly selective and none of the saturated product is formed. Furthermore, the reduction is completely stereospecific and the only product from a disubstituted acetylene is the corresponding *E*-alkene (7.65). This method thus complements the formation of *Z*-alkenes by catalytic hydrogenation of acetylenes discussed on p. 418. It has been thought that the

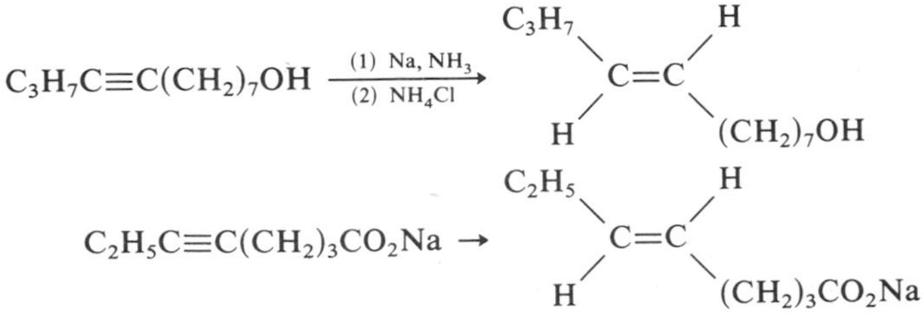

$$C_3H_7C{\equiv}C(CH_2)_7OH \xrightarrow[(2)\ NH_4Cl]{(1)\ Na,\ NH_3} (C_3H_7)(H)C{=}C(H)((CH_2)_7OH)$$

$$C_2H_5C{\equiv}C(CH_2)_3CO_2Na \rightarrow (C_2H_5)(H)C{=}C(H)((CH_2)_3CO_2Na) \quad (7.65)$$

reaction takes place by stepwwise addition of two electrons to the triple bond to form a dianion, the more stable *E*-configuration of which, with maximum separation of the two negatively charged sp^2-orbitals, is protonated to form the *E*-alkene. It appears, however, that solutions of alkali metals in liquid ammonia are not powerful enough reducing agents to convert alkynes into dianions and it may be that with sodium-ammonia organosodium compounds (59) are the true intermediates.

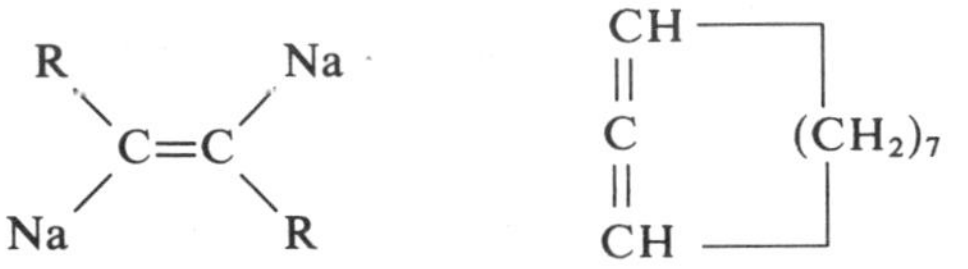

$$\underset{(59)}{\text{R(Na)C=C(Na)R}} \qquad \underset{(60)}{\text{cyclo-}[\text{CH=C=CH}(\text{CH}_2)_7]} \tag{7.66}$$

An interesting exception to the general rule that reduction of alkynes with sodium and ammonia gives the *E*-alkene is found in the reduction of medium-ring cyclic alkynes where considerable amounts of the corresponding *Z*-alkene are often produced. Cyclodecyne, for example, gives a mixture of *cis*- and *trans*-cyclodecene in the ratio 47 : 1. A *trans* dianion, or disodium salt, is obviously formed with difficulty from compounds of this type and the *Z*-alkene is thought to arise by an alternative mechanism involving the corresponding allene (60), formed by isomerisation of the alkyne by the accumulating sodamide.

Terminal triple bonds may be reduced or not as required. Under ordinary conditions, reaction of the alkyne with the metal solution leads to the formation of the metallic acetylide which resists further reduction because of the negative charge on the ethynyl carbon atom. In presence of ammonium sulphate, however, the free ethynyl group is preserved and on reduction yields the corresponding alkene. Thus 1,6-heptadiyne is converted into 1,6-heptadiene in high yield under these conditions. This is a better method for preparing 1-alkenes than the alternative catalytic hydrogenation of 1-alkynes, which sometimes gives small amounts of the saturated hydrocarbons which may be difficult to separate from the alkene. On the other hand, reduction of a terminal alkyne can be suppressed by converting it into its sodium salt by reaction with sodamide, thus allowing the selective reduction of an internal triple bond in the same molecule. 1,7-Undecadiyne, for example, is directly converted into *trans*-7-undecen-1-yne in high yield (7.67).

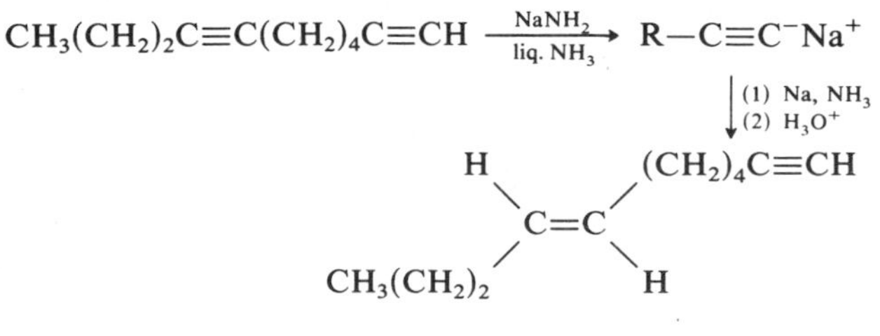

$$CH_3(CH_2)_2C{\equiv}C(CH_2)_4C{\equiv}CH \xrightarrow[\text{liq. NH}_3]{\text{NaNH}_2} R{-}C{\equiv}C^-Na^+ \xrightarrow{(1)\ \text{Na, NH}_3;\ (2)\ H_3O^+} \underset{\text{H}}{\overset{CH_3(CH_2)_2}{}}\!C{=}C\!\underset{\text{H}}{\overset{(CH_2)_4C\equiv CH}{}} \tag{7.67}$$

(product: *trans* $CH_3(CH_2)_2CH{=}CH(CH_2)_4C{\equiv}CH$, with H on each alkene carbon trans to each other)

Reductive fission

Metal amine reducing agents and other dissolving metal systems can bring about a variety of reductive fission reactions, some of which are useful in synthesis.

Most of these reactions probably proceed by direct addition of two electrons from the metal to the bond which is broken. The anions produced may be protonated by an acid in the reaction medium, or may survive until work-up. In some cases it may be desirable to maintain the fission products as anions to prevent further reduction. Reaction is

$$A{-}B \xrightarrow{2e} A^- + B^- \xrightarrow{2H^+} AH + BH$$

facilitated when the anions are stabilised by resonance or by an electronegative atom. As expected, therefore, bonds between hetero-atoms or between a hetero-atom and an unsaturated system which can stabilise a negative charge by resonance, are particularly easily cleaved. Thus allyl and benzyl halides, ethers and esters, and sometimes even the alcohols themselves, are readily cleaved by metal–amine systems and other dissolving metal reagents, as well as by catalytic hydrogenation (see p. 412). Fission reactions of this type, particularly with solutions of an alkali metal in liquid ammonia, have been widely used in structural studies, and also for the reductive removal of unsaturated groups used as protecting agents for amino, imino, hydroxyl and thiol groups. They are of particular value in peptide and nucleotide synthesis, and have an obvious general advantage over hydrolytic or catalytic methods for compounds which are labile or contain sulphur. Benzyl, *p*-toluenesulphonyl and benzyloxycarbonyl groups are all efficiently replaced by hydrogen, and cystinyl-peptides are cleaved to cysteinyl derivatives (7.68). For example

$$\begin{aligned}
RSCH_2C_2H_5 &\rightarrow RSH + CH_3C_6H_5\\
RNHSO_2C_6H_4CH_3\text{-}p &\rightarrow RNH_2 + HSC_6H_4CH_3\text{-}p\\
RNHCO_2CH_2C_6H_5 &\rightarrow RNH_2 + CO_2 + CH_3C_6H_5\\
RSSR &\rightarrow 2RSH\\
ROCH_2C_6H_5 &\rightarrow ROH + CH_3C_6H_5
\end{aligned} \qquad (7.68)$$

in the synthesis of the naturally occurring peptide, glutathione, *N*-benzyloxycarbonyl-γ-glutamyl-*S*-benzylcysteinylglycine was prepared as the key intermediate in which the benzyloxycarbonyl group was used for protection of the amino group and the benzyl group for protection of the thiol. The protective groups were finally removed to yield glutathione by cleavage with sodium in liquid ammonia (7.69).

Reductive fission has been of great assistance in the elucidation of the structures of naturally occurring allyl and benzyl alcohols, ethers and

$$\begin{array}{l} \qquad\qquad\qquad\qquad CH_2SCH_2C_6H_5 \\ \qquad\quad CH_2CONHCHCONHCH_2CO_2H \\ \qquad\quad CH_2 \\ C_6H_5CH_2OCONHCHCO_2H \end{array}$$

$$\downarrow \text{Na, } NH_3 \qquad (7.69)$$

$$\begin{array}{l} \qquad\qquad\qquad CH_2SH \\ CH_2CONHCHCONHCH_2CO_2H \\ CH_2 \\ H_2NCHCO_2H \end{array}$$

esters. Thus, the structure of the alcohol lanceol was neatly confirmed by reduction with sodium and alcohol in liquid ammonia to the known sesquiterpene hydrocarbon bisabolene (7.70).

CH_3 ... CH_2OH $\xrightarrow[C_2H_5OH]{\text{Na, } NH_3}$ CH_3 ... (7.70)

The reductive fission of allyl or benzyl alcohols can be prevented, if necessary, by converting them into the corresponding alkoxide ions by reaction with sodamide or other strong base, as in (7.71).

$$C_6H_5CH{=}CH{-}C(CH_3)_2OH \xrightarrow[\text{(2) Na-liq. } NH_3]{\text{(1) } NaNH_2} C_6H_5CH_2CH_2C(CH_3)_2OH \qquad (7.71)$$

In contrast to benzyl alcohols, benzylamines are not readily reduced with metal–ammonia reagents, because nitrogen is less electronegative than oxygen, but quaternary salts are easily cleaved as in the well-known Emde reaction of quaternary ammonium salts (7.72).

$\overset{+}{N}(CH_3)_2$, CH_3 $\xrightarrow{\text{Na, } NH_3}$ $N(CH_3)_2$, CH_2, CH_3 (7.72)

Solutions of sodium in liquid ammonia also cleave aryl ethers. Diaryl ethers react particularly readily. Thus phenyl *p*-tolyl ether gives a mixture of *p*-cresol and phenol. Alkyl aryl ethers are cleaved only slowly, but aryl diethylphosphates, which are easily obtained from the phenol and diethylphosphorochloridate or tetraethyl pyrophosphate, give the hydrocarbon in high yield, and this is one of the few available methods for removal of the oxygen from a phenolic ring (Kenner and Williams, 1955). Enols are similarly converted into alkenes. This has been exploited in a method, illustrated in (7.73), for converting $\alpha\beta$-unsaturated ketones into structurally specific alkenes (Ireland *et al.*, 1972).

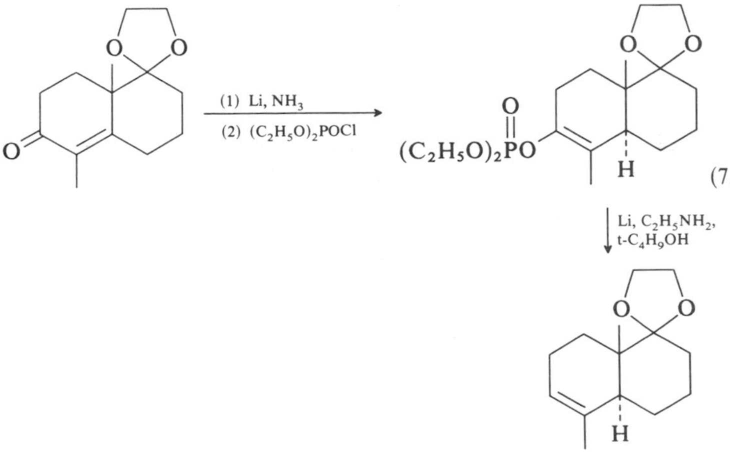

(1) Li, NH_3
(2) $(C_2H_5O)_2POCl$

$(C_2H_5O)_2\overset{O}{\overset{\|}{P}}O$

Li, $C_2H_5NH_2$, t-C_4H_9OH

(7.73)

Alcoholic hydroxyl groups can also be replaced by hydrogen by reaction of the corresponding methanesulphonates with sodium and an alcohol in liquid ammonia, or by cleavage of the diethyl phosphates with lithium and t-butanol in ethylamine; even tertiary alcohols are readily de-oxygenated by the latter procedure (Ireland, Muchmore and Hengartner, 1972; see also Trost and Renaut, 1982). However, these methods might not always be suitable, especially with polyfunctional compounds, and to meet such cases alternative methods have been developed (Hartwig, 1983). In one procedure secondary alcohols are conveniently deoxygenated by reaction of the *O*-cycloalkyl thiobenzoates or *S*-methyldithiocarbonates with tributylstannane in refluxing toluene or xylene (Barton and McCombie, 1975). These reactions proceed by a free-radical chain mechanism (7.74) and have the resultant advantage that rearrangements common in reactions involving carbonium ions are avoided. Lanosterol, for example, is converted into the oxygen-free compound without rearrangement. This reaction has already proved useful in the synthesis and modification of a number of natural products (see, for example, Robins, Wilson and Hansske, 1983). Thus in the carbohydrate series, the *S*-methyl dithiocarbonate (61) gave the 3-deoxy compound in 80–90 per cent yield. By suitable modification of the experimental conditions the reaction can be extended to primary (Barton, Motherwell and Stange, 1981) and tertiary (Barton and Crich, 1984) alcohols as well.

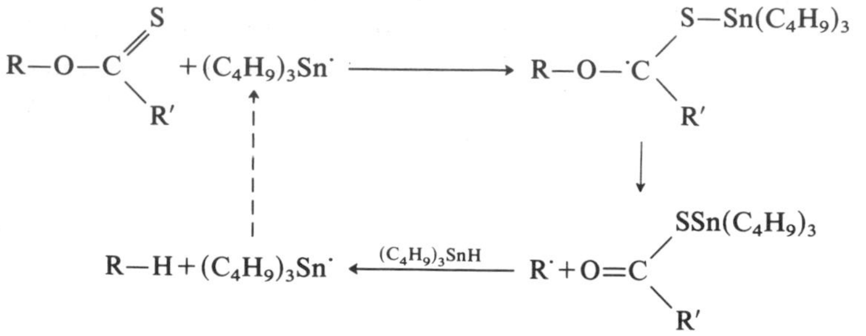

$$R{-}O{-}C(=S)R' + (C_4H_9)_3Sn^{\cdot} \longrightarrow R{-}O{-}\dot{C}(S{-}Sn(C_4H_9)_3)R'$$
$$\downarrow$$
$$R{-}H + (C_4H_9)_3Sn^{\cdot} \xleftarrow{(C_4H_9)_3SnH} R^{\cdot} + O{=}C(SSn(C_4H_9)_3)R'$$

$(R' = \text{e.g. } SCH_3)$ (7.74)

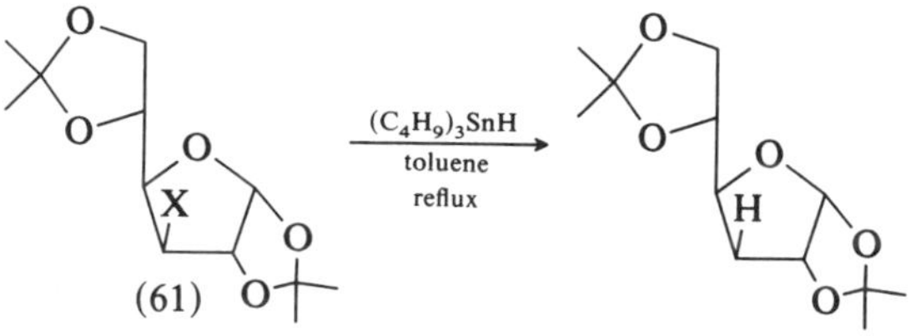

(61) $\xrightarrow[\text{toluene, reflux}]{(C_4H_9)_3SnH}$

$X = OC(=S){-}SCH_3$

Primary and secondary alcohols can also be deoxygenated by reduction of the derived phenyl alkyl selenides with tri-n-butylstannane. The phenyl-selenides themselves are readily prepared from the alcohol by reaction with phenyl selenocyanate in the presence of tributylphosphine (Grieco *et al.*, 1976). Primary alcohols are more easily converted into the selenides than secondary allowing selective removal of hydroxyl groups in some cases. The diol (62), for example, was selectively deoxygenated to the alcohol (63) (Grieco *et al.*, 1982).

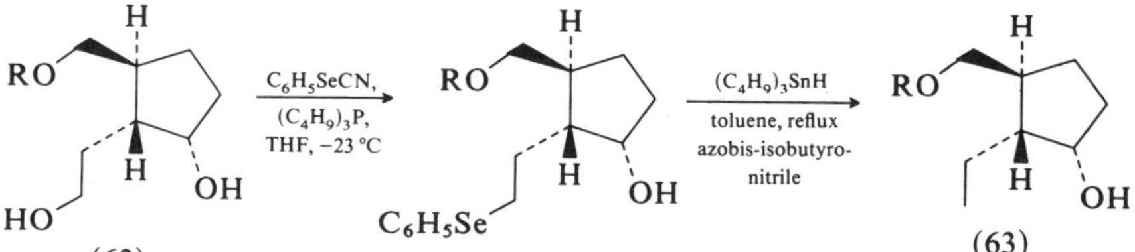

(62) $\xrightarrow[\text{THF, } -23\,°C]{C_6H_5SeCN,\ (C_4H_9)_3P,}$ C_6H_5Se derivative $\xrightarrow[\text{toluene, reflux, azobis-isobutyronitrile}]{(C_4H_9)_3SnH}$ (63)

$R = C_6H_5CH_2$ (7.75)

Organic halides can be converted into the halogen-free compounds with dissolving metal reagents, but, since the highly electropositive metals such as sodium induce Wurtz coupling, metal–amine systems are not often used. Good results are often obtained with magnesium or zinc in a protic solvent such as dilute mineral acid, acetic acid or an alcohol. The mechanism of these reactions is uncertain but it is thought that they may proceed by way of short-lived organometallic intermediates. This is supported by the observation that if a substituent which can be lost as a stable anion (e.g. —OH, —OR, —OCOR, halogen) is present on the adjacent carbon atom elimination rather than simple reductive cleavage is observed (7.76).

$$C_3H_7CH(Br)-CH(OCH_3)C_3H_7 \xrightarrow[C_2H_5OH]{Zn} C_3H_7CH{=}CHC_3H_7 \quad (55\%\ Z\text{-}, 45\%\ E\text{-}) \qquad (7.76)$$

Reductive cleavage of alkyl halides may or may not occur with retention of configuration. With alkenyl halides cleavage with sodium in liquid ammonia usually gives mixtures of the *Z*- and *E*-alkenes formed by interconversion of the intermediate radicals or anions (7.77).

$$(C_2H_5)(H)C{=}C(C_2H_5)(Cl) \xrightarrow[THF]{Na,\ NH_3} (C_2H_5)(H)C{=}C(C_2H_5)(H)\ (52\%) + (C_2H_5)(H)C{=}C(H)(C_2H_5)\ (48\%) \qquad (7.77)$$

Other useful reagents for reduction of carbon–halogen bonds are trialkyltin hydrides and chromiumII salts; other hydride reducing agents may also be employed (see pp. 475, 477). Reduction with trialkyltin hydrides is a general method and can be applied to both alkyl and aryl halides (Kuivila, 1968). It proceeds by a free-radical chain process. From the synthetic viewpoint, stepwise reduction of geminal dihalides is a useful application (7.78). As ordinarily employed the reagent is expensive

$$\text{2,2-dimethyl-3-methylene-1,1-dibromocyclopropane} \xrightarrow{(C_4H_9)_3SnH} \text{(H, Br)} \xrightarrow{(C_4H_9)_3SnH} \text{(H, H)} \qquad (7.78)$$

and isolation of the product from trialkyltin halide formed in the reaction may be troublesome. A procedure using catalytic amounts of a trialkyltin halide with sodium borohydride has been reported by Corey and Suggs (1975*b*). Here the trialkyltin hydride is continuously regenerated in the catalytic cycle shown in (7.79). The diborane produced is trapped by the

$$\begin{aligned} 2R_3SnX + 2NaBH_4 &\rightarrow 2R_3SnH + 2NaCl + B_2H_6 \\ R_3SnH + R'X &\rightarrow R_3SnX + R'H \end{aligned} \qquad (7.79)$$

ethanol used as solvent. Using this procedure (R)-(+)-dimethyl malate was prepared from (*R,R*)-(+)-dimethyl tartrate as shown in (7.80) (Corey, Marfat and Hoover, 1981).

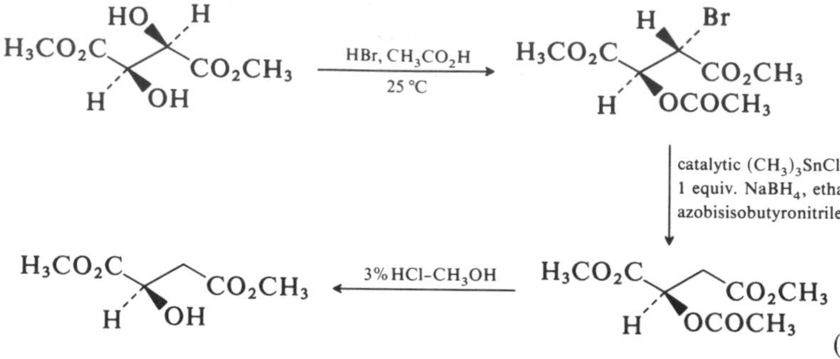

HBr, CH_3CO_2H, 25 °C; catalytic $(CH_3)_3SnCl$, 1 equiv. $NaBH_4$, ethanol, azobisisobutyronitrile; 3% HCl-CH_3OH

(7.80)

Reduction of organic halides by metal hydrides such as tributyltin hydride can go wrong when the intermediate carbon radical does something else faster than it abstracts a hydride ion from the tin hydride. For example, it may cyclise if there is an appropriately situated carbon-carbon double bond (cf. Beckwith, Lawrence and Serelis, 1980). This difficulty can be overcome by using the anionic vanadium hydride $[(\eta^5\text{-}C_5H_5)V(CO)_3H]^-Na^+$. This reagent effects the same kinds of reduction as tributyltin hydride, but with a much faster transfer of hydride from vanadium to the carbon radical (Kenney, Jones and Bergman, 1978).

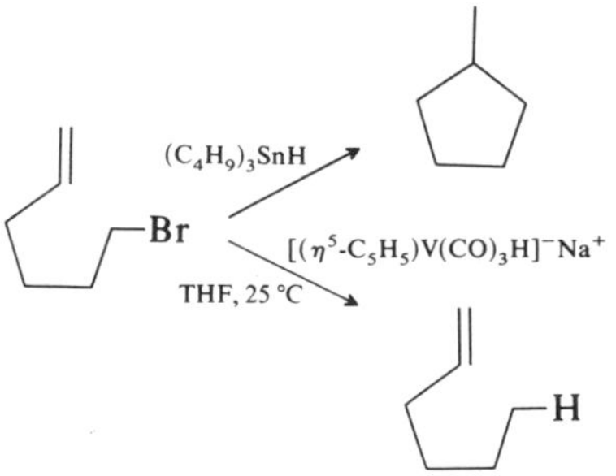

$(C_4H_9)_3SnH$; $[(\eta^5\text{-}C_5H_5)V(CO)_3H]^-Na^+$, THF, 25 °C

(7.81)

Reduction of simple alkyl or aralkyl halides with chromiumII salts usually effects replacement of halogen by hydrogen (Hanson, 1974). Reaction is thought to proceed by two one-electron transfer steps as in (7.82): Compounds carrying a vicinal substituent which can be eliminated

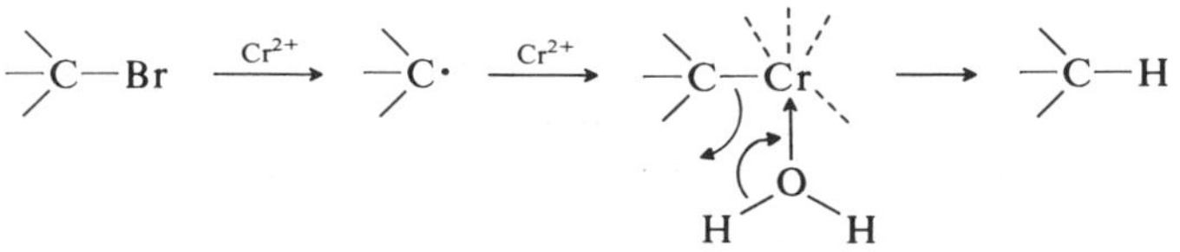

$$\text{C—Br} \xrightarrow{Cr^{2+}} \text{C}\cdot \xrightarrow{Cr^{2+}} \text{C—Cr} \longrightarrow \text{C—H}$$

(7.82)

as an anion (e.g. OH, OR, Cl) usually give alkenes (7.83). Elimination can be circumvented by effecting the reduction with chromiumII acetate

$$CH_2BrCHBrCH_2Cl \xrightarrow[\text{aq. dimethylformamide}]{CrSO_4} CH_2{=}CHCH_2Cl \qquad (7.83)$$

in presence of a hydrogen atom transfer agent such as butanethiol, thus providing, for example, an excellent route from a bromohydrin to the corresponding alcohol as in the route to 11-β-hydroxysteroids (7.84).

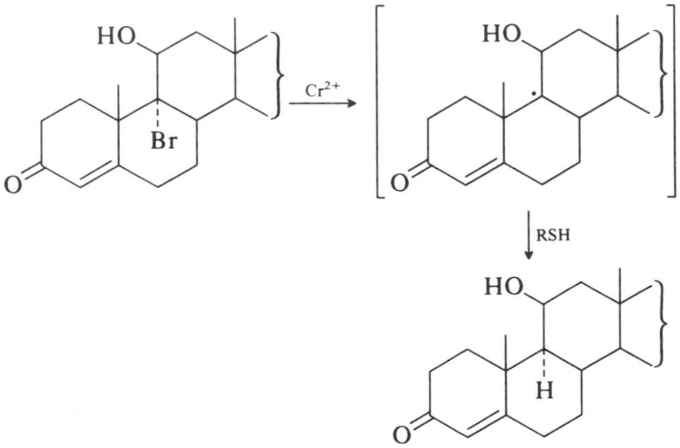

HO, Br, O — Cr^{2+} → [HO, O] — RSH → HO, H, O (7.84)

Lithium in alkylamines

Solutions of lithium in aliphatic amines of low molecular weight are powerful reducing agents, but are less selective than metal–ammonia reagents (Kaiser, 1972). The amines have the advantage that they are better solvents for organic substances than is ammonia, and have higher boiling points which makes for easier handling. The higher working temperatures facilitate the initial stages of the reduction, but they also favour conjugation (and therefore further reduction) of the initial product, and unless precautions are taken reduction of an aromatic compound leads to the tetrahydro or even the hexahydro derivative. Under suitable conditions yields of 1,4-dihydro compounds comparable to those formed under Birch conditions can be obtained.

Solutions of lithium in alkylamines have been used successfully for reductive cleavage of allylic ethers and esters. Reduction of steroid epoxides leads to the axial alcohols, and in this reaction they are more powerful and specific reagents than lithium aluminium hydride.

7.3. Reduction by hydride-transfer reagents

Reactions which proceed by transfer of hydride ions are widespread in organic chemistry, and they are important also in biological systems. Reductions involving the reduced forms of coenzymes I and II, for example, are known to proceed by transfer of a hydride ion from a 1,4-dihydropyridine system to the substrate. In the laboratory the most useful reagents of this type in synthesis are aluminium isopropoxide and the various metal hydride reducing agents.

Aluminium alkoxides

The reduction of carbonyl compounds to alcohols with aluminium isopropoxide has long been known under the name of the Meerwein–Pondorff–Verley reduction (Wilds, 1944). The reaction is easily effected by heating the components together in solution in isopropanol. An equilibrium is set up and the product is obtained by using an excess of the reagent or by distilling off the acetone as it is formed. The reaction is thought to proceed by transfer of hydride ion from the isopropoxide to the carbonyl compound through a six-membered cyclic transition state (7.85). Aldehydes are reduced to primary alcohols, and ketones give secondary alcohols, often in high yield. The reaction owes its usefulness to the fact that carbon–carbon double bonds and many other unsaturated groups are unaffected, thus allowing selective reduction of carbonyl groups. Cinnamaldehyde, for example, is converted into cinnamyl alcohol, *o*-nitrobenzaldehyde gives *o*-nitrobenzyl alcohol, and phenacyl bromide gives styrene bromohydrin.

$$R^1R^2C{=}O + Al(OPr^i)_2{-}O{-}CH(CH_3)_2 \rightleftharpoons R^1R^2CH{-}O{-}Al{<} + O{=}C(CH_3)_2 \xrightarrow{H_3O^+} R^1R^2C(H)OH \qquad (7.85)$$

Reductions of a similar type can be brought about by other metallic alkoxides, but aluminium alkoxide is particularly effective because it is soluble in both alcohols and hydrocarbons and, being a weak base, it shows little tendency to bring about wasteful condensation reactions of the carbonyl compounds.

Lithium aluminium hydride and sodium borohydride

A number of metal hydrides have been employed as reducing agents in organic chemistry, but the most commonly used are lithium aluminium hydride and sodium borohydride, both of which are commercially available. Another useful reagent, borane, is discussed on p. 478.

The anions of the two complex hydrides can be regarded as derived from lithium or sodium hydride and either aluminium hydride or borane.

$$LiH + AlH_3 \rightarrow Li^+AlH_4^-$$
$$NaH + BH_3 \rightarrow Na^+BH_4^-$$

These anions are nucleophilic reagents and as such they normally attack polarised multiple bonds such as C=O or C≡N by transfer of hydride ion to the more positive atom. They do not usually reduce isolated carbon–carbon double or triple bonds.

With both reagents all four hydrogen atoms may be used for reduction, being transferred in a stepwise manner as illustrated in (7.86) for the reduction of a ketone with lithium aluminium hydride. There is evidence that in borohydride reductions a more complex path may be followed (cf. Wigfield, 1979). For reductions with lithium aluminium hydride (but not with sodium borohydride) each successive transfer of hydride ion takes place more slowly than the one before, and this has been exploited for the preparation of modified reagents which are less reactive and more selective than lithium aluminium hydride itself by replacement of two or three of the hydrogen atoms of the anion by alkoxy groups (see p. 470).

$$H_3\bar{Al}{-}H + (CH_3)_2C{=}O \rightarrow H{-}C(CH_3)_2{-}O{-}\bar{Al}H_3 \xrightarrow{CH_3COCH_3} H_2\bar{Al}[-OCH(CH_3)_2]_2$$
$$\xrightarrow{CH_3COCH_3} H\bar{Al}[OCH(CH_3)_2]_3 \xrightarrow{CH_3COCH_3} \bar{Al}[OCH(CH_3)_2]_4 \xrightarrow{H_3O^+} 4\,H{-}C(CH_3)_2{-}OH \qquad (7.86)$$

Lithium aluminium hydride is a more powerful reducing agent than sodium borohydride and reduces most of the commonly encountered organic functional groups (see Table 7.3). It reacts readily with water and other compounds which contain active hydrogen atoms and must be used under anhydrous conditions in a non-hydroxylic solvent; ether and tetrahydrofuran are commonly employed. Sodium borohydride reacts only slowly with water and most alcohols at room temperature and reductions with this reagent are often effected in ethanol solution. Being less reactive than lithium aluminium hydride it is more discriminating in its action (cf. Walker, 1976). At room temperature in ethanol it readily reduces aldehydes and ketones but it does not generally attack esters or amides and it is normally possible to reduce aldehydes and ketones selectively with sodium borohydride in the presence of a variety of other functional groups. Some typical examples are shown in (7.87). Lithium borohydride is also sometimes used. It is a more powerful reducing agent than sodium borohydride and less selective in its action, but it has the advantage that it is soluble in ether and tetrahydrofuran (Brown and Krishnamurthy, 1979).

$$CH_2{=}CH{-}CH{=}CH{-}CHO \xrightarrow[or\ NaBH_4]{LiAlH_4} CH_2{=}CH{-}CH{=}CH{-}CH_2OH$$

$$\text{4-(ethoxycarbonyl)cyclohexanone} \xrightarrow[C_2H_5OH]{NaBH_4} \text{cis- and trans-4-hydroxycyclohexane}CO_2C_2H_5$$

$$\text{N-methyl-2-piperidone} \xrightarrow[ether]{LiAlH_4} \text{N-methylpiperidine} \qquad (7.87)$$

$$O_2N\text{(CH}_2)_4\text{CHO} \xrightarrow{NaBH_4,\ C_2H_5OH} O_2N\text{(CH}_2)_4\text{CH}_2OH$$

An exception to the general rule that carbon–carbon double bonds are not attacked by hydride reducing agents is found in the reduction of β-aryl-$\alpha\beta$-unsaturated carbonyl compounds with lithium aluminium hydride, where the carbon–carbon double bond is often reduced as well as the carbonyl group. Even in these cases, however, selective reduction of the carbonyl group can generally be achieved by working at low temperatures or by using sodium borohydride or aluminium hydride (p. 473) as reducing agent (7.88). This type of reduction of the double

$$C_6H_5CH{=}CH{\cdot}CHO \xrightarrow{\text{excess }LiAlH_4,\text{ ether, }35\,^\circ C} C_6H_5CH_2CH_2CH_2OH$$

$$C_6H_5CH{=}CH{\cdot}CHO \xrightarrow{NaBH_4\ or\ LiAlH_4\text{ ether, }-10\,^\circ C} C_6H_5CH{=}CHCH_2OH \qquad (7.88)$$

bond of allylic alcohols is thought to proceed through a cyclic organoaluminium compound (64), for it is found experimentally that only one of the two hydrogen atoms added to the double bond is derived from the hydride, and acidification with a deuterated solvent leads to the deuterated alcohol shown in (7.89).

$$\text{(64)} \xrightarrow{RO^2H} C_6H_5CH^2HCH_2CH_2OH \qquad (7.89)$$

(64): cyclic Al compound with two rings $C_6H_5{-}CH{-}CH_2{-}CH_2{-}O{-}Al$ (Al bonded to CH and O of each chain)

(64)

Table 7.3. *Common functional groups reduced by lithium aluminium hydride*

Functional group	Product
$>C=O$	$>CH-OH$
$-CO_2R$	$-CH_2OH+ROH$
$-CO_2H$	$-CH_2OH$
$-CONHR$	$-CH_2NHR$
$-CONR_2$	$-CH_2NR_2$ or $[-CH(OH)-NR_2] \longrightarrow -CHO+R_2NH$
$-C\equiv N$	$-CH_2NH_2$ or $[-CH=NH] \xrightarrow{H_2O} -CHO$
$>C=NOH$	$>CH-NH_2$
$-\overset{\vert}{\underset{\vert}{C}}-NO_2$ (aliphatic)	$-\overset{\vert}{\underset{\vert}{C}}-NH_2$
$ArNO_2$	$ArNHNHAr$ or $ArN=NAr$
$-CH_2OSO_2C_6H_5$ or $-CH_2Br$	$-CH_3$
$>CHOSO_2C_6H_5$ or $>CH-Br$	$>CH_2$
$-CH-C<$ (epoxide, bridged by O)	$-CH_2-\overset{\vert}{\underset{OH}{C}}-$

H.O. House, *Modern synthetic reactions*, copyright 1965, 1972, W. A. Benjamin, Inc., Menlo Park, California.

A similar type of aluminium compound is thought to be involved in the well-known reduction of the triple bond of propargylic alcohols to a *trans* double bond with lithium aluminium hydride (cf. Grant and Djerassi, 1974). Only triple bonds flanked by hydroxyl groups are reduced, as illustrated in the example (7.90). The reaction can be used for the preparation of labelled allylic alcohols.

$$\mathrm{CH{\equiv}C(CH_2)_2C{\equiv}CCO_2C_2H_5 \xrightarrow{LiAlH_4} CH{\equiv}C(CH_2)_2CH{=}CHCH_2OH}\ \textit{trans}$$

$$\mathrm{HC{\equiv}CCHOHC_4H_9 \xrightarrow[(2)\ ^2H_2O]{(1)\ LiAlH_4} (H)(^2H)C{=}C(H)(CHOHC_4H_9)} \tag{7.90}$$

Lithium aluminium hydride and sodium borohydride have probably found their most widespread use in the reduction of carbonyl compounds. Aldehydes, ketones, carboxylic acids, esters and lactones can all be reduced smoothly to the corresponding alcohols under mild conditions (for examples and experimental conditions see Gaylord, 1956; Pizey, 1974). Reaction with lithium aluminium hydride is the method of choice for the reduction of carboxylic esters to primary alcohols. Substituted amides are converted into amines or aldehydes, depending on the experimental conditions (see p. 471). For compounds containing more than one kind of carbonyl group reaction conditions have to be properly chosen to effect the reduction required. When necessary, the carbonyl group of an aldehyde or ketone may be protected against reduction by conversion into the acetal. For example ethyl acetoacetate which contains both an ester and a ketone functional group, on reduction with lithium aluminium hydride, affords 1,3-butandiol, by reduction of both groups. With the milder sodium borohydride, however, only the keto group is reduced to give ethyl 3-hydroxybutanoate. To effect selective reduction of the ester, the keto group must be protected as its acetal, and the ester reduced with lithium aluminium hydride. Mild acid hydrolysis then regenerates the ketone to give the β-keto-alcohol.

$$\mathrm{CH_3COCH_2CO_2C_2H_5 \xrightarrow{LiAlH_4,\ ether} CH_3CHOHCH_2CH_2OH}$$

$$\mathrm{CH_3COCH_2CO_2C_2H_5 \xrightarrow[C_2H_5OH]{NaBH_4} CH_3CHOHCH_2CO_2C_2H_5}$$

$$\mathrm{CH_3COCH_2CO_2C_2H_5 \xrightarrow{HOCH_2CH_2OH,\ H^+} CH_3C(OCH_2CH_2O)CH_2CO_2C_2H_5 \xrightarrow[(2)\ H_3O^+]{(1)\ LiAlH_4} CH_3COCH_2CH_2OH} \tag{7.91}$$

The reducing properties of sodium borohydride are substantially modified in the presence of metal salts, and particularly useful in this respect are lanthanide salts. In the presence of cerium(IV) chloride, for example, sodium borohydride can discriminate between different ketonic and aldehydic carbonyl groups, effecting the selective reduction of the *less* reactive carbonyl group in a dicarbonyl compound. For example, $\alpha\beta$-unsaturated ketones are reduced selectively in the presence of saturated ketones (Gemal and Luche, 1979) and ketones can be reduced in the presence of aldehydes by reaction with one equivalent of sodium borohydride in aqueous ethanol in the presence of cerium chloride. It is believed that, under the reaction conditions the aldehyde group is protected as the hydrate, which is stabilised by complexation with the cerium ion, and is regenerated during the isolation of the product (Luche and Gemal, 1979).

0.5 equiv. $NaBH_4$, CH_3OH

$CeCl_3.6H_2O$ (1 equiv.)

HO (7.92)

O CHO → H OH CHO

(1) $NaBH_4-CeCl_3.6H_2O$ $C_2H_5OH-H_2O$

(2) H_3O^+

Cerium chloride is also an efficient catalyst for the regio-selective 1,2-reduction of $\alpha\beta$-unsaturated ketones to allylic alcohols in cases where the uncatalysed reaction leads to reduction of the double bond as well (Gemal and Luche, 1981).

O CHO → OH CHO (64%)

$NaBH_4$, $CeCl_3$

C_2H_5OH, −15 °C

(7.93)

Selective reduction of aldehydes in the presence of ketones can be effected with tetra-n-butylammonium triacetoxyborohydride (Nutaitis and Gribble, 1983), lithium tris[(3-ethyl-3-pentyl)oxy]aluminium hydride (Krishnamurthy, 1981) and other reagents.

Ordinarily, reduction of the carbonyl group of an unsymmetrical ketone such as ethyl methyl ketone leads to the racemic carbinol. With ketones which contain an asymmetric centre, however, the two forms of the carbinol may not be produced in equal amount. Thus, in the reduction of the ketone (65) with lithium aluminium hydride the *threo* (*anti*, see p. 52) form of the alcohol predominates in the product (7.94). The main

CH_3, H, C_6H_{11}, O, CH_3 (65) → CH_3, H, C_6H_{11}, CH_3, OH, H (72% of product) + CH_3, H, C_6H_{11}, CH_3, H, OH (28% of product) (7.94)

product formed in these reactions can be predicted on the basis of Cram's rule (Cram and Abd Elhafez, 1952) according to which that diastereomer predominates in the product which is formed by approach of the reagent to the less hindered side of the carbonyl group when the rotational conformation of the molecule is such that the carbonyl group is flanked by the two least bulky groups on the adjacent chiral centre. This may be represented, using Newman projection formulae, as in (7.95) where S,

transition state of least energy → predominant stereoisomer (7.95)

or as

M and L represent small, medium and large substituents. Thus, for the reduction of the ketone (65) the predominant *threo-* (*anti-*) carbinol arises by attack of the metal hydride on the less hindered side of the carbonyl group in the conformation shown. A somewhat different model for the transition state which is said to predict results more in agreement with experiment has been proposed (Cherest, Felkin and Prudent, 1968) but the essential feature of Cram's hypothesis, involving approach of the hydride reagent to the less hindered side of the carbonyl group, is preserved. The selectivity obtained in these reactions increases with the bulk of the reducing agent and some highly stereoselective reductions have been achieved using complex hydride reducing agents specifically designed for the purpose (see p. 478).

When the reducing agent itself is chiral and optically active asymmetric reduction of unsymmetrical (prochiral) ketones to optically active secondary alcohols can often be achieved even when the ketones themselves are achiral. A number of reagents have been used (see ApSimon and Seguin, 1979). One of the best is *B*-3-pinanyl-9-borabicyclo[3,3,1]nonane (66), which is readily prepared from (+)-α-pinene and 9-borabicyclononane (see p. 291). It is an excellent reagent for the asymmetric reduction of ketones and of 1-deuteroaldehydes. Thus, benzaldehyde-1-d was reduced to essentially pure (*S*)-(+)-benzyl-1-d alcohol and 1-octyn-

3-one gave (R)-1-octyn-3-ol with 92 per cent optical purity (7.96). In the latter reaction it is supposed that the transition state has the form (67) in which the pentyl group is lying over the pinene ring (Midland *et al.*, 1979; Midland *et al.*, 1980; Brown and Pai, 1982). Other reagents have been developed. The complex hydride (68) has been used for the asymmetric reduction of aliphatic ketones (Midland and Kazubski, 1982),

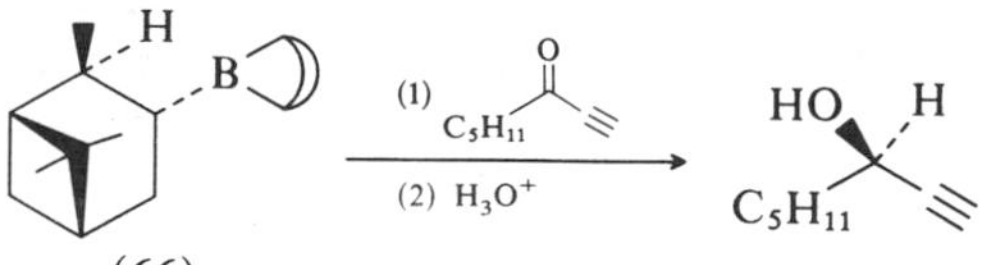

(66) (1) $C_5H_{11}COC{\equiv}CH$ (2) H_3O^+ → HO H C_5H_{11}

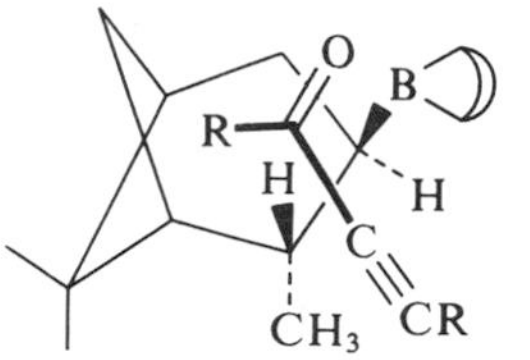

(67)

(7.96)

and high selectivity has been obtained in the reduction of prochiral aromatic ketones with the borane (69) prepared from borane and (S)-(−)-2-amino-3-methyl-1,1-diphenylbutan-1-ol (Itsuno, *et al.*, 1983) and with the complex aluminium hydride (70) which contains a disymmetric 1,1′-binaphthyl moiety as chiral auxiliary ligand (Noyori, Tomino and Tanimoto, 1979).

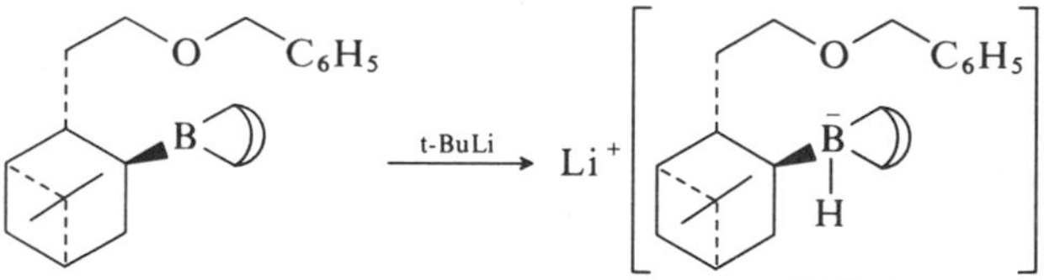

t-BuLi → Li^+ (68)

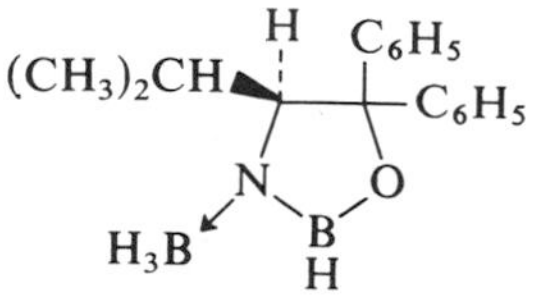

(69) 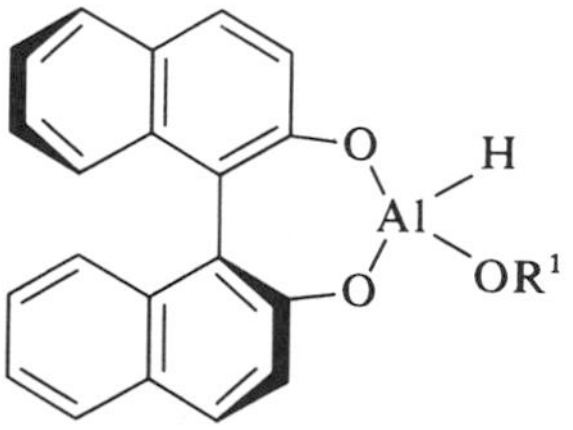 (70)

(7.97)

Enzymes are, of course, chiral reagents *par excellence*, and enzymic reduction of benzaldehyde derivatives using the enzyme liver alcohol dehydrogenase and the reduced form of its co-enzyme, nicotinamide adenine dinucleotide (NADH), has been used to prepare a number of specifically labelled benzyl alcohols required for biosynthetic work. In these reactions transfer of hydride takes place from one face of NADH to the *re* face of the aldehyde, that is to the top face as it is written in (7.98). Thus, reaction of deuterated benzaldehyde (71) with liver alcohol dehydrogenase and NADH, generated *in situ* from NAD^+ and ethanol gives specifically the (*S*)-[1-^{2}H] benzyl alcohol (72), whereas using benzaldehyde itself and the deuterated co-enzyme, prepared *in situ* from NAD^+ and CH_3CD_2OH, the (*R*)-alcohol is obtained (compare Battersby and Staunton, 1974; Battersby, 1984).

^{2}H O (71) liver alcohol dehydrogenase H H $CONH_2$ N ^{2}H H OH (72)

H O liver alcohol dehydrogenase H ^{2}H $CONH_2$ N H ^{2}H OH

(7.98)

The ketonic carbonyl group of β-keto-esters and cyclic ketones are also very conveniently reduced with high stereoselectivity by fermenting bakers' yeast. Thus, ethyl acetoacetate gave ethyl (*S*)-(+)-3-hydroxybutyrate with 87 per cent optical purity (Mori, 1981), 2-chlorocyclohexanone gave optically pure (1*S*,2*R*)-2-chlorocyclohexanol (Crumbie, Ridley and Simpson, 1977) and the prochiral diketone (73) was converted almost entirely into the chiral ketol (74) (Brooks, Grothaus and Irwin, 1982).

Interestingly, reduction of ethyl γ-chloroacetoacetate under the same conditions gave predominantly the corresponding (*S*)-alcohol (i.e. the opposite configuration from that of the alcohol from ethyl acetoacetate itself), but the corresponding octyl ester gave almost entirely the (*R*)-alcohol (Zhou *et al.*, 1983); that is the stereochemistry of the reduction appears to depend on the dimensions of the molecule. If this effect should be general it will be possible to obtain either enantiomer of a β-hydroxy ester at will by enzymic reduction of the appropriate β-keto ester. It is thought that this effect owes its origin to the fact that the yeast contains

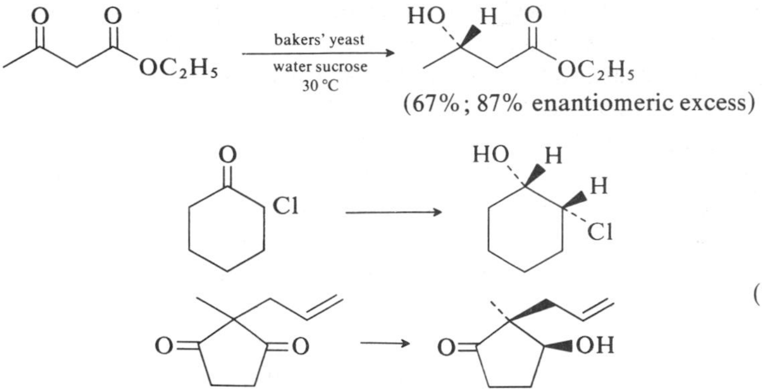

bakers' yeast, water sucrose, 30 °C

OC_2H_5 → OC_2H_5 (67%; 87% enantiomeric excess)

(73) (74) (7.99)

at least two different oxidoreductase enzymes which produce the two enantiomeric alcohols at different rates.

In cases where there is a polar group on the carbon atom adjacent to the carbonyl group Cram's rule may not be followed, because the conformation of the carbonyl compound in the transition state is no longer determined solely by steric factors. In α-hydroxy and α-amino ketones, for example, reaction is thought to proceed through a relatively rigid chelate compound of type (75). Because of this, reduction of α-hydroxy and α-amino ketones usually proceeds with a comparatively high degree of stereoselectivity by attack on the chelate from the less hindered side, but not necessarily in the Cram sense. Again, where the adjacent carbon atom carries a halogen substituent, the most reactive conformation of the molecule appears to be the one in which the polar halogen atom and the polar carbonyl are *anti* disposed, to minimise dipole–dipole repulsion. The predominant product is then formed by approach of the metal hydride anion to the less hindered side of this conformation.

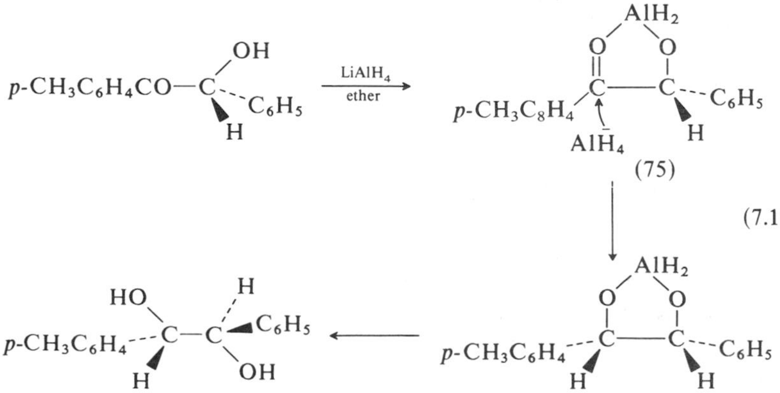

p-$CH_3C_6H_4CO$—C(OH)(C_6H_5)(H) $\xrightarrow[\text{ether}]{LiAlH_4}$ (75) → p-$CH_3C_6H_4$–CH(OH)–CH(OH)–C_6H_5 (85% of product) (7.100)

With strongly hindered cyclic ketones the main product is again formed by approach of the reagent to the less hindered side of the carbonyl group. Thus, reduction of camphor with lithium aluminium hydride leads mainly to the *exo* alcohol (isoborneol), whereas norcamphor, in which approach of the hydride anion is now easier from the side of the methylene bridge, leads mainly to the *endo* alcohol.

With other less rigid cyclohexanones the stereochemical course of the reduction is not so easy to predict. In general, a mixture of products is obtained in which, with comparatively unhindered ketones the more stable equatorial alcohol predominates; with hindered ketones the axial alcohol is the main product (Barton, 1953). Thus, reduction of 4-t-butylcyclohexanone with lithium aluminium hydride affords the equatorial *trans*-4-t-butylcyclohexanol in 92 per cent yield, whereas the hindered 3,3,5-trimethylcyclohexanone affords a mixture containing mainly the axial alcohol (76). The latter is practically the only product

$LiAlH_4$; $LiAl(OC_4H_9\text{-t})_3$

(76) (7.101)

when the more selective reducing agent lithium hydridotri-t-butoxyaluminate, $LiAlH(OBu^t)_3$, (p. 470) is employed. There have been numerous attempts to rationalise these effects (see Wigfield, 1979). It is supposed that in general there is a tendency for approach of the reagent to the carbonyl group in an axial direction as in (77), leading to the equatorial alcohol. In hindered ketones such as (78), however, axial approach may

(77) $\xrightarrow{LiAlH_4,\ ether}$ (88–90% of product) + (10–12% of product)

(7.102)

(78) $\xrightarrow{LiAlH_4,\ ether}$ (45% of product) + (55% of product)

be hampered by steric factors , favouring equatorial approach of reagent and formation of the axial alcohol.

To obtain the axial alcohol from an unhindered ketone we employ not borohydride or lithium aluminium hydride but one of the new highly hindered reducing agents such as (L)-selectride (p. 478). With this reagent 4-t-butylcyclohexanone is reduced to the axial alcohol in 96 per cent yield. To obtain the equatorial alcohol from a hindered cyclohexanone is more difficult, but it can sometimes be achieved by equilibration of the axial–equatorial alcohol product mixture with, for example, aluminium alkoxides.

Reductive fission of epoxides and of alkyl halides or sulphonates with lithium aluminium hydride or, better, lithium triethylborohydride (p. 477) proceeds by S_N2 substitution by hydride ion, with formation of a new C—H bond. Epoxides are thereby reduced to alcohols. With unsymmetrical epoxides reaction takes place at the less substituted carbon atom to give the more substituted alcohol. Thus, 1,2-epoxybutane gives mainly 2-butanol and 1-methylcyclohexene epoxide gives 1-methylcyclohexanol. In the presence of Lewis acids, however, the direction of ring opening may be reversed. Thus, reduction of 1-methylcyclohexene epoxide with sodium cyanoborohydride in the presence of boron trifluoride-etherate gives, not 1-methylcyclohexanol, but *cis*-2-methylcyclohexanol by backside attack of hydride on the epoxide–boron trifluoride complex. The direction of ring-opening is now dictated by the formation of the more stable carbonium ion (Hutchins, Taffer and Burgoyne, 1981).

CH_3 / O; $LiAlH_4$ or $Li(C_2H_5)_3BH$ → CH_3, OH; $NaBH_3CN$, $BF_3{\cdot}(C_2H_5)_2O$, THF → CH_3, H, H, OH

(7.103)

In accordance with the S_N2 mechanism reduction of epoxides with hydride reagents takes place with inversion of configuration at the carbon atom attacked. Thus, 1,2-dimethylcyclohexene epoxide affords *trans*-1,2-dimethylcyclohexanol (7.104).

CH_3 O CH_3 $\xrightarrow[\text{ether}]{LiAlH_4}$ CH_3 OH CH_3 H (7.104)

Primary and secondary alkyl halides are reductively cleaved to the corresponding hydrocarbons with lithium aluminium hydride, although better results are obtained with sodium cyanoborohydride (p. 475) or with lithium triethylborohydride (p. 477). Tertiary halides react only slowly and give mostly alkenes. Aryl iodides and bromides may also be reduced to the hydrocarbons with lithium aluminium hydride in boiling tetrahydrofuran (Brown and Krishnamurthy, 1969) and other supposedly inert compounds such as vinyl and bridgehead halides can also be converted into the parent hydrocarbons under appropriate conditions. α-Bromo-α-methystyrene, for example, gives α-methylstyrene and 1-bromoadamantane is easily converted into adamantane. Better results are said to be obtained with clear solutions of lithium aluminium hydride than with the slurries usually employed (Krishnamurthy and Brown, 1982). There is evidence that reduction of alkyl halides, particularly iodides, with lithium aluminium hydroxide proceeds by a single electron transfer pathway (Ashby, De Priest, Goel, Wenderoth and Pham, 1984). Sulphonate esters of primary and secondary alcohols are also readily reduced with lithium aluminium hydride and this reaction has been widely employed in synthesis for the replacement of an alcoholic hydroxyl group by hydrogen. Again even better results are obtained with lithium triethylborohydride (cf. Brown and Krishnamurthy, 1979).

Carboxylic acids, aldehydes and ketones are very conveniently converted into methyl or methylene groups by reduction of the corresponding acyl toluene-*p*-sulphonylhydrazide or the toluene-*p*-sulphonylhydrazone of the aldehyde or ketone, with lithium aluminium hydride. Thus, 2-naphthaldehyde is converted into 2-methylnaphthalene by this procedure and 1-naphthylacetic acid gives 1-ethylnaphthalene (Cagliotti, 1972). Better results are obtained with sodium cyanoborohydride (p. 476).

Lithium hydridoalkoxyaluminates

Lithium aluminium hydride itself is a powerful and versatile reducing agent. More selective reagents can be obtained by modification of lithium aluminium hydride by treatment with alcohols or with aluminium chloride. One such reagent is the sterically bulky lithium hydridotri-t-butoxyaluminate, which is readily prepared by action of the stoichiometric amount of t-butyl alcohol on lithium aluminium hydride.

$$LiAlH_4 + 3ROH \rightarrow LiAlH(OR)_3 + 3H_2$$

Analogous reagents are obtained in the same way from other alcohols, and by replacement of only one or two of the hydrogen atoms of the hydride by alkoxy groups, affording a range of reagents of graded activities (Rerick, 1968; Brown and Krishnamurthy, 1979).

Lithium hydrotri-t-butoxyaluminate is a much milder reducing agent than lithium aluminium hydride itself. Thus, although aldehydes and ketones are reduced normally to alcohols, carboxylic esters and epoxides react only slowly, and halides, nitriles and nitro groups are not attacked. Aldehydes and ketones can thus be selectively reduced in presence of these groups. A good example is the first reaction shown in equation (7.105).

$$\text{OHC, } CH_3O_2C\text{-substituted bicyclic lactone epoxide (CO—O, H, O)} \xrightarrow[\text{THF, }-10\,°C]{LiAlH(OC_4H_9\text{-t})_3} \text{HOCH}_2\text{, } CH_3O_2C\text{-substituted bicyclic lactone epoxide (CO—O, H, O)}$$

$$N{\equiv}C\text{–}C_6H_4\text{–}COCl \xrightarrow[\text{diglyme, }-78\,°C]{LiAlH(OC_4H_9\text{-t})_3} N{\equiv}C\text{–}C_6H_4\text{–}CHO \quad (80\%) \qquad (7.105)$$

One of the most useful applications of the alkoxy reagents is in the preparation of aldehydes from carboxylic acids by partial reduction of the acid chlorides or dimethylamides. Acid chlorides are readily reduced with lithium aluminium hydride itself or with sodium borohydride to the corresponding alcohols, but with one equivalent of the tri-t-butoxy compound high yields of the aldehyde can be obtained in the presence of a range of other functional groups (7.105).

Although esters, in general, are reduced only slowly, phenyl esters are converted into the aldehyde with $LiAlH(OC_4H_9)_3$. Thus, phenyl cyclohexanecarboxylate gives formylcyclohexane in 60 per cent yield.

Reduction of tertiary amides with excess of lithium aluminium hydride affords the corresponding amines in good yield (7.106). Reaction is believed to proceed through an aldehyde-ammonia derivative of the type (79). With the less active $LiAlH(OC_2H_5)_3$ (the tri-t-butoxy compound is ineffective in this case) reaction stops at the aldehyde-ammonia stage and hydrolysis of the product affords the corresponding aldehyde. Similarly, reduction of nitriles with lithium aluminium hydride affords a primary amine by way of the imine salt. With lithium hydridotriethoxyaluminate, however, reaction stops at the imine stage, and hydrolysis gives the aldehyde. By this procedure trimethylacetonitrile was converted into trimethylacetaldehyde in 75 per cent yield.

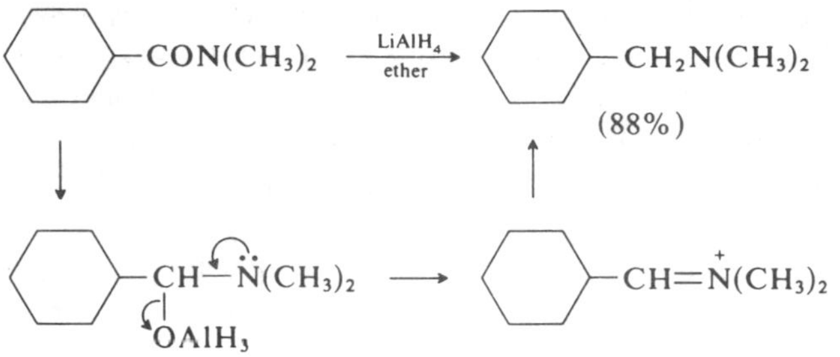

$$\text{C}_6\text{H}_{11}\text{-CON(CH}_3)_2 \xrightarrow[\text{ether}]{\text{LiAlH}_4} \text{C}_6\text{H}_{11}\text{-CH}_2\text{N(CH}_3)_2 \ (88\%)$$

$$\text{C}_6\text{H}_{11}\text{-CH(OAlH}_3\text{)-}\ddot{\text{N}}\text{(CH}_3)_2 \ (79) \longrightarrow \text{C}_6\text{H}_{11}\text{-CH=}\overset{+}{\text{N}}\text{(CH}_3)_2 \quad (7.106)$$

$$CH_2{=}CH(CH_2)_8CON(CH_3)_2 \xrightarrow[(2)\ H_3O^+]{(1)\ LiAlH(OC_2H_5)_3} CH_2{=}CH(CH_2)_8CHO \ (85\%)$$

$$R{-}C{\equiv}N \xrightarrow[\text{ether}]{LiAlH_4} [R{-}CH{=}N{-}\bar{A}lH_3] \xrightarrow[(2)\ H_3O^+]{(1)\ LiAlH_4} RCH_2NH_2$$

Because of the steric effect of the alkoxy groups the hydrido-alkoxyaluminates are more stereoselective in their action than is lithium aluminium hydride itself. But there is no very clear correlation between the size of the alkoxy groups and the selectivity of the reductions. The situation is complicated by the fact that some of the alkoxy compounds disproportionate readily, with regeneration of the reactive tetrahydro-aluminate ion. Concentration effects may also have to be considered.

$$2H_3\bar{A}lOCH(CH_3)_2 \rightleftharpoons AlH_4^- + H_2\bar{A}l[OCH(CH_3)_2]_2$$

Thus, in the reduction of 3,3,5-trimethylcyclohexanone, lithium hydridotrimethoxyaluminate $LiAlH(OCH_3)_3$ was, as expected, more stereoselective than lithium aluminium hydride, but it was also more selective than the apparently more bulky tri-t-butoxy compound, $LiAlH(OC_4H_9\text{-t})_3$. The reason is that in tetrahydrofuran, the usual solvent, the trimethoxy compound is dimeric or trimeric, whereas the tri-t-butoxy compound is monomeric. The selectivity obtained with the trimethoxy compound depends on its concentration.

Table 7.4. *Reduction of* 3,3,5-*trimethylcyclohexanone*

CH_3, CH_3, O, CH_3 → CH_3, H, OH, CH_3, CH_3 + CH_3, OH, H, CH_3, CH_3

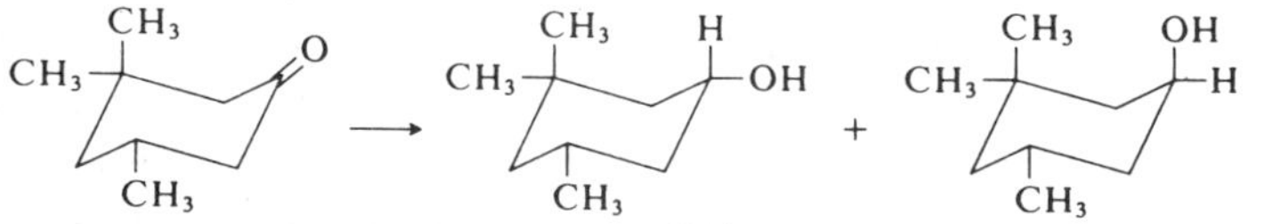

	(%)	(%)
$NaBH_4$, $(CH_3)_2CHOH$	36–45	55–64
$LiAlH_4$, ether	37–48	52–63
$LiAlH(OCH_3)_3$, THF	2–8	92–98
$LiAlH(OC_4H_9\text{-t})_3$, THF	4–12	88–96

Mixed lithium aluminium hydride–aluminium chloride reagents

Further useful modification of the properties of lithium aluminium hydride is achieved by addition of aluminium chloride in various proportions. This serves to release mixed chloride–hydrides of aluminium as shown in (7.107). The general effect of the addition of

$$\begin{aligned} 3LiAlH_4 + AlCl_3 &\rightarrow 3LiCl + 4AlH_3 \\ LiAlH_4 + AlCl_3 &\rightarrow LiCl + 2AlH_2Cl \\ LiAlH_4 + 3AlCl_3 &\rightarrow LiCl + 4AlHCl_2 \end{aligned} \qquad (7.107)$$

aluminium chloride is to lower the reducing power of lithium aluminium hydride and in consequence to produce reagents which are more specific for particular reactions (Rerick, 1968). For example, the carbon–halogen bond is often inert to the mixed hydride reagents. Advantage is taken of this in the reduction of polyfunctional compounds in which retention of halogen is desired, as in the conversion of methyl 3-bromopropionate into 3-bromopropanol; lithium aluminium hydride alone produces propanol (7.108).

$$BrCH_2CH_2CO_2CH_3 \xrightarrow[\text{ether}]{LiAlH_4\text{–}AlCl_3(1:1)} BrCH_2CH_2CH_2OH \qquad (7.108)$$

Similarly, nitro groups are not so easily reduced as with lithium aluminium hydride itself, and *p*-nitrobenzaldehyde can be converted into *p*-nitrobenzyl alcohol in 75 per cent yield. Aldehydes and ketones are reduced to carbinols, and there is no advantage in the use of mixed hydrides in these cases, although it should be noted that the stereochemical result obtained in the reduction of cyclic ketones may not be the same as with lithium aluminium hydride itself. With diaryl ketones and with aryl alkyl ketones however the carbonyl group is reduced to methylene in high yield, and this procedure offers a useful alternative to the Clemmensen or Huang–Minlon methods for reduction of this type of ketone.

Reduction with lithium aluminium hydride–aluminium chloride (3:1) also provides an excellent route from $\alpha\beta$-unsaturated carbonyl compounds to unsaturated alcohols which are difficult to prepare with lithium aluminium hydride alone because of competing reduction of the carbon–carbon double bond. The effective reagent is thought to be aluminium hydride formed *in situ* from lithium aluminium hydride and aluminium chloride (7.109).

$$C_6H_5CH{=}CHCO_2C_2H_5 \xrightarrow[\text{ether}]{3LiAlH_4\text{–}AlCl_3} \underset{(90\%)}{C_6H_5CH{=}CHCH_2OH} \qquad (7.109)$$

Di-isobutylaluminium hydride, DIBAL

This derivative of aluminium hydride is available commercially as a solution in toluene or hexane. It is a versatile reducing agent (cf. Winterfeldt, 1975). At ordinary temperatures esters and ketones are reduced to alcohols, nitriles give amines and epoxides are cleaved to alcohols. It finds its greatest use, probably, in the preparation of aldehydes. At low temperatures esters and lactones are reduced directly to aldehydes, and nitriles and amides give aldimines which are readily converted into the aldehydes by hydrolysis (7.110).

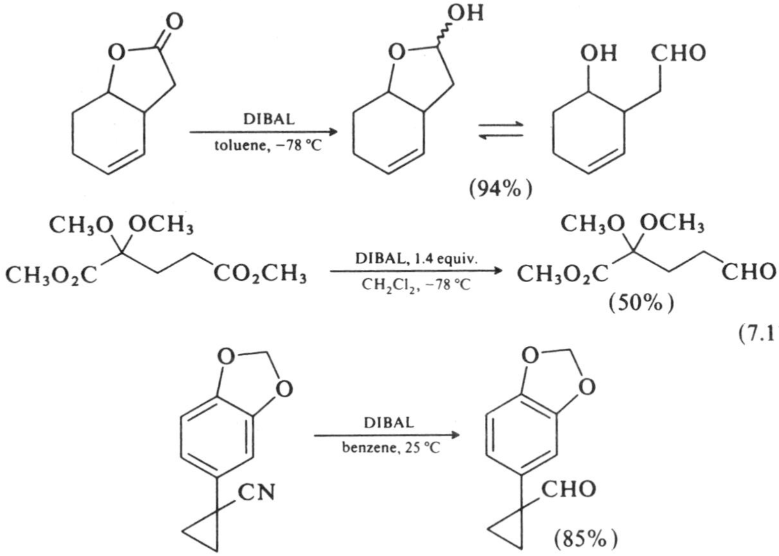

DIBAL, toluene, −78 °C (94%)

$CH_3O\ OCH_3$ CH_3O_2C CO_2CH_3 —DIBAL, 1.4 equiv., CH_2Cl_2, −78 °C→ $CH_3O\ OCH_3$ CH_3O_2C CHO (50%)

(7.110)

CN —DIBAL, benzene, 25 °C→ CHO (85%)

It is also a useful reagent for the reduction of disubstituted alkynes to *cis*-alkenes, and for selective 1,2-reduction of $\alpha\beta$-unsaturated carbonyl compounds to allylic alcohols. Thus, the enedione (81) gave (82) and, in a synthesis of linoleic acid the diyne (80) was reduced to the *cis,cis*-diene in 82 per cent yield without cleavage of the C—Cl bond (7.111).

$CH_3(CH_2)_5C{\equiv}CCH_2C{\equiv}C(CH_2)_5CH_2Cl$

—(1) DIBAL, 0 °C; (2) CH_3OH→ $CH_3(CH_2)_5C{\equiv}CCH_2C{\equiv}C(CH_2)_5CH_2Cl$ (80)

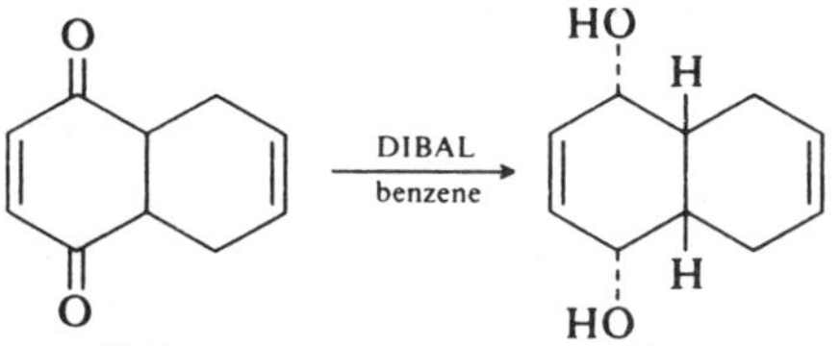

(81) —DIBAL, benzene→ (82) (7.111)

Sodium cyanoborohydride

A number of reagents derived from sodium borohydride by replacement of one or more of the hydrogen atoms by other groups allow more selective reduction than with sodium borohydride itself. Among the most useful are sodium cyanoborohydride and some trialkylborohydrides (Walker, 1976; Brown and Krishnamurthy, 1979).

Because of the electron-withdrawing effect of the cyano group, sodium cyanoborohydride is a weaker and more selective reagent than sodium borohydride (see Lane, 1975). It has the further advantage that it is stable in acid to pH = 3 and can thus be employed to effect reductions in the presence of functional groups which are sensitive to the more basic conditions of reduction with sodium borohydride. In neutral solution in hexamethylphosphoramide it is a useful reagent for reductive removal of iodo, bromo and tosyloxy groups. Under these conditions carbonyl groups and other reducible groups are not affected. 1-Iododecane gives decane in 90 per cent yield, and 3-bromo-1,2-epoxy-1-phenylpropane is converted into the halogen-free compound without cleavage of the epoxide group (7.112).

$$C_6H_5\overset{O}{CH{-}CH}CH_2Br \xrightarrow[\text{hexamethylphosphoramide, 70 °C}]{NaBH_3CN} C_6H_5\overset{O}{CH{-}CH}CH_3 \quad (63\%)$$

$$C_6H_5CHO + C_2H_5NH_2 \overset{pH6}{\rightleftharpoons} [C_6H_5CH{=}\overset{+}{N}HC_2H_5] \xrightarrow{NaBH_3CN,\ CH_3OH} C_6H_5CH_2NHC_2H_5 \quad (80\%) \qquad (7.112)$$

$$C_6H_5CH_2COCO_2Na \xrightarrow[CH_3OH]{NH_4Br,\ NaBH_3CN} C_6H_5CH_2CH(NH_2)CO_2H \quad (49\%)$$

Aldehydes and ketones are unaffected by sodium cyanoborohydride in neutral solution, but they are readily reduced to the corresponding alcohol at pH = 3–4 by way of the protonated carbonyl group. By previous exchange of the hydrogens of the borohydride for deuterium or tritium, by reaction with ^2H_2O or tritiated water, an efficient and economical route is available for deuteride or tritiide reduction of aldehydes and ketones.

Iminium groups are even more easily reduced than carbonyl groups in acid solution, and this has been exploited in a method for reductive amination of aldehydes and ketones by way of the iminium salts formed from the carbonyl compounds and a primary or secondary amine at pH = 6; at this pH the carbonyl compounds themselves are unaffected.

The reaction has been adapted for the preparation of α-amino acids from α-keto acids (7.112).

A well-known method for the conversion of carbonyl compounds into the corresponding hydrocarbons involves reduction of the derived toluene-*p*-sulphonyl-(tosyl-) hydrazones with sodium borohydride (Cagliotti, 1972). Very much better yields are obtained using sodium cyanoborohydride in acidic dimethylformamide; the reactions are more selective and alkenes and other unwanted wide products are not formed. The reaction is specific for aliphatic and alicyclic carbonyl compounds; aromatic compounds are unaffected. The tosylhydrazone need not be isolated and is prepared and reduced *in situ*. With αβ-unsaturated carbonyl compounds reduction of the tosylhydrazone is accompanied by migration of the double bond to the carbon atom which originally carried the oxygen. For example, cinnamaldehyde tosylhydrazone gives 3-phenyl-1-propene in 98 per cent yield, and β-ionone is converted into the diene (83) with deconjugation of the double bonds. The double bond migrations are stereoselective, giving predominantly the *trans* alkene.

$$CH_3(CH_2)_8CHO \xrightarrow[\text{(2) } NaBH_3CN,\ \text{dimethylformamide, 100 °C}]{\text{(1) Tosylhydrazide}} CH_3(CH_2)_8CH_3 \quad (95\%)$$

(7.113)

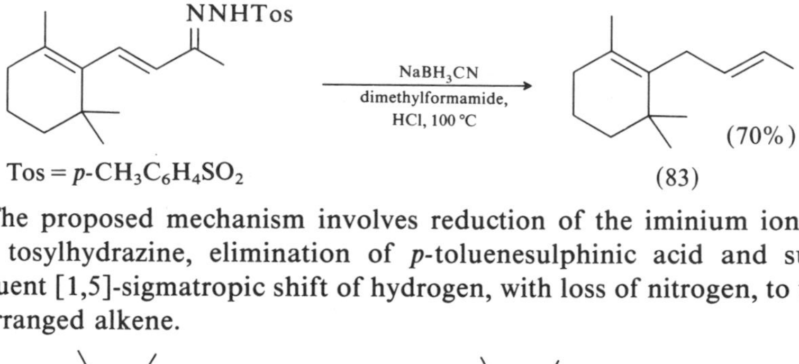

NNHTos

NaBH$_3$CN, dimethylformamide, HCl, 100 °C

(70%)

Tos = *p*-$CH_3C_6H_4SO_2$ (83)

The proposed mechanism involves reduction of the iminium ion to the tosylhydrazine, elimination of *p*-toluenesulphinic acid and subsequent [1,5]-sigmatropic shift of hydrogen, with loss of nitrogen, to the rearranged alkene.

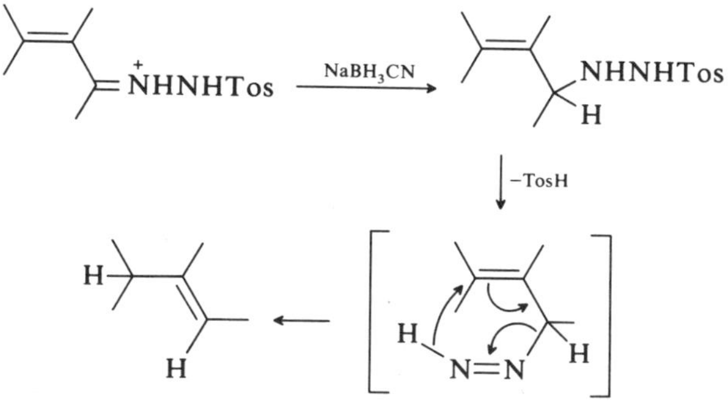

$\overset{+}{N}$HNHTos —NaBH$_3$CN→ NHNHTos, H

↓ −TosH (7.114)

H, H ← [H, N=N, H]

Tos = *p*-$CH_3C_6H_4SO_2$

Trialkylborohydrides

The lithium trialkylborohydrides shown in [7.115] are more powerful reducing agents than lithium borohydride itself (aldehydes, ketones and esters, for example, are readily reduced to the corresponding alcohols) but their reactions show special features which give them value in synthesis (Brown and Krishnamurthy, 1979). Because of the inductive effect of the ethyl groups and possibly also because of their effect in reducing solvation of the anion, lithium triethylborohydride (84), 'Superhydride', is a very powerful nucleophile. It is an excellent reagent for

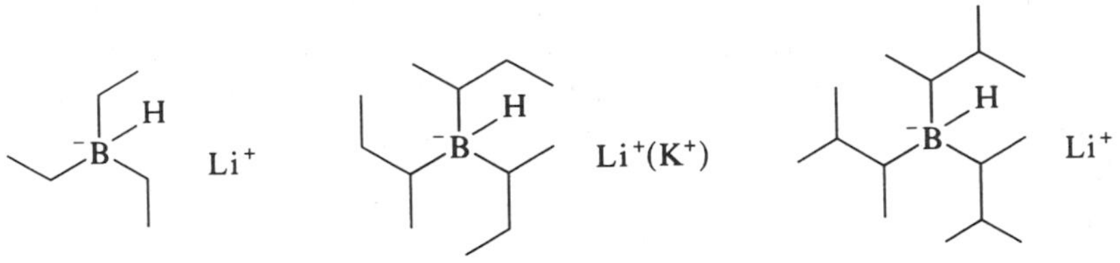

(84) (85) (86) (7.115)

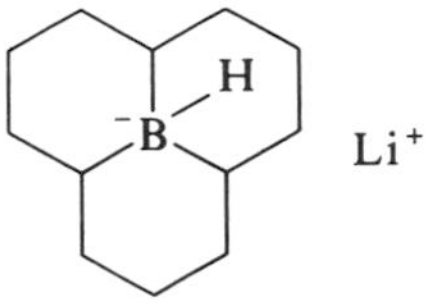

the reductive fission of primary and secondary alkyl bromides and tosylates and is far superior to lithium aluminium hydride in this respect. Aryl halides are not affected, allowing the possibility of selective reduction in a molecule containing both aryl and alkyl halide groups. Cycloheptyl bromide was converted into cycloheptane in 99 per cent yield in a smooth reaction at room temperature and cyclopentyl tosylate gave cyclopentane quantitatively. Reduction of the halides with the deuterated reagent, which is easily obtained by hydrogen exchange in D_2O at pH = 3, provides a simple means for introducing deuterium into an alkane, with stereochemical inversion at the site of substitution.

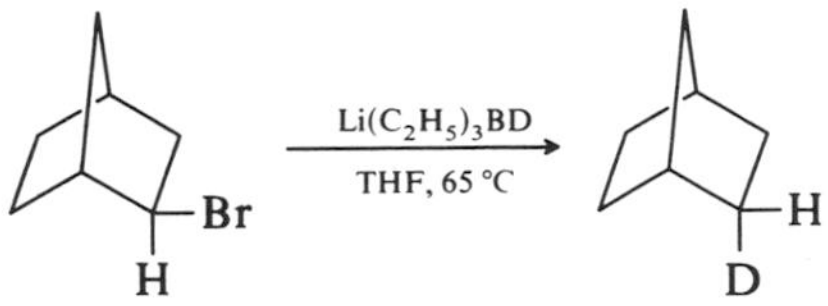

$Li(C_2H_5)_3BD$, THF, 65 °C (7.116)

Epoxides also are smoothly reduced to the alcohols in high yield, by attack at the less substituted carbon. Even highly hindered epoxides react readily without rearrangement.

The other trialkylborohydrides shown in (7.115) have been used chiefly for the stereoselective reduction of ketones. The lithium and potassium tri-s-butylborohydrides (85), commercially available as 'Selectrides', are among the best reagents yet available for the stereoselective reduction of ketones. Even better results have been obtained with the reagent (86) (Krishnamurthy and Brown, 1976). Introduction of bulky alkyl substituents into the borohydride anion drastically alters the direction of attack on a cyclic ketone, and cyclohexanones are now attacked predominantly from the equatorial side to give the axial alcohol. Thus,

$NaBH_4$ $\quad$ $Li(s\text{-}C_4H_9)_3BH$ $\quad$ (7.117)

reduction of 2-methylcyclohexanone with the tri-s-butyl reagent (85) gives *cis*-2-methylcyclohexanol in 99 per cent yield, and even 4-t-butylcyclo-hexanone, in which the substituent is several carbon atoms removed from the carbonyl group, gives 93 per cent of *cis*-4-t-butylcyclohexanol by equatorial attack. The reagent is prepared *in situ* from $LiAlH(OCH_3)_3$ and tri-s-butylborane, and reductions proceed rapidly at 0 °C or below in tetrahydrofuran.

The trialkylborohydride (87) containing the asymmetric α-pinanyl group, has been prepared from 9-BBN (see p. 291) and (+)-α-pinene followed by t-butyl lithium and used to reduce a number of ketones. The alcohols obtained were optically active and useful optical yields were obtained in some cases (Brown and Pai, 1982).

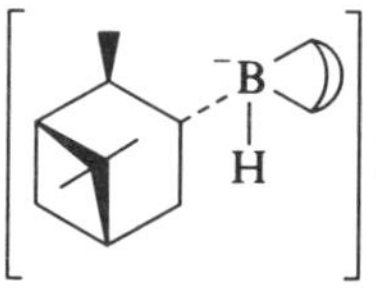

(87)

7.4. Reductions with borane and dialkylboranes

Borane, BH_3 (which exists as the gaseous dimer diborane B_2H_6) is a powerful reducing agent and attacks a variety of unsaturated groups. It can be prepared by reaction of boron trifluoride etherate with sodium borohydride in diglyme solution (diglyme is the dimethyl ether of diethylene glycol) but for most purposes the commercially available solutions in tetrahydrofuran, which contain the borane-tetrahydrofuran complex, BF_3.THF, are suitable.

Table 7.5. *Reduction of functional groups with diborane*

Reactant	Product
$-CO_2H$	$-CH_2OH$
$-CH{=}CH-$	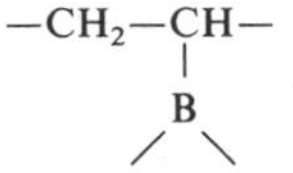 $-CH_2-CH(B<)-$ (before hydrolysis)
$-CHO$	$-CH_2OH$
$-CO-$ (with one more bond)	$-CHOH-$ (with one more bond)
$-C{\equiv}N$	$-CH_2NH_2$
$-CONR_2$	$-CH_2NR_2$
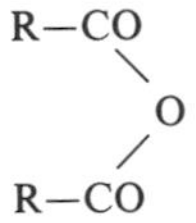 $(R-CO)_2O$ (R–CO–O–CO–R)	$R-CH_2OH$
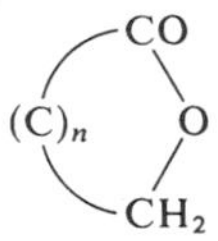 Lactone ring: $(C)_n$ with CO–O–CH$_2$	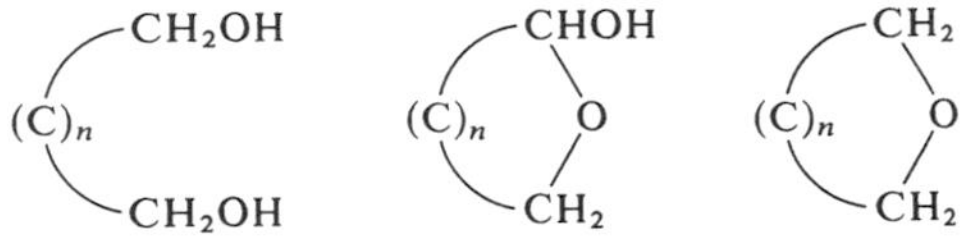 $(C)_n$ bearing CH_2OH and CH_2OH; ring $(C)_n$ with CHOH–O–CH$_2$; ring $(C)_n$ with CH$_2$–O–CH$_2$
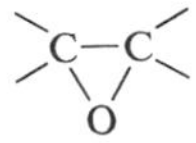 Epoxide $>C-C<$ bridged by O	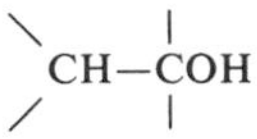 $>CH-C(OH)<$
$-CO_2R$	$-CH_2OH+ROH$
$-COCl$	no reaction
$-NO_2$	no reaction

Reaction of borane with unsaturated groups takes place readily at room temperature, and the products shown in Table 7.5 are isolated in high yield after hydrolysis of the intermediate boron compound (Walker, 1976; Brown and Krishnamurthy, 1979). Borane reacts rapidly with water, and reactions must be effected under anhydrous conditions, best under nitrogen, since borane itself and the lower alkylboranes may ignite in air.

Reductions with borane do not simply parallel those with sodium borohydride. This is because sodium borohydride is a nucleophile and reacts by addition of hydride ion to the more positive end of a polarised multiple bond, whereas borane is a Lewis acid and attacks electron-rich centres. For example, while sodium borohydride very rapidly reduces acid chlorides to primary alcohols, the reaction being facilitated by the electron-withdrawing effect of the halogen, borane does not react with acid chlorides under the usual mild conditions. Reduction of carbonyl groups by borane takes place by addition of the electron-deficient borane to the oxygen atom, followed by the irreversible transfer of hydride ion from boron to carbon (7.118). The inertness of acid chlorides can be ascribed to the decreased basic properties of the carbonyl oxygen resulting from the electron-withdrawing effect of the halogen. For a similar reason esters are reduced only slowly by borane.

$$\rangle C=O + BH_3 \rightleftharpoons \rangle C=\overset{+}{O}-\overset{-}{B}H_3 \rightarrow -\underset{H}{\overset{|}{C}}-OBH_2 \qquad (7.118)$$

Remarkable is the ready reduction of carboxylic acids to primary alcohols with borane, which can often be selectively effected in presence of other unsaturated groups. *p*-Nitrobenzoic acid, for example, is reduced to *p*-nitrobenzyl alcohol in 79 per cent yield. Reduction of carboxylic acids is believed to proceed by way of a triacyloxyborane, the carbonyl groups of which are rapidly reduced by further reaction with borane.

The reduction of epoxides with borane is also noteworthy since it gives rise to the *less* substituted carbinol in preponderant amount, in contrast to reduction with complex hydrides (see p. 469). The reaction is catalysed by small amounts of sodium or lithium borohydride and high yields of the alcohol are obtained. With 1-alkylcycloalkene epoxides the 2-alkylcycloalkanols produced are entirely *cis*, and this reaction thus complements the hydroboration–oxidation of cycloalkenes described on p. 295, which leads to *trans*-2-alkylcycloalkanols. Reaction with borane in the presence of boron trifluoride has also been used for the reduction of epoxides and for the conversion of lactones and some esters into ethers.

$$CH_3-\overset{CH_3}{\overset{|}{C}}\underset{O}{\diagdown\diagup}CH-CH_3 \xrightarrow[\text{THF, 0°C}]{B_2H_6,\ LiBH_4} (CH_3)_2\overset{OH}{\overset{|}{C}}CH_2CH_3 + (CH_3)_2CHCHOHCH_3$$

(25% of product) (75% of product)

$$\text{1-methylcyclopentene oxide} \xrightarrow[\text{THF, 0 °C}]{B_2H_6,\ LiBH_4} \textit{cis}\text{-2-methylcyclopentanol} + \text{1-methylcyclopentanol} \qquad (7.119)$$

(72% of product) (28% of product)

Several borane complexes have come into use in place of borane itself. The borane–dimethyl sulphide complex has the advantage that it is stable and is soluble in several organic solvents. It is effective for the reduction of a range of functional groups (Brown, Choi and Narasimhan, 1982). The stable ammonia–borane and t-butylamine–borane complexes are excellent reagents for the reduction of aldehydes and ketones to alcohols

(Andrews and Crawford, 1980) and the borane–pyridine complex is a cheap and readily available alternative to sodium cyanoborohydride for reductive amination of carbonyl compounds (Pelter, Rosser and Mills, 1984). Catecholborane is another versatile reducing agent with milder action than borane (Lane and Kabalka, 1976).

The difference in the relative reactivities of lithium borohydride and borane in the reduction of carboxylic acid and ester groups is nicely illustrated in the syntheses of (*R*)- and (*S*)-mevalonolactone from the same starting material shown in (7.120) (Huang *et al.*, 1975).

H_3C, HO, CO_2H, CO_2CH_3 → (1) $(CH_3CO)_2O$ (2) $BH_3.THF$ → H_3C, HO, CH_2OH, CO_2CH_3 → lactone

H_3C, HO, CO_2H, CO_2CH_3 → $LiBH_4$ *or* $Na-NH_3$ → H_3C, HO, CO_2H, CH_2OH → lactone

(7.120)

Reducing agents more selective than borane itself are obtained by replacing one or two of the hydrogen atoms of the borane by bulky alkyl groups. Particularly useful reagents of this type are di(3-methyl-2-butyl)borane (disiamylborane) and 2,3-dimethyl-2-butylborane (thexylborane) which are easily obtained by action of borane on 2-methyl-2-butene and 2,3-dimethyl-2-butene (see p. 289). These are milder and more selective reducing agents than borane, and because of the large steric requirements of the alkyl groups the rate of reaction is strongly influenced by the structure of the compound being reduced. Aldehydes and ketones are converted into the corresponding alcohols, although the reactivity of ketones varies widely with their structure. Acid chlorides, acid anhydrides, and esters do not react, and epoxides are reduced only very slowly. Carboxylic acids are not reduced, in contrast to the rapid reaction with borane; they simply form the dialkylboron carboxylate which on hydrolysis regenerates the acid. Presumably the large size of the dialkylboron group in the carboxylate prevents further attack by the reagent on the carbonyl group.

Two useful reactions of disiamylborane, which are not paralleled in the reactions of borane itself, are the reductions of lactones to hydroxy-aldehydes and of dimethylamides to aldehydes (7.121). High yields of

$$\text{(γ-butyrolactone)} + (C_5H_{11})_2BH \xrightarrow{\text{THF, 0 °C}} \text{(2-tetrahydrofuranyl, H, }OB(C_5H_{11})_2\text{)} \xrightarrow{H_2O} \underset{CH_2OH}{CH_2}-\underset{CHO}{CH_2} \ (73\%) \tag{7.121}$$

$$RCON(CH_3)_2 \xrightarrow[\text{THF, 0 °C}]{(C_5H_{11})_2BH} RC(H)(OB(C_5H_{11})_2)N(CH_3)_2 \xrightarrow{H_2O} RCHO$$

aldehyde are obtained, and reduction with disiamylborane supplements the reactions with hydridoalkoxyaluminates described on p. 470 for the conversion of carboxylic acid derivatives into aldehydes. Another reagent, 9-borabicyclo[3,3,1]nonane (p. 291), is highly selective for the reduction of $\alpha\beta$-unsaturated carbonyl compounds to allylic alcohols in the presence of other functional groups; nitro and ester groups, for example, react only slowly under the reaction conditions.

BH ≡ H—B (88) $)_2BH$ (89)   (7.122)

A valuable feature of the alkylboranes is the high stereoselectivity which they show in the reduction of cyclic ketones to alcohols (Brown and Krishnamurthy, 1979). With 2-substituted cycloalkanones and bicyclic ketones reaction takes place from the less hindered side of the carbonyl group to give the less stable of the two possible epimers. Thus, whereas reduction of 2-methylcyclohexanone with borane gave mainly the more stable *trans*-2-methylcyclohexanone, with disiamylborane the preponderant product was the *cis*-isomer, and using the bulky optically active terpenoid derivative diisopinocampheylborane (89) as reducing agent, the yield of the *cis* isomer rose to 94 per cent. The product of the last reaction was optically active.

7.5. Other methods

Wolff–Kishner reduction

The Wolff–Kishner reduction provides an excellent method for the reduction of the carbonyl group of many aldehydes and ketones to methyl or methylene. As originally described the reaction involved heating the semicarbazone of the carbonyl compound with sodium ethoxide or other base at 200 °C in a sealed tube, but it is now more conveniently effected by heating a mixture of the carbonyl compound, hydrazine hydrate and sodium or potassium hydroxide in a high-boiling solvent (diethylene glycol is often used) at 180–200 °C for several hours. Excellent yields of the reduced product are often obtained. With potassium t-butoxide in dimethyl sulphoxide, reduction can often be effected at room temperature (Cram *et al.*, 1962). The mechanism of the reaction has not been widely studied, but it is believed that the hydrazone initially formed is transformed by reactions similar to those shown in the equations (7.123). A wide variety of aldehydes and ketones has been reduced by this method. Sterically hindered ketones, such as 11-keto steroids, are resistant, but they too can be reduced by use of anhydrous hydrazine and sodium metal as base.

$$RR'C{=}O + H_2NNH_2 \rightleftharpoons RR'C{=}NNH_2 + H_2O$$
$$RR'C{=}NNH_2 \xrightarrow{OH^-} RR'C{=}N\ddot{N}H^- \rightarrow RR'CH{-}N{=}\ddot{N}^- \rightarrow RR'CH^- + N_2 \rightarrow R'CH_2R \qquad (7.123)$$

Reduction of conjugated unsaturated ketones is sometimes accompanied by a shift in the position of the double bond (7.124). In other cases pyrazoline derivatives may be formed which decompose yielding cyclopropanes isomeric with the expected alkene (7.124).

$$\text{(structure, CHO)} \rightarrow \text{(structure, CH}_2\text{)} \quad (80\%) \qquad (7.124)$$

$$(CH_3)_2C{=}CHCOCH_3 \xrightarrow{H_2NNH_2} (CH_3)_2\text{(pyrazoline, } CH_3\text{, N, NH)} \rightarrow \text{(cyclopropane: } H_3C, H_3C, CH_3\text{)}$$

With many ketones carrying a leaving group in the α-position elimination accompanies reduction. For example the sterol (90) affords the alkene (91) Of particular importance is the reductive opening of $\alpha\beta$-epoxyketones which are converted into allylic alcohols. (Wharton and Bohlen, 1961; see also Barton *et al.*, 1981*b*).

CH_3COO OH O (90) → CH_3COO (91) (82%) (7.125)

Raney nickel desulphurisation of thio-acetals

Another method for reduction of the carbonyl group of aldehydes and ketones to methyl or methylene which is useful on occasion is desulphurisation of the corresponding thio-acetals with Raney nickel in boiling ethanol (Pettit and van Tamelen, 1962). Hydrogenolysis is effected by the hydrogen adsorbed on the nickel during its preparation (7.126).

CH_3 O → CH_2SH–CH_2SH, BF_3 etherate → CH_3 S S ↓ Raney Ni, C_2H_5OH (7.126)

CH_3 + CH_3

Reaction is effected under fairly mild conditions, but the method suffers from the disadvantage that large amounts of Raney nickel are required, and other unsaturated groups in the compound may also be reduced. Lithium and ethylamine may be used instead of Raney nickel, to bring about desulphurisation but again unsaturated groups in the molecule may be affected.

An illustration of the way in which desulphurisation can be used in the synthesis of a complex natural product is given in (7.127), which shows steps from a synthesis of *Cecropia* juvenile hormone (94). The dihydrothiopyran (92) is used to build up the precursor (93) in which the stereochemistry of two of the double bonds in the hormone is controlled by incorporating them in rings. Desulphurisation gives the open chain compound with the correct configuration of the double bonds, and at the same time unmasks the ethyl substituents, to give a product which is converted by further steps into the hormone.

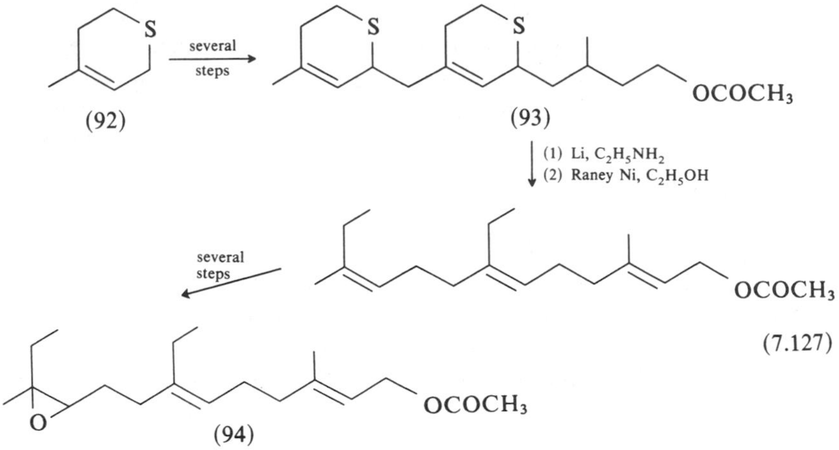

(92) —several steps→ (93) $OCOCH_3$

(1) Li, $C_2H_5NH_2$
(2) Raney Ni, C_2H_5OH

$OCOCH_3$ —several steps→

(94) $OCOCH_3$

(7.127)

Another method for effecting desulphurisation which is helpful in the formation of carbon–carbon double bonds is described on p. 141.

Reductions with di-imide

It has long been known that isolated carbon–carbon bonds can be reduced with hydrazine in presence of oxygen or an oxidising agent. Thus, it was found as early as 1914 that oleic acid is reduced to stearic acid by this method. The actual reducing agent in these reactions is in fact the highly active species di-imide, HN=NH, formed *in situ* by oxidation of hydrazine, and it has been found that this compound is a highly selective reducing agent which in many cases offers a useful alternative to catalytic hydrogenation for the reduction of carbon–carbon multiple bonds (Miller, 1965).

The reagent is not isolated but is prepared *in situ*, usually by oxidation of hydrazine with oxygen or an oxidising agent (e.g. hydrogen peroxide); by thermal decomposition of *p*-toluenesulphonylhydrazide; or from azodicarboxylic acid (7.128).

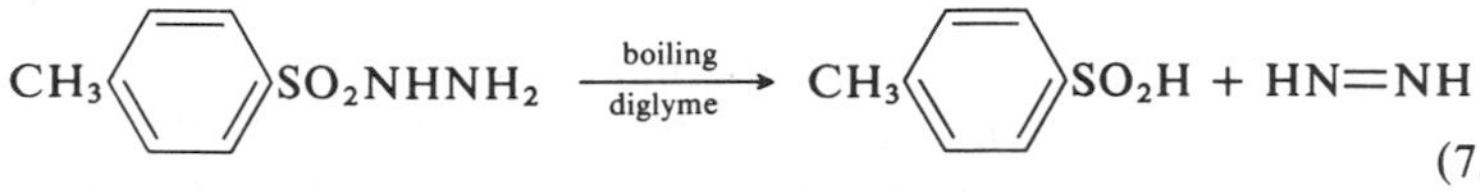

$$CH_3C_6H_4SO_2NHNH_2 \xrightarrow[\text{diglyme}]{\text{boiling}} CH_3C_6H_4SO_2H + HN{=}NH \quad (7.128)$$

$$HO_2CN{=}NCO_2H \rightarrow HN{=}NH + CO_2$$

Di-imide is a highly selective reagent. In general, it is found that symmetrical double bonds such as C≡C, C=C, N=N, O=O are readily reduced, but unsymmetrical, more polar, bonds (C≡N, $N\begin{smallmatrix}\nearrow O \\ \searrow O\end{smallmatrix}$, C=N, S=O, S—S, C—S) are not. For example, under conditions where oleic acid and azobenzene were readily reduced, methyl cyanide, nitrobenzene and dibenzyl sulphide were unaffected, and the selectivity of the reagent is strikingly demonstrated by the reduction of diallyl disulphide to dipropyl disulphide in almost quantitative yield (7.129).

$$(CH_2{=}CHCH_2)_2S_2 \xrightarrow[\text{boiling glycol}]{\text{tosylhydrazide,}} \underset{(93\text{–}100\%)}{(CH_3CH_2CH_2)_2S_2} \qquad (7.129)$$

$$C_6H_5N{=}NC_6H_5 \xrightarrow[\text{boiling methanol}]{\text{azodicarboxylic acid,}} \underset{\text{quantitative}}{C_6H_5NHNHC_6H_5}$$

The reactions are highly stereoselective and take place by *cis*-addition of hydrogen in all cases. Thus, reduction of fumaric acid with tetradeuterohydrazine or with potassium azodicarboxylate and deuterium oxide affords (±)-dideuterosuccinic acid exclusively, while maleic acid gives the *meso*-isomer.

Reduction of acetylenes is also a *cis* process. Thus, partial reduction of diphenylacetylene gave, besides starting material and diphenylethane, only (*Z*)-stilbene; no (*E*)-stilbene was detected.

In sterically hindered molecules, addition takes place to the less hindered side of the double bond, but in examples where steric influences are moderate, much less stereochemical discrimination is observed.

The reactions are regarded as taking place by synchronous transfer of a pair of hydrogen atoms through a cyclic six-membered transition state (7.130). This mechanism explains the high stereospecificity of the reaction, and couples the driving force of nitrogen formation with the addition reaction. Concerted *cis*-transfer of hydrogen is symmetry allowed for the ground-state reaction (Woodward and Hoffman, 1969).

$$\underset{HN{=}NH}{>C{=}C<} \longrightarrow \left[\begin{matrix} >C{\equiv}C< \\ H \qquad H \\ N{\equiv}N \end{matrix}\right] \longrightarrow \underset{N{\equiv}N}{\begin{matrix} >C{-}C< \\ H \quad H \end{matrix}} \qquad (7.130)$$

Reduction with low-valent titanium species

Some synthetically useful reductions with low-valent titanium species have been reported (McMurry, 1974). Titanium(III), conveniently available as titanium chloride, is a mild reducing agent in aqueous solution.

Table 7.6. *Some reductions effected with aqueous Ti(III)*

Reactant	Product
$ArNO_2$	$ArNH_2$
R_2CHNO_2	$R_2CH{=}NH \rightarrow R_2C{=}O$
$R_2S{=}O$	R_2S
$R_3N{\rightarrow}O$	R_3N
2-bromocyclohexanone (O, Br)	cyclohexanone (O)
$CH_3COCH{=}CHCO_2C_2H_5$ (O, $CO_2C_2H_5$)	$CH_3COCH_2CH_2CO_2C_2H_5$ (O, $CO_2C_2H_5$)
bicyclic ene-dione (O, O)	bicyclic dione (O, O)

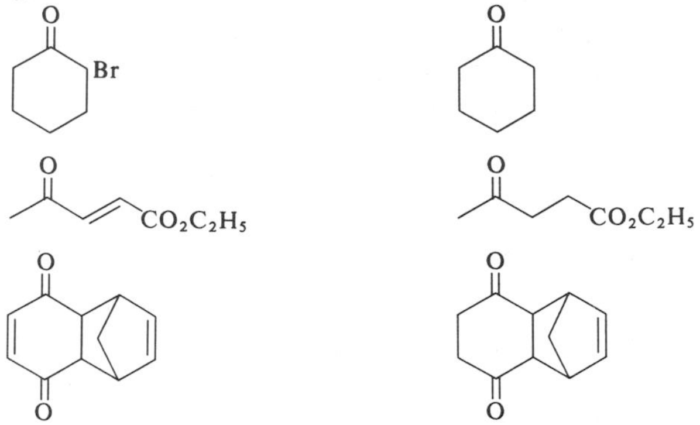

Table 7.6 shows some reductions which can be effected with the reagent. One of the most useful reactions synthetically is the conversion of aliphatic nitro compounds into imines, which are directly hydrolysed to carbonyl compounds under the conditions of the reaction. Aliphatic $-\overset{|}{C}HNO_2$ groups thus become equivalent to carbonyl groups. This greatly extends the usefulness of nitro compounds in synthesis. They are easily converted into anions which serve, for example, as excellent donors in Michael addition to electrophilic alkenes (p. 47), and the sequence Michael addition followed by reaction with titaniumIII chloride provides an excellent route to 1,4-dicarbonyl compounds and thence to cyclopentenones. *cis*-Jasmone, for example was readily obtained as shown in (7.131).

$$CH_3(CH_2)_4CH_2NO_2 \xrightarrow[\text{dimethoxyethane}]{TiCl_3,\ H_2O} CH_3(CH_2)_4CHO \quad (74\%)$$

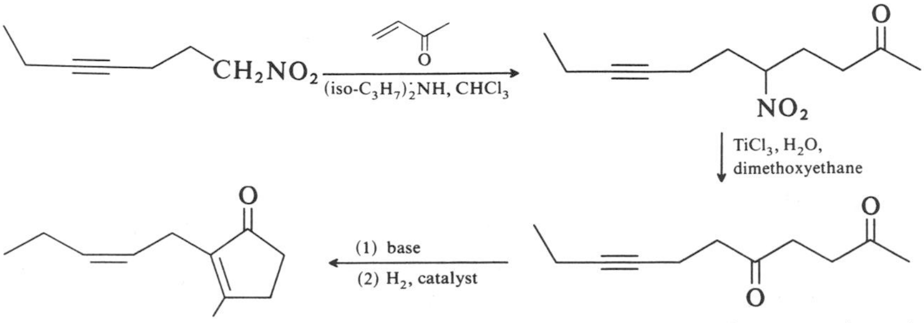

CH_2NO_2 — (O) / $(iso\text{-}C_3H_7)_2NH$, $CHCl_3$ → O, NO_2

↓ $TiCl_3$, H_2O, dimethoxyethane

cis-Jasmone (O) ← (1) base (2) H_2, catalyst — O, O (85%) (7.131)

The reagent is also very effective for reduction of α-halo ketones to the halogen-free compounds, and is more convenient than chromium(II) salts for this, and for the reduction of ene-diones to the saturated diones.

Trialkyltin hydrides

Trialkyltin hydrides (in practice, the commercially available tri-n-butyltin hydride is frequently employed) are useful reagents for the reductive cleavage of carbon–halogen bonds (Kuivila, 1968, 1970). Alkyl, aryl and alkenyl halides are all readily converted into the corresponding halogen-free compounds.

$$R_3SnH + R'X \rightarrow R_3SnX + R'H$$

The reactions proceed by a free-radical chain mechanism and in practice are generally effected in the presence of a free-radical initiator, such as azobisisobutyronitrile or by irradiation with light.

$$\begin{aligned} R_3SnH + In^{\cdot} &\rightarrow R_3Sn^{\cdot} + In{-}H \\ R_3Sn^{\cdot} + R'X &\rightarrow R'^{\cdot} + R_3SnX \\ R'^{\cdot} + R_3SnH &\rightarrow R'H + R_3Sn^{\cdot} \end{aligned} \qquad (7.132)$$

In agreement bromides are reduced more readily than chlorides and for a series of alkyl bromides the order of activity is

tertiary alkyl > secondary > primary

Reactions which proceed through a stabilised radical, such as an allyl or benzyl radical take place particularly readily. Bromoadamantane was converted into adamantane in quantitative yield by irradiation in hexane solution in the presence of tri-n-butyltin hydride and cyclohexyl bromide gave cyclohexane with azo*bis*isobutyronitrile as the initiator. 1,1-Dibromocyclopropanes, readily available by addition of dibromocarbene to alkenes, can be reduced in a stepwise manner to give successively the monobromocyclopropane and the halogen-free compound. With the chlorobromo derivative (95) the chlorine is removed first. A useful synthetic application is in the removal of iodine from iodolactones, themselves commonly prepared from unsaturated acids (7.133).

Simple carbon free-radicals lose their stereochemical identity either because they are planar or undergo extremely rapid inversion. Consequently, reduction of optically active halogen compounds with tin hydrides affords racemic products. Optically active 1-chloro-1-phenylethane and triphenyltin deuteride gave racemic α-deuterioethylbenzene, and in the reduction of (Z)- and (E)-2-bromo-2-butene a mixture of (Z)- and (E)-2-butene was obtained.

(95) [Cl,Br-cyclopropane fused to cyclohexane] $\xrightarrow[0\ ^\circ C,\ (no\ solvent)]{1\ mol.\ (C_4H_9)_3SnH}$ [H, Br product] 2.5 + [Br, H product] 1 (97%)

[1-chloro-4-bromobenzene] $\xrightarrow[150\ ^\circ C]{(C_6H_5)_3SnH}$ [bromobenzene] (75%) (7.133)

[bicyclic acid, H, CH_2CO_2H] $\xrightarrow{I_2,\ NaHCl_3,\ H_2O}$ [iodolactone] $\xrightarrow[(2)\ ^-OH,\ H_2O]{(1)\ (C_4H_9)_3SnH,\ benzene}$ [HO, H, CH_2CO_2H product]

Tributyltin hydride is also an excellent reagent for the replacement of tertiary and some secondary aliphatic nitro groups by hydrogen. Most other commonly encountered functional groups such as keto, ester, cyano or organosulphur groups are unaffected under the reaction conditions. This possibility greatly extends the synthetic applications of nitro compounds which are well known as versatile and reactive partners in alkylation and Michael addition reactions.

$$C_6H_5CH_2NO_2 + CH_2{=}CHCO_2CH_3 \rightarrow C_6H_5CH(NO_2)CH_2CH_2CO_2CH_3 \xrightarrow[\text{azobisisobutyronitrile, benzene, reflux}]{(C_4H_9)_3SnH,} C_6H_5CH_2CH_2CH_2CO_2CH_3 \quad (64\%) \qquad (7.134)$$

Earlier methods for dehalogenation with tin hydrides used stoichiometric amounts of the reagent, and there was often difficulty in separating the product from the trialkyltin halide formed at the same time. In an improved procedure reduction is effected with catalytic tri-n-butyltin hydride in the presence of sodium borohydride, which effects continuous regeneration of the tin hydride from the trialkyltin halide (Corey and Suggs, 1975*b*). Reactions are effected under irradiation with ultraviolet light and take place rapidly at or below room temperatures.

Free radicals produced from alkyl and alkenyl halides with tributyltin hydride undergo synthetically useful intramolecular addition to suitably placed double and triple bonds to form ring compounds, as discussed on p. 97.

Reduction with trialkylsilanes

The addition of Si—H to unsaturated substrates is a useful method of reduction in some cases, and is also an important route to complex organosilanes. Addition can be brought about under catalytic or ionic conditions.

Silanes will reduce a variety of functional groups in the presence of transition metal catalysts (Colvin, 1978; Fleming, 1979). Alkynes undergo *cis* addition to give vinylsilanes, ketones give silyl ethers of the corresponding secondary alcohol and aromatic Schiff bases are readily reduced to secondary amines. A useful reaction is the conversion of acid chlorides into aldehydes, which provides an alternative to the Rosenmunde reduction and reducton with complex hydrides (cf. p. 471) (Citron, 1969) Reaction of $\alpha\beta$-unsaturated aldehydes and ketones with triethylsilane in the presence of $[(C_6H_5)_3P]_3RhCl$ gives the silyl enol ether of the corresponding saturated compound, and thence, on hydrolysis, the saturated carbonyl compound. Isolated carbon–carbon double bonds are unaffected (7.135) (Liu and Browne, 1981).

$$C_6H_5CH{=}NCH_3 \xrightarrow[\text{benzene, } [(C_6H_5)_3P]_3RhCl]{(C_2H_5)_2SiH_2,} C_6H_5CH_2N(CH_3)SiH(C_2H_5)_2 \xrightarrow{CH_3OH} C_6H_5CH_2NHCH_3 \quad (95\%)$$

$$C_7H_{15}COCl \xrightarrow{(C_2H_5)_3SiH,\ Pd{-}C} C_7H_{15}CHO \quad (80\%) \qquad (7.135)$$

O, CO_2CH_3, H $\xrightarrow[[(C_6H_5)_3P]_3RhCl]{(C_2H_5)_3SiH}$ $(C_2H_5)_3SiO$, CO_2CH_3, H $\longrightarrow$ O, CO_2CH_3, H (88%)

Silanes are hydride donors and ionic hydrogenation with silanes involves stepwise addition to the substrate of H^+ and H^-. A combination of triethylsilane and trifluoroacetic acid is frequently used (Kursanov, Parnes and Loin, 1974) and the procedure provides a non-catalytic method for hydrogenation of C=C, C=O and C=N double bonds and for hydrogenolysis of some single bonds such as C—OH, and C—Hal. Alkenes give saturated hydrocarbons, but only if the double bond is at least trisubstituted, allowing the possibility of selective hydrogenation in a compound containing different types of double bond. For example, a mixture of cyclohexene and methylcyclohexene gave only methylcyclohexane; the cyclohexene was unaffected. In trifluoroacetic acid

$$>C=C< \xrightarrow{H^+} >CH-\overset{+}{C}< \xrightarrow{R_3SiH} >CH-CH< \qquad (7.136)$$

alcohols undergo hydrogenolysis with replacement of OH by H, presumably by way of the corresponding carbonium ion, and aromatic imines give secondary amines. Probably the most useful application is in the selective reduction of ketones. Trialkylsilanes selectively reduce the carbonyl group of arylcarbonyl compounds to CH_2. Other potentially reducible groups, NO_2, CN, ester halogen are unaffected. The ketone (96) for example gave the reduced compound (97) in 81 per cent yield under mild conditions (Maurer *et al.*, 1984). Even better results are said to be

$$(96) \xrightarrow[CF_3CO_2H]{(C_2H_5)_3SiH} (97)\ (81\%)$$

$$m\text{-}NO_2C_6H_4COCH_3 \xrightarrow[CH_2Cl_2,\ 0° \rightarrow 25\ °C]{(C_2H_5)_3SiH,\ BF_3} m\text{-}NO_2C_6H_4C_2H_5\ (91\%) \qquad (7.137)$$

obtained with triethylsilane and boron trifluoride. Under these conditions *para*-methoxybenzaldehyde gave *para*-methoxytoluene in quantitative yield, 2-undecanone gave undecane in 80 per cent yield and *meta*-nitroacetophenone was converted into *meta*-nitroethylbenzene at room temperature in 91 per cent yield, without attack on the generally easily reducible nitro group (Fry and Silverman, 1981). Under the same conditions many alcohols are rapidly converted into the corresponding hydrocarbon, without rearrangement (Adlington, Orfanopoulos and Fry, 1976). During the reaction the organosilicon hydride is converted into an organosilicon fluoride and the reaction is facilitated by co-ordination of gaseous boron trifluoride with oxygen and the formation of the very strong silicon–fluorine bond.

References

Abramovitch, R. A. and Davis, B. A. (1964). *Chem. Rev.*, **64**, 149.

Achmatowicz, O. and Pietraszkiewicz, P. (1981). *J. Chem. Soc. Perkin I*, 2680.

Adam, W., Baeza, J. and Liu, J. C. (1972). *J. Am. Chem. Soc.*, **94**, 2000.

Adams, W. R. (1971) in *Oxidation* Vol. 2, pp. 65–112 ed. R. L. Augustine, (New York: Marcel Dekker).

Adlington, R. M., Baldwin, J. E., Bottaro, J. C. and Perry, M. W. D. (1983). *J. Chem. Soc. Chem. Commun.*, 1040.

Adlington, M. G., Orfanopoulos, M. and Fry, J. L. (1976). *Tetrahedron Lett.*, 2955.

Adlington, R. M. and Barrett, A. G. M. (1983). *Acc. Chem. Res.*, **16**, 55.

Ager, D. J. (1982). *Chem. Soc. Rev.*, **11**, 493.

Ager, D. J. (1984). *Synthesis*, 384.

Ager, D. J. and Fleming, I. (1978). *J. Chem. Soc. Chem. Commun.*, 177.

Akashi, K., Palermo, R. E. and Sharpless, K. B. (1978). *J. Org. Chem.*, **43**, 2063.

Akhrem, A. A., Reshetova, I. G. and Titov, Yu. A. (1972). *Birch Reduction of Aromatic Compounds* (New York: Plenum Press).

Akhtar, A. and Barton, D. H. R. (1964). *J. Am. Chem. Soc.*, **86**, 1528.

Albright, J. D. (1983). *Tetrahedron*, **39**, 3207.

Allen, J. A., Boar, R. B., McGhie, J. F. and Barton, D. H. R. (1973). *J. Chem. Soc. Perkin I*, 2402.

Alston, P. V. and Ottenbrite, R. M. (1975). *J. Org. Chem.*, **40**, 111.

Alston, P. V., Ottenbrite, R. M. and Cohen, T. (1978). *J. Org. Chem.* **43**, 1864.

Ando, W. (1977). *Acc. Chem. Res.*, **10**, 179.

Andrews, C. C. and Crawford, T. C. (1980). *Tetrahedron Lett.*, 693, 697.

Ansell, M. J. *et al.* (1971). *J. Chem. Soc.* (*C*), 1401, 1414, 1423, 1429.

ApSimon, J. W. and Seguin, R. P. (1979). *Tetrahedron*, **35**, 2797.

Arco, M. J., Trammell, M. H. and White, J. D. (1976). *J. Org. Chem.*, **41**, 2075.

Arias, L. A., Adkins, S., Nagel, C. J. and Bach, R. D. (1983). *J. Org. Chem.*, **48**, 888.

Arigoni, D., Vasella, A., Sharpless, K. B. and Jensen, H. P. (1973). *J. Am. Chem. Soc.*, **95**, 7917.

Arnold, B. J., Mellows, S. M. and Sammes, P. G. (1973). *J. Chem. Soc. Perkin I*, 1266.

Arnold, B. J., Sammes, P. G. and Wallace, T. W. (1974). *J. Chem. Soc. Perkin I*, 409, 415.

Ashby, E. C., De Priest, R. N., Goel, A. B., Wenderoth, B. and Pham, T. N. (1984). *J. Org. Chem.*, **49**, 3545.

Ashby, E. C., De Priest, R. N., Tuncay, A. and Srivastava, S. (1982). *Tetrahedron Lett.*, **23**, 5251.

Augustine, R. L. (1965). *Catalytic Hydrogenation* (London: Arnold).
Augustine, R. L. (1968). *Reduction* (London: Arnold).

Bach, R. D. and Knight, J. W. (1981). *Organic Syntheses*, **60**, 63.
Bachman, W. E. and Struve, W. C. (1942). *Organic Reactions*, Vol. 1, p. 38. (New York: Wiley).
Back, T. G., Barton, D. H. R., Britten-Kelley, M. R. and Guziec, F. S. (1975). *J. Chem. Soc. Chem. Commun.*, 539.
Baggiolini, E. G., Lee, H. L., Pizzolato, G. and Uskokovic, M. R. (1982). *J. Am. Chem. Soc.*, **104**, 6460.
Bailey, P. S. (1978, 1982). *Ozonisation in Organic Chemistry*, Vol. 1 and Vol. 2 (New York: Academic Press).
Baker, R. (1973). *Chem. Rev.*, **73**, 487.
Baldwin, J. E., Bailey, P. D., Gallacher, G., Singleton, K. A. and Wallace, P. M. (1983). *J. Chem. Soc. Chem. Commun.*, 1049.
Baldwin, J. E. and Bottaro, J. C. (1981). *J. Chem. Soc. Chem. Commun.*, 1121.
Baldwin, J. E., Kelly, D. R. and Ziegler, C. B. (1984). *J. Chem. Soc. Chem. Commun.*, 133.
Baldwin, J. E. and Lopez, R. C. G. (1982). *J. Chem. Soc. Chem. Commun.*, 1029.
Baldwin, J. E. and Walker, J. A. (1972). *J. Chem. Soc. Chem. Commun.*, 354.
Bartlett, P. A. (1980). *Tetrahedron*, **36**, 2.
Bartlett, P. A. and Barstow, J. F. (1982). *J. Org. Chem.*, **47**, 3933.
Bartlett, P. A., Green, F. R. and Webb, T. R. (1977). *Tetrahedron Lett.*, 331.
Bartlett, P. A. and Jernstedt, K. K. (1977). *J. Am. Chem. Soc.*, **99**, 4829.
Bartlett, P. A. and Myerson, J. (1978). *J. Am. Chem. Soc.*, **100**, 3950.
Bartlett, P. A., Tanzella, D. J. and Barstow, J. F. (1982). *Tetrahedron Lett.*, **23**, 619.
Bartlett, P. D. (1970). *Quart. Rev. Chem. Soc. Lond.*, **24**, 473.
Bartlett, P. D. (1971). *Pure and Appl. Chem.*, **27**, 597.
Barton, D. H. R. (1953). *J. Chem. Soc.*, 1027, Note 23.
Barton, D. H. R. (1959). *Helv. Chim. Acta*, **42**, 2604.
Barton, D. H. R., Beaton, J. M., Geller, L. E. and Pechet, M. M. (1961). *J. Am. Chem. Soc.*, **83**, 4076.
Barton, D. H. R. and Crich, D. (1984*a*). *J. Chem. Soc. Chem. Commun.*, 774.
Barton, D. H. R. and Crich, D. (1984*b*). *Tetrahedron Lett.*, **25**, 2787.
Barton, D. H. R., Gastiger, M. J. and Motherwell, W. B. (1983). *J. Chem. Soc. Chem. Commun.*, 41, 731.
Barton, D. H. R., Guziec, F. S. and Shahak, I. (1974). *J. Chem. Soc. Perkin I*, 1794.
Barton, D. H. R., Haynes, R. K., Leclerc, G., Magnus, P. D. and Menzies, I. D. (1975). *J. Chem. Soc. Perkin I*, 2055.
Barton, D. H. R., Hui, R. A. H. F. and Ley, S. V. (1982). *J. Chem. Soc. Perkin I*, 2179.
Barton, D. H. R., Kitchin, J. P., Lester, D. L., Motherwell, W. B. and Papoula, M. T. B. (1981). *Tetrahedron*, **37**, Suppl. 9, 73.
Barton, D. H. R., Lamotte, G., Motherwell, W. B. and Narang, S. C. (1979). *J. Chem. Soc. Perkin I*, 2030.
Barton, D. H. R. and McCombie, S. W. (1975). *J. Chem. Soc. Perkin I*, 1574.
Barton, D. H. R., Motherwell, R. S. H. and Motherwell, W. B. (1981). *J. Chem. Soc. Perkin I*, 2363.
Barton, D. H. R., Motherwell, R. S. H. and Motherwell, W. B. (1983). *Tetrahedron Lett.*, **24**, 1979.
Barton, D. H. R., Motherwell, W. B. and Stange, A. (1981). *Synthesis*, 743.

Barton, D. H. R. and Robinson, C. H. (1954). *J. Chem. Soc.*, 3045.
Battersby, A. R. (1984). *Chemistry in Britain*, 611.
Battersby, A. R. and Staunton, J. (1974). *Tetrahedron*, **30**, 1707.
Beak, P. and Reitz, D. B. (1978). *Chem. Rev.*, **78**, 275.
Becker, K. B. (1980). *Tetrahedron*, **36**, 1717.
Beckwith, A. L. J. (1981). *Tetrahedron*, **37**, 3073.
Beckwith, A. L. J. and Duong, T. (1979). *J. Chem. Soc. Chem. Commun.*, 690.
Beckwith, A. L. J., Lawrence, T. and Serelis, A. K. (1980). *J. Chem. Soc. Chem. Commun.*, 484.
Bednarski, M., Maring, C. and Danishefsky, S. (1983). *Tetrahedron Lett.*, **24**, 3451.
Behan, J. M., Johnstone, R. A. W. and Wright, M. J. (1975). *J. Chem. Soc. Perkin I*, 1216.
Behrens, C. H. and Sharpless, K. B. (1983). *Aldrichimica Acta*, **16**, 67.
Belleville, D. J. and Bauld, N. L. (1982). *J. Am. Chem. Soc.*, **104**, 2665.
Bergstein, W. and Martens, J. (1981). *Synthesis*, 76.
Berson, J. A., Hamlet, Z. and Mueller, W. A. (1962). *J. Am. Chem. Soc.*, **84**, 297.
Berson, J. A., Wall, R. G. and Perlmutter, H. D. (1966). *J. Am. Chem. Soc.*, **88**, 187.
Berti, G. (1973). In *Topics in Stereochemistry*, Vol. 7, ed. N. L. Allinger and E. L. Eliel (London: Wiley).
Bestman, H. J., Armsen, R. and Wagner, H. (1969). *Chem. Ber.*, **102**, 2259.
Bhalerao, U. T. and Rapoport, H. (1971). *J. Am. Chem. Soc.*, **93**, 4835, 5311.
Biellmann, J. F. (1968). *Bull. Soc. Chim. Fr.*, 3055.
Billington, D. C. (1985). *Chem. Soc. Rev.*, **14**, 93.
Birch, A. J. (1959). *Quart. Rev. Chem. Soc. 4*, 69.
Birch, A. J., Hinde, A. L. and Radom, L. (1980). *J. Am. Chem. Soc.*, **102**, 3370, 4074, 6430, See also *ibid*, **103**, 284.
Birch, A. J. and Nasipuri, D. (1959). *Tetrahedron*, **6**, 148.
Birch, A. J. and Slobbe, J. (1976*a*). *Tetrahedron Lett.*, 2079.
Birch, A. J. and Slobbe, J. (1976*b*). *Heterocycles*, **5**, 905.
Birch, A. J. and Smith, H. (1958). *Quart. Rev. Chem. Soc. Lond.*, **12**, 17.
Birch, A. J. and Subba Rao, G. (1972). In *Advances in Organic Chemistry: Methods and Results*, Vol. 8, p. 1, ed. E. C. Taylor (New York: Wiley-Interscience).
Blackburn, E. V. and Timmons, C. J. (1969). *Quart. Rev. Chem. Soc. Lond.*, **23**, 482.
Blatcher, P., Grayson, J. I. and Warren, S. (1978). *J. Chem. Soc. Chem. Commun.*, 657.
Bloch, R. and Abecassis, J. (1982). *Tetrahedron Lett.*, **23**, 3277.
Bloch, R. and Abecassis, J. (1983). *Tetrahedron Lett.*, **24**, 1247.
Block, E. (1978). *Reactions of Organosulphur Compounds.* (New York: Academic Press).
Bloomfield, J. J. (1968). *Tetrahedron Lett.*, 587.
Bloomfield, J. J., Owsley, D. C., Ainsworth, C. and Robertson, R. E. (1975). *J. Org. Chem.*, **40**, 393.
Blumenkopf, T. A. and Heathcock, C. H. (1983). *J. Am. Chem. Soc.*, **105**, 2354.
Bluthe, N., Malacria, M. and Gore, J. (1982). *Tetrahedron Lett.*, **23**, 4263.
Bluthe, N., Malacria, M. and Gore, J. (1983). *Tetrahedron Lett.*, **24**, 1157.
Bluthe, N., Malacria, M. and Gore, J. (1984). *Tetrahedron*, **40**, 3277.
Boeckman, R. K. (1983). *Tetrahedron*, **39**, 925.
Boeckman, R. K. and Alessi, T. R. (1982). *J. Am. Chem. Soc.*, **104**, 3216.
Boeckman, R. K. and Bruza, K. T. (1974). *Tetrahedron Lett.*, 3365.
Boeckman, R. K. and Demko, D. M. (1982). *J. Org. Chem.*, **47**, 1789.
Boeckman, R. K. and Ganem, B. (1974). *Tetrahedron Lett.*, 913.

Boeckman, R. K. and Ko, S. S. (1982). *J. Am. Chem. Soc.*, **104**, 1033.
Bogdanović, B. (1973). *Agnew. Chem. internat. edn*, **12**, 954.
Bogdanowicz, M. J. and Trost, B. M. (1974). *Organic Syntheses*, **54**, 27.
Boger, D. L. (1983). *Tetrahedron*, **39**, 2869.
Bongini, A., Cardillo, G., Orena, M., Porzi, G. and Sandri, S. (1982). *J. Org. Chem.*, **47**, 4626.
Bosworth, N. and Magnus, P. D. (1972). *J. Chem. Soc. Perkin I*, 943.
Boyd, J., Epstein, W. and Frater, G. (1976). *J. Chem. Soc. Chem. Commun.*, 380.
Breslow, R. and Heyer, D. (1982). *J. Am. Chem. Soc.*, **104**, 2045.
Breslow, R., Corcoran, R. J., Snider, B. B., Doll, R. J., Khanna, P. L. and Kaleya, R. (1977). *J. Am. Chem. Soc.*, **99**, 905.
Breslow, R., Baldwin, S., Flechtner, T., Kalicky, P., Liu, S. and Washburn, W. (1973). *J. Am. Chem. Soc.*, **95**, 3251.
Breslow, R. (1972). *Chem. Soc. Rev.*, **1**, 553.
Briger, G. and Bennett, J. N. (1980). *Chem. Rev.*, **80**, 63.
Bridges, A. J. and Whitham, G. H. (1974). *J. Chem. Soc. Chem. Commun.*, 142.
Brooks, D. W., Grothaus, P. G. and Irwin, W. L. (1982). *J. Org. Chem.*, **47**, 2820.
Brown, H. C. (1972). *Boranes in Organic Chemistry.* (Ithaca: Cornell University Press).
Brown, H. C. (1981). In *Organic Synthesis Today and Tomorrow.* ed. B. M. Trost and C. R. Hutchinson, Oxford: Pergamon, p. 121.
Brown, H. C. (1983). In *Current Trends in Organic Synthesis.* ed. H. Nozaki, Pergamon, p. 247.
Brown, H. C., Basavaiah, D. and Kulkarni, S. U. (1982). *J. Org. Chem.*, **47**, 171.
Brown, H. C. and Basavaiah, D. (1982). *J. Org. Chem.*, **47**, 3806.
Brown, H. C. and Brown, C. A. (1962). *J. Am. Chem. Soc.*, **84**, 2827.
Brown, H. C. and Chandrasekharan, J. (1983). *J. Org. Chem.*, **48**, 644.
Brown, H. C., Choi, Y. M. and Narasimhan, S. (1982). *J. Org. Chem.*, **47**, 3153.
Brown, H. C., Desai, M. C. and Jadhav, P. K. (1982). *J. Org. Chem.*, **47**, 5065.
Brown, H. C., Ford, T. M. and Hubbard, J-L. (1980). *J. Org. Chem.*, **45**, 4067.
Brown, H. C. and Imai, T. (1983). *J. Am. Chem. Soc.*, **105**, 6285.
Brown, H. C., Jadhav, P. K. and Desai, M. C. (1982). *J. Am. Chem. Soc.*, **104**, 6844.
Brown, H. C., Jadhav, P. K. and Desai, M. C. (1984). *Tetrahedron*, **40**, 1325.
Brown, H. C., Kramer, G. W., Levy, A. B. and Midland, M. M. (1975). In *Organic Syntheses via Boranes.* (New York: Wiley).
Brown, H. C. and Krishnamurthy, S. (1969). *J. Org. Chem.*, **34**, 3918.
Brown, H. C. and Krishnamurthy, S. (1979). *Tetrahedron*, **35**, 567.
Brown, H. C., Mandal, A. K. and Kulkarni, S. M. (1977). *J. Org. Chem.*, **42**, 1392.
Brown, H. C., Mandal, A. K., Yoon, N. M., Singaram, B., Schwier, J. R. and Jadhav, P. K. (1982*a*). *J. Org. Chem.*, **47**, 5069, 5074.
Brown, H. C. and Midland, M. M. (1972). *Angew. Chem. internat. edn*, **11**, 692.
Brown, H. C., Midland, M. M. and Levy, A. B. (1972). *J. Am. Chem. Soc.*, **94**, 3662.
Brown, H. C. and Molander, G. A. (1981). *J. Org. Chem.*, **46**, 645.
Brown, H. C., Nambu, H. and Rogic, M. (1969). *J. Am. Chem. Soc.*, **91**, 6852.
Brown, H. C. and Pai, G. G. (1982). *J. Org. Chem.*, **47**, 1606.
Brown, H. C., Ravindran, N. and Kulkarni, S. U. (1979). *J. Org. Chem.*, **44**, 2417.
Brown, H. C., Sikorski, J. A., Kulkarni, S. U. and Lee, H. D. (1982*b*). *J. Org. Chem.*, **47**, 863.
Brown, H. C. and Singaram, B. (1984). *J. Am. Chem. Soc.*, **106**, 1797.
Brownbridge, P. (1983). *Synthesis*, 1.
Bryce-Smith, D., Deshpande, R. R. and Gilbert, A. (1975). *Tetrahedron Lett.*, 1627.

Bryson, T. A. and Donelson, D. M. (1977). *J. Org. Chem.*, **42**, 2930.
Buchanan, J. G. St.C. and Woodgate, P. D. (1969). *Quart. Rev. Chem. Soc. Lond.*, **23**, 522.
Buchi, G. and Chu, P-S. (1979). *J. Am. Chem. Soc.*, **101**, 6767. See also Buchi, G. and Chu, P-S. (1978). *J. Org. Chem.*, **43**, 3717.
Burford, C., Cooke, F., Roy, G. and Magnus, P. (1983). *Tetrahedron*, **39**, 867.
Burke, S. D. and Grieco, P. A. (1979). *Organic Reactions*, **26**, 361.
Buse, C. T. and Heathcock, C. H. (1978). *Tetrahedron Lett.*, 1685.
Buss, A. D. and Warren, S. (1981). *J. Chem. Soc. Chem. Commun.*, 100.
Butler, R. N. (1977). In *Synthetic Reagents*. Ed. J. S. Pizey, Vol. 3 (London: Wiley).
Butz, L. W. and Rytina, A. W. (1949). *Organic Reactions*, **5**, 136.

Cacchi, S., La Torre, F. and Misiti, D. (1979). *Synthesis*, 356.
Cagliotti, L. (1972). *Organic Syntheses*, **52**, 122.
Caine, D. (1976). *Organic Reactions*, **23**, 1.
Cambie, R. C., Hayward, R. C., Roberts, J. L. and Rutledge, P. S. (1974). *J. Chem. Soc. Perkin I*, 1858.
Campbell, J. B. and Brown, H. C. (1980). *J. Org. Chem.*, **45**, 549.
Čaplar, V., Comisso, G. and Sunjić, V. (1981). *Synthesis*, 85.
Caputo, R., Mangoni, L., Neri, O. and Palumbo, G. (1981). *Tetrahedron Lett.*, **22**, 3551.
Carless, H. A. J., Atkins, R. and Fekarurhobo, G. K. (1985). *Tetrahedron Lett.*, **26**, 803.
Carlson, P. H. J., Katsuki, T., Martin, V. S. and Sharpless, K. B. (1981). *J. Org. Chem.*, **46**, 3936.
Carr, R. V. C. and Paquette, L. A. (1980). *J. Am. Chem. Soc.*, **102**, 853.
Carruthers, W. (1982). In *Comprehensive Organometallic Chemistry*, eds Wilkinson, G., Stone, F. G. A. and Abel, E. W. Vol. 7, p. 716 (Oxford: Pergamon Press).
Carter, M. J., Fleming, I. and Percival, A. (1981). *J. Chem. Soc. Perkin I*, 2415.
Čeković, Z. and Green, M. M. (1974). *J. Am. Chem. Soc.*, **96**, 3000.
Čeković, Z. and Srnic, T. (1976). *Tetrahedron Lett.*, 561.
Cha, J. K., Christ, W. J. and Kishi, Y. (1983). *Tetrahedron Lett.*, **24**, 3943, 3947.
Chamberlin, A. R., Dezube, M. and Dussault, P. (1981). *Tetrahedron Lett.*, **22**, 4611.
Chamberlin, A. R., Liotta, E. L. and Bond, F. T. (1983). *Organic Syntheses*, **61**, 141.
Chamberlin, A. R., Stemke, J. E. and Bond, F. T. (1978). *J. Org. Chem.*, **43**, 147.
Chan, T. H. and Fleming, I. (1979). *Synthesis*, 761.
Chan, T. H., Paterson, I. and Pinsonnault, J. (1977). *Tetrahedron Lett.*, 4183.
Chapleo, C. B., Hallett, P., Lythgoe, B., Waterhouse, I. and Wright, P. W. (1977). *J. Chem. Soc. Perkin I*, 1211, 1218.
Cherest, M., Felkin, H. and Prudent, N. (1968). *Tetrahedron Let.*, 2199.
Chmielewski, M. and Jurczak, J. (1981). *J. Org. Chem.*, **46**, 2230.
Chong, A. O., Oshima, K. and Sharpless, K. B. (1977). *J. Am. Chem. Soc.*, **99**, 3420.
Chow, H-F. and Fleming, I. (1984). *J. Chem. Soc. Perkin I*, 1815.
Choy, W., Reed, L. A. and Masamune, S. (1983). *J. Org. Chem.*, **48**, 1137.
Ciganek, E. (1967). *Tetrahedron Lett.*, 3321.
Citron, J. D. (1969). *J. Org. Chem.*, **34**, 1977.
Clark, R. D. and Heathcock, C. H. (1973). *J. Org. Chem.*, **38**, 3658.
Clive, D. L. J. (1978). *Tetrahedron*, **34**, 1049.
Clive, D. L. J. and Denyer, C. V. (1973). *J. Chem. Soc. Chem. Commun.*, 253.

Clough, J. M. and Pattenden, G. (1981). *Tetrahedron*, **37**, 3911.
Coates, R. M., Shah, S. K. and Mason, R. W. (1982). *J. Am. Chem. Soc.*, **104**, 2198.
Cohen, T. and Kosarych, Z. (1982). *J. Org. Chem.*, **47**, 4005.
Collins, J. C., Hess, W. W. and Frank, F. J. (1968). *Tetrahedron Lett.*, 3363.
Colonge, J. and Descotes, G. (1967). In *1,4-Cycloaddition Reactions*, ed. J. Hamer (New York: Academic Press).
Colonna, F. P., Fattuta, S., Resaliti, A. and Russo, C. (1970). *J. Chem. Soc. (C)*, 2377.
Colvin, E. W. (1978). *Chem. Soc. Rev.*, **7**, 15.
Colvin, E. W., Beck, A. K. and Seebach, D. (1981). *Helv. Chim. Acta*, **64**, 2264.
Concepción, J. I., Francisco, C. G., Hernández, J. A., Salazar, J. A. and Suárez, E. (1984). *Tetrahedron Lett.*, **25**, 1953.
Conia, J-M. and Blanco, L. (1983). *Current Trends in Organic Synthesis.* ed. H. Nozaki, p. 331, (Oxford: Pergamon).
Conia, J-M. and Le Perchec, P. (1975). *Synthesis*, 1.
Cook, A. G. (1968). Ed. *Enamines: Their Synthesis, Structure and Reactions.* (New York: Marcel Dekker).
Cooke, F. and Magnus, P. (1977). *J. Chem. Soc. Chem. Commun.*, 513.
Cooke, F., Schwindemann, J. and Magnus, P. (1979). *Tetrahedron Lett.*, 1995.
Cope, A. C. and Trumbull, E. R. (1960). *Organic Reactions*, **11**, 317.
Corey, E. J., Arnett, J. F. and Widiger, G. N. (1975). *J. Am. Chem. Soc.*, **97**, 430.
Corey, E. J., Crouse, D. N. and Anderson, T. E. (1975). *J. Org. Chem.*, **40**, 2140.
Corey, E. J., Danheiser, R. L. and Chandrasekaran, S. (1976). *J. Org. Chem.*, **41**, 260.
Corey, E. J. and Enders, D. (1976). *Tetrahedron Lett.*, 11.
Corey, E. J. and Engler, T. A. (1984). *Tetrahedron Lett.*, **25**, 149.
Corey, E. J., Gilman, N. W. and Ganem, B. E. (1968). *J. Am. Chem. Soc.*, **90**, 5616.
Corey, E. J. and Hertler, W. R. (1959). *J. Am. Chem. Soc.*, **81**, 5209.
Corey, E. J. and Hertler, W. R. (1960). *J. Am. Chem. Soc.*, **82**, 1657.
Corey, E. J. and Hopkins, P. B. (1982). *Tetrahedron Lett.*, **23**, 1979.
Corey, E. J., Kang, J. and Kyler, K. (1985). *Tetrahedron Lett.*, **26**, 555.
Corey, E. J., Katzenellenbogen, J. A., Gilman, N. W., Roman, S. A. and Erickson, B. W. (1968). *J. Am. Chem. Soc.*, **90**, 5618.
Corey, E. J., Katzenellenbogen, J. A. and Posner, G. H. (1967). *J. Am. Chem. Soc.*, **89**, 4245.
Corey, E. J. and Kim, C. U. (1972). *J. Am. Chem. Soc.*, **94**, 7587.
Corey, E. J. and Kim, C. U. (1974). *Tetrahedron Lett.*, 287.
Corey, E. J. and Kwiatkowsky, G. T. (1968). *J. Am. Chem. Soc.*, **90**, 6816.
Corey, E. J. and Lansbury, P. T. (1983). *J. Am. Chem. Soc.*, **105**, 4093.
Corey, E. J., Marfat, A., Falck, J. R. and Albright, J. O. (1980). *J. Am. Chem. Soc.*, **102**, 1433.
Corey, E. J., Marfat, A. and Hoover, J. (1981). *Tetrahedron Lett.*, **22**, 1587.
Corey, E. J., Mitra, R. B. and Uda, U. (1964). *J. Am. Chem. Soc.*, **86**, 485.
Corey, E. J. and Schmidt, G. (1979). *Tetrahedron Lett.*, 399.
Corey, E. J., Schmidt, G. and Shimoji, K. (1983). *Tetrahedron Lett.*, **24**, 3169.
Corey, E. J. and Suggs, J. W. (1975*a*). *Tetrahedron Lett.*, 2647.
Corey, E. J. and Suggs, J. W. (1975*b*). *J. Org. Chem.*, **40**, 2554.
Corey, E. J. and Venkateswarlu, A. (1972). *J. Am. Chem. Soc.*, **94**, 6190.
Cornforth, J. W., Cornforth, R. H. and Mathew, K. K. (1959). *J. Chem. Soc.*, 112.
Cossey, A. L., Gunter, M. J. and Mander, L. N. (1980). *Tetrahedron Lett.*, **21**, 3309.
Cowell, G. W. and Ledwith, A. (1970). *Quart. Rev. Chem. Soc. Lond.*, **24**, 119.

Cox, M. T., Heaton, D. W. and Horbury, J. (1980). *J. Chem. Soc. Chem. Commun.*, 799.
Crabtree, R. H. and Morris, G. E. (1977). *J. Organomet. Chem.*, **135**, 395.
Cragg, G. L. (1973). *Organoboranes in Organic Synthesis* (New York: Marcel Dekker).
Cram, D. J. and Abd Elhafez, F. A. (1952). *J. Am. Chem. Soc.*, **74**, 4828, 5828, 5851.
Cram, D. J., Sahyun, M. R. V. and Knox, G. R. (1962). *J. Am. Chem. Soc.*, **84**, 1734.
Crandall, J. K. and Luan-Ho Chang. (1967). *J. Org. Chem.*, **32**, 435.
Criegee, R. (1975). *Agnew. Chem. internat. edn*, **14**, 745.
Crumbie, R. L., Ridley, D. D. and Simpson, G. W. (1977). *J. Chem. Soc. Chem. Commun.*, 315.
Curran, D. P. (1982). *J. Am. Chem. Soc.*, **104**, 4024.

Danishefsky, S. (1981). *Acc. Chem. Res.*, **14**, 400.
Danishefsky, S., Chackalamannil, S. and Uang, B-J. (1982). *J. Org. Chem.*, **47**, 2231.
Danishefsky, S., Funk, R. L. and Kerwin, J. F. (1980). *J. Am. Chem. Soc.*, **102**, 6889, 6891.
Danishefsky, S., Harayama, T. and Singh, R. K. (1979). *J. Am. Chem. Soc.*, **101**, 7008.
Danishefsky, S. and Hershenson, F. M. (1979). *J. Org. Chem.*, **44**, 1180.
Danishefsky, S., Kato, N., Askin, D. and Kerwin, J. F. (1982). *J. Am. Chem. Soc.*, **104**, 360.
Danishefsky, S. and Kerwin, J. F. (1982). *J. Org. Chem.*, **47**, 3803, 3183.
Danishefsky, S., Kerwin, J. F. and Kobayashi, S. (1982). *J. Am. Chem. Soc.*, **104**, 358.
Danishefsky, S., Kobayashi, S. and Kerwin, J. F. (1982). *J. Org. Chem.*, **47**, 1981.
Danishefsky, S., Morris, J. and Clizbe, L. A. (1981). *J. Am. Chem. Soc.*, **103**, 1602.
Danishefsky, S., Singh, R. K. and Gammill, R. B. (1978). *J. Org. Chem.*, **43**, 379.
Daub, G. W., Sanchez, M. G., Cromer, R. A. and Gibson, L. L. (1982). *J. Org. Chem.*, **47**, 743.
Dauben, W. G., Kessel, C. R. and Takemura, K. H. (1980). *J. Am. Chem. Soc.*, **102**, 6893.
Dauben, W. G. and Krabbenhoft, H. O. (1977). *J. Org. Chem.*, **42**, 282.
David, S. and Eustache, J. (1979). *J. Chem. Soc. Perkin I*, 2230.
Davis, A. P. and Whitham, G. H. (1980). *J. Chem. Soc. Chem. Commun.*, 639.
Deem, M. L. (1972). *Synthesis*, 675.
de Mayo, P. (1971). *Acc. Chem. Res.*, **4**, 41.
Demole, E., Demole, C. and Berthet, D. (1973). *Helv. Chim. Acta*, **56**, 265.
Demoulin, A., Gorissen, H., Hesbain-Prizque, A-M. and Ghosez, L. (1975). *J. Am. Chem. Soc.*, **97**, 4409.
Denmark, S. E. and Jones, T. K. (1982). *J. Am. Chem. Soc.*, **104**, 2642.
Denny, R. W. and Nickon, A. (1973). *Organic Reactions*, **20**, 133.
Deuchert, K., Hertenstein, U., Hürig, S. and Wehner, G. (1979). *Chem. Ber.*, **112**, 2045.
Dewar, M. J. S. (1971). *Angew. Chem. internat. edn*, **10**, 761.
Dewar, M. J. S. (1984). *J. Am. Chem. Soc.*, **106**, 209.
Dewar, M. J. S. and Pierini, A. B. (1984). *J. Am. Chem. Soc.*, **106**, 203.
Dilling, W. L. (1966). *Chem. Rev.*, **66**, 373.
Djerassi, C. (1951). *Organic Reactions*, **6**, 207.
Djerassi, C. (1963). *Steroid Reactions*, p. 267. (San Francisco: Holden-Day, Inc.).

Doering, W. von E., Birlandeanu, L., Andrews, D. W. and Pagnotta, M. (1985). *J. Am. Chem. Soc.*, **107**, 428.
Dubs, P. and Stüssi, R. (1978). *Helv. Chim. Acta*, **61**, 990.
Duncia, J. V., Lansbury, P. T., Miller, T. and Snider, B. (1982). *J. Am. Chem. Soc.*, **104**, 1930.
Dupius, J., Giese, B., Rüegge, D., Fischer, H., Korth, H-G. and Sustmann, R. (1984). *Agnew. Chem. internat. edn*, **23**, 896.

Eaborn, C., Pant, B. C., Peeling, E. R. A. and Taylor, S. C. (1969). *J. Chem. Soc.* (*C*), 2823.
Eastmond, R., Johnson, T. R. and Walton, D. R. M. (1972). *Tetrahedron*, **28**, 4601.
Eaton, P. E. (1968). *Acc. Chem. Res.*, **1**, 50.
Edwards, M. P., Ley, S. V., Lister, S. G., Palmer, B. D. and Williams, D. J. (1984). *J. Org. Chem.*, **49**, 3503.
Ehlinger, E. and Magnus, P. (1980). *J. Am. Chem. Soc.*, **102**, 5004.
Emde, H., Domsch, D., Feger, H., Frick, U., Götz, A., Hergott, H., Hofmann, K., Kober, W., Krägeloh, K., Oesterle, T., Steppan, W., West, W. and Simchen, G. (1982). *Synthesis*, 1.
Enders, D., Eichenauer, H., Baus, U., Schubert, H. and Kremer, K. A. M. (1984). *Tetrahedron*, **40**, 1345.
Epstein, W. W. and Sweat, F. W. (1967). *Chem. Rev.*, **67**, 247.
Erdik, E. (1984). *Tetrahedron*, **40**, 641.
Eschenmoser, A. (1970). *Quart. Rev. Chem. Soc. Lond.*, **24**, 366.
Evans, D. A. and Andrews, G. C. (1974). *Acc. Chem. Res.*, **7**, 147.
Evans, D. A. and Bartroli, J. (1982). *Tetrahedron Lett.*, **23**, 807.
Evans, D. A., Bartroli, J. and Godel, T. (1982). *Tetrahedron Lett.*, **23**, 4577.
Evans, D. A., Bartroli, J. and Shih, T. L. (1981). *J. Am. Chem. Soc.*, **103**, 2127.
Evans, D. A., Bryan, C. A. and Sims, C. L. (1972). *J. Am. Chem. Soc.*, **94**, 2891.
Evans, D. A., Carroll, G. L. and Truesdale, L. K. (1974). *J. Org. Chem.*, **39**, 914.
Evans, D. A., Ennis, M. D. and Mathre, D. J. (1982). *J. Am. Chem. Soc.*, **104**, 1737.
Evans, D. A. and Golob, A. M. (1975). *J. Am. Chem. Soc.*, **97**, 4765.
Evans, D. A. and Hoffmann, J. M. (1976). *J. Am. Chem. Soc.*, **98**, 1983.
Evans, D. A. and McGee, L. R. (1981). *J. Am. Chem. Soc.*, **103**, 2876.
Evans, D. A. and Morrissey, M. M. (1984). *J. Am. Chem. Soc.*, **106**, 3866.
Evans, D. A. and Nelson, J. V. (1980). *J. Am. Chem. Soc.*, **102**, 774.
Evans, D. A., Nelson, J. V. and Taber, T. R. (1982). In *Topics in Stereochemistry*, Vol. 13, p. 1, eds N. L. Allinger, E. L. Eliel and S. H. Wilen (London: Interscience).
Evans, D. A., Nelson, J. V., Vogel, E. and Taber, T. R. (1981). *J. Am. Chem. Soc.*, **103**, 3099.
Evans, D. A., Scott, W. L. and Truesdale, L. K. (1972). *Tetrahedron Lett.*, 121.
Evans, D. A. and Takacs, J. M. (1980). *Tetrahedron Lett.*, **21**, 4233.
Evans, R. M. (1959). *Quart. Rev. Chem. Soc. Lond.*, **13**, 61.

Fabre, J-L. and Julia, M. (1983). *Tetrahedron Lett.*, **24**, 4311.
Fabre, J-L., Julia, M. and Verpeaux, J-N. (1982). *Tetrahedron Lett.*, **23**, 2469.
Fallis, A. G. (1984). *Canad. J. Chem.*, **62**, 183.
Felix, D., Gschwend-Steen, K., Wick, A. E. and Eschenmoser, A. (1969). *Helv. Chim. Acta*, **52**, 1030.
Fétizon, M. and Golfier, M. (1968). *C.r. hebd. Séanc. Acad. Sci., Paris*, **267** (C), 900.

Fiaud, J-C. and Malleron, J-C. (1981). *J. Chem. Soc. Chem. Commun.*, 1159.
Fieser, L. F. and Fieser, M. (1967). *Reagents for Organic Synthesis*, Vol. 1, p. 759 (London: Wiley).
Fleming, I. (1976). *Frontier Orbitals and Organic Chemical Reactions* (London: Wiley).
Fleming, I. (1979). In *Comprehensive Organic Chemistry*, Vol. 3, p. 572 (Oxford: Pergamon).
Fleming, I. (1981). *Bull. Soc. Chim. France*, (II), 7; *Chem. Soc. Rev.* **10**, 83.
Fleming, I. and Goldhill, J. (1980). *J. Chem. Soc. Perkin I*, 1493.
Fleming, I., Iqbal, J. and Krebs, E-P. (1983). *Tetrahedron*, **39**, 841.
Fleming, I. and Kargar, M. H. (1967). *J. Chem. Soc.* (*C*), 226.
Fleming, I. and Michael, J. P. (1981). *J. Chem. Soc. Perkin I*, 1549.
Fleming, I. and Paterson, I. (1979). *Synthesis*, 736.
Fleming, I. and Pearce, A. (1981). *J. Chem. Soc. Perkin I*, 251.
Fleming, I. and Terrett, N. K. (1983). *Tetrahedron Lett.*, **24**, 4151.
Fleming, I. and Waterson, D. (1984). *J. Chem. Soc. Perkin I*, 1809.
Foote, C. S. (1968). *Acc. Chem. Res.*, **1**, 104.
Foote, C. S. (1971). *Pure and Appl. Chem.*, **27**, 635.
Fortunato, J. M. and Ganem, B. (1976). *J. Org. Chem.*, **41**, 2194.
Franck, R. W., John, T. V., Olejniczak, K. and Blount, J. F. (1982). *J. Am. Chem. Soc.*, **104**, 1106.
Fräter, G. (1979). *Helv. Chim. Acta*, **62**, 2825, 2829.
Fry, J. L. and Silverman, S. B. (1981). *Organic Synthesis*, **60**, 108.
Fuks, R. and Viehe, H. G. (1969). In *Chemistry of Acetylenes* ed. H. G. Viehe, p. 477 (New York: Marcel Dekker).
Fukuyama, T., Vranesic, B., Negri, D. P. and Kishi, Y. (1978). *Tetrahedron Lett.*, 2741.
Funk, R. L. and Vollhardt, K. P. C. (1979). *J. Am. Chem. Soc.*, **101**, 215.
Funk, R. L. and Vollhardt, K. P. C (1980). *Chem. Soc. Rev.*, **9**, 41.

Ganem, B. and Boeckman, R. K. (1974). *Tetrahedron Lett.*, 917.
Gardner, J. N., Carlon, F. E. and Gnoj, O. (1968). *J. Org. Chem.*, **33**, 3294.
Gassman, P. G. *et al.* (1974). *J. Am. Chem. Soc.*, **96**, 3002, 5487, 5495, 5512.
Gaylord, N. G. (1956). *Reduction with Complex Metal Hydrides* (New York: Wiley).
Gemal, A. L. and Luche, J-L. (1979). *J. Org. Chem.*, **44**, 4187.
Gemal, A. L. and Luche, J-L. (1981). *J. Am. Chem. Soc.*, **103**, 5454.
Genet, J. P. and Ficini, J. (1979). *Tetrahedron Lett.*, 1499.
Giering, W. P., Rosenblum, M. and Tancredi, J. (1972). *J. Am. Chem. Soc.*, **94**, 7170.
Giese, B. (1983). *Angew. Chem. internat. edn*, **22**, 753.
Giese, B. and Dupuis, J. (1983). *Agnew. Chem. internat. edn*, **22**, 622.
Giese, B. and Heuck, K. (1979). *Chem. Ber.*, **112**, 3759.
Giese, B. and Heuck, K. (1981). *Chem. Ber.*, **114**, 1572.
Gilchrist, T. L. and Storr, R. C. (1979). *Organic Reactions and Orbital Symmetry*, 2nd edn p. 35. (Cambridge: Cambridge University Press).
Gillis, B. T. (1967). In *1,4-Cycloaddition Reactions*, ed. J. Hamer (New York: Academic Press).
Girard, C. and Bloch, R. (1982). *Tetrahedron Lett.*, **23**, 3683.
Gollnick, K. (1968). *Adv. Photochem.*, **6**, 1.

Gollnick, K. and Schenck, O. (1967). In *1,4-Cycloaddition Reactions*, ed. J. Hamer (New York: Academic Press).
Gorman, A. A. and Rodgers, M. A. J. (1981). *Chem. Soc. Rev.*, **10**, 205.
Gosney, I. and Rowley, A. G. (1979). In *Organophosphorus Reagents in Organic Synthesis*, ed. J. I. G. Cadogan p. 17 (New York: Academic Press).
Grant, B. and Djerassi, C. (1974). *J. Org. Chem.*, **39**, 968.
Gray, R., Harruff, L. G., Krymowski, J., Peterson, J. and Boekelheide, V. (1978). *J. Am. Chem. Soc.*, **100**, 2892.
Grieco, P. A., Gilman, S. and Nishizawa, M. (1976). *J. Org. Chem.*, **41**, 1485.
Grieco, P. A., Inanaga, J., Lin, N-H. and Yanani, T. (1982). *J. Am. Chem. Soc.*, **104**, 5781.
Grieco, P. A., Jaw, J. Y., Claremon, D. A. and Nicolaou, K. C. (1981). *J. Org. Chem.*, **46**, 1215.
Grieco, P. A. and Pogonowski, C. S. (1974). *J. Org. Chem.*, **39**, 732.
Grieco, P. A., Yokoyama, Y., Gilman, S. and Nishizawa, M. (1977). *J. Org. Chem.*, **42**, 2035.
Grob, C. A. and Schiess, P. W. (1967). *Angew. Chem. internat. edn*, **6**, 1.
Gröbel, B. T. and Seebach, D. (1977). *Synthesis*, 357.
Gupta, R. C., Harland, P. A. and Stoodley, R. J. (1983). *J. Chem. Soc. Chem. Commun.*, 754.
Gupta, R. C., Jackson, D. A. and Stoodley, R. J. (1982). *J. Chem. Soc. Chem. Commun.*, 929. See also Liu, H–J. and Browne, E. N. C. (1981). *Canad. J. Chem.*, **59**, 601.
Gupta, D., Soman, R. and Dev. S. (1982). *Tetrahedron*, **38**, 3013.
Gupta, I. and Yates, P. (1982). *J. Chem. Soc. Chem. Commun.*, 1227.

Hamer, J. and Ahmad, M. (1967). In *1,4-Cycloaddition Reactions*, ed. J. Hamer (New York: Academic Press).
Hammond, G. S., Turro, N. J. and Liu, R. S. H. (1963). *J. Org. Chem.*, **28**, 3297.
Hanessian, S., Bargiotti, A. and La Rue, M. (1978). *Tetrahedron Lett.*, 737.
Hanson, J. R. (1974). *Synthesis*, 1.
Harmon, R. E., Gupta, S. K. and Brown, D. J. (1973). *Chem. Rev.*, **73**, 21.
Harris, T. M. and Harris, C. M. (1969). *Organic Reactions*, **17**, 155.
Hart, D. J., Cain, P. A. and Evans, D. A. (1978). *J. Am. Chem. Soc.*, **100**, 1548.
Hartwig, W. (1983). *Tetrahedron*, **39**, 2609.
Hasan, I. and Kishi, Y. (1980). *Tetrahedron Lett.*, **21**, 4229.
Hase, T. A. and Koskimies, J. K. (1982). *Aldrichimica Acta*, **15**, 33.
Hassall, C. H. (1957). *Organic Reactions*, **9**, 73.
Heathcock, C. H. (1981). *Science*, **214**, 395.
Heathcock, C. H. and Flippin, L. A. (1983). *J. Am. Chem. Soc.*, **105**, 1667.
Heathcock, C. H., Pirrung, M. C., Lampe, J., Buse, C. T. and Young, S. D. (1981). *J. Org. Chem.*, **46**, 2290.
Heck, R. F. (1979). *Acc. Chem. Res.*, **12**, 146.
Heck, R. F. (1982). *Organic Reactions*, **27**, 345.
Hegedus, L. S. (1984). *Tetrahedron*, **40**, 2415.
Helmchen, G. and Schmierer, R. (1981). *Angew. Chem. internat. edn*, **20**, 205.
Hentges, S. G. and Sharpless, K. B. (1980). *J. Am. Chem. Soc.*, **102**, 4263.
Heusler, K. and Kalvoda, J. (1964). *Angew. Chem. internat. edn*, **3**, 525.
Heyns, K. and Paulsen, H. (1963). In *Newer Methods of Preparative Organic Chemistry*, Vol. 2, ed. W. Foerst, p. 303 (New York: Academic Press).
Hickmott, P. W. (1982). *Tetrahedron*, **38**, 1975, 3363.

Hill, R. K. and Carlson, R. M. (1965). *J. Org. Chem.*, **30**, 2414.
Hill, R. K. and Rabinovitz, M. (1964). *J. Am. Chem. Soc.*, **86**, 965.
Hines, J. N., Peagram, M. J., Thomas, E. J. and Whitham, G. H. (1973). *J. Chem. Soc. Perkin I*, 2332.
Hiranuma, H. and Miller, S. I. (1982). *J. Org. Chem.*, **47**, 5083.
Hiyama, T., Kimura, K. and Nozaki, H. (1981). *Tetrahedron Lett.*, **22**, 1037.
Hiyama, T. and Nozaki, H. (1973). *Bull. Chem. Soc. Japan*, **46**, 2248.
Hoffmann, H. M. R. (1969). *Angew. Chem. internat. edn*, **8**, 556.
Hoffmann, H. M. R. (1984). *Angew. Chem. internat. edn*, **23**, 1.
Hoffmann, H. M. R., Henning, R. and Lalko, O. R. (1982). *Angew. Chem. internat. edn*, **21**, 442. See also Henning, R. and Hoffmann, H. M. R. (1982). *Tetrahedron Lett.*, **23**, 2305.
Hoffmann, R. and Woodward, R. B. (1965). *J. Am. Chem. Soc.*, **87**, 2046, 4388.
Hoffmann, R. W. (1967). *Dehydrobenzene and Cycloalkynes* (New York: Academic Press).
Hoffmann, R. W. (1982). *Angew. Chem. internat. edn*, **21**, 555.
Hoffmann, R. W. and Zeiss, H. J. (1981). *J. Org. Chem.*, **46**, 1309.
Holmes, H. L. (1948). *Organic Reactions*, **4**, 60.
Hooz, J., Gunn, D. M. and Kono, H. (1971). *Can. J. Chem.*, **49**, 2371.
Hori, T. and Sharpless, K. B. (1978). *J. Org. Chem.*, **43**, 1689.
Houk, K. N. (1975). *Acc. Chem. Res.*, **8**, 361.
Houk, K. N., Domelsmith, L. N., Strozier, K. W. and Patterson, R. W. (1978). *J. Am. Chem. Soc.*, **100**, 6531.
House, H. O. (1965). *Modern Synthetic Reactions* (New York: Benjamin).
House, H. O. (1976). *Acc. Chem. Res.*, **9**, 59.
House, H. O. (1967). *Rec. Chem. Prog.*, **28**, 99.
House, H. O. (1972). *Modern Synthetic Reactions*, 2nd edn, p. 175 (Menlo Park, California: Benjamin).
Hsiao, C-N. and Shechter, H. (1982). *Tetrahedron Lett.*, **23**, 1963.
Huang, F. C., Lee, L. F. H., Mittal, R. S. D., Ravikumar, P. R., Chan, J. A. and Sih, C. J. (1975). *J. Am. Chem. Soc.*, **97**, 4144.
Huckin, S. N. and Weiler, L. (1974). *J. Am. Chem. Soc.*, **90**, 1082.
Hudlicky, T., Kossyk, F. J., Dochwat, D. M. and Cantrell, G. L. (1981). *J. Org. Chem.*, **46**, 2911.
Hudrlik, P. F. and Kulkarni, A. K. (1981). *J. Am. Chem. Soc.*, **103**, 6251.
Hudrlik, P. F. and Peterson, D. (1975). *J. Am. Chem. Soc.*, **97**, 1464.
Huffmann, J. W. (1983). *Acc. Chem. Res.*, **16**, 399.
Huisgen, R. (1963). *Angew. Chem. internat. edn*, **2**, 565.
Huisgen, R. (1964). In *The Chemistry of Alkenes*, ed. S. Patai (London: Interscience).
Huisgen, R. (1968*a*). *Angew. Chem. internat. edn*, **7**, 321.
Huisgen, R. (1968*b*). *J. Org. Chem.*, **33**, 2291.
Huisgen, R., Grashey, R. and Sauer, J. (1964). In *The Chemistry of Alkenes*, ed. S. Patai, p. 739 (London: Interscience).
Hünig, S. and Wehner, G. (1979). *Chem. Ber.*, **112**, 2062.
Hutchins, R. O. and Natale, N. R. (1978). *J. Org. Chem.*, **43**, 2299.
Hutchins, R. O., Taffer, I. M. and Burgoyne, W. (1981). *J. Org. Chem.*, **46**, 5214.
Huynh, C., Derguini-Boumechal, F. and Linstrumelle, G. (1979). *Tetrahedron Lett.*, 1503.

Ireland, R. E., Anderson, R. C., Badoud, R., Fitzsimmons, B. J., McGarvey, G. J., Thaisrivongs, S. and Wilcox, C. S. (1983). *J. Am. Chem. Soc.*, **105**, 1988.

Ireland, R. E. and Daub, J. P. (1983). *J. Org. Chem.*, **48**, 1303.
Ireland, R. E., Muchmore, D. C. and Hengartner, U. (1972). *J. Am. Chem. Soc.*, **94**, 5098.
Ireland, R. E., Mueller, R. H. and Willard, A. K. (1976). *J. Am. Chem. Soc.*, **98**, 2868.
Ireland, R. E., O'Neil, T. H. and Tolman, G. L. (1983). *Organic Syntheses*, **61**, 116.
Ishida, T. and Wada, K. (1979). *J. Chem. Soc. Perkin I*, 323.
Ito, Y., Hirao, T. and Saegusa, T. (1978). *J. Org. Chem.*, **43**, 1011.
Ito, Y., Nakasuka, M. and Saegusa, T. (1981). *J. Am. Chem. Soc.*, **103**, 476.
Ito, Y., Nakasuka, M. and Saegusa, T. (1982). *J. Am. Chem. Soc.*, **104**, 7609.
Itsuno, S., Hirao, A., Nakahama, S. and Yamasaki, N. (1983). *J. Chem. Soc. Perkin I*, 1673.
Iwata, C., Miyashita, K., Ida, Y. and Yamada, M. (1981). *J. Chem. Soc. Chem. Commun.*, 461.

Jabri, N., Alexakis, A. and Normant, J. F. (1981). *Tetrahedron Lett.*, **22**, 959, 3851.
Jabri, N., Alexakis, A. and Normant, J. F. (1982). *Tetrahedron Lett.*, **23**, 1589.
Jakovac, I. J., Ng, G., Lok, K. P. and Jones, J. B. (1980). *J. Chem. Soc. Chem. Commun.*, 515.
Januszkiewicz, K. and Alper, H. (1983). *Tetrahedron Lett.*, **24**, 5159, 5163.
Jerussi, R. A. (1970). In *Selective Organic Transformations*, Vol. 1, ed. B. S. Thyagarajan (London: Wiley-Interscience).
Johnson, A. W. (1966). *Ylid Chemistry* (London: Academic Press).
Johnson, C. R. (1973). *Acc. Chem. Res.*, **6**, 341.
Johnson, C. R., Mori, K. and Nakanishi, A. (1979). *J. Org. Chem.*, **44**, 2065.
Johnson, M. R., Nakata, T. and Kishi, Y. (1979). *Tetrahedron Lett.*, **20**, 4343. But see Hasan, I. and Kishi, Y. (1980).
Johnson, W. S., Bannister, B., Bloom, B. M., Kemp, A. D., Pappo, R., Rogier, E. R. and Szmuszkovicz, J. (1953). *J. Am. Chem. Soc.*, **75**, 2275.
Johnstone, R. A. W. and Wilby, A. H. (1981). *Tetrahedron*, **37**, 3667.
Johnstone, R. A. W., Wilby, A. H. and Entwistle, I. D. (1985). *Chem. Rev.*, **85**, 129.
Julia, M. (1971). *Acc. Chem. Res.*, **4**, 386.
Julia, M. (1974). *Pure and Appl. Chem.*, **40**, 553.
Julia, M. and Paris, J-M. (1973). *Tetrahedron Lett.*, 4833.
Jung, M. E. and Gaede, B. (1979). *Tetrahedron*, **35**, 621.
Jung, M. E., McCombs, C. A., Takeda, Y. and Pan, Y-G. (1981). *J. Am. Chem. Soc.*, **103**, 6677.
Jurczak, J. and Zamojski, A. (1972). *Tetrahedron*, **28**, 1505.

Kabalka, G. W. and Hedgecock, H. C. (1975). *J. Org. Chem.*, **40**, 1776.
Kaiser, E. M. (1972). *Synthesis*, 391.
Kakushima, M. (1979). *Canad. J. Chem.*, **57**, 2564.
Kalvoda and Heusler, K. (1971). *Synthesis*, 501.
Kametani, T. (1979). *Pure and Appl. Chem.*, **51**, 747.
Kametani, T. and Nemoto, H. (1981). *Tetrahedron*, **37**, 3.
Karpf, M., Huguet, J. and Dreiding, A. S. (1982). *Helv. Chim. Acta*, **65**, 13.
Kasai, M. and Ziffer, H. (1983). *J. Org. Chem.*, **48**, 2346.
Katsuki, T. and Sharpless, K. B. (1980). *J. Am. Chem. Soc.*, **102**, 5974.
Kauffmann, T. (1974). *Angew. Chem. internat. edn*, **13**, 627.
Kay, I. T. and Bartholomew, D. (1984). *Tetrahedron Lett.*, **25**, 2035.
Kearns, D. R. (1971). *Chem. Rev.*, **71**, 395.
Keck, G. E. and Yates, J. B. (1982). *J. Am. Chem. Soc.*, **104**, 5829.

Kende, A. S. and Toder, B. H. (1982). *J. Org. Chem.*, **47**, 167.
Kenner, G. W. and Williams, N. R. (1955). *J. Chem. Soc.*, 522.
Kenney, R. J., Jones, W. D. and Bergman, R. G. (1978). *J. Am. Chem. Soc.*, **100**, 635.
Kerdesky, F. A. J., Ardecky, R. J., Lakshmikantham, M. V. and Cava, M. P. (1981). *J. Am. Chem. Soc.*, **103**, 1992.
Kerwin, J. F. and Danishefsky, S. (1982). *Tetrahedron Lett.*, **23**, 3739.
Keung, E. C. and Alper, H. (1972). *J. Chem. Educ.*, **49**, 97.
Khatri, N. A., Schmitthenner, H. F., Shringarpure, J. and Weinreb, S. M. (1981). *J. Am. Chem. Soc.*, **103**, 6387.
Kirby, G. W. (1977). *Chem. Soc. Rev.*, **6**, 1.
Kirby, G. W. and Sweeny, J. G. (1981). *J. Chem. Soc. Perkin I*, 3250.
Kishi, Y., Aratani, M., Tanino, H., Fukuyama, T. and Goto, T. (1972). *J. Chem. Soc. Chem. Commun.*, 64.
Klebe, J. F. (1972). In *Advances in Organic Chemistry*, ed. E. C. Taylor, Vol. 8, p. 97 (London: Wiley-Interscience).
Knowles, W. S. (1983). *Acc. Chem. Res.*, **16**, 106.
Kocienski, P. J., Lythgoe, B. and Ruston, S. (1978). *J. Chem. Soc. Perkin I*, 829.
Kodama, M., Takahashi, T., Kurihara, T. and Ito, S. (1980). *Tetrahedron Lett.*, **21**, 2811.
Kofron, W. G. and Yeh, M-K. (1976). *J. Org. Chem.*, **41**, 439.
Kolonko, K. J. and Shapiro, R. H. (1978). *J. Org. Chem.*, **43**, 1404.
Koreeda, M. and Ciufolini, M. A. (1982). *J. Am. Chem. Soc.*, **104**, 2308.
Kornblum, N., Erickson, A. E., Kelly, W. J. and Henngler, B. (1982). *J. Org. Chem.*, **47**, 4534.
Kosarych, Z. and Cohen, T. (1982). *Tetrahedron Lett.*, **23**, 3019.
Kozikowski, A. P. (1984). *Acc. Chem. Res.*, **17**, 410.
Kozikowski, A. P. and Ghosh, A. K. (1982). *J. Am. Chem. Soc.*, **104**, 5788.
Kozikowski, A. P. and Stein, P. D. (1982). *J. Am. Chem. Soc.*, **104**, 4023.
Krapcho, A. P. (1978). *Synthesis*, 77.
Krapcho, A. P. (1982). *Synthesis*, 805, 893.
Krebs, A. and Ruger, W. (1979). *Tetrahedron Lett.*, 1305.
Kreiser, W., Haumesser, W. and Thomas, A. F. (1974). *Helv. Chim. Acta*, **57**, 164.
Kreiser, W. and Wurziger, H. (1975). *Tetrahedron Lett.*, 1669.
Kresze, G. and Dittel, W. (1981). *Annalen*, 610.
Krishnamurthy, S. (1981). *J. Org. Chem.*, **46**, 4628.
Krishnamurthy, S. and Brown, H. C. (1976). *J. Am. Chem. Soc.*, **98**, 3383.
Krishnamurthy, S. and Brown, H. C. (1982). *J. Org. Chem.*, **47**, 276.
Krow, G. R. (1981). *Tetrahedron*, **37**, 2697.
Kuczkowski, R. L. (1983). *Acc. Chem. Res.*, **16**, 42.
Kuhn, H. J. and Gollnick, K. (1972). *Tetrahedron Lett.*, 1909.
Kuivila, H. G. (1968). *Acc. Chem. Res.*, **1**, 299. *Synthesis* (1970), 499.
Kumada, M. (1980). *Pure and Appl. Chem.*, **52**, 669.
Kursanov, D. N., Parnes, Z. N. and Loin, N. M. (1974). *Synthesis*, 633.
Kuwajima, I., Nakamura, E. and Hashimoto, K. (1983). *Organic Syntheses*, **61**, 122.
Kuwajima, I., Nakamura, E. and Shimizu, M. (1982). *J. Am. Chem. Soc.*, **104**, 1025.
Kwart, H. and King, K. (1968). *Chem. Rev.*, **68**, 415.

Laarhoven, W. H. (1983*a*). *Rev. Trav. Chim.*, **102**, 185.
Laarhoven, W. H. (1983*b*). *Rev. Trav. Chim.*, **102**, 241.
Lahima, N. J. and Levy, A. B. (1978). *J. Org. Chem.*, **43**, 1279.

Lane, C. F. (1974). *J. Org. Chem.*, **39**, 1437.
Lane, C. F. (1975). *Synthesis*, 135.
Lane, C. F. and Kabalka, G. W. (1970). *Tetrahedron*, **32**, 981.
Larson, E. R. and Danishefsky, S. (1982). *J. Am. Chem. Soc.*, **104**, 6458.
Lee, D. G. (1982). In *Oxidation in Organic Chemistry*, ed. W. S. Trahanovsky, (New York: Academic Press), Part D.
Lee, D. G., Lamb, S. E. and Chang, V. C. (1981). *Organic Syntheses*, **60**, 11.
Lee, D. G. and van den Engh, M. (1973). In *Oxidation in Organic Chemistry*, ed. W. S. Trahanovsky, Part B, Chapter 4 (New York: Academic Press).
Lee, J. B. and Uff, B. C. (1967). *Quart. Rev. Chem. Soc. Lond.*, **21**, 429.
Lee, R. A., McAndrews, C., Patel, K. M. and Reusch, W. (1973). *Tetrahedron Lett.*, 965.
Lee, T. V. and Toczek, J. (1982). *Tetrahedron Lett.*, **23**, 2917.
Lever, O. W. (1976), *Tetrahedron*, **32**, 1943.
Levy, A. B., Angelastro, R. and Marinelli, E. R. (1980). *Synthesis*, 945.
Leyendecker, F., Drouin, J. and Conia, J. M. (1974). *Tetrahedron Lett.*, 2931.
Liotta, D. (1984). *Acc. Chem. Res.*, **17**, 28.
Lipshutz, B. H., Parker, D., Kozlowski, J. A. and Miller, R. D. (1983). *J. Org. Chem.*, **48**, 3334.
Lipshutz, B. H., Wilhelm, R. S. and Floyd, D. M. (1981). *J. Am. Chem. Soc.*, **103**, 7672.
Lipton, M. F. and Shapiro, R. H. (1978). *J. Org. Chem.*, **43**, 1409.
Liu, H-J. and Browne, E. W. C. (1981). *Canad. J. Chem.*, **59**, 601.
Lu, L. D-L., Johnson, R. A., Finn, M. G. and Sharpless, K. B. (1984). *J. Org. Chem.*, **49**, 728.
Luche, J-L. and Gemal, A. L. (1979). *J. Am. Chem. Soc.*, **101**, 5848.
Lwowski, W. (1967). *Angew. Chem. internat. edn*, **6**, 897.
Lythgoe, B., Moran, T. A., Nambudiry, M. E. N. and Ruston, S. (1976). *J. Chem. Soc. Perkin I*, 2386.

Ma, P., Martin, V. S., Masamune, S., Sharpless, K. B. and Viti, S. M. (1983). *J. Org. Chem.*, **47**, 1378.
McCormick, J. P., Tomasik, W. and Johnson, M. W. (1981). *Tetrahedron Lett.*, **22**, 607.
McIntosh, J. and Sieler, R. A. (1978). *J. Org. Chem.*, **43**, 4431.
McKillop, A., Hunt, J. D., Kienzle, F., Bingham, E. and Taylor, E. C. (1973*a*). *J. Am. Chem. Soc.*, **95**, 3635.
McKillop, A., Oldenziel, O. H., Swann, B. P., Taylor, E. C. and Robey, R. L. (1973*b*). *J. Am. Chem. Soc.*, **95**, 1296.
McKillop, A., Swann, B. P. and Taylor, E. C. (1973). *J. Am. Chem. Soc.*, **95**, 3340.
McKillop, A. and Taylor, E. C. (1973). *Chem. in Britain*, **9**, 4.
McKillop, A. and Young, D. W. (1979). *Synthesis*, 401.
McMurry, J. E. (1974). *Acc. Chem. Res.*, **7**, 281.
McMurry, J. E. (1983). *Acc. Chem. Res.*, **16**, 405.
McMurry, J. E. and Fleming, M. P. (1975). *J. Org. Chem.*, **40**, 2555.
McMurry, J. E. and Scott, W. J. (1980). *Tetrahedron Lett.*, **21**, 4313; **24**, 979.
McQuillin, F. J. (1973). In *Progress in Organic Chemistry*, eds W. Carruthers and J. K. Sutherland, Vol. 8, p. 314 (London: Butterworth).
Madge, N. C. and Holmes, A. B. (1980). *J. Chem. Soc. Chem. Commun.*, 956.
Maercker, A. (1965). *Organic Reactions*, **14**, 270.
Magnus, P. D. (1977). *Tetrahedron*, **33**, 2019.

Magnus, P., Gallagher, T., Brown, P. and Pappalardo, P. (1984). *Acc. Chem. Res.*, **17**, 35.
Maignan, C. and Raphael, R. A. (1983). *Tetrahedron Lett.*, **39**, 3245.
Majetich, G., Casares, A. M., Chapman, D. and Behnke, M. (1983). *Tetrahedron Lett.*, **24**, 1909.
Makosza, M. and Jonćzyk, A. (1976). *Organic Syntheses*, **55**, 91.
Mancuso, A. J., Huang, S-L. and Swern, D. (1978). *J. Org. Chem.*, **43**, 2480.
Mancuso, A. J. and Swern, D. (1981). *Synthesis*, 165.
Marfat, A., McGuirk, P. R. and Helquist, P. (1979). *J. Org. Chem.*, **44**, 3888.
Marino, J. P. and Jaén, J. C. (1982). *J. Am. Chem. Soc.*, **104**, 3165.
Marshall, J. A. (1969). *Rec. Chem. Prog.*, **30**, 3.
Marshall, J. A. (1971). *Synthesis*, 229.
Martin, J. G. and Hill, R. K. (1961). *Chem. Rev.*, **61**, 537.
Martin, S. F. and Chih-yun Tu. (1981). *J. Org. Chem.*, **46**, 3763.
Martin, S. F., Tu, C., Kimura, M. and Simonsen, S. H. (1982). *J. Org. Chem.*, **47**, 3634.
Martin, V. S., Woodward, S. S., Katsuki, T., Yamada, Y., Ikeda, M. and Sharpless, K. B. (1981). *J. Am. Chem. Soc.*, **103**, 6237.
Masamune, S. (1981). *Organic Synthesis Today and Tomorrow*, eds B. M. Trost and C. R. Hutchinson, p. 197 (Oxford: Pergamon).
Masamune, S., Ali, Sk.A., Snitman, D. L. and Garvey, D. S. (1980). *Angew. Chem. internat. edn*, **19**, 557.
Masamune, S. and Choy, W. (1982). *Aldrichimica Acta*, **15**, 47.
Masamune, S., Choy, W., Petersen, J. S. and Sita, L. R. (1985). *Angew. Chem. internat. edn*, **24**, 1.
Masamune, S., Kaiho, T. and Garvey, D. S. (1982). *J. Am. Chem. Soc.*, **104**, 5521.
Matteson, D. S. and Sadhu, K. M. (1983). *J. Am. Chem. Soc.*, **105**, 2077.
Maurer, P. J., Takahata, H. and Rapoport, H. (1984). *J. Am. Chem. Soc.*, **106**, 1095.
Mazur, Y. (1975). *Pure and Appl. Chem.*, **41**, 145.
Meinwald, J. and Meinwald, Y. C. (1966). In *Advances in Alicyclic Chemistry*, ed. H. Hart and G. J. Karabatsos, Vol. 1, p. 1 (New York: Academic Press).
Meyers, A. I. (1978). *Acc. Chem. Res.*, **11**, 375.
Meyers, A. I., Nabeya, A., Adickes, H. W., Politzer, I. R., Malone, G. R., Kovelevsky, A. C., Nolen, R. L. and Portnoy, R. C. (1973). *J. Org. Chem.*, **38**, 36.
Meyers, A. I. Edwards, P. D., Rieker, W. F. and Bailey, T. R. (1984). *J. Am. Chem. Soc.*, **106**, 3270.
Meyers, A. I. and Fuentes, L. M. (1983). *J. Am. Chem. Soc.*, **105**, 117.
Meyers, A. I. and Mihelich, E. D. (1976). *Angew. Chem. internat. edn*, **15**, 270.
Meyers, A. I. and Smith, R. K. (1979). *Tetrahedron Lett.*, 2749.
Meyers, A. I., Tait, T. A. and Commins, D. L. (1978). *Tetrahedron Lett.*, 4657.
Meyers, A. I., Williams, D. R., Erickson, G. W., White, S. and Druelinger, M. (1981). *J. Am. Chem. Soc.*, **103**, 3081.
Meyers, A. I., Williams, D. R., White, S. and Erickson, G. W. (1981). *J. Am. Chem. Soc.*, **103**, 3088.
Midland, M. M., Greer, S., Tramontano, A. and Zderic, S. A. (1979). *J. Am. Chem. Soc.*, **101**, 2352.
Midland, M. M. and Kazubski, A. (1982). *J. Org. Chem.*, **47**, 2495.
Midland, M. M., McDowell, D. C., Hatch, R. L. and Tramontano, A. (1980). *J. Am. Chem. Soc.*, **102**, 867.
Midland, M. M., Sinclair, J. A. and Brown, H. C. (1974). *J. Org. Chem.*, **39**, 731.

Mihailović, M. Lj., Gojković, S. and Konstantinović, S. (1973). *Tetrahedron*, **29**, 3675.
Mihelich, E. D., Daniels, K. and Eickhoff, D. J. (1981). *J. Am. Chem. Soc.*, **103**, 7690.
Mikami, K., Kishi, N. and Nakai, T. (1983). *Tetrahedron Lett.*, **24**, 795.
Miller, C. E. (1965). *J. Chem. Educ.*, **42**, 254.
Miller, R. D. and McKean, D. R. (1979). *Synthesis*, 730.
Mimoun, H., Charpentier, R., Mitschler, A., Fischer, J. and Weiss, R. (1980). *J. Am. Chem. Soc.*, **102**, 1047.
Minami, N., Ko, S. S. and Kishi, Y. (1982). *J. Am. Chem. Soc.*, **104**, 1109.
Mitchell, R. H. and Boekelheide, V. (1974). *J. Am. Chem. Soc.*, **96**, 1547, 1558.
Morgans, D. J. (1981). *Tetrahedron Lett.*, **22**, 3721.
Mori, K. (1981). *Tetrahedron*, **37**, 1341.
Moriarty, R. M. and Kwang Chung Hou. (1984). *Tetrahedron Lett.*, **25**, 691.
Mukaiyama, T. (1977). *Angew. Chem. internat. edn*, **16**, 817.
Mukaiyama, T. (1982). *Organic Reactions*, **28**, 203.
Mukaiyama, T. (1983). *Pure and Appl. Chem.*, **55**, 1749.
Mulzer, J., Pointer, A., Chucholanski, A. and Brüntrup, G. (1979). *J. Chem. Soc. Chem. Commun.*, 52.
Murray, R. W. (1968). *Acc. Chem. Res.*, **1**, 313.

Nace, H. R. (1962). *Organic Reactions*, **12**, 57.
Nader, B., Bailey, T. R., Franck, R. W. and Weinreb, S. M. (1981). *J. Am. Chem. Soc.*, **103**, 7573.
Nader, B., Franck, R. W. and Weinreb, S. M. (1980). *J. Am. Chem. Soc.*, **102**, 1153.
Nakabayashi, T. (1960). *J. Am. Chem. Soc.*, **82**, 3900, 3906, 3909.
Negishi, E. (1982). *Acc. Chem. Res.*, **15**, 340.
Nehishi, E., Luo, F-T. and Rand, C. L. (1982). *Tetrahedron Lett.*, **23**, 27.
Negishi, E. and Yoshida, T. (1973). *J. Chem. Soc. Chem. Commun.*, 606.
Newton, R. F., Reynolds, D. P., Finch, M. A. W., Kelly, D. R. and Roberts, S. M. (1979). *Tetrahedron Lett.*, 3981.
Newton, P. F. and Whitham, G. H. (1979). *J. Chem. Soc. Perkin I*, 3067.
Nicolaou, K. C., Clarenion, D. A., Barnette, W. E. and Seitz, S. P. (1979). *J. Am. Chem. Soc.*, **97**, 3704.
Nicolaou, K. C. and Magolda, R. L. (1981). *J. Org. Chem.*, **46**, 1506.
Nicolaou, K. C., Seitz, S. R. and Pavia, M. R. (1981). *J. Am. Chem. Soc.*, **103**, 1222. See also **104**, 2030.
Nielson, A. T. and Houlihan, W. J. (1968). *Organic Reactions*, **16**, 1.
Normant, J. F. (1978). *Pure and Appl. Chem.*, **50**, 709
Normant, J. F., Cahiez, G., Bourgain, M., Chuit, C. and Villieras, J. (1974). *Bull. Soc. Chim. France*, 1656.
Noureldin, N. A. and Lee, D. G. (1981). *Tetrahedron Lett.*, **22**, 4889.
Noyori, R., Murata, S. and Suzuki, M. (1981). *Tetrahedron*, **37**, 3899.
Noyori, R., Tomino, I. and Tanimoto, Y. (1979). *J. Am. Chem. Soc.*, **101**, 3129.
Nozake, H., Kato, H. and Noyori, R. (1969). *Tetrahedron*, **25**, 1661.
Nussbaum, A. L. and Robinson, C. H. (1962). *Tetrahedron*, **17**, 35.
Nutaitis, C. F. and Gribble, G. V. (1983). *Tetrahedron Lett.*, **24**, 4287.

Ohta, H., Tetsukawa, H. and Noto, N. (1982). *J. Org. Chem.*, **47**, 2400.
Olah, G. A., Husain, A., Gupta, B. G. B., Salem, G. F. and Narang, S. C. (1981). *J. Org. Chem.*, **46**, 5212.

Olah, G. A., Husain, A., Singh, B. P. and Mehrotra, A. K. (1983). *J. Org. Chem.*, **48**, 3667.
Olah, G. A., Narang, S. C., Gupta, B. G. B. and Malhotra, R. (1979). *Synthesis*, 61.
Olofson, R. A. and Dougherty, C. M. (1973). *J. Am. Chem. Soc.*, **95**, 582.
Onischenko, A. S. (1964). *Diene Synthesis* (New York: Davey).
Ono, N., Miyaki, H. and Kaji, A. (1982). *J. Chem. Soc. Chem. Commun.*, 33.
Oppolzer, W. (1972). *Angew. Chem. internat. edn*, **11**, 1031.
Oppolzer, W. (1973). *Helv. Chim. Acta*, **56**, 1812.
Oppolzer, W. (1974) *Tetrahedron Lett.*, 1001.
Oppolzer, W. (1977). *Angew. Chem. internat. edn*, **16**, 10.
Oppolzer, W. (1978). *Synthesis*, 793.
Oppolzer, W. (1981). *Pure and Appl. Chem.*, **53**, 1181.
Oppolzer, W. (1982). *Acc. Chem. Res.*, **15**, 135.
Oppolzer, W. (1984). *Angew. Chem. internat. edn*, **23**, 876.
Oppolzer, W. and Bättig, K. (1982). *Tetrahedron Lett.*, **23**, 4669.
Oppolzer, W., Bieber, L. and Francotti, E. (1979). *Tetrahedron Lett.*, 4537.
Oppolzer, W., Chapius, C., Dao, G. M., Reichlin, D. R. and Godel, T. (1982). *Tetrahedron Lett.*, **23**, 4781.
Oppolzer, W. and Flashkamp, E. (1977). *Helv. Chim. Acta*, **60**, 204.
Oppolzer, W., Francotti, E. and Bättig, K. (1981). *Helv. Chim. Acta*, **64**, 478.
Oppolzer, W., Moretti, R., Godel, T., Meunier, A. and Löher, H. (1983). *Tetrahedron Lett.*, **24**, 4971.
Oppolzer, W. and Petrzilka, M. (1978). *Helv. Chim. Acta*, **61**, 2755. See also Wookulich, P. M. and Uskoković, M. R. (1981). *J. Am. Chem. Soc.*, **103**, 3956; Baggiolini, E. G., Lee, H. L., Pizzolato, G. and Usković, M. R. (1982). *J. Am. Chem. Soc.*, **104**, 6460.
Oppolzer, W. and Pitteloud, R. (1982). *J. Am. Chem. Soc.*, **104**, 6478.
Oppolzer, W. and Roberts, D. N. (1980). *Helv. Chim. Acta*, **63**, 1703.
Oppolzer, W. and Thirring, K. (1982). *J. Am. Chem. Soc.*, **104**, 4978.
Orban, I., Schaffner, K. and Jeger, O. (1963). *J. Am. Chem. Soc.*, **85**, 3033.
Overman, L. E. (1980). *Acc. Chem. Res.*, **13**, 218.
Overman, L. E. and Burk, R. M. (1984). *Tetrahedron Lett.*, **25**, 5739.
Overman, L. E., Freerks, R. L., Petty, C. B., Clizbe, L. A., Ono, R. K., Taylor, G. F. and Jessup, P. J. (1981*a*). *J. Am. Chem. Soc.*, **103**, 2816.
Overman, L. E. and Fukaya, C. (1980). *J. Am. Chem. Soc.*, **102**, 1454.
Overman, L. E. and Jessup, P. J. (1978). *J. Am. Chem. Soc.*, **100**, 5179.
Overman, L. E., Kakimoto, M., Okazaki, M. E. and Meier, G. P. (1983). *J. Am. Chem. Soc.*, **105**, 6622.
Overman, L. E. and Renaldo, A. F. (1980). *Tetrahedron Lett.*, **24**, 3757.
Overman, L. E., Sworin, M., Bass, L. S. and Clardy, J. (1981*b*). *Tetrahedron*, **37**, 4041.

Padwa, A. (1976). *Angew. Chem. internat. edn*, **15**, 123.
Paquette, L. A. (1977). *Organic Reactions*, **35**, 1.
Paquette, L. A. and Crouse, G. D. (1983). *J. Org. Chem.*, **48**, 141.
Paquette, L. A., Moerck, R. E., Harichian, B. and Magnus, P. D. (1978). *J. Am. Chem. Soc.*, **100**, 1597.
Paquette, L. A. and Williams, R. V. (1981). *Tetrahedron Lett.*, **22**, 4643.
Parham, W. E. and Schweizer, E. E. (1963). *Organic Reactions*, **13**, 55.
Parker, A. J. (1962). *Quart. Rev. Chem. Soc. Lond.*, **16**, 163.
Parker, R. E. and Isaacs, N. S. (1959). *Chem. Rev.*, **59**, 737.

Parshall, G. W. (1980). *Homogeneous Catalysis. The Applications and Chemistry of Catalysis by Soluble Transition Metal Complexes* (London: Wiley).
Paterson, I. (1979). *Tetrahedron Lett.*, 1519.
Paterson, I. and Fleming, I. (1979). *Tetrahedron Lett.*, 993, 995, 2179.
Pearson, D. E. and Buehler, C. A. (1974). *Chem. Rev.*, **74**, 45.
Pelter, A., Bentley, T. W., Harrison, C. R., Subrahmanyam, C. and Laub, R. J. (1976*a*). *J. Chem. Soc. Perkin I*, 2419, 2428.
Pelter, A., Hughes, R., Smith, K. and Tabata, M. (1976*b*). *Tetrahedron Lett.*, 4385.
Pelter, A., Hutchings, M. G., Smith, K. and Williams, D. J. (1975*a*). *J. Chem. Soc. Perkin I*, 145.
Pelter, A. and Rao, J. M. (1981). *J. Chem. Soc. Chem. Commun.*, 1149.
Pelter, A., Rosser, R. M. and Mills, S. (1984). *J. Chem. Soc. Perkin I*, 717.
Pelter, A., Smith, K., Hutchings, M. G. and Rowe, K. (1975*b*). *J. Chem. Soc. Perkin I*, 129, 138.
Peterson, D. J. (1968). *J. Org. Chem.*, **33**, 781.
Petrzilka, M. and Grayson, J. I. (1981). *Synthesis*, 753.
Pettit, G. R. and van Tamelen, E. E. (1962). *Organic Reactions*, **12**, 356. (London: Wiley.)
Piancatelli, G., Scettri, A. and D'Auria, M. (1982). *Synthesis*, 245.
Pizey, S. S. (1974). *Synthetic Reagents*, Vol. 1, p. 101 (London: Wiley).
Politzer, I. R. and Meyers, A. I. (1971). *Organic Syntheses*, **51**, 24.
Posner, G. H. (1975*a*). *Organic Reactions*, **19**, 1.
Posner, G. H. (1975*b*). *Organic Reactions*, **22**, 253.
Posner, G. H. and Lenz, C. M. (1979). *J. Am. Chem. Soc.*, **101**, 934.
Pyne, S. G., Hensel, M. J. and Fuchs, P. L. (1982). *J. Am. Chem. Soc.*, **104**, 5721.

Ranganathan, D., Rao, C. B., Ranganathan, S., Mehrotra, A. K. and Iyengar, R. (1980). *J. Org. Chem.*, **45**, 1185.
Rao, A. S., Paknikar, S. K. and Kirtane, J. G. (1983). *Tetrahedron*, **39**, 2323.
Rao, C. G., Kulkarni, S. U. and Brown, H. C. (1979). *J. Organomet. Chem.*, **172**, C20.
Rasmussen, J. K. (1977). *Synthesis*, 91.
Ratcliffe, R. W. (1976). *Organic Syntheses*, **55**, 84.
Reetz, M. T. (1982). *Angew. Chem. internat. edn*, **21**, 96.
Reetz, M. T., Chatziiosifidis, I., Löwe, U. and Maier, W. F. (1979). *Tetrahedron Lett.*, 1427.
Reich, H. J. (1975). *J. Org. Chem.*, **40**, 2571.
Reich, H. J. (1979). *Acc. Chem. Res.*, **12**, 22.
Reich, H. J., Renga, J. M. and Reich, I. L. (1974). *J. Org. Chem.*, **39**, 2133.
Rerick, M. N. (1968). In *Reduction*, ed. R. L. Augustine, p. 1. (London: Arnold).
Reucroft, J. and Sammes, P. G. (1971). *Quart. Rev. Chem. Soc. Lond.*, **25**, 135.
Rhoads, S. J. and Raulins, N. R. (1975). *Organic Reactions*, **22**, 1.
Rickborn, B. and Thummel, R. P. (1969). *J. Org. Chem.*, **34**, 3583.
Ripoll, J. L., Rouessac, A. and Rouessac, F. (1978). *Tetrahedron*, **34**, 19.
Robins, M. J., Wilson, J. S. and Hansske, F. (1983). *J. Am. Chem. Soc.*, **105**, 4059.
Rossiter, B. E., Katsuki, T. and Sharpless, K. B. (1981). *J. Am. Chem. Soc.*, **103**, 464.
Rossiter, B. E., Verhoeven, T. R. and Sharpless, K. B. (1979). *Tetrahedron Lett.*, 4733.
Roush, W. R., Gillis, H. R. and Ko, A. I. (1982). *J. Am. Chem. Soc.*, **104**, 2269.
Roush, W. R. and Myers, A. G. (1981). *J. Org. Chem.*, **46**, 1506, 1509.

Roussel, M. and Mimoun, H. (1980). *J. Org. Chem.*, **45**, 5387.
Ruden, R. A. and Bonjouklian, R. (1974). *Tetrahedron Lett.*, 2095.
Ruhlmann, K., Seefluth, H. and Becker, H. (1967). *Chem. Ber.*, **100**, 3820.
Rylander, P. N. (1967). *Catalytic Hydrogenation over Platinum Metals* (London: Academic Press).

Saavedra, J. E. (1983). *J. Org. Chem.*, **48**, 2388.
Sainte, F., Serckx-Poncin, B., Hesbain-Frisque, A-M. and Ghosez, L. (1982). *J. Am. Chem. Soc.*, **104**, 1428.
Sakurai, H. (1982). *Pure and Appl. Chem.*, **54**, 1.
Sala, T. and Sargent, M. V. (1978). *J. Chem. Soc. Chem. Commun.*, 253.
Salomon, R. G., Coughlin, D. J., Ghosh, S. and Zagorski, M. G. (1982). *J. Am. Chem. Soc.*, **104**, 998.
Sammes, P. G. (1970). *Quart. Rev. Chem. Soc. Lond.*, **24**, 37.
Sammes, P. G. (1976). *Tetrahedron*, **32**, 405.
Sauer, J. (1966). *Angew. Chem. internat. edn*, **5**, 211.
Sauer, J. (1967). *Angew. Chem. internat. edn*, **6**, 16.
Sauer, J. and Kredel, J. (1966). *Tetrahedron Lett.*, 731, 6359.
Sauer, J. and Sustmann, R. (1980). *Angew. Chem. internat. edn*, **19**, 779.
Schiess, P., Heitzmann, M., Rutschmann, S. and Staheli, R. (1978). *Tetrahedron Lett.*, 4569.
Schiller, G. (1955). In Houben-Weyl, *Methoden der organischen Chemie*, ed. E. Müller, Vol. 4(2), p. 241. (Stuttgart: Georg Thieme Verlag).
Schlosser, M. (1970). In *Topics in Stereochemistry*, ed. E. L. Eliel and N. L. Allinger, Vol. 5, p. 1. (New York: Interscience).
Schlosser, M. (1972). In Houben-Weyl, *Methoden der organischen Chemie*, p. 134, ed. E. Müller, Vol. 5(1b). (Stuttgart: Georg Thieme Verlag).
Schmidt, A. H. (1981). *Aldrichimica Acta*, **14**, 31.
Schmidt, R. R. (1973). *Angew. Chem. internat. edn*, **12**, 212.
Schmittenner, H. F. and Weinreb, S. M. (1980). *J. Org. Chem.*, **45**, 3372.
Schröder, M. (1980). *Chem. Rev.*, **80**, 187.
Schröder, M. and Constable, E. C. (1982). *J. Chem. Soc. Chem. Commun.*, 734.
Schultz, A. G. (1983). *Acc. Chem. Res.*, **16**, 210.
Scott, J. W., Keith, D. D., Nix, G., Parrish, D. R., Remington, S., Roth, G. P., Townsend, J. M., Valentine, D. and Yang, R. (1981). *J. Org. Chem.*, **46**, 5086.
Seebach, D. (1979*a*). *Angew. Chem. internat. edn*, **18**, 239.
Seebach, D. (1979*b*). *Chimia*, **33**, 1.
Seebach, D. and Aebi, J. D. (1983). *Tetrahedron Lett.*, **24**, 3311.
Seebach, D., Beck, A. K., Mukhepadhyay, T. and Thomas, E. (1982). *Helv. Chim. Acta*, **65**, 1101.
Seebach, D., Boes, M., Naef, R. and Schweizer, W. B. (1983). *J. Am. Chem. Soc.*, **105**, 5390.
Seebach, D. and Enders, D. (1975). *Angew. Chem. internat. edn*, **14**, 15.
Seebach, D., Henning, R. and Lehr, F. (1978). *Angew. Chem. internat. edn*, **17**, 458.
Seebach, D., Henning, R., Lehr, F. and Gonnermann, J. (1977). *Tetrahedron Lett.*, 1161.
Seebach, D. and Weber, T. (1983). *Tetrahedron Lett.*, **24**, 3315.
Semmelhack, M. F. (1972). *Organic Reactions*, **19**, 115.
Semmelhack, M. F., Helquist, P., Jones, L. D. Keller, L., Mendelson, L., Ryono, L. S., Smith, J. G. and Stauffer, R. D. (1981). *J. Am. Chem. Soc.*, **103**, 6460.
Serckx-Poncin, B., Hesbain-Frisque, A-M. and Ghosez, L. (1982). *Tetrahedron Lett.*, **23**, 3261.

Sharpless, K. B., Behrens, C. H., Katsuki, T., Lee, A. W. M., Martin, V. S., Takatani, M., Viti, S. M., Walker, F. J. and Woodward, S. S. (1983). *Pure and Appl. Chem.*, **55**, 589.
Sharpless, K. B., Chong, A. O. Oshima, K. (1976). *J. Org. Chem.*, **41**, 177.
Sharpless, K. B. and Gordon, K. M. (1976). *J. Am. Chem. Soc.*, **98**, 300.
Sharpless, K. B. and Hori, T. (1976). *J. Org. Chem.*, **41**, 176.
Sharpless, K. B. and Michaelson, R. C. (1973). *J. Am. Chem. Soc.*, **95**, 6136.
Sharpless, K. B., Umbreit, M. A., Nieh, M. T. and Flood, T. C. (1972). *J. Am. Chem. Soc.*, **94**, 6538.
Sharpless, K. B. and Verhoeven, T. R. (1979). *Aldrichimica Acta*, **12**, 63.
Sharpless, K. B., Woodward, S. S. and Finn, M. G. (1983). *Pure and Appl. Chem.*, **55**, 1823.
Shea, K. J. and Gilman, J. W. (1983). *Tetrahedron Lett.*, **24**, 657.
Shea, K. J. and Wada, E. (1982). *J. Am. Chem. Soc.*, **104**, 5715.
Shea, K. J., Wise, S., Burke, L. D., Davis, P. D., Gilman, J. W. and Greeley, A. C. (1982). *J. Am. Chem. Soc.*, **104**, 5708.
Sheldon, R. A. and Kochi, J. K. (1972). *Organic Reactions*, **19**, 279.
Shenoi, S. and Stille, J. K. (1982). *Tetrahedron Lett.*, **23**, 627.
Sicher, J. (1972). *Angew. Chem. internat. edn*, **11**, 200.
Sinclair, J. H. and Brown, H. C. (1976). *J. Org. Chem.*, **41**, 1078.
Simmons, H. E., Cairns, T. L., Vladuchick, S. A. and Hoiness, C. M. (1973). *Organic Reactions*, **20**, 1.
Smith, A. B. and Dieter, R. K. (1981). *Tetrahedron*, **37**, 2407.
Smith, A. B. and Richmond, R. E. (1983). *J. Am. Chem. Soc.*, **105**, 575.
Smith, A. B. and Scarborough, R. M. (1980). *Synthetic Commun.*, 205.
Smith, J. G. (1984). *Synthesis*, 629.
Smith, J. K., Newcomb, M., Bergbreiter, D. E., Williams, D. R. and Meyers, A. I. (1983). *Tetrahedron Lett.*, **24**, 3559.
Smith, P. A. S. (1963). In *Molecular Rearrangements*, ed. P. de Mayo, Vol. 1, p. 571 (London: Wiley-Interscience).
Snider, B. B. (1980). *Acc. Chem. Res.*, **13**, 426.
Snider, B. B. and Phillips, G. B. (1983). *J. Org. Chem.*, **48**, 464.
Snider, B. B., Rodini, D. J., Kirk, T. C. and Cordova, R. (1982). *J. Am. Chem. Soc.*, **104**, 555.
Soderquist, J. A. and Brown, H. C. (1981). *J. Org. Chem.*, **44**, 4599.
Stang, P. J. (1982). *Acc. Chem. Res.*, **15**, 348.
Stephenson, L. M., Grdina, J. J. and Orfanopoulos, M. (1980). *Acc. Chem. Res.*, **13**, 419.
Sternbach, D., Shibuya, M., Jaisli, F., Bonetti, M. and Eschenmoser, A. (1979). *Angew. Chem. internat. edn*, **18**, 634.
Stevens, R. V. (1977). *Acc. Chem. Res.*, **10**, 193.
Stevens, R. V., Chapman, K. T., Stubbs, C. A., Tam, W. W. and Albizati, K. F. (1982). *Tetrahedron Lett.*, **23**, 4647.
Stewart, R. (1965). In *Oxidation in Organic Chemistry*, ed. K. B. Wiberg, part A, p. 42 (London: Academic Press).
Still, W. C. (1977). *J. Am. Chem. Soc.*, **99**, 4186.
Still, W. C. (1979). *J. Am. Chem. Soc.*, **101**, 2493.
Still, W. C. and Barrish, A. (1983). *J. Am. Chem. Soc.*, **105**, 2487.
Still, W. C. and Darst, K. P. (1980). *J. Am. Chem. Soc.*, **102**, 7385.
Still, W. C. and Novack, V. J. (1981). *J. Am. Chem. Soc.*, **103**, 1283.
Still, W. C. and Ohmizu, H. (1981). *J. Org. Chem.*, **46**, 5243.

Still, W. C. and Shaw, K. R. (1981). *Tetrahedron Lett.*, **22**, 3725.
Still, W. C. and Tsai, M-Y. (1980). *J. Am. Chem. Soc.*, **102**, 3654.
Stork, G. (1983). *Current Trends in Organic Synthesis*, ed. H. Nozaki, p. 359 (Oxford: Pergamon).
Stork, G. and Baine, N. H. (1982). *J. Am. Chem. Soc.*, **104**, 2321.
Stork, G. and Benaim, J. (1971). *J. Am. Chem. Soc.*, **93**, 5938.
Stork, G., Brizzolana, A., Landesman, H., Szmuszkovicz, J. and Terrell, R. (1963). *J. Am. Chem. Soc.*, **85**, 207.
Stork, G., Cama, L. D. and Coulson, D. R. (1974). *J. Am. Chem. Soc.*, **96**, 5268, 5270.
Stork, G. and Dowd, S. R. (1963). *J. Am. Chem. Soc.*, **85**, 2178.
Stork, G. and Dowd, S. R. (1974). *Organic Syntheses*, **54**, 46.
Stork, G. and Ganem, B. (1973). *J. Am. Chem. Soc.*, **95**, 6152.
Stork, G. and Kahn, M. (1985). *J. Am. Chem. Soc.*, **107**, 500.
Stork, G. and Kahne, D. E. (1983). *J. Am. Chem. Soc.*, **105**, 1072.
Stork, G. and Mook, R. (1983). *J. Am. Chem. Soc.*, **105**, 3720.
Stork, G., Mook, R., Biller, S. A. and Rychnovsky, S. D. (1983). *J. Am. Chem. Soc.*, **105**, 3741.
Stork, G. and Nakamura, E. (1979). *J. Org. Chem.*, **44**, 4010.
Stork, G. and Raucher, S. (1976). *J. Am. Chem. Soc.*, **98**, 1583.
Stork, G. and Singh, J. (1974). *J. Am. Chem. Soc.*, **96**, 6181.
Sustmann, R. (1974). *Pure and Appl. Chem.*, **40**, 569.
Suzuki, M., Takada, H. and Noyori, R. (1982). *J. Org. Chem.*, **47**, 902.

Taber, D. F., Gunn, B. P. and Ching Chiu, I. (1983). *Organic Syntheses*, **61**, 59.
Takaya, H., Makino, S., Hayakawa, Y. and Noyori, R. (1978). *J. Am. Chem. Soc.*, **100**, 1765.
Tanigawa, Y., Nishimura, K., Kawasaki, A. and Murahashi, S-I. (1982). *Tetrahedron Lett.*, **23**, 5549.
Tedder, J. M. and Walton, J. C. (1980). *Tetrahedron*, **36**, 701.
Thummel, R. P. and Rickborn, B. (1970). *J. Am. Chem. Soc.*, **92**, 2064.
Tomioka, H., Oshima, K. and Nozaki, H. (1982). *Tetrahedron Lett.*, **23**, 539.
Trost, B. M. (1974). *Acc. Chem. Res.*, **7**, 85.
Trost, B. M. (1978). *Acc. Chem. Res.*, **11**, 453; *Chem. Rev.*, **78**, 363.
Trost, B. M. (1980). *Acc. Chem. Res.*, **13**, 385.
Trost, B. M. (1981). *Aldrichimica Acta*, **14**, 43.
Trost, B. M. and Coppola, B. P. (1982). *J. Am. Chem. Soc.*, **104**, 6879.
Trost, B. M., Ippen, J. and Vladuchik, W. C. (1977). *J. Am. Chem. Soc.*, **99**, 8117.
Trost, B. M. and Klein, T. P. (1979). *J. Am. Chem. Soc.*, **101**, 6756; **103**, 1864
Trost, B. M. and Lunn, G. (1977). *J. Am. Chem. Soc.*, **99**, 7079.
Trost, B. M. and Molander, G. A. (1981). *J. Am. Chem. Soc.*, **103**, 5969.
Trost, B. M., O'Krongly, D. and Bellatire, J. L. (1980). *J. Am. Chem. Soc.*, **102**, 7595.
Trost, B. M. and Renaut, P. (1982). *J. Am. Chem. Soc.*, **104**, 6668.
Trost, B. M. and Tamaru, Y. (1975). *J. Am. Chem. Soc.*, **97**, 3528.
Trost, B. M., Vladuchick, W. C. and Bridges, A. J. (1980). *J. Am. Chem. Soc.*, **102**, 3554.
Tsuji, J. (1982). *Pure and Appl. Chem.*, **54**, 197
Tufariello, J. J. (1979). *Acc. Chem. Res.*, **12**, 396.
Turro, N. J. (1969). *Acc. Chem. Res.*, **2**, 25.

Ullrich, V. (1972). *Angew. Chem. internat. edn*, **11**, 701.
Umbreit, M. A. and Sharpless, K. B. (1977). *J. Am. Chem. Soc.*, **99**, 5526.
Umbreit, M. A. and Sharpless, K. B. (1981). *Organic Syntheses*, **60**, 29.
Utimoto, K. (1983). *Pure and Appl. Chem.*, **55**, 1845
Utimoto, K., Obayashi, M. and Nozaki, H. (1976). *J. Org. Chem.*, **41**, 2940.

van Tamelen, E. E., Pappas, S. P. and Kirk, K. L. (1971). *J. Am. Chem. Soc.*, **93**, 6092.
van Tamelen, E. E. and Sharpless, K. B. (1967). *Tetrahedron Lett.*, 2655.
Vedejs, E. (1975). *Organic Reactions*, **22**, 401.
Vedejs, E. (1984). *Acc. Chem. Res.*, **17**, 358.
Vedejs, E., Eberlein, T. H. and Varie, D. L. (1982). *J. Am. Chem. Soc.*, **104**, 1445.
Vedejs, E., Engler, D. A. and Telschow, J. E. (1978). *J. Org. Chem.*, **43**, 188.
Vollhardt, K. P. C. (1984). *Angew. Chem. internat. edn*, **23**, 539.

Wadsworth, W. S. (1977). *Organic Reactions*, **25**, 73.
Wagner, P. J. and Hammond, G. S. (1968). In *Advances in Photochemistry*, eds. W. A. Noyes, G. S. Hammond and J. N. Pitts. Vol. 5, p. 136 (London: Interscience).
Walker, B. J. (1979). *Organophosphorus Reagents in Organic Synthesis*, ed. J. I. G. Cadogan, p. 155 (New York: Academic Press).
Walker, E. R. H. (1976). *Chem. Soc. Rev.*, **5**, 23.
Walling, C. and Padwa, A. (1963). *J. Am. Chem. Soc.*, **85**, 1597.
Wasserman, H. H. and Ives, J. L. (1981). *Tetrahedron*, **37**, 1825.
Wasserman, H. H. and Ives, J. L. (1982). *Tetrahedron*, **37**, 1825.
Wasserman, H. H. and Lipshutz, B. H. (1975). *Tetrahedron Lett.*, 1731.
Wehrli, R., Bellus, D., Hansen, H-J. and Schmidt, H. (1976). *Chimia*, **30**, 416.
Weinreb, S. M. (1985). *Acc. Chem. Res.*, **18**, 16.
Weinreb, S. M., Khatri, N. A. and Shringarpure, J. (1979). *J. Am. Chem. Soc.*, **101**, 5073.
Weinreb, S. M. and Levin, J. J. (1979). *Heterocycles*, **12**, 949.
Weinreb, S. M. and Staib, R. R. (1982). *Tetrahedron*, **39**, 3087.
Wender, P. A., Erhardt, J. M. and Letendre, L. J. (1981). *J. Am. Chem. Soc.*, **103**, 2114.
Wender, P. A. and Schaus, J. M. (1978). *J. Org. Chem.*, **43**, 782.
Wenkert, E. (1968). *Acc. Chem. Res.*, **1**, 78.
Westmijze, H., Meyer, J., Bos, H. J. J. and Vermeer, P. (1976). *Rec. Trav. Chim.*, **95**, 299, 304.
Wharton, P. S. and Bohlen, D. H. (1961). *J. Org. Chem.*, **26**, 3615.
White, J. D. and Sheldon, B. G. (1981). *J. Org. Chem.*, **46**, 2273.
Whitesell, J. K., Bhattacharya, A., Aguilar, D. A. and Henke, K. (1982). *J. Chem. Soc. Chem. Commun.*, 989.
Whitesell, J. W. and Whitesell, M. A. (1983). *Synthesis*, 517.
Wigfield, D. C. (1979). *Tetrahedron*, **35**, 449.
Wilds, A. L. (1944). *Organic Reactions*, **2**, 178.
Wimmer, K. (1955). In *Methoden der Organischer Chemie* (*Houben-Weyl*), Vol. 4(2), p. 137, ed. E. Müller (Stuttgart: Georg Thieme Verlag).
Wiseman, J. R., Pendery, J. J., Otto, C. A. and Chiong, K. G. (1980). *J. Org. Chem.*, **45**, 516.
Wittig, G. and Frommeld, H-D. (1964). *Chem. Ber.* **97**, 3548.
Wittig, G. (1962). *Angew. Chem. internat. edn*, **1**, 415.

Woggon, W-D, Ruther, F. and Egli, H. (1980). *J. Chem. Soc. Chem. Commun.*, 706.
Wolff, M. E. (1963). *Chem. Rev.* **63**, 55.
Woodward, R. B. (1963). The Harvey Lectures, Vol. 31. (Compare Fleming, I (1973). *Selected Organic Syntheses*, p. 202 (London: Wiley).
Woodward, P. B., Freary, E. R. J., Nestler, G., Raman, H., Sitrin, R., Suter, Ch. and Whitesell, J. K. (1973). *J. Am. Chem. Soc.*, **95**, 6853.
Woodward, R. B. and Hoffmann, R. (1969). *Angew. Chem. internat. edn*, **8**, 781.
Woodward, R. B. and Hoffmann, R. (1970). *The Conservation of Orbital Symmetry* (New York: Academic Press).
Woodward, R. B., Pachter, I. J. and Scheinbaum, M. L. (1974). *Organic Syntheses*, **54**, 39.

Yamaguchi, K., Yabushita, S., Fueno, T. and Houk, K. N. (1981). *J. Am. Chem. Soc.*, **103**, 5043.
Yamamoto, Y., Yamamoto, S., Yatagai, H., Ishihara, Y. and Maruyama, K. (1982). *J. Org. Chem.*, **47**, 119.
Yasuda, A., Yamamoto, H. and Nozaki, H. (1979). *Bull. Chem. Soc. Japan*, **52**, 1705.
Yates, P. and Eaton, P. (1960). *J. Am. Chem. Soc.*, **82**, 4436.

Zhou, B., Gopalan, A. S., Van Middlesworth, F., Shieh, W-R. and Sih, C. J. (1983). *J. Am. Chem. Soc.*, **105**, 5925.
Ziegler, F. E. (1977). *Acc. Chem. Res.*, **10**, 227.
Ziegler, F. E. and Spitzer, E. B. (1970). *J. Am. Chem. Soc.*, **92**, 3492.
Zurflüh, R., Wall, E. N., Siddall, J. B. and Edwards, J. A. (1968). *J. Am. Chem. Soc.*, **90**, 6224.
Zweifel, G. and Pearson, N. R. (1980). *J. Am. Chem. Soc.*, **102**, 5919.
Zweifel, G. and Steele, R. B. (1967*a*). *J. Am. Chem. Soc.*, **89**, 2754.
Zweifel, G. and Steele, R. B. (1967*b*). *J. Am. Chem. Soc.*, **89**, 5085.

INDEX